Actinide Nanoparticle Research

Stepan N. Kalmykov • Melissa A. Denecke
Editors

Actinide Nanoparticle Research

Springer

Editors
Prof. Dr. Stepan N. Kalmykov
Lomonosov Moscow State University
Dept. Chemistry
Leninskie Gory 1/3
119991 Moscow
Russia
stepan@radio.chem.msu.ru

Dr. Melissa A. Denecke
Forschungszentrum Karlsruhe
Inst. Nukleare
Entsorgungstechnik
Postfach 36 40
76021 Karlsruhe
Germany
melissa@ine.fzk.de

ISBN 978-3-642-11431-1 e-ISBN 978-3-642-11432-8
DOI 10.1007/978-3-642-11432-8
Springer Heidelberg Dordrecht London New York

Library of Congress Control Number: 2011924898

Cover design: eStudio Calamar S.L.

Printed on acid-free paper

Springer is part of Springer Science+Business Media (www.springer.com)

Foreword

A challenge for most of the countries that developed nuclear fuel cycles is the problem of safe and economically reasonable methods for management of spent nuclear fuel (SNF) and radioactive wastes. In Russia, the government plans to increase the share of nuclear power from 15% up to 25% by 2030. This would increase the volume of SNF that have to be processed and nuclear wastes that have to be disposed into the geological repositories. On these grounds, research in different fields including behavior of SNF and vitrified nuclear wastes under repository conditions, interaction of radionuclides with engineered and natural barriers, migration and speciation of radionuclides in the environment, and so on, is highly appreciated. Another challenging problem that many countries face is the remediation of contaminated sites as a legacy of such diverse activities as weapon programs, SNF reprocessing, or uranium mining and milling. No effective remediation strategies could be developed without knowledge on the speciation of radionuclides.

It was demonstrated by research groups in many countries that actinides, as well as other radionuclides, tend to form colloid particles of different nature that are responsible for their environmental behavior. Structurally, most of these colloids are hierarchic particles that consist of agglomerated nanoparticles. In order to understand and predict their environmental behavior, their characterization at nano- and subnano scale is required.

This is the first book that covers different aspects of actinide nanoresearch: formation, structure, and properties of actinide nanoparticles and thin layers, environmental behavior of actinide colloids including modeling approaches. Several chapters are devoted to the new innovative spectroscopic and microscopic techniques of actinide nanoparticle characterization.

This book is of interest both for fundamental research into chemistry and physics of actinides and for applied researchers. The chapters are written and reviewed by top-level specialists in the field of actinide chemistry, allowing one to recommend this book to broad audience including scientists, engineers, and students who study nuclear chemistry.

Moscow, Russia
February 2011

Boris F. Myasoedov
Academician, Professor, Honorary chairman of
Russian-German Symposium on Actinide Nanoparticles

Preface

The idea for this book on actinide nanoparticle research was born during the "Russian–German Symposium on Actinide Nanoparticles" held in May 2009 in Moscow and organized by the institutes Department of Chemistry of the Lomonosov Moscow State University (MSU) and the Institute of Nuclear Waste Disposal at the Karlsruhe Institute of Technology (KIT-INE). The symposium itself is a milestone in a cooperation between MSU and KIT-INE, the Helmholtz-Russian Joint Research Group entitled "Actinide Nanoparticles: Formation, Stability, and Properties Relevant to the Safety of Nuclear Waste Disposal" funded by the German Helmholtz Society and the Russian Foundation for Basic Research.

The research interest in actinide nanoparticles is multifaceted. An obvious interest lies in their importance for safety research related to nuclear waste disposal and assessment of actinide-contaminated sites, as the formation and transport of actinide nanoparticles is an environmental concern that is vital not only to our generation but also to future generations. There are other aspects of more basic character, which are a driving force behind actinide nanoparticle research. This research belongs to a distinct realm of nanoscience, but with a history longer than other areas of today's conventional nanoscience. Modern nanoscience, as a presently well-funded, "sexy" research area, has actually joined the ranks of the area of nanoscience referred to for decades as colloid chemistry or colloid research, including macro- and supramolecular chemistry.

Our endeavors in actinide nanoparticle research entail much more than being simply classified as a sector of nanoscience, however. We are clearly tackling the grand challenge of understanding the behavior of these heavy elements as nanoclusters and nanoparticles with inherent special behavior, reactivity, and structure. Part of this challenge lies in the radioactivity of actinide-containing systems investigated, requiring specialized expertise and infrastructure. It is not always a trivial task equipping instrumentation for nanoparticle characterization to make safe studies of radioactive systems possible and only a limited number of laboratories worldwide have the capability and licensing for working with transuranium actinide elements. Certainly an exceptionally challenging aspect is the theoretical description of heavy actinide elements. Experimental observation and interpretation on the nanoscale is driving instrumentation and method development and the thrust to couple experiments on actinide systems with quantum chemical calculations is promoting advances in theory. Theoretical, computational description of 5f element systems must deal with numerous electrons in (5f, 6d, 7s, and 6p) orbitals close in energy, requiring scalar and often spin-orbit relativistic effects or electron correlation effects.

Many chapters of this book, but not all, are an extension or overview of presentations given at the Russian–German Symposium on Actinide Nanoparticles. The first chapter Actinide Nanoparticles – Generation, Stability, and Mobility is a summary of the presentation given by Professor Horst Geckeis at the symposium and is an excellent discourse on the state of knowledge and the importance of environmental actinide nanoparticle behavior. The remaining chapters of this book are divided into four sections dealing with different aspects of actinide nanoparticle research, in a manner similar to the structure of the symposium: Methods, Modeling, Surface Reactivity and Environmental Behavior. In the Methods section, we have included imaging, spectroscopic and separation techniques covering overviews of previous studies (STXM, FFF, and MS), newer results (STEM), and a chapter on a newer spectroscopy offering great potential for direct characterization of actinide nanoparticle bulk electronic structure (RIXS). The Modeling section includes not only a treatise on modeling of colloid-mediated actinide transport in the environment but also a chapter on the outlook of quantum chemical strategies for computing actinide nanoparticle structures. This is followed by a section on Interfacial Phenomena and Formation of Actinide Nanoparticles. The interfacial phenomena are addressed in three separate chapters, one reporting new results from an investigation of actinide ion interaction with silica nanoparticles, the second chapter on well-defined thin films produced under controlled sputtering conditions as models for reactive surfaces, especially spent fuel but potentially applicable for advancing our understanding of actinide nanoparticles with inherent large surface to bulk ratios. The third chapter in this section deals with novel structures of uranium selenate clusters, which can serve as models for formation and reactivity of certain classes of actinide nanoparticles. The last section presents results from previous and new investigations of actinide nanoparticle environmental behavior. Four of this section's chapters deal with anthropogenic actinide contamination legacies; the last chapter reports new results of X-ray tomography investigations with nanoscale resolution and techniques offering a potentially encouraging outlook for future research activities.

Ultimate goal in actinide nanoparticle research of attaining a level of understanding, which permits us to predict and ultimately control actinide behavior in systems covering the entire size-scale range – from ion, to nanoclusters, to nanoparticles and microparticles, up to bulk, condensed systems – is not only a noble ambition necessary for providence of modern society, it is at the forefront of research. Potential applications include synthesis of actinide systems targeted toward novel structures, as well as selected chemical and physical characteristics, and reliable prediction of actinide mobility in geochemical milieus on a long-term timescale necessary for safety assessment of nuclear waste disposal concepts and strategies for eliminating/minimizing risks associated with contamination legacies. This is indeed a grand challenge. This humble endeavor, as the first book published on the subject of actinide nanoparticle research, represents one step in our exciting journey into the fascinating domain of the multifarious actinide nanoscopic world.

Karlsruhe, Germany
Moscow, Russia
February 2011

Melissa A. Denecke
Stepan N. Kalmykov

Contents

Abbreviations

AEM	Analytical electron microscopy
AES	Auger electron spectroscopy
AFM	Atomic force microscopy
ALS	Advanced Light Source, Lawrence Berkeley National laboratory, USA
AME	Actinide Migration Evaluation
An	Actinide
AsFlFFF	Asymmetric flow field flow fractionation
BE	Binding energy
BEI	Backscattered electron imaging
BET	Brunauer, Emmett, Teller (model or technique)
BF	Bright-field
CA	Cellular automata
CCC	Critical coagulation concentration
CE	Counter electrode
CFF	Cross-flow (ultra) filtration
CLSM	Confocal laser scanning microscopy
CRL	Compound refractive lenses
CSH	Ca-silicate hydrate phases
CTEM	Conventional transmission electron microscopy
CTF	Contrast transfer function
CV	Cyclovoltammetry
CZE	Capillary zone electrophoresis
CZP	Condenser zone plate
DC	Direct current
DFT	Density functional theory
DGT	Diffusive gradients in thin film
DLM	Diffuse layer model
DLS	Dynamic light scattering
DLVO	Derjaguin, Landau, Verwey, Overbeek (theory)

DOC	Dissolved organic carbon
DOE	U.S. Department of Energy
DOM	Dissolved organic matter
dpa	Displacement per atom
DU	Depleted uranium
ECAFM	Electrochemical atomic force microscopy
ECP	Effective core potentials
Ecorr	Corrosion potential
EDS	Energy dispersive X-ray spectroscopy
EELS	Electron-energy-loss spectroscopies
EF	Fermi-level
EFTEM	Energy filtered transmission electron microscopy
EPA	US Environmental Protection Agency
EPR	Electron paramagnetic resonance spectroscopy
EPS	Extracellular polymer substances
EQCM	Electrochemical quartz crystal microbalance
ESEM	Environmental scanning electron microscopes
ESI	Electrospray ionization
ESI MS	Electrospray ionization mass spectroscopy
ESRF	European Synchrotron Radiation Facility, Grenoble, France
EXAFS	Extended X-ray absorption fine structure
FA	Fulvic acids
FBB	Fundamental building block
FCS	Fluorescence correlation spectroscopy
FFF	Flow-field flow fractionation
FFT	Fast Fourier transformed
FIB	Focussed ion beam
FlFFF	Flow field-flow fractionation
FT	Fourier transform
FT ICR	Fourier transform ion-cyclotron-resonance
FWHM	Full width at half maximum
FZP	Fresnel zone plates
GSL	Great Salt Lake
GTS	Grimsel Test Site
HA	Humic acids
HAADF-STEM	High-angle annular dark-field scanning transmission electron microscopy
HAS	Hydroxy aluminosilicate
HDCH	Hydrodynamic chromatography
HFO	Hydrous ferric oxide
HLW	High level wastes
HREELS	High resolution electron energy loss spectroscopy
HR-ICP-MS	High resolution inductively coupled plasma mass spectrometry
HRL	Hard rock laboratory

HS	Humic substances
ICP-MS	Inductively coupled plasma mass spectrometry
IR	Infrared
KB	Kirkpatrick-Baez (mirrors)
Kd	Distribution coefficient
KE	Kinetic energy
KIT	Karlsruhe Institute of Technology
LA	Laser-ablation
LAM	Linear additive model
LCD	Ligand charge distribution
LEED	Low-energy electron diffraction
LIBD	Laser-induced breakdown detection
LMM	Low molecular mass
LOD	Limits of detection
LS	Light scattering
LSLM	Laser scanning luminescence microscopy
LSV	Linear sweep voltammetry
LUMO	Lowest unoccupied molecular orbitals
MALDI	Matrix-assisted laser-desorption ionization
MALLS	Multiple-angle laser light scatterings
MCL	Maximum contaminant level
MCP	Multi-channel plates
MDA	Multi detector analysis
MES	Molecular environmental science
ML	Monolayers
MLEM	Maximum likelihood expectation maximization
MS	Mass spectrometry
MWCO	Molecular weight cut-off
μ-XRF	Micro- X-ray fluorescence analysis
μ-XRD	Micro- X-ray diffraction
nano-SIMS	Secondary ion mass spectrometry with nanoscale resolution
NEXAFS	Near edge X-ray absorption fine structure
NMR	Nuclear magnetic resonance
NORM	Naturally occurring radioactive material
NP	Nanoparticle
NSLS	National Synchrotron Light Source, Brookhaven, New York, USA
NTS	Nevada Test Site, Nevada, USA
PAA	Polyacrylic acid
PBU	Primary building units
PCBM	(6,6)-phenyl C_{61} butyric acid methyl ester
PCS	Photon correlation spectroscopy
pHpzc	Point of zero charge
PP	Pseudo potentials

μ-PIXE	Micro-particle (or proton) induced X-ray emission
PNBU	Prenucleation building units
QP	Quadrupol
RBS	Rutherford backscattering spectrometry
RE	Reference electrode
Rf	Retardation factor
RF	Radiofrequency
RFETS	Rocky Flats Environmental Technology Site
RHEED	Reflection high energy electron diffraction
RI	Refractive index
RIMS	Resonance Ionization
RIXS	Resonant inelastic x-ray scattering
RT	Room temperature
RTG	Radioisotope thermoelectric generators
RW	Radioactive waste
SAED	Selected area electron diffraction
SANS	Small angle neutron scattering
SAXS	Small angle X-ray scattering
SBU	Secondary building units
SCM	Surface complexation modeling
SEC	Size exclusion chromatography
SEM	Scanning electron microscopy
SEM-EDX	Scanning electron microscopy combined with energy-dispersive X-ray spectroscopy
SEXAFS	Surface extended X-ray absorption fine structure
SF	Sector field (analyzers)
SIMS	Secondary ion mass spectrometry
SKB	Swedish Nuclear Fuel and Waste Management Company
SLS	Swiss Light Source
SNF	Spent nuclear fuel
2SPNE SC/CE	Two-site protolysis, nonelectrostatic surface complexation and cation exchange model
SR	Synchrotron radiation
SRS	Savannah River Site
SSCP	Scanning stripping chronopotentiometry
STM	Scanning tunneling microscopy
STEM	Scanning transmission electron microscopy
STS	Semipalatinsk Test Site
STXM	Scanning transmission X-ray microscopy
SWIFT	Stored-waveform inverse-Fourier-transform
SymFlFFF	Symmetric flow field flow fractionation
TDA	Taylor dispersions analysis
TDS	Thermal diffuse scattering
TEM	Transmission electron microscopy

TEM/EDX	Transmission electron microscopy coupled to X-ray energy dispersive spectroscopy
TENORM	Technologically enhanced NORM
TFF	Tangential-flow filtration
TIMS	Thermal ionization mass spectrometry
TOF	Time-of-flight
TRLFS	Time-resolved laser fluorescence spectroscopy
TXM	Transmission X-ray microscopy
UF	Ultrafiltration
UHV	Ultra-high vacuum
URL	Underground rock laboratory
UPS	Ultraviolet photoelectron spectroscopy
UV	Ultraviolet
UV/Vis	Ultra violet/visible
VLS	Variable-line-spacing-grating (spectrograph)
WE	Working electrode
WIPP	Waste Isolation Pilot Plant
XANES	X-ray absorption near edge spectroscopy
XAS	X-ray absorption spectroscopy
XPS	X-ray photoelectron spectroscopy
XRD	X-ray diffraction
XRM	X-ray microscopy
XSW	X-ray standing wave
YAG	Yttrium aluminum garnet (laser)

Actinide-Nanoparticle Interaction: Generation, Stability and Mobility

Horst Geckeis, Thomas Rabung, and Thorsten Schäfer

Abstract Actinide ions are widely considered immobile under environmental conditions relevant to their disposal notably under reducing conditions prevalent in subsoil systems. Colloid formation is, however, known to increase their mobility. Observations of this in environmental systems are regularly reported. Different processes may lead to actinide colloid formation, by either generation of polymeric oxide/hydroxide nanoparticles or sorption to already existing or neo-formed ground water colloids. In order to assess actinide mobilization by colloidal species, the cascade of individual colloidal processes under given geochemical conditions must be considered and quantified: colloid formation, actinide–colloid interaction, colloid stabilization and migration. This chapter describes and discusses actinide colloid reactions with a focus on the underlying mechanisms.

1 Introduction

Naturally occurring or "environmental"colloids/nanoparticles can be defined as solids of organic or inorganic nature in the size range from ~1 nm to ~1 μm (upper size limit depending on their density) that remain suspended in water. Inorganic colloids/nanoparticles are formed by chemical alteration or physical erosion of minerals, or by direct precipitation from ground water, and typically comprise silica, clay minerals, calcite or metal-oxyhydroxides. Two types of

H. Geckeis (✉) and T. Rabung
Karlsruhe Institute of Technology (KIT), Institute for Nuclear Waste Disposal (INE), P.O. Box 3640, 76021 Karlsruhe, Germany
e-mail: horst.geckeis@kit.edu

T. Schäfer
Karlsruhe Institute of Technology (KIT), Institute for Nuclear Waste Disposal (INE), P.O. Box 3640, 76021 Karlsruhe, Germany
Freie Universität Berlin, Institute of Geological Sciences, Hydrogeology Group, 12249 Berlin, Germany

S.N. Kalmykov and M.A. Denecke (eds.), *Actinide Nanoparticle Research*,
DOI 10.1007/978-3-642-11432-8_1, © Springer-Verlag Berlin Heidelberg 2011

inorganic nanoparticles are usually distinguished. The so-called intrinsic- or eigen-colloids/nanoparticles consisting essentially of polymerized complexes of a particular element (e.g., Pu(IV)-polymer) formed by condensation of hydrolyzed Pu(IV) ions and consisting of cations linked by anions. These types of polymeric actinide species have recently attracted attention, as they may dominate the solubility of the tetravalent actinides and play a significant role in the complex redox chemistry of plutonium (Neck et al. 2007a; Soderholm et al. 2008) (see, e.g., Chaps. 5 and 13). The carrier- or pseudo- colloids/nanoparticles consist of mineral fragments of crystalline or amorphous solids to which a particular element/radionuclide may be attached. Organic nanoparticles may comprise fragments of degrading organic material, or they may be organic macromolecules (i.e., humic or fulvic acids) or microbes (biocolloids). Organic materials, especially humic and fulvic acids may form coatings on inorganic nanoparticles, increasing their electrostatic stability.

Numerous publications have demonstrated that the classical two phase flow concept of transport (stationary phase: sediment; mobile phase: ground water) has to be expanded to include a third phase, namely the nanoparticles, which are ubiquitous in aquifer systems (Kretzschmar and Schäfer 2005; McCarthy and Zachara 1989; Wilkinson and Lead 2006). Neglecting the nanoparticulate phase in ground water systems can drastically underestimate the mobility of strong sorbing contaminants (e.g., plutonium) (Kersting et al. 1999; Kim 1994; Kim et al. 1984). The discussion of clay colloids washed out from the bentonite barrier in a granitic nuclear repository due to low salinity glacial melt water intrusion (Neretnieks and Liu 2006) has focused new attention to this issue. Geochemically or hydraulically disturbed systems especially might generate a high concentration of colloidal phases, and therefore can enhance the mobility of strong sorbing contaminants (Ryan and Gschwend 1994). On the other hand, colloid destabilization via aggregation can also lead to an additional retention mechanism for strong sorbing contaminants.

Transport in far-field environments/geological formations has to take into account saturated and unsaturated flow regimes (Robinson et al. 2003). Nanoparticle mobilization and transport in unsaturated host rocks and from soils to the ground water table (vadose zone) is an important research topic, which changes nanoparticle migration mechanism (DeNovio et al. 2004). Such effects must be considered, e.g., in the far-field system of the U.S. Department of Energy (DOE) Yucca Mountain (Nevada) site, the underground waste tank leakages at the U.S. DOE Washington State Hanford Site (Zhuang et al. 2007) and potential gas migration pathways. Understanding colloid transport in unsaturated media is complicated by the presence of dynamic air–water interfaces and the increasing discontinuity of water-filled pores with decreasing water content. Additional mechanisms of nanoparticle retention have been suggested such as film-straining and deposition of nanoparticles onto the air–water interface.

Attachment of actinides to colloids does not necessarily mean high mobility. Colloid attachment to geosphere surfaces, colloid agglomeration, sedimentation and filtration in nanoporous rock such as clay rock, and actinide desorption from colloids followed by sorption to minerals contribute to actinide retention even in the presence of colloids.

In general, the importance of colloidal phases or nanoparticles on transport processes depends on (a) their presence, (b) their mobility and stability and (c) the slow reversibility or "irreversibility" of metal/radionuclide interaction with these nanoparticles, which are discussed in detail below.

2 Background Nanoparticles in Geological Systems

The sampling of colloids/nanoparticles in ground water systems is not a trivial issue: a number of artefacts such as oxygen intrusion can bias the sampling procedure and, consequently, the reliable measurement of background nanoparticle concentration. The exact determination of nanoparticle background concentration is motivated by the fact that this concentration will give an estimation of the maximum transport enhancement for strongly sorbing tracers. In porous and fractured rocks the sampling methology is well documented (Backhus et al. 1993; Barcelona et al. 2005; Degueldre et al. 1996b). A general procedure for low-flow passive sampling technique has been established for different geological formations, as documented in, e.g., an US Environmental Protection Agency (USEPA) workshop summary (Puls 1995). Discussions on the advantages of different analytical techniques that can be used for background nanoparticle sampling are also found elsewhere (Barcelona et al. 2005; Nilsson et al. 2008). Much progress has also been made in the field of analytic development concerning the detection of nanoparticles (see other contributions in this book). The community clearly benefits from developments made, especially in the field of nanoparticle size distribution analysis (e.g., via laser-induced breakdown detection, LIBD) and flow-field flow fractionation (see Chap. 4), and the possibility to detect size dispersion/chromatography effects in nanoparticle migration experiments (Delos et al. 2008; Walther et al. 2006).

2.1 *Inorganic Nanoparticles*

Different colloid field studies are available in the literature. Degueldre et al. (1996a, b, 2000) collected colloid data and derived stability criteria. A mobile LIBD equipment (Hauser et al. 2002) has been used in field investigations at the Grimsel Test Site, in the Swedish site investigation program and other site investigations (e.g., the Ruprechtov site, Czech Republic) and provided detailed information on inorganic nanoparticle concentration in several observation boreholes and correlated these to geochemical parameters. The data show that inorganic nanoparticle concentration in advective transport dominated far-field environments (crystalline and rock salt overburden) is mostly limited to concentrations <1 mg/L and strongly depends on the ionic strength, notably the alkaline earth element concentration of the ground water. Higher colloid concentrations usually can be explained by high flow fields in disturbed systems. Lower concentrations of inorganic nanoparticles are expected if the ground water composition exceeds an ionic strength of 0.1 M, where nanoparticle concentrations decrease even down to the detection limit of the LIBD system (~10 ppt) (SKB 2005). In the SR-Can report (SKB 2006b) the highest nanoparticle concentration in ground water is expected during a glacial period, especially during the advance and retreat of an ice sheet when ground water velocities are highest. The concentration in this case is assumed to be the highest measured nanoparticle concentrations in ground waters at repository depths, $\approx$1 mg/L. The data obtained within the site investigations mentioned above [see, e.g., Fig. 3 in Hauser et al. (2007)] give further confidence in this upper concentration limit.

2.2 Organic Nanoparticles

One of the best investigated systems concerning natural organic nanoparticle migration and actinide interaction in field and laboratory studies is the Gorleben site (Lower Saxony, Germany) (Artinger et al. 1998b, 2000; Dearlove et al. 1991; Kim et al. 1990). This site has been continuously investigated within European Commission projects HUMICS, HUPA and FUNMIG (Buckau 2000, 2006). The strata covering the Permian salt dome at Gorleben, which is considered as a possible nuclear waste repository, consists of sedimentary rocks with a distinct aquifer system; sandy sediments of high permeability are found from the surface down to the top of the salt dome (BGR 1991). Due to local Miocene brown coal and Pleistocene peat deposits, dissolved organic carbon (DOC) in ground water is present locally up to 200 mg/L (Buckau 1991). In these ground waters DOC mainly consists of colloidal fulvic and humic acids (Kim et al. 1984; Schäfer et al. 2005).

For far-field environments in diffusion controlled claystone systems, model calculations based on pore size distribution and connectivity argue nanoparticles $<<$10 nm fitting through pore necks, and nanoparticle-related transport is negligible (Nagra 2002). Several in situ collected water samples from the Opalinus Clay (Mont Terri) were studied with focus on the size range 1–10 nm. A concentration of inorganic nanoparticles around 10–20 ppt with organic nanoparticles (humic substances) below the detection limit were determined (Degueldre et al. 1998). Detailed studies on the organic inventory in indurated clay (Opalinus and Callovo-Oxfordian) pore water have shown rather high DOC concentrations of 56.7 mg/L in the Callovo-Oxfordian and from 1.2 to 15.8 $\pm$ 0.5 mg/L in the Opalinus Clay pore water (Courdouan et al. 2007a, b). However, a significant DOC fraction belongs to low molecular weight compounds and not nanoparticles with a molecular size $<$500 Da (Courdouan et al. 2007a, b). For plastic clays (Boom Clay formation) the pore water contains ~100 mg/L DOC (Liu et al. 2008; Maes et al. 2006). Comparison between batch-type studies, where the entire organic inventory was brought into suspension, and the pore water sampled in the HADES underground laboratory (Ionescu et al. 2008), indicates that the pore-size cut-off of the mobile fraction is $\leq$~300 kDa. Taking the size distribution of humic and fulvic acids analyzed from the Gorleben aquifer system as reference (Buckau 2004; Kim et al. 1990), one has to assume that both humic and fulvic acids can also be mobile in the Boom Clay system.

Concerning the potential role of organic nanoparticles in crystalline environments, studies on ground water of the Äspö Hard Rock Laboratory (Sweden) have shown that fulvic acids represent a significant fraction of the DOC inventory (Buckau 2005; SKB 2005). In general, for the Grimsel Test Site, Switzerland, a DOC concentration of around 1 mg L^{-1} is determined, ranging from 0.24 mg L^{-1} (Degueldre et al. 1996a) to 1.2 $\pm$ 0.2 mg L^{-1} (Schäfer et al. 2004). Äspö ground water has a higher DOC content of 10–22 mg L^{-1} (Tullborg and Gustafsson 1999), with maximum values found in the CHEMLAB2 borehole of 42.3 mg L^{-1} (Kienzler et al. 2003). For the Whiteshell Research Area (Canada) average DOC concentration of 0.5 mg L^{-1} are reported (Vilks and Bachinski 1996).

3 Repository Near-Field Nanoparticles

3.1 Canister and Glass Corrosion

Early work by Bates and co-workers had already identified insoluble Pu- and Am- bearing nanoparticles formed during simulated weathering of a high-level nuclear waste glass, and pointed to the need of a nanoparticle-trapping mechanism for a waste repository in order to meet long-term performance specifications. Buck and co-workers identified plutonium enriched layers on corroded spent uranium oxide fuel by analytical transmission electron microscopy (TEM), which pose increased potential for generation of Pu-bearing colloids (Buck et al. 2004). UO_2 nanoparticles were identified as corrosion products of metallic uranium nuclear fuel and predominantly clay nanoparticles were released from corroded uranium–aluminium fuels (Kaminski et al. 2005) and identified in glass corrosion products (Buck and Bates 1999), which shows the dependency of the colloid source term on the corroded waste forms. The generation mechanism of nanoparticular hydroxyl-aluminosilicates (HAS), which can be considered precursor phases of clay minerals and the entrapping of actinides, has been the research focus of Kim and Panak, showing the potential structural incorporation by spectroscopic methods (Kim et al. 2005, 2007; Panak et al. 2005). Additional work includes classical thermodynamic data generation by measuring solubility limits of tetravalent elements through the onset of nanoparticle formation (Neck et al. 2007b). The results indicate that especially tetravalent actinides form nanoparticles in the near-field.

As a consequence of these near-field nanoparticle observations, the bentonite buffer as repository component in the e.g. KBS-3 disposal system has to filter nanoparticles as a safety function (SKB 2006b). As a quantitative limit or safety function indicator criterion to fulfill the role of nanoparticle filtration, a buffer density >1,650 kg/m^3 is required. This requirement on the buffer ensures it being dense enough to immobilize fuel nanoparticles escaped from a defective canister.

The release of several key radionuclides will thereby be limited purely by their solubility. The criterion is based on 15 nm gold nanoparticle transport experiments showing that the bentonite microstructure with a dry density of 1,000 kg/m^3 (saturated clay density of about 1,640 kg/m^3) effectively filter these nanoparticles (Kurosawa and Ueta 2001). Homogenization analysis indicates that interstitial pores sizes of <1 nm may be achieved for montmorillonite partial densities in excess of 1,200 kg/m^3 (Kurosawa et al. 2006).

3.2 *Bentonite Backfill and Buffer Erosion*

Bentonite as backfill and buffer material is an important component of engineered-barrier systems (Gates et al. 2009) and will be used in its pure form, or as bentonite/sand mixture in different repository concepts for spent fuel/high level waste (Hummel 2008). Upon glacial melt water intrusion, clay colloid matter can be released during bentonite erosion (Liu and Neretnieks 2006). Using a simplified zero order model to predict the erosion rates (fracture aperture $\delta = 1 \times 10^{-4}$ m; outer periphery of the intruding bentonite as circular boundary with a diameter $d = 2$ m), stages of the glacial melt water intrusion can be differentiated. In the initial stage of glacial water intrusion lasting a few years, when the bentonite becomes equilibrated with typical fresh glacial ground water without re-supply of calcium, erosion rates of 4.7×10^{-4}–6.6×10^{-2} mg cm^{-2} s^{-1} are predicted. In the intermediate stage the bentonite erosion rate is dependent on the presence or absence of readily soluble calcium phases such as gypsum/anhydrite. In the presence of gypsum and associated calcium supply erosion rates of 4.7×10^{-5}–6.8×10^{-3} mg cm^{-2} s^{-1} are predicted, whereas in the absence of gypsum erosion rates of 8.5×10^{-3}–1.2 mg cm^{-2} s^{-1} are calculated. Experimental studies show that colloid release is controlled by the ground water composition (pH, ionic strength) and by the supply of Ca^{2+} ions acting as coagulant, coming either from exchangeable cations and/or gypsum/calcite impurities in the bentonite (Alonso et al. 2008; Seher et al. 2009). Furthermore, investigations of Baik et al. under advective flow conditions using ground water from the Yuseong Site revealed highest erosion rates of 6.8×10^{-7} mg cm^{-2} s^{-1} for a flow velocity of 1.5×10^{-5} m s^{-1} (Baik et al. 2007). Taking the highest experimental derived erosion rates of 1.2×10^{-6} mg cm^{-2} s^{-1} of Na-montmorillonite in contact with bi-distilled waters the lowest erosion rate predicted by the zero order model overestimates bentonite erosion rate by at least one order of magnitude.

Additionally, nanoparticles might also be generated through local oversaturation by the mixing of bentonite pore water with granite ground water. Studies on the neo-formation of nanoparticles in synthetic Febex porewater, resembling the saturated bentonite state with natural ground water from the Grimsel Test Site (GTS) revealed a limited nanoparticle concentration increase (~100 ppt, average diameter ~ 30 nm) which showed an influence on trivalent actinide speciation most likely due to the formation of polysilicilic acid nanoparticles (Kunze et al. 2008).

4 Nanoparticle Stability

Colloidal stability in *sensu stricto* is related to agglomeration reactions which are controlled both by the colloid surface charge and the solution composition, i.e., pH and ionic strength. Beside charge repulsion, steric stabilization processes can prevent colloid aggregation (mainly known for organic and silica particles). In a wider sense, colloids are "stable" or "instable" depending on their solubility, under given geochemical conditions or their interaction with mineral surfaces. This latter aspect is discussed in Sect. 6 below. Actinide(IV) oxihydroxide polymers are for instance found stable even at high ionic strengths for months. However, these readily dissolve if the pH is varied or the solution is diluted so that concentrations fall below solubility limits (Neck et al. 2007a). The polynuclear oxide/hydroxide colloids of tetravalent actinide ions are assumed to be thermodynamically stable species when in contact with the respective solid. Presently, knowledge about their exact nature, colloidal stability and their mobility is severely limited; further studies are required. Nanoparticulate crystalline actinide oxides should behave in a classical sense and dispersions are destabilized if the surface charge approaches zero or is screened at high ionic strength. The point of zero charge (pH_{pzc}) for PuO_2 is reported to lie at 8–9, for ThO_2 at 7–9.5 and for UO_2 even lower at 5.2 (Kosmulski 2009).

If actinides are bound to aquatic colloids, the stability of those colloids will determine actinide mobility. As a safety function indicator for the stability of the bentonite barrier (and thus also for the stability of suspended clay particles), a concentration of divalent cations of 1 mM was set as a criterion in the Swedish nuclear waste disposal safety report (SKB 2006b). Independent comprehensive studies confirmed a pH independent critical coagulation concentration Ca-CCC of 1 mmol L^{-1} for montmorillonite colloids prepared from bentonite (Seher et al. 2010). In the case of NaCl background electrolyte pH dependent Na-CCC values could be determined by the same authors. The determined Na-CCC values are in quite good agreement with published data for similar clay minerals (Chheda et al. 1992; Hetzel and Doner 1993; Lagaly and Ziesmer 2003; Tombácz and Szekeres 2004). There are, however, deviating data found in the literature which may be due to organic matter coating (Schäfer et al. 2007), or to different isomorphic substitutions or to different sizes leading to variable charge and/or steric stabilization.

For the stability of humic/fulvic acids in natural water no strict colloid stability criterion as for inorganic colloids can be derived. While it is well known that humic acid flocculation is favored at decreasing pH and increasing concentration of polyvalent cations (in natural waters Ca^{2+} plays an important role) (Sholkovitz 1976), fulvic acids are found to be stable over a wide range of pH and ionic strength ($\leq$0.5 M). Only limited aggregation (possibly formation of dimers and trimers) could be demonstrated, e.g., for the hydrophilic Suwannee River fulvic acid (Lead et al. 2000). Other studies performed in the Gorleben aquifer system showed the presence of fulvic acids even in high salinity brines (Buckau et al. 2000).

5 Colloid–Metal-Ion-Interaction

Actinide chemistry in deep aquifers is dominated by complexation, redox reactions, colloid formation and interactions with mineral surfaces (Kim 2006). Notably, the radionuclide reactions at the colloid-ground water interface are not yet sufficiently understood, and it is thus imperative to have compelling evidence of reaction mechanisms and mobilization/immobilization processes. Spectroscopic investigations on actinide interaction with inorganic and organic colloids are available in the literature and a summary of them are compiled in the following section.

5.1 *Inorganic Colloids*

To gain insight into the nature of colloid-bound actinide species, various spectroscopic techniques have been applied. The trivalent actinide ion Cm^{3+} can be used as a luminescent probe at concentrations down to 10^{-9} mol/L and investigated by the time-resolved laser fluorescence spectroscopy (TRLFS) as speciation tool (Chung et al. 1998; Simoni 2002; Wimmer et al. 1992). Variations in the first coordination sphere of Cm(III) compared to the Cm-aquo species in general induce a red shift in the emission spectrum. The fluorescence lifetime provides insight into the composition of the first coordination sphere. The empirical Horrock's relationship is usually taken to derive the number of coordinated H_2O/OH^- units (Kimura et al. 1998). Cm(III) sorption onto different clay mineral colloids dispersed in solution studied by TRLFS led to the following conclusions with regard to sorption mechanisms and the nature of the surface complexes:

1. At low pH and low ionic strength outer sphere sorption takes place (Stumpf et al. 2001). This can be quantified by comparing TRLFS with batch sorption data on Cm(III). Under these conditions, the first hydration sphere of the metal ion remains unaffected and in principle Cm(III) sorption is only visible in batch sorption data but not directly detectable by TRLFS. By observing a long ranging quench effect of structurally abundant iron in clay minerals, outer sphere bound Cm(III) was identified and quantified by TRLFS (Hartmann et al. 2011).
2. Analysis of Cm(III) TRLFS spectra with subsequent peak de-convolution allows differentiation of three different inner-sphere surface complexes (Stumpf et al. 2002; Rabung et al. 2005).
3. TRLFS fluorescence lifetimes indicate displacement of ~4 H_2O molecules from the first coordination sphere upon inner-sphere surface complex formation. Lifetime does not vary significantly upon further pH increase even though the emission band shift to longer wavelengths, which indicates the formation of new species.
4. These experimental observation is compatible with the postulation of the following reaction scheme:

$$\equiv S - OH + Cm^{3+} \Leftrightarrow (\equiv S - O - Cm)^{2+} + H^+$$

$$(\equiv S - O - Cm)^{2+} + H_2O \Leftrightarrow (\equiv S - O - Cm(OH))^+ + H^+$$

$$(\equiv S - O - Cm(OH))^+ + H_2O \Leftrightarrow (\equiv S - O - Cm(OH)_2) + H^+$$

These conclusions are supported by extended X-ray absorption fine structure (EXAFS) studies on Am(III) sorption onto smectite, kaolinite (Stumpf et al. 2004) and marl colloids (Degueldre et al. 2004) and recently for trivalent yttrium and lanthanides on hectorite (Schlegel 2008) using polarized EXAFS. These studies suggest that the coordination number of An/Y/Ln(III) hardly changes upon sorption, while TRLFS shows that 4 H_2O are removed from the sorbed Cm ion hydration sphere. The very similar chemistry of trivalent Am, Gd and Cm together with information from EXAFS and TRLFS leads to the conclusion that the metal cation is coordinated with about four oxygen atoms from the solid surface (Stumpf et al. 2004).

Spectroscopic studies on sorption of other actinides to clay minerals have appeared as well including uranyl and thorium (Dähn et al. 2002; Schlegel and Descostes 2009; Sylwester et al. 2000). Like the trivalent actinide ions, ${UO_2}^{2+}$ undergoes ion exchange binding at low pH and ionic strength, while inner-sphere complexation prevails at higher pH and higher ionic strength. Ternary surface complexes including OH^- (Kowal-Fouchard et al. 2004) and ${CO_3}^{2-}$ (Catalano and Brown 2005) as further ligands have been proposed to form under certain conditions, also shown for trivalent actinide ions (Fernandes et al. 2010). TRLFS (Chisholm-Brause et al. 2004; Kowal-Fouchard et al. 2004), X-ray photoelectron spectroscopy (XPS) (Kowal-Fouchard et al. 2004) and EXAFS (Catalano and Brown 2005; Schlegel and Descostes 2009; Sylwester et al. 2000) have been successfully applied to follow clay surface sorbed uranyl speciation. For the tetravalent thorium, inner-sphere surface complexation was found at even low pH where the tetravalent metal ion is bound in a bidentate fashion to the oxygen atoms of Si-tetrahedra and Al-octahedra. EXAFS points to different possible surface complexes with slightly varying bond lengths and angles (Dähn et al. 2002). The interesting point is that apparently metal ion species in different oxidation states occupy different sites and to some extent there seems to exist some non-competitive sorption between metal ions with different chemical properties (Bradbury and Baeyens 2005). Nevertheless, there is a broad dataset available for the speciation of actinide surface complexes on montmorillonite and illite, derived from experimental data and a linear free energy relationship (Bradbury and Baeyens 2005, 2009). The underlying two-site protolysis, nonelectrostatic surface complexation and cation exchange (2SPNE SC/CE) model is basically in agreement with spectroscopic findings (see Fig. 1).

All information compiled above indicate sorption reactions being constrained to the mineral surface and should, thus, be fully reversible during laboratory time scales. TRLFS studies on Cm(III) sorption to clay minerals at pH $\geq$ 12 shows

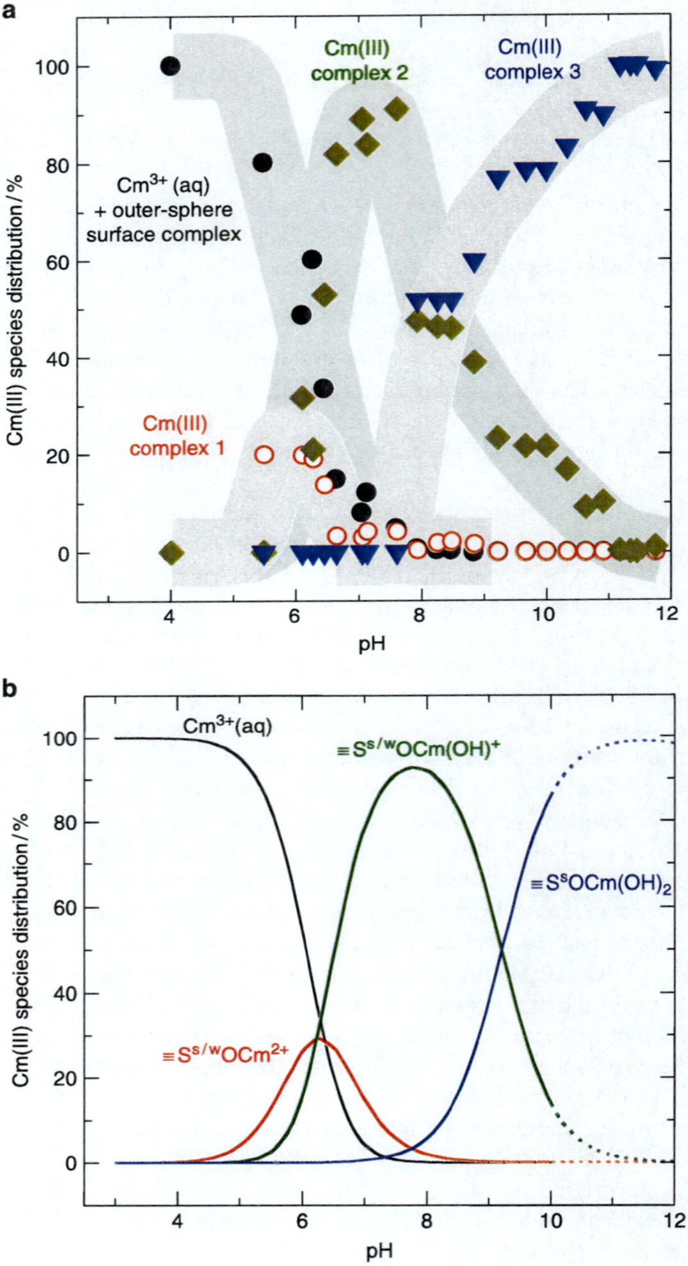

Fig. 1 Comparison of TRLFS speciation of montmorillonite bound Cm with the prediction made by using the 2SPNE/CE model [taken from Rabung et al. (2005)]

a dramatic shift of the emission band to 620 nm and an increase of the lifetime to 753 ± 88 μs (Rabung et al. 2005). This must be due to a dramatic change in the first coordination sphere and an almost complete loss of H_2O/OH^- from the first metal coordination sphere. According to existing thermodynamic data, portlandite (Ca $(OH)_2$) or Ca-silicate hydrate phases (CSH) are supposed to precipitate around this pH and may coprecipitate Cm(III). Such processes may lead to partial irreversibility of actinide sorption at least for some time. Indications for mineralization reactions of actinide ions are also reported for colloidal silica (Chung et al. 1998), calcite (Schmidt et al. 2009; Stumpf and Fanghänel 2002; Stumpf et al. 2006) and gibbsite (Huittinen et al. 2010, accepted; Huittinen et al. 2009). Tri- and tetravalent actinides are also found to be integrated into the oxo-bridge structure of hydroxy aluminosilicate (HAS) colloids once they are introduced in a mixture solution of Si and Al (Kim et al. 2003, 2005; Panak et al. 2003). The metal cations behave structurally similar to Al and thus become incorporated into a stable colloid-borne species. This could be very relevant in natural aquifer systems, in which colloids, including aluminosilicate colloids, are always present. Aquatic colloids, such as aluminosilicate colloids may therefore facilitate the migration of trivalent actinides as colloid-borne species, with little interaction on geomatrices surfaces. Actinide incorporation into colloid structure has been observed in repository relevant ground water mixing zones as discussed above (Kunze et al. 2008). Kinetic data on desorption from colloids, under conditions where spectroscopy indicates actinide incorporation are widely missing. However, in the case of Th(IV) adsorption onto montmorillonite in ground water kinetically hindered desorption or partially irreversible binding of Th over 3 years is found (Bouby et al. 2010). Which surface phenomenon is responsible for the partial irreversible sorption is not understood.

5.2 *Organic Colloids*

Humic/fulvic acids (HA/FA) represent another colloid type being frequently discussed as potential radionuclide carriers in the far field of a repository (Kim 1991). Actinide interaction with humic/fulvic acid is described by inner-sphere complexation to carboxylate and possibly phenolic functional groups. The tetravalent actinide Th(IV) studied by XPS and EXAFS indicate thorium interaction with carboxylic groups, coordinated by 9–10 oxygen at a distance of 2.43 A (Denecke et al. 1999). Similar results have been reported for trivalent lanthanide ions (Monsallier et al. 2003). Formation of binary humate and ternary hydroxo and carbonato humate complexes are found for the interaction of trivalent actinides such as Am(III) and Cm (III) by UV/Vis absorption and TRLFS (Buckau et al. 1992; Panak et al. 1996). Hydroxo-fulvate-Np(IV) complexes were also identified with EXAFS (Denecke et al. 2002). Such ternary species are rarely taken into account by most model approaches up to now, but appear to be even dominant in the near-neutral range in typical ground water.

A number of geochemical model approaches have been developed to predict actinide speciation in HA/FA containing ground water (Benedetti et al. 1995; Kim and Czerwinski 1996; Tipping 1998). Radionuclide-HA/FA association is rapid and is followed by rearrangement to kinetically more stable species (Schuessler et al. 2000). Dissociation is considerably slower and rates decrease with increasing ageing (Geckeis et al. 2002). Different kinetic modes are populated with dissociation time constants varying from seconds to weeks (Choppin and Clark 1991; Schuessler et al. 2000). The metal ion is believed to either migrate to stronger binding sites or to move to inner sites of the humic macromolecular structure. In view of the now generally accepted small size of the humic acid molecules the existence of a significant inner space becomes somewhat implausible (Leenheer and Croue 2003). It is more likely that intermolecular bridging by polyvalent cations leads to the interaction of metal ions with more than one humic acid entity and observed time dependent decrease of dissociation rates. A comprehensive review on humic/fulvic acid interaction with actinide/lanthanide ions and respective kinetics is available from Bryan et al. (submitted).

Recent combined spectro-microscopy studies on Eu(III)-HA agglomerates provides interesting insight into the polyvalent metal humic acid interaction (Naber et al. 2006; Plaschke et al. 2002). Enrichment of Eu(III) in those agglomerates is correlated to a local increase in carboxylic acid group density. In addition, different HA binding sites for Eu(III) could be identified, being attributed to stronger and weaker binding functional groups. Both findings suggest a certain degree of functional group heterogeneity in HA with regard to their distribution and chemical nature (Rabung and Geckeis 2010). Such heterogeneity is not found for Eu(III) polyacrylic acid interaction, which is sometimes taken as a model for HA. While HA-agglomeration at increased metal ion loading is not surprising, the existence of metal induced humic agglomeration has been visualized even at trace metal concentrations by flow-field flow fractionation (FFF) studies (Geckeis et al. 2002; Suteerapataranon et al. 2006). FFF allows the analysis of colloid hydrodynamic size distributions. Th(IV) humate complexes have their maximum in size distribution at almost double the hydrodynamic diameter for HA alone. The degree of agglomeration appears to depend on the metal ion charge. Plutonium induced humic acid agglomeration is also observed in recent FFF experiments and described in Chap. 4. The existence of different chemical binding modes and agglomeration phenomena can thus be responsible for the delayed dissociation of aged metal ion-HA/FA complexes. In natural systems organic coatings of inorganic colloids can lead to an increased overall colloidal stability and metal ions bound to such kind of colloids are found to form ternary surface complexes (Wang et al. 2004). Concerning prediction of metal distribution in the ternary system (mineral surface – organic matter – metal) the linear additive model (LAM) approach based on the binary systems often fails (Heidmann et al. 2005). Spectroscopic investigations probing both the metal environment (Eu^{III}-TRLFS) and the organic functionality (carbon K-edge X-ray absorption near-edge structure, XANES) showed that fractionation can occur depending on the mineral surface site occupancy (Claret et al. 2008). Developments on the ligand charge distribution (LCD) model were made to explain the differences in sorption behavior of different humic substance size fractions to overcome the limitations of LAM models (Weng et al. 2007).

In many cases redox reactions are involved into the metal ion–humic interaction. Such processes can also be followed by EXAFS and XPS, as has been shown for the reactions of Np(V), Pu(VI) and Tc(VII) (Maes et al. 2004; Marquardt et al. 2004; Zeh et al. 1999). Humic bound Fe(II) and/or quinone groups are thought to act as electron donors. In all cases precipitating oxihydroxides of the reduced tetravalent species are kept in solution to some extent as colloidal species stabilized by humic acid coatings. The differentiation of Me–HA complexes and colloid/HA agglomerates becomes challenging in such systems, as pronounced long-term kinetic effects are involved in such redox reactions.

6 Actinide Colloid Migration

In general, polyvalent actinide ions are considered widely immobile under reducing repository conditions due to their low solubility and strong tendency to be retained at mineral surfaces. It is recognized, however, already since the 1980's that colloid formation may dominate actinide speciation in natural environments (Kim 1986). In fact colloid borne actinide migration has been observed in various laboratory and field studies. The mobility of colloids/nanoparticles is mainly driven by their stability and retardation processes during transport (attachment, matrix diffusion, size cut-off), which are usually lumped together in the term *filtration*. The following sections summarize the state of experimental observations and their interpretation in terms of colloid mediated actinide transport assessment for nuclear waste disposal and contaminated sites.

6.1 Actinide Colloid Observations at Contaminated Sites

Colloid bound radionuclides have been found in various aquifer systems adjacent to contaminated sites (Kersting et al. 1999; Novikov et al. 2006; Penrose et al. 1990; Roberts et al. 2004; Utsunomiya et al. 2009; Zachara et al. 2007). Cs and Pu colloidal migration has been observed to act at the US DOE Hanford site at various places. Release of high solid containing liquid waste of temperatures >100°C from single shell tanks containing nuclear waste from earlier Pu production facilities was confirmed already in 1959 (Zachara et al. 2007) and may undergo variable geochemical reactions including colloid formation. In the vadose zone below the leaking tanks colloids experience retention in variable water films in the soil. Under saturated conditions Cs is found to desorb from colloids (Chen and Flury 2005). Plutonium released from surface basins is found to migrate into the ground water and 7–29% of the Pu exists associated with colloids (1 kDa–0.2 μm) in the oxidation states III and IV (Dai et al. 2005). However, Pu concentrations and fractions of colloidal species decrease downstream. Thus, no clear evidence for the colloid enhanced Pu migration over longer distances could be derived from observations. Actinide migration at the Savannah River Site is also controversially discussed. At some sample points, e.g., Pu colloid species are found but of less relevance compared to dissolved Pu species (Dai et al. 2002). Pu detected 3,400 m downstream from Los Alamos National Laboratory has initially been assigned to colloid transport (Penrose et al. 1990). This was later questioned; the arguments being that one cannot rule out Pu found at some distance away from the source may come from other sources, e.g., from windborne particulates (Marty et al. 1997).

Other studies, however, find evidence for relevant colloid mediated actinide transport in natural environments. Actinides discharged into seawater from the Sellafield reprocessing facilities are mainly attached to fine sediments, which only can be mobilized under resuspension conditions. However minor fractions of Pu(IV) and Am(III) have been found to persist as colloids in sea water (Leonard et al. 1995; Salbu et al. 1993). It is also likely that DOC of high molecular size is responsible for Pu and Am transport in river water over long distances from the Chernobyl Nuclear Power plant (Matsunaga et al. 2004). At the Rocky flats Environmental Technology Site actinides and notably plutonium are found in colloidal (but also in particulate) fractions in surface, pond and near surface ground water. The Pu is mainly associated with organic matter, which is considered as an important vector for actinide transport (Roberts et al. 2004). Kersting et al. (1999) found trace concentrations of plutonium, activation and fission products in colloid fractions in a distance of 1.3 km downstream of a nuclear weapon test site in Nevada (see Chap. 15). Most likely inorganic colloids (zeolite and sheet silicate nanoparticles) act as actinide carriers. Novikov et al. (2006) found colloid-mediated plutonium transport over a distance of 3 km from the assumed source in ground water at the Mayak Production Association, Urals, Russia. Colloidal actinide carriers can be hydrous ferric oxide (HFO) species where the contribution of natural organic matter, e.g., present as HFO coating, cannot be ruled out (Malkovsky et al. 2009).

A clear quantitative assessment of colloid mediated actinide migration under given conditions in contaminated sites or in any water intrusion scenario for a high level nuclear waste repository is, however, still not possible (Honeyman 1999). Findings on colloidal species in natural water samples have to be considered very carefully, as, e.g., high flow rate water sampling (Puls et al. 1990) or artifacts of sampling (Buesseler et al. 2003; Kaplan et al. 1993; Ryan and Gschwend 1990) may counterfeit their existence. Re-suspension of larger actinide containing particulate matter is relevant in contaminated surface water, but hard to quantify and to predict, and certainly less relevant in deep ground water. Another point of discussion related to the assessment of colloidal matter for the possible long term actinide transport in the vicinity of nuclear waste repository in deep geology is the question of kinetics. This question is notably important when contaminated sites are investigated, where strong chemical gradients are active and variations in redox potential, pH and ionic strength are very favorable to colloid generation and mobilization. This might not be necessarily expected in the far field of nuclear waste repository or a contaminated site where such strong gradients are absent.

6.2 *Actinide Colloid Mobility Studied in Laboratory and Field Experiments*

Extensive experimental studies have been carried out to elucidate and quantify the humic/fulvic colloid facilitated migration of actinides in a sandy aquifer system relevant to the tentative high level nuclear waste disposal site in Germany at Gorleben. Migration of inorganic colloid bound actinides has been almost exclusively studied in systems related to radionuclide transport in fractured rock. Such studies aim to provide relevant data for the safety assessment of nuclear waste repositories in granite host rock. In most laboratory column and field migration studies, the colloid/nanoparticle breakthrough curve maximum arrives slightly earlier (retardation factor $R_f < 1$) or is similar to that of the conservative tracer ($R_f = 1$), with the latter usually observed under near-natural (low) flow conditions.

6.2.1 Actinide-Humic Colloid Migration

Mobility of humic/fulvic colloid-borne tri-, tetra-, penta- and hexavalent actinide ions has been investigated already since the late eighties in sand column experiments (Artinger et al. 1998a, 2002a, b, c, 2003; Kim 1994). In all these experiments, sandy sediment columns were firstly equilibrated with humic/fulvic acid containing ground water and radionuclides of interest were spiked to the ground water for different contact times. Humic colloid facilitated actinide migration was observed with retention factors slightly below 1, which is attributed to a size or charge exclusion of humic colloids in the porous media. Quite early it was recognized that the experimental results could not be quantitatively described by assuming establishment of equilibrium between dissolved humate complexes and surface sorbed species. Trace radionuclide recoveries were found to vary depending on the contact time of the tracer in the ground water prior to injection on the column, on humic acid concentration in the solution, on the type of sediment, and mostly as a function of the residence time of the tracer in the column. Quite a number of studies focused on the kinetics of actinide dissociation from humate/fulvate complexes, and at least two kinetic modes of dissociation were postulated. A complete understanding is still missing, and the discussion whether or not species with very slow dissociation modes can form is still open. The state of the art in understanding kinetics of actinide/fission product interaction with humic matter is given in (Bryan et al. submitted). Combined experiments with radiolabelled humic acid allowed distinguishing between sorption of actinide-humate complexes and actinide-humate dissociation followed by actinide sorption. Even though slight retention of labeled humic acid is observed, the decreasing actinide recovery in column experiments is mainly due to dissociation. Similar migration studies were performed in plastic clay considered in Belgium as the reference host rock for a high-level nuclear waste repository (Boom clay). The porewater of this rock contains relatively high levels of natural organic matter. Upon contacting Am(III) labeled humate complexes with Boom Clay, the major part of Am becomes almost instantaneously dissociated and retained on the clay. Only a small fraction persists to migrate along with the mobile organic matter at even long contact times. Dissociation of Am(III) from this fraction was fit to a first order kinetic model with a rate of $8 \times 10^{-8}\ s^{-1}$ only (Maes et al. 2006). This value lies in-line with dissociation rates determined by Bryan et al. (2006) and King et al. (2001) [summarized and collected in Bryan et al. (submitted)]. Experimental results consequently can be quantitatively described to some extent by kinetic model approaches (Schuessler et al. 2001; Van De Weerd 1997). Even though there are various parameters controlling actinide-humic colloid mobility, there is a striking similarity in slow dissociation rates observed in many studies (see Fig. 2). A straightforward assignment of variable actinide speciation or rearrangements in humic molecular structure to the different kinetic dissociation modes is currently not available (see discussion in Sect. 5.2). Still we cannot exclude that "irreversibly" humic-colloid bound actinide species can be transported over longer distances. Possible mechanisms for such process could be the incorporation of actinide ions into inorganic nanoparticles coated and thus stabilized in ground water by humic/fulvic acids. A proof of the principle that such processes can indeed enhance actinide ion mobility has been demonstrated by coprecipitation of Am(III) with ferrihydrite and hematite nanoparticles and subsequent observation of actinide and colloid migration in humic rich ground water passing through a sandy sediment column (Schäfer et al. 2003).

For nanoparticle mobility in clay stone formations or compacted bentonite, where diffusive ground water transport dominates and rock porosity is in the nm range, mostly indirect information is available. Diffusion experiments with

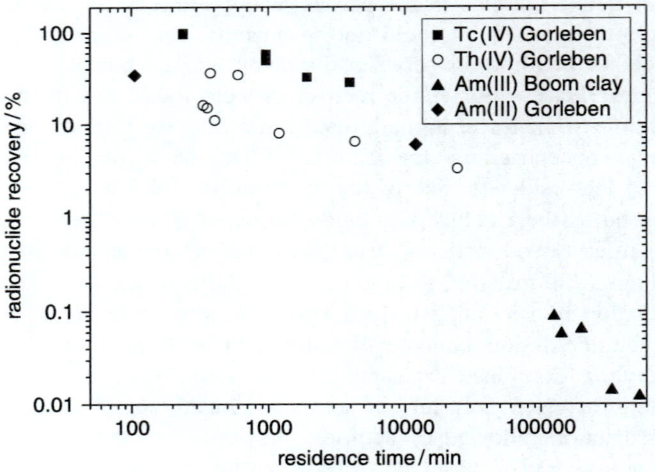

Fig. 2 Radionuclide recoveries in dynamic migration experiments with humic/fulvic acid containing ground/porewater in sandy sediments and plastic clay as a function of tracer residence time [Gorleben data are taken from Artinger et al. (1998, 2003) and Boom Clay data from Maes et al. (2006)]

lignosulfonate and humic acid nanoparticles through MX-80 bentonite (dry densities from 600 to 1,800 kg/m^3) with an average size of 80 and <10 nm, respectively, showed diffusivities of these organic nanoparticles to be in the same range as negatively charged ions like Cl^- and I^-, which was unexpected. The effect of humic nanoparticles on both Co and Eu sorption, as well as on their diffusion, was significant. The apparent diffusivity (D_a) increased significantly for both Co(II) and Eu(III) when humic nanoparticles were present and the distribution coefficient (K_d) values decreased (SKB 2006a; Wold and Eriksen 2003, 2007). Explanations for the lack of nanoparticle filtration could be the presence of larger pores in the compacted bentonite, or transformation of the molecules during the course of the experiment.

6.2.2 Actinide-Inorganic Colloid Migration

Earlier studies on actinide behavior in granitic fractures of a block scale experiment under defined geochemical conditions (pH = 8.3; Ionic strength: 0.013 mol/L) (Vilks and Baik 2001) showed that Am(III) was carried only to a small extent by natural colloids (19.3 mg/L). The colloids were extracted from natural ground water and consisted primarily of organic particles and aluminosilicates. Montmorillonite-type clay nanoparticles are investigated as most relevant inorganic colloidal radionuclide carriers. They are inherently present in all disposal concepts in crystalline rock via introduction of bentonite backfill material. Clay colloids might then be mobilized and in turn may mobilize radionuclides, e.g., after intrusion of low mineralized glacial melt water into the repository near field (Neretnieks 2006), as discussed above. The relevance of inorganic colloid mediated actinide transport is closely coupled to colloid stability and thus to ionic strength and pH of the ground water. Those stability boundary conditions are much stricter than for colloidal humic/fulvic matter, which can exist under much wider geochemical conditions. Under the relatively high ionic strength conditions of ground water ($I > 0.5$ mol/L) taken from the Swedish underground rock laboratory (URL), colloids can be even a sink for actinide migration due to their agglomeration and subsequent sedimentation (Bouby et al. 2008). In the low ionic strength high pH (pH = 9.6, $I = 10^{-3}$ mol/L) ground water of the Swiss URL at the GTS, however, high mobility for bentonite colloids has been reported in various in situ migration studies. The colloid-borne lanthanides and actinides Tb(III), Am(III) and Th(IV), Pu(IV) were found to be transported in dipole experiments over distances up to 5 m with high colloid and actinide recoveries ranging from 30 to 80% (Geckeis et al. 2004; Hauser et al. 2002; Möri et al. 2003). Various laboratory and field experiments (Bouby et al. 2010; Missana et al. 2008; Schäfer et al. 2004) were subsequently carried out to gain more detailed insight into the long-term mobility of clay colloids under given conditions. By decreasing ground water flow rates and thus increasing lanthanide/actinide-colloid tracer residence times in the fracture reduced tracer recoveries were observed (see Fig. 3). Recoveries decline for colloidal matter and lanthanide/actinide ions to a different extent with residence times (Albarran et al. 2009; Schäfer et al. 2004, 2009). The chemical conditions favored colloidal stability and "unfavorable"

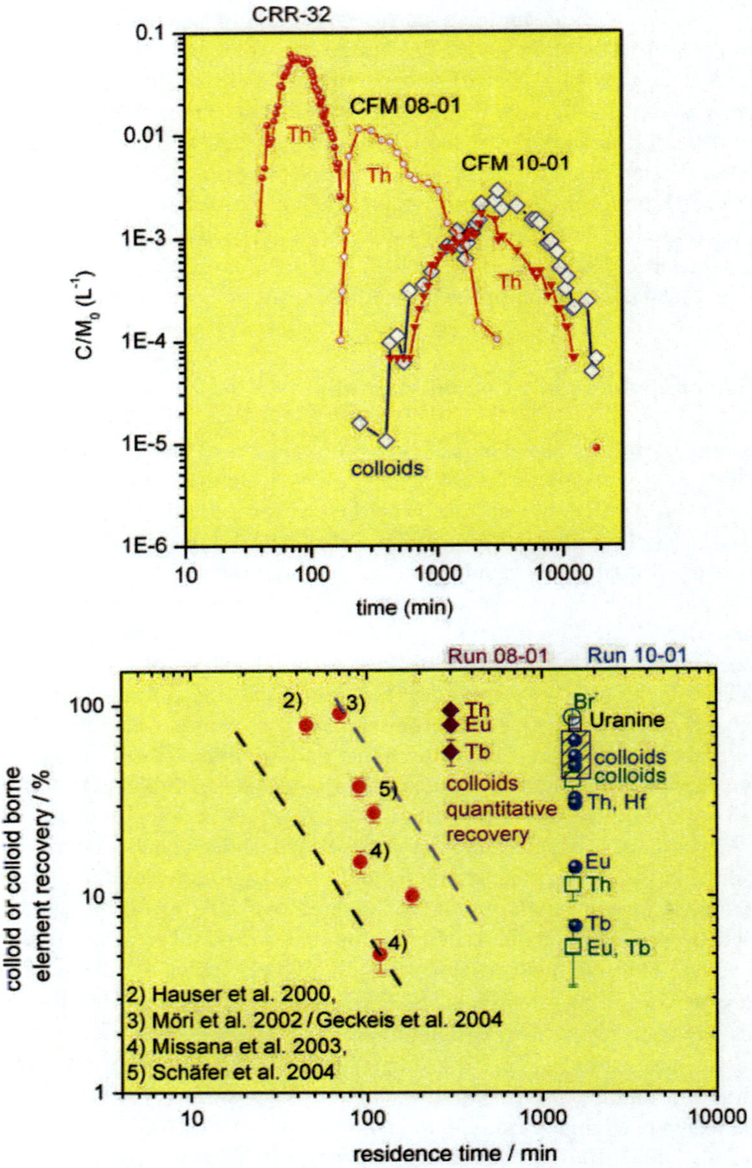

Fig. 3 (**a**) Th(IV) and clay colloid breakthrough curves recorded in migration experiments carried out at the GTS; c/m$_0$: measured Th(IV) and colloid concentration normalized to the total injected tracer mass; (**b**) Colloid and actinide/lanthanide recoveries obtained in various laboratory and field studies on GTS relevant conditions (GTS ground water and GTS fractures or borecores) (Bouby et al. 2010)

electrostatic conditions for colloid-rock attachment existed. Similar results of decreasing nanoparticle recovery were obtained when the roughness of the fracture surface was increased, as predicted by microscale modeling approaches implementing surface roughness (Boutt et al. 2006) and known from membrane fouling (Hoek and Agarwal 2006; Hoek et al. 2003). How retention of inorganic colloids in GTS fractures works precisely is presently not clear. In contrast to experiments on humic colloid facilitated actinide transport, most dynamic experiments with inorganic colloids were carried out without equilibrating the stationary phase with colloid-containing ground water. There are indications that clay colloids are subject to slow agglomeration and surface attachment over time, i.e., they are not stable in the ground water on the long term even though ionic strength is low and pH is high.

The quantification of colloid/nanoparticle retention at mineral or rock surfaces under conditions both unfavorable and favorable to electrostatic attachment was also analyzed by μPIXE from 2 to 100 nm diameter gold nanoparticles (Alonso et al. 2009), fluorescence microscopy using 25 nm carboxylated latex microspheres (Filby et al. 2008) and AFM force spectroscopy using the so-called "colloid probe" technique (Filby et al. 2009). Under favorable conditions, electrostatic attraction prevailed: strong adsorption of negatively charged colloids was in fact observed at pH values close or below the point of zero charge (pH_{PCZ}) of the respective mineral surface. The adsorption of polyvalent cations as Eu^{3+} (concentration: 10^{-5} mol L^{-1}) increased nanoparticle attachment to minerals (Filby et al. 2008). Under unfavorable conditions nanoparticle attachment was nevertheless still observed (Alonso et al. 2009), even in the alkaline regime typical for the Grimsel ground waters (Filby 2009). In this case, nanoparticle (carboxylated microspheres) adsorption can be expected on apatite. In the presence of Ca(II) or with Ca(II) containing Grimsel ground water, the repulsion forces clearly decrease and thus surface attachment becomes easier. Overall, nanoparticle retention on granite was found to be strongly dependent on small scale chemical effects, roughness or non-uniform distribution of charge sites or even the combination of the effects.

Beside the above mentioned surface retention on fracture walls, possible matrix diffusion is an additional process that can retain or eliminate colloids/nanoparticles from the mobile aqueous phase. Precise determination of bentonite nanoparticle diffusion coefficient in the granite rock matrix by Rutherford backscattering spectrometry (RBS) revealed diffusion coefficients of 10^{-17} m^2 s^{-1} for 250 ± 50 nm diameter bentonite nanoparticles (Alonso et al. 2007a). Surprisingly, studies on the size dependency of the diffusion coefficient performed with gold nanoparticles in the size range from 2 nm up to 100 nm by the same group revealed lower D_{max} values of 8×10^{-18} m^2 s^{-1} and D_{min} values of 8×10^{-19} m^2 s^{-1}, which might indicate additional sorption taking place (Alonso et al. 2007b). The role and the extent of matrix diffusion in granite rock on colloid retention is still controversially discussed (Schäfer et al. submitted). Even though clay colloid recoveries vary in many experiments, dissociation from clay nanoparticles seems to control lanthanide/actinide mobility under given conditions. In accordance with laboratory batch type studies, tetravalent actinide (Th(IV)) desorption is slower than that of trivalent elements (Tb/Eu/Am(III)). As already discussed for natural organic colloids, the impact and relevance of "irreversible" actinide colloid binding is still unclear and subject of further investigations.

7 Conclusions and Outlook

In order to obtain a complete picture of the colloid-mediated radionuclide migration processes, more knowledge on the interaction of the colloids with sediment and rock surfaces is required. The currently available theories on colloid filtration still appear to be inappropriate to describe the processes in heterogeneous natural systems (Ryan and Elimelech 1996). The existence of chemical and physical heterogeneity of natural solids is invoked as a main reason for the failure of classical electrostatic surface interaction models such as the DLVO theory describing attractive and repulsive forces between surfaces. The elucidation of colloid-rock interaction processes would certainly enhance the reliability of colloid transport modeling efforts. Currently available approaches follow the classical filter theory (Rajagopalan and Tien 1976) or recently modified methods have also appeared (Tufenkji and Elimelech 2005). Further developments take physical and chemical heterogeneities into account (Bekhit and Hassan 2005; Severino et al. 2007; Song et al. 1994). Metal ion colloid interaction kinetics is observed in most laboratory and field experiments and has to be considered in respective transport models. Reaction rates have thus been implemented in various codes (Contardi et al. 2001; Ibaraki and Sudicky 1995; Lührmann et al. 1998). However, in order to separate the effects of flowpath geometry, chemical heterogeneity, and kinetics of actinide reactions as much as possible, a detailed characterization of the systems is required, e.g., using a combination of tomographic analysis of flowpaths, post mortem analysis, direct visualization of contaminant transport and development of respective modeling approaches.

Acknowledgments Part of the studies described here were funded by the Federal Ministry of Economics and Technology (BMWi) under the joint KIT-INE, GRS research project "KOLLORADO & KOLLORADO 2" and the SKB/KTH "Colloid project". We also want to thank all organizations contributing to the funding of the Colloid Formation and Migration (CFM) project, namely: ANDRA (France), BMWi (Germany), JAEA (Japan), CRIEPI (Japan), AIST (Japan), NAGRA (Switzerland), POSIVA (Finland) and SKB (Sweden).

References

Albarran N., Missana T., Alonso U., Garcia-Gutierrez M., and Lopez-Torrubia T. (2009) Dynamic study of bentonite colloid retention in a smooth granite fracture under unfavorable electrostatic conditions. *4th Annual Workshop Proceeding of integrated project "Fundamental processes of Radionuclide Migration" – 6th EC FP IP FUNMIG, FZKA report 7461*, 421–428.

Alonso U., Missana T., Patelli A., and Rigato V. (2007a) Bentonite colloid diffusion through the host rock of a deep geological repository. *Physics and Chemistry of the Earth* **32**(1–7), 469–476.

Alonso U., Missana T., Patelli A., Rigato V., and Ravagnan J. (2007b) Colloid diffusion in crystalline rock: An experimental methodology to measure diffusion coefficients and evaluate colloid size dependence. *Earth and Planetary Science Letters* **259**(3–4), 372–383.

Alonso U., Missana T., and Garcia-Gutierrez M. (2008) Experimental approach to study the bentonite colloid generation source term in different geochemical conditions. *3rd Annual workshop Proceedings of IP FUNMIG, NDA report*, 329–336.

Alonso U., Missana T., Patelli A., Ceccato D., Albarran N., Garcia-Gutierrez M., Lopez-Torrubia T., and Rigato V. (2009) Quantification of Au nanoparticles retention on a heterogeneous rock surface. *Colloids and Surfaces A: Physicochemical and Engineering Aspects* **347**(1–3), 230–238.

Artinger R., Kienzler B., Schuessler W., Kim J.I. (1998) Effects of humic substances on the Am-241 migration in a sandy aquifer: column experiments with Gorleben groundwater/sediment systems. *J. Contam. Hydrol.*, **35**, 261–275.

Artinger R., Buckau G., Geyer S., Wolf M., Kim J. I., and Fritz P. (1998a) Characterization of humic and fulvic acids from the Gorleben aquifer system by GPC with UV/Vis and fluorescence detection. *9th international meeting of the International Humic Substance Society.*

Artinger R., Kienzler B., Schüssler W., and Kim J. I. (1998b) Effects of humic substances on the 241Am migration in a sandy aquifer: Column experiments with Gorleben groundwater/sediment systems. *Journal of Contaminant Hydrology* **35**, 261–275.

Artinger R., Buckau G., Geyer S., Fritz P., Wolf M., and Kim J. I. (2000) Characterization of groundwater humic substances: Influence of sedimentary organic carbon. *Applied Geochemistry* **15**(1), 97–116.

Artinger R., Rabung T., Kim J. I., Sachs S., Schmeide K., Heise K. H., Bernhard G., and Nitsche H. (2002a) Humic colloid-borne migration of uranium in sand columns. *Journal of Contaminant Hydrology* **58**(1–2), 1–12.

Artinger R., Schuessler W., Schäfer T., and Kim J. I. (2002b) A kinetic study of Am(III)/humic colloid interactions. *Environmental Science & Technology* **36**(20), 4358–4363.

Artinger R., Schuessler W., Scherbaum F., Schild D., and Kim J. I. (2002c) Am-241 migration in a sandy aquifer studied by long-term column experiments. *Environmental Science & Technology* **36**(22), 4818–4823.

Artinger R., Buckau G., Zeh P., Geraedts K., Vancluysen J., Maes A., and Kim J. I. (2003) Humic colloid mediated transport of tetravalent actinides and technetium. *Radiochimica Acta* **91**(12), 743–750.

Backhus D. A., Ryan J. N., Groher D. M., MacFarlane J. K., and Gschwend P. M. (1993) Sampling colloids and colloid associated contaminants in ground water. *Ground Water* **31**(3), 466–479.

Baik M. H., Cho W. J., and Hahn P. S. (2007) Erosion of bentonite particles at the in-terface of a compacted bentonite and a fractured granite. *Engineering Geology* **91**(2–4), 229–239.

Barcelona M. J., Varljen M. D., Puls R. W., and Kaminski D. (2005) Ground water-purging and sampling methods: History vs. hysteria. *Ground water Monitoring and Remediation* **25**, 52–62.

Bekhit H. M., Hassan A. E. (2005) Two-dimensional modeling of contaminant transport in porous media in the presence of colloids. *Adv. in Water Res.*, **28**, 1320–1335.

Benedetti M. F., Milne C. J., Kinniburgh D. G., Vanriemsdijk W. H., and Koopal L. K. (1995) Metal-ion binding to humic substances - application of the nonideal competitive adsorption model. *Environmental Science & Technology* **29**(2), 446–457.

BGR (1991) Untertägige geowissenschaftliche Erkundung des Standortes Gorleben. Bundesanstalt für Geowissenschaften und Rohstoffe.

Bouby M., Geckeis H., Schäfer T., Luützenkirchen J., Seher H., Bauer A., Plaschke M., Hauser W., and Kienzler B. (2008) Laboratory study on colloid stability and radionuclide-colloid interaction under Äspö groundwater conditions. *3rd Annual workshop Proceedings of IP FUNMIG, NDA report*, 391–400.

Bouby M., Filby A., Geckeis H., Geyer F., Götz R., Hauser W., Huber F., Kienzler B., Kunze P., Küntzel M., Lützenkirchen J., Noseck U., Plaschke M., Pudewills A., Schäfer T., Seher H., and Walther C. (2010) Colloid/Nanoparticle formation and mobility in the context of deep geological nuclear waste disposal (Project KOLLORADO-1; Final report). In *FZKA report 7515*, Vol. FZKA report 7515 (ed. T. Schäfer and U. Noseck), pp. 210. Forschungszentrum Karlsruhe.

Boutt D. F., Grasselli G., Fredrich J. T., Cook B. K., Williams J. R. (2006) Trapping zones: The effect of fracture roughness on the directional anisotropy of fluid flow and colloid transport in a single fracture. *Geophys. Res. Letters*, **33**, L21402.

Bradbury M. H. and Baeyens B. (2005) Modelling the sorption of Mn(II), Co(II), Ni(II), Zn(II), Cd(II), Eu(III), Am(III), Sn(IV), Th(IV), Np(V) and U(VI) on montmorillonite: Linear free energy relationships and estimates of surface binding constants for some selected heavy metals and actinides. *Geochimica Et Cosmochimica Acta* **69**, 875–892.

Bradbury M. H. and Baeyens B. (2009) Sorption modelling on illite. Part II: Actinide sorption and linear free energy relationships. *Geochimica Et Cosmochimica Acta* **73**(4), 1004–1013.

Bryan N. D., Jones D. M., Keepax R. E., Farrelly D. H. (2006) Non-exchangeable binding of radionuclides by humic substances: a chemical analogue. In *Advances in Actinide Science*, (eds. I. May, N. D. Bryan and R. Alvarez), Royal Society of Chemistry, Cambridge, UK.

Bryan N. D., Abrahamsen L., Evans N., Warwick P., Buckau G., Wenig L., Van Riemsdijk W. H. (2011) The effects of humic substances on the transport of radionuclides: recent improvements in the prediction of behavior and understanding of mechanisms. *Appl. Geochem., accepted.*

Buck E. C. and Bates J. K. (1999) Microanalysis of colloids and suspended particles from nuclear waste glass alteration. *Applied Geochemistry* **14**(5), 635–653.

Buck E. C., Finn P. A., and Bates J. K. (2004) Electron energy-loss spectroscopy of anomalous plutonium behavior in nuclear waste materials. *Micron* **35**(4), 235–243.

Buckau G. (1991) Komplexierung von Americium (III) mit Huminstoffen in natürlichen Grundwässern. PhD Freie Universität Berlin.

Buckau G. (2000) Effects of Humic Substances on the Migration of Radionuclides: Complexation and Transport of Actinides (HUMICS), Final Report, Report EUR 19610 EN.

Buckau G. (2004) Humic substances in performance assessment of nuclear waste disposal: Actinide and iodine migration in the far-field: 2nd technical progress report pp. 350. Forschungszentrum Karlsruhe.

Buckau G. (2005) Origin, stability and mobility of humic colloids in Äspö groundwater: Feasibility study. *The Colloid investigations conducted at the Äspö Hard Rock laboratory during the years 2000–2004*.

Buckau G. (2006) Humic Substances in Performance Assessment of Nuclear Waste Disposal: Actinide and Iodine Migration in the Far-Field (HUPA), Final Report, Report EUR 21928.

Buckau G., Kim J. I., Klenze R., Rhee D. S., and Wimmer H. (1992) A comparative spectroscopy study of the fulvate complexation of trivalent transuraniums ions. *Radiochimica Acta* **57**, 105–111.

Buckau G., Artinger R., Fritz P., Geyer S., Kim J. I., and Wolf M. (2000) Origin and mobility of humic colloids in the Gorleben aquifer system. *Applied Geochemistry* **15**(2), 171–179.

Buesseler K. O., Hassellov M., and Dai M. (2003) Comment on "Trace metal levels in uncontaminated groundwater of a coastal watershed: Importance of colloidal forms". *Environmental Science & Technology* **37**(3), 657–658.

Catalano J. G. and Brown G. E. (2005) Uranyl adsorption onto montmorillonite: Evaluation of binding sites and carbonate complexation. *Geochimica Et Cosmochimica Acta* **69**(12), 2995–3005.

Chen G. and Flury M. (2005) Retention of mineral colloids in unsaturated porous media as related to their surface properties. *Colloids and Surfaces A: Physicochemical and Engineering Aspects* **256**(2–3), 207–216.

Chheda P., Grasso D., and van Oss C. J. (1992) Impact of ozone on the stability of montmorillionite suspensions. *Journal of Colloid and Interface Science* **153**, 226–236.

Chisholm-Brause C. J., Berg J. M., Little K. M., Matzner R. A., and Morris D. E. (2004) Uranyl sorption by smectites: spectroscopic assessment of thermodynamic modeling. *Journal of Colloid and Interface Science* **277**(2), 366–382.

Choppin G. R. and Clark S. B. (1991) The kinetic interactions of metal-ions with humic acids. *Marine Chemistry* **36**(1–4), 27–38.

Chung K. H., Klenze R., Park K. K., Paviet-Hartmann P., and Kim J. I. (1998) A study of the surface sorption process of Cm(III) on silica by time-resolved laser fluorescence spectroscopy (I). *Radiochimica Acta* **82**, 215–219.

Claret F., Schäfer T., Brevet J., and Reiller P. E. (2008) Fractionation of Suwannee river fulvic acid and aldrich humic acid on α-Al2O3: Spectroscopic evidence. *Environmental Science & Technology* **42**(23), 8809–8815.

Contardi J. S., Turner D. R., and Ahn T. M. (2001) Modeling colloid transport for performance assessment. *Journal of Contaminant Hydrology* **47**(2–4), 323–333.

Courdouan A., Christl I., Meylan S., Wersin P., and Kretzschmar R. (2007a) Characterization of dissolved organic matter in anoxic rock extracts and in situ pore water of the Opalinus Clay. *Applied Geochemistry* **22**(12), 2926–2939.

Courdouan A., Christl I., Meylan S., Wersin P., and Kretzschmar R. (2007b) Isolation and characterization of dissolved organic matter from the Callovo-Oxfordian formation. *Applied Geochemistry* **22**, 1537–1548.

Dähn R., Scheidegger A. M., Manceau A., Curti E., Baeyens B., Bradbury M. H., and Chateigner D. (2002) Th uptake on montmorillonite: A powder and polarized extended X-ray absorption fine structure (EXAFS) study. *Journal of Colloid and Interface Science* **249**(1), 8–21.

Dai M., Kelley J. M., and Buesseler K. O. (2002) Sources and migration of plutonium in groundwater at the Savannah River Site. *Environmental Science & Technology* **36**(17), 3690–3699.

Dai M., Buesseler K. O., and Pike S. M. (2005) Plutonium in groundwater at the 100K-Area of the U.S. DOE Hanford Site. *Journal of Contaminant Hydrology* **76**(3–4), 167–189.

Dearlove J. P. L., Longworth G., Ivanovich M., Kim J. I., Delakowitz B., and Zeh P. (1991) A study of groundwater-colloids and their geochemical interactions with natural radionuclides in Gorleben aquifer systems. *Radiochimica Acta* **52–53**, 83–89.

Degueldre C., Grauer R., Laube A., Oess A., and Silby H. (1996a) Colloid properties in granitic groundwater systems.2. Stability and transport study. *Applied Geochemistry* **11**(5), 697.

Degueldre C., Pfeiffer H.-R., Alexander W., Wernli B., and Bruetsch R. (1996b) Colloid properties in granitic groundwater systems. I: Sampling and characterisation. *Applied Geochemistry* **11**, 677–695.

Degueldre C., Scholtis A., and Thomas B. (1998) Opalinus clay groundwaters and colloids. A sampling exercise at Mt. Terri (June/July 197). Analytical results.

Degueldre C., Triay I., Kim J.I., Vilks P., Laaksoharju M., Miekeley N. (2000) Groundwater colloid properties: a global approach. *Appl. Geochem.*, **15**, 1043–1051.

Degueldre C., Reed D., Kropf A. J., and Mertz C. (2004) XAFS study of americium sorbed onto groundwater colloids. *Journal of Synchrotron Radiation* **11**, 198–203.

Delos A., Walther C., Schäfer T., and Buchner S. (2008) Size dispersion and colloid mediated radionuclide transport in a synthetic porous media. *Journal of Colloid and Interface Science* **324**(1–2), 212–215.

Denecke M. A., Bublitz D., Kim J. I., Moll H., and Farkes I. (1999) EXAFS investigation of the interaction of hafnium and thorium with humic acid and Bio-Rex70. *Journal of Synchrotron Radiation* **6**, 394–396.

Denecke M. A., Marquardt C. M., Rothe J., Dardenne K., and Jensen M. P. (2002) XAFS studies of actinide coordination structure in Np(IV)-fulvates. *J. Nuclear Science Technology* **3**(Suppl.), 410–413.

DeNovio N. M., Saiers J. E., and Ryan J. N. (2004) Colloid movement in unsaturated porous media: recent advances and future directions. *Vadose Zone Journal* **3**, 338–351.

Fernandes M. M., Stumpf T., Baeyens B., Walther C., and Bradbury M. H. (2010) Spectroscopic identification of ternary Cm-carbonate surface complexes. *Environmental Science & Technology* **44**(3), 921–927.

Filby A. (2009) Interaction of colloids with mineral surfaces: a microscopical and nanoscopical approach. PhD thesis, Universität Fridericiana Karlsruhe (TH).

Filby A., Plaschke M., Geckeis H., and Fanghänel T. (2008) Interaction of latex colloids with mineral surfaces and Grimsel granodiorite. *Journal of Contaminant Hydrology* **102**(3–4), 273–284.

Filby A., Plaschke M., Geckeis H., and Bosbach D. (2009) Interaction of latex colloids with mineral surfaces and Grimsel granodiorite. *4th Annual Workshop Proceeding of integrated project "Fundamental processes of Radionuclide Migration" – 6th EC FP IP FUNMIG, FZKA report 7461*, 285–290.

Gates W. P., Bouazza A., and Churchman G. J. (2009) Bentonite clay keeps pollutants at Bay. *Elements* **5**(2), 105–110.

Geckeis H., Rabung T., Manh T. N., Kim J. I., and Beck H. P. (2002) Humic colloid-borne natural polyvalent metal ions: Dissociation experiment. *Environmental Science & Technology* **36**(13), 2946–2952.

Geckeis H., Schäfer T., Hauser W., Rabung T., Missana T., Degueldre C., Möri A., Eikenberg J., Fierz T., and Alexander W. R. (2004) Results of the Colloid and Radionuclide Retention experiment (CRR) at the Grimsel Test Site (GTS), Switzerland – Impact of reaction kinetics and speciation on radionuclide migration-. *Radiochimica Acta* **92**(9–11), 765–774.

Hartmann E., Brendebach B., Polly R., Geckeis H., Stumpf Th. (2011) Characterization and quantification of Sm(III)/ and Cm(III)/clay mineral outer-sphere species by TRLFS in D_2O and EXAFS studies. *J. Coll. Interface Sci.*, **353**, 562–568.

Hauser W., Geckeis H., Kim J. I., and Fierz T. (2002) A mobile laser-induced breakdown detection system and its application for the in situ-monitoring of colloid migration. *Colloids and Surfaces A: Physicochemical and Engineering Aspects* **203**(1–3), 37–45.

Hauser W., Geckeis H., Götz R., Noseck U., and Laciok A. (2007) Colloid detection in natural ground water from Ruprechtov by laser-induced breakdown detection. In *2nd Annual Workshop Proceeding of integrated project "Fundamental processes of Radionuclide Migration" – 6th EC FP IP FUNMIG* (ed. G. Buckau, B. Kienzler, L. Duro, and V. Montoya), pp. 367–373.

Heidmann I., Christl I., and Kretzschmar R. (2005) Sorption of Cu and Pb to kaolinite-fulvic acid colloids: Assessment of sorbent interactions. *Geochimica Et Cosmochimica Acta* **69**(7), 1675–1686.

Hetzel F. and Doner H. E. (1993) Some colloidal properties of beidellite – Comparison with low and high charge montmorillonites. *Clays and Clay Minerals* **41**(4), 453–460.

Hoek E. M. V. and Agarwal G. K. (2006) Extended DLVO interactions between spherical particles and rough surfaces. *Journal of Colloid and Interface Science* **298**(1), 50–58.

Hoek E. M. V., Bhattacharjee S., and Elimelech M. (2003) Effect of membrane surface roughness on colloid-membrane DLVO interactions. *Langmuir* **19**(11), 4836–4847.

Honeyman B. D. (1999) Colloidal culprits in contamination. *Nature* **397**, 23–24.

Huittinen N., Rabung T., Lutzenkirchen J., Mitchell S. C., Bickmore B. R., Lehto J., and Geckeis H. (2009) Sorption of Cm(III) and Gd(III) onto gibbsite, alpha-Al(OH)(3): A batch and TRLFS study. *Journal of Colloid and Interface Science* **332**(1), 158–164.

Huittinen N., Rabung T., Andrieux P., Lehto J., Geckeis H. (2010) A comparative batch sorption and time-resolved laser fluorescence spectroscopy study on the sorption of Eu(III) and Cm(III) on synthetic and natural kaolinite. *Radiochim. Acta*, **98**, 613–620.

Hummel W. (2008) Radioactive contaminants in the subsurface: The influence of complexing ligands on trace metal speciation. *Monatshefte Fur Chemie* **139**(5), 459–480.

Ibaraki M. and Sudicky E. A. (1995) Colloid-facilitated contaminant transport in discretely fractured porous media .1. Numerical formulation and sensitivity analysis. *Water Resources Research* **31**(12), 2945–2960.

Ionescu A., Maes N., and Mallants D. (2008) Modelling transport of C-14-labelled Natural Organic Matter (NOM) in Boom Clay. *Scientific Basis for Nuclear Waste Management Xxxi* **1107**, 629–636.

Kaminski M. D., Goldberg M. M., and Mertz C. J. (2005) Colloids from the aqueous corrosion of aluminium-based nuclear fuel. *Journal of Nuclear Materials* **347**(1–2), 88–93.

Kaplan D., Bertsch P., Adriano D., and Miller W. P. (1993) Soil-borne mobile colloids as influenced water flow and organic carbon. *Environmental Science & Technology* **27**(6), 1193–1200.

Kersting A. B., Efurd D. W., Finnegan D. L., Rokop D. J., Smith D. K., and Thompson J. L. (1999) Migration of plutonium in ground water at the Nevada Test Site. *Nature* **397**, 56–59.

Kienzler B., Vejmelka P., Romer J., Fanghänel E., Jansson M., Eriksen T. E., and Wikberg P. (2003) Swedish-German actinide migration experiment at ASPO hard rock laboratory. *Journal of Contaminant Hydrology* **61**(1–4), 219–233.

Kim J. I. (1986) Chemical behaviour of transuranic elements in natural aquatic systems. In *Handbook on the Physics and Chemistry of the Actinide* (ed. A. I. Freeman and C. Keller), pp. Chap. 8. Elsevier Science Publishers.

Kim J. I. (1991) Actinide colloid generation in groundwater. *Radiochimica Acta* **52/53**, 71.

Kim J. I. (1994) Actinide colloids in natural aquifer systems. *Materials Research Society Bulletin* **19**(12), 47–53.

Kim J. I. (2006) Significance of actinide chemistry for the long-term safety of waste disposal. *Nuclear Engineering and Technology* **38**, 459–482.

Kim J. I. and Czerwinski K. R. (1996) Complexation of metal ions with humic acid: Metal ion charge neutralization model. *Radiochimica Acta* **73**, 5–10.

Kim J. I., Buckau G., Baumgärtner F., Moon H., and Lux D. (1984) Colloid generation and the actinide migration in Gorleben groundwaters. *Materials Research Society Symposium Proceedings* **26**, 31–40.

Kim J. I., Buckau G., Li G. H., Duschner H., and Psarros N. (1990) Characterization of humic and fulvic-acids from Gorleben groundwater. *Fresenius Journal of Analytical Chemistry* **338**(3), 245–252.

Kim M. A., Panak P. J., Yun J. I., Kim J. I., Klenze R., and Kohler K. (2003) Interaction of actinides with aluminosilicate colloids in statu nascendi Part I: generation and characterization of actinide(III)-pseudocolloids. *Colloids and Surfaces A: Physicochemical and Engineering Aspects* **216**(1–3), 97–108.

Kim M. A., Panak P. J., Yun J. I., Priemyshev A., and Kim J. I. (2005) Interaction of actinides(III) with aluminosilicate colloids in "statu nascendi" Part III. Colloid formation from monosilanol and polysilanol. *Colloids and Surfaces A: Physicochemical and Engineering Aspects* **254**(1–3), 137–145.

Kim M. A., Panak P. J., Breban D. C., Priemyshev A., Yun J. I., Mansel A., and Kim J. I. (2007) Interaction of actinides(III) with aluminosilicate colloids. Part IV. Influence of humic acid. *Colloids and Surfaces A: Physicochemical and Engineering Aspects* **296**(1–3), 206–215.

Kimura T., Kato Y., Takeishi H., and Choppin G. R. (1998) Comparative study on the hydration states of Cm(III) and Eu(III) in solution and in cation exchange resin. *Journal of Alloys and Compounds* **271**, 719–722.

King S. J., Warwick P., Hall A., Bryan N. D. (2001) The dissociation kinetics of dissolved metal-humic complexes. *Phys. Chem. Chem. Phys.*, **3**, 2080–2085.

Kosmulski M. (2009) pH-dependent surface charging and points of zero charge. IV. Update and new approach. *Journal of Colloid and Interface Science* **337**(2), 439–448.

Kowal-Fouchard A., Drot R., Simoni E., and Ehrhardt J. J. (2004) Use of spectroscopic techniques for uranium (VI)/montmorillonite interaction modeling. *Environmental Science & Technology* **38**(5), 1399–1407.

Kretzschmar R. and Schäfer T. (2005) Metal retention and transport on colloidal particles in the environment. *Elements* **1**(4), 205–210.

Kunze P., Seher H., Hauser W., Panak P. J., Geckeis H., Fanghänel T., and Schäfer T. (2008) The influence of colloid formation in a granite groundwater bentonite pore water mixing zone on radionuclide speciation. *Journal of Contaminant Hydrology* **102**, 263–272.

Kurosawa S. and Ueta S. (2001) Effect of colloids on radionuclide migration for performance assessment of HLW disposal in Japan. *Pure and Applied Chemistry* **73**(12), 2027–2037.

Kurosawa S., Tanaka S., James S. C., and Yui M. (2006) The effect of montmorillonite partial density on the role of colloid filtration by a bentonite buffer. *Journal of Nuclear Science and Technology* **43**(5), 605–609.

Lagaly G. and Ziesmer S. (2003) Colloid chemistry of clay minerals: the coagulation of montmorillonite dispersions. *Advances in Colloid and Interface Science* **100–102**, 105–128.

Lead J. R., Wilkinson K. J., Balnois E., Cutak B. J., Larive C. K., Assemi S., and Beckett R. (2000) Diffusion coefficients and polydispersities of the Suwannee River fulvic acid: Comparison of fluorescence correlation spectroscopy, pulsed-field gradient nuclear magnetic resonance, and flow field-flow fractionation. *Environmental Science & Technology* **34**(16), 3508–3513.

Leenheer J. A. and Croue J. P. (2003) Characterizing aquatic dissolved organic matter. *Environmental Science & Technology* **37**(1), 18A–26A.

Leonard K. S., McCubbin D., and Lovett M. B. (1995) Physico-chemical characterisation of radionuclides discharged from a nuclear establishment. *Science of the Total Environment* **175**(1), 9–24.

Liu J. and Neretnieks I. (2006) Physical and chemical stability of the bentonite buffer. Svensk Kärnbränslehantering AB.

Liu D. J., Bruggeman C., and Maes N. (2008) The influence of natural organic matter on the speciation and solubility of Eu in Boom Clay porewater. *Radiochimica Acta* **96**(9–11), 711–720.

Lührmann L., Noseck U., and Tix C. (1998) Model of contaminant transport in porous media in the presence of colloids applied to actinide migration in column experiments. *Water Resources Research* **34**(3), 421–426.

Maes A., Geraedts K., Bruggeman C., Vancluysen J., Rossberg A., and Hennig C. (2004) Evidence for the interaction of technetium colloids with humic substances by X-ray absorption spectroscopy. *Environmental Science & Technology* **38**(7), 2044–2051.

Maes N., Wang L., Hicks T., Bennett D., Warwick P., Hall T., Walker G., and Dierckx A. (2006) The role of natural organic matter in the migration behaviour of americium in the Boom Clay – Part I: Migration experiments. *Physics and Chemistry of the Earth, Parts A/B/C* **31**(10–14), 541–547.

Malkovsky V. I., Dikov Y. P., Kalmykov S. N., and Buleev M. I. (2009) Structure of colloid particles in groundwaters on the territory of the Mayak production association and its impact on the colloid transport of radionuclides in subsoil environments. *Geochemistry International* **47** (11), 1100–1106.

Marquardt C. M., Seibert A., Artinger R., Denecke M. A., Kuczewski B., Schild D., and Fanghänel T. (2004) The redox behaviour of plutonium in humic rich groundwater. *Radiochimica Acta* **92** (9–11), 617–623.

Marty R. C., Bennett D., and Thullen P. (1997) Mechanism of plutonium transport in a shallow aquifer in Mortandad canyon, Los Alamos National Laboratory, New Mexico. *Environmental Science & Technology* **31**(7), 2020–2027.

Matsunaga T., Nagao S., Ueno T., Takeda S., Amano H., and Tkachenko Y. (2004) Association of dissolved radionuclides released by the Chernobyl accident with colloidal materials in surface water. *Applied Geochemistry* **19**(10), 1581–1599.

McCarthy J. F. and Zachara J. M. (1989) Subsurface transport of contaminants: Mobile colloids in the subsurface environment may alter the transport of contaminants. *Environmental Science & Technology* **23**(5), 497–502.

Missana T., Alonso U., Garcia-Gutierrez M., and Mingarro M. (2008) Role of bentonite colloids on europium and plutonium migration in a granite fracture. *Applied Geochemistry* **23**(6), 1484–1497.

Monsallier J. M., Schüssler W., Buckau G., Rabung T., Kim J. I., Jones D., Keepax R., and Bryan N. (2003) Kinetic investigation of Eu(III)-humate interactions by ion exchange resins. *Analytical Chemistry* **75**(13), 3168–3174.

Möri A., Alexander W. R., Geckeis H., Hauser W., Schäfer T., Eikenberg J., Fierz T., Degueldre C., and Missana T. (2003) The colloid and radionuclide retardation experiment at the Grimsel Test Site: influence of bentonite colloids on radionuclide migration in a fractured rock. *Colloids and Surfaces A: Physicochemical and Engineering Aspects* **217**(1–3), 33–47.

Naber A., Plaschke M., Rothe J., Hofmann H., and Fanghänel T. (2006) Scanning transmission X-ray and laser scanning luminescence microscopy of the carboxyl group and Eu(III) distribution in humic acid aggregates. *Journal of Electron Spectroscopy and Related Phenomena* **153**(3), 71–74.

Nagra. (2002) Stability and mobility of colloids in Opalinus Clay (ed. A. Voegelin and R. Kretzschmar), pp. 33. Nagra.

Neck V., Altmaier M., Seibert A., Yun J. I., Marquardt C. M., and Fanghanel T. (2007a) Solubility and redox reactions of Pu(IV) hydrous oxide: Evidence for the formation of PuO2+x(s, hyd). *Radiochimica Acta* **95**(4), 193–207.

Neck V., Altmaier M., Seibert A., Yun J. I., Marquardt C. M., and Fanghänel T. (2007b) Solubility and redox reactions of Pu(IV) hydrous oxide: Evidence for the formation of PuO2+x(s, hyd). *Radiochimica Acta* **95**(4), 193–207.

Neretnieks I. (2006) Flow and transport through a damaged buffer - exploration of the impact of a cemented and an eroded buffer, pp. 43. SKB Technical Report 06-33, Svensk Kärnbränslehantering AG, Stockholm, Sweden.

Neretnieks I. and Liu J. (2006) Physical and chemical stability of the bentonite buffer. SKB Technical Report 06-103, Svensk Kärnbränslehantering AG, Stockholm, Sweden.

Nilsson A. C., Hedqvist I., and Degueldre C. (2008) Granitic groundwater colloids sampling and characterisation: The strategy for artefact elimination. *Analytical and Bioanalytical Chemistry* **391**(4), 1327–1333.

Novikov A. P., Kalmykov S. N., Utsunomiya S., Ewing R. C., Horreard F., Merkulov A., Clark S. B., Tkachev V. V., and Myasoedov B. F. (2006) Colloid transport of plutonium in the far-field of the Mayak Production Association, Russia. *Science* **314**(5799), 638–641.

Panak P., Klenze R., and Kim J. I. (1996) A study of ternary complexes of Cm(III) with humic acid and hydroxide or carbonate in neutral pH range by time-resolved laser fluorescence spectroscopy. *Radiochimica Acta* **74**, 141–146.

Panak P. J., Kim M. A., Yun J. I., and Kim J. I. (2003) Interaction of actinides with aluminosilicate colloids in statu nascendi. Part II: spectroscopic speciation of colloid-borne actinides(III). *Colloids and Surfaces A: Physicochemical and Engineering Aspects* **227**(1–3), 93–103.

Panak P. J., Kim M. A., Klenze R., Kim J. I., and Fanghänel T. (2005) Complexation of Cm(III) with aqueous silicic acid. *Radiochimica Acta* **93**(3), 133–139.

Penrose W. R., Polzer W. L., Essington E. H., Nelson D. M., and Orlandi K. A. (1990) Mobility of plutonium and americium through a shallow aquifer in a semiarid region. *Environmental Science & Technology* **24**(2), 228–234.

Plaschke M., Rothe J., Schäfer T., Denecke M. A., Dardenne K., Pompe S., and Heise K. H. (2002) Combined AFM and STXM in situ study of the influence of Eu(III) on the agglomeration of humic acid. *Colloids and Surfaces A: Physicochemical and Engineering Aspects* **197**(1–3), 245–256.

Puls R. W. (1995) Use of low-flow or passive sampling techniques for sampling ground water. In *Ground Water Sampling- A Workshop Summary.* U.S.E.P.A. United States Environmental Protection Agency, report number: EPA/600/R-94/205.

Puls R. W., Eychaner J. H., and Powell R. M. (1990) Colloidal-facilitated transport of inorganic contaminants in ground water: Part I. Sampling considerations. *USEPA Environmental Research Brief* EPA/600/M-90/023.

Rabung T. and Geckeis H. (2010) Influence of pH and metal ion loading on the Cm(III) humate complexation: A time resolved laser fluorescence spectroscopy study. *Radiochimica Acta* **97**, 265–271.

Rabung T., Pierret M. C., Bauer A., Geckeis H., Bradbury M. H., and Baeyens B. (2005) Sorption of Eu(III)/Cm(III) on Ca-montmorillonite and Na-illite. Part 1: Batch sorption and time-resolved laser fluorescence spectroscopy experiments. *Geochimica Et Cosmochimica Acta* **69** (23), 5393–5402.

Rajagopalan R. and Tien C. (1976) Trajectory analysis of deep-bed filtration with the sphere-in-cell porous media model. *AIChE Journal* **22**, 523–533.

Roberts K. A., Santschi P. H., Leppard G. G., and West M. M. (2004) Characterization of organic-rich colloids from surface and ground waters at the actinide-contaminated Rocky Flats Environmental Technology Site (RFETS), Colorado, USA. *Colloids and Surfaces A: Physicochemical and Engineering Aspects* **244**(1–3), 105–111.

Robinson B. A., Li C. H., and Ho C. K. (2003) Performance assessment model development and analysis of radionuclide transport in the unsaturated zone, Yucca Mountain, Nevada. *Journal of Contaminant Hydrology* **62–63**, 249–268.

Ryan J. N. and Elimelech M. (1996) Colloid mobilization and transport in groundwater. *Colloids and Surfaces A* **107**, 1–56.

Ryan J. N. and Gschwend P. M. (1990) Colloid mobilization in two Atlantic Coastal Plain aquifers: Field studies. *Water Resources Research* **26**(2), 307–322.

Ryan J. N. and Gschwend P. M. (1994) Effect of solution chemistry on clay colloid release from an iron oxide-coated aquifer sand. *Environmental Science & Technology* **28**(9), 1717–1726.

Salbu B., Bjornstad H. E., Svaren I., Prosser S. L., Bulman R. A., Harvey B. R., and Lovett M. B. (1993) Size distribution of radionuclides in nuclear-fuel reprocessing liquids after mixing with seawater. *Science of the Total Environment* **130**, 51–63.

Schäfer T., Artinger R., Dardenne K., Bauer A., Schuessler W., and Kim J. I. (2003) Colloid-borne Americium migration in Gorleben groundwater: Significance of iron secondary phase transformation. *Environmental Science & Technology* **37**(8), 1528–1534.

Schäfer T., Geckeis H., Bouby M., and Fanghänel T. (2004) U, Th, Eu and colloid mobility in a granite fracture under near-natural flow conditions. *Radiochimica Acta* **92**, 731–737.

Schäfer T., Buckau G., Artinger R., Kim J. I., Geyer S., Wolf M., Bleam W. F., Wirick S., and Jacobsen C. (2005) Origin and mobility of fulvic acids in the Gorleben aquifer system: implications from isotopic data and carbon/sulfur XANES. *Organic Geochemistry* **36**(4), 567–582.

Schäfer T., Chanudet V., Claret F., and Filella M. (2007) Spectromicroscopy mapping of colloidal/particulate organic matter in Lake Brienz, Switzerland. *Environmental Science & Technology* **41**(22), 7864–7869.

Schäfer T., Seher H., Hauser W., Walther C., Degueldre C., Yamada M., Suzuki M., Missana T., Alonso U., Trick T., and Blechschmidt I. (2009) The Colloid Formation and Migration (CFM) project at the Grimsel Test Site (Switzerland): Results from the homologue tests. *Geochimica Et Cosmochimica Acta* **73**(13), A1168.

Schäfer T., Huber F., Seher H., Missana T., Alonso U., Kumke M. U., Eidner S., Claret F., Enzmann F. (2011) Nanoparticles and their influence on radionuclide mobility in deep geological formations. *Appl. Geochem., accepted.*

Schlegel M. L. and Descostes M. (2009) Uranium uptake by hectorite and montmorillonite: A solution chemistry and polarized EXAFS study. *Environmental Science & Technology* **43**(22), 8593–8598.

Schmidt M., Stumpf T., Walther C., Geckeis H., and Fanghänel T. (2009) Incorporation versus adsorption: Substitution of Ca^{2+} by Eu^{3+} and Cm^{3+} in aragonite and gypsum. *Dalton Transactions* (33), 6645–6650.

Schuessler W., Artinger R., Kienzler B., and Kim J. I. (2000) Conceptual modeling of the humic colloid-borne Americium(III) migration by a kinetic approach. *Environmental Science & Technology* **34**, 2608–2611.

Schuessler W., Artinger R., Kim J. I., Bryan N. D., and Griffin D. (2001) Numerical modeling of humic colloid borne Americium(III) migration in column experiments using the transport/speciation code K1D and the KICAM model. *Journal of Contaminant Hydrology* **47**(2–4), 311–322.

Seher H., Albarran N., Hauser W., Götz R., Missana T., Geckeis H., Fanghänel T., and Schäfer T. (2009) Colloid generation by erosion of compacted bentonite under different geochemical conditions. *4th Annual Workshop Proceeding of integrated project "Fundamental processes of Radionuclide Migration" – 6th EC FP IP FUNMIG, FZKA report 7461*, 143–150.

Seher H., Hauser W., Geckeis H., Fanghänel T., and Schäfer T. (2010) Bentonite nanoparticle stability and effect of fulvic acids: experiments and modelling. *Colloids and Surfaces A.*

Seher H. (2010) Der Einfluss von Kolloiden auf die Migration von Radionukliden (The influence of colloids on radionuclide migration), Thesis, University of Heidelberg, 2011.

Severino G., Cvetkovic V., and Coppola A. (2007) Spatial moments for colloid-enhanced radionuclide transport in heterogeneous aquifers. *Advances in Water Resources* **30**(1), 101–112.

Sholkovitz E. R. (1976) Flocculation of dissolved organic and inorganic matter during mixing of river water and seawater. *Geochimica Et Cosmochimica Acta* **40**(7), 831–845.

Simoni E. (2002) Radionuclides retention: from macroscopic to microscopic. *Comptes Rendus Physique* **3**(7–8), 987–997.

SKB (2005) The colloid investigations conducted at the Äspö Hard Rock Laboratory during 2000–2004, Technical report TR-05-20 (ed. M. Laaksoharju and S. Wold), pp 620. Svensk Kärnbränslehantering AB.

SKB (2006a) Buffer and backfill process report for the safety assessment SR-Can, pp. 620. Svensk Kärnbränslehantering AB.

SKB (2006b) Long-term safety for KBS-3 repositories at Forsmark and Laxemar – a first evaluation; Main report of the SR-Can project. SKB Technical report TR-06-09, pp. 620. Svensk Kärnbränslehantering AB.

Soderholm L., Almond P. M., Skanthakumar S., Wilson R. E., and Burns P. C. (2008) The structure of the plutonium oxide nanocluster [Pu38O56Cl54(H2O)(8)](14-). *Angewandte Chemie-International Edition* **47**(2), 298–302.

Song L., Johnson P. R., and Elimelech M. (1994) Kinetics of colloid deposition onto heterogeneously charged surfaces in porous media. *Environmental Science & Technology* **28**, 1164–1171.

Stumpf T. and Fanghänel T. (2002) A time-resolved laser fluorescence spectroscopy (TRLFS) study of the interaction of trivalent actinides (Cm(III)) with calcite. *Journal of Colloid and Interface Science* **249**(1), 119–122.

Stumpf T., Bauer A., Coppin F., and Il Kim J. (2001) Time-resolved laser fluorescence spectroscopy study of the sorption of Cm(III) onto smectite and kaolinite. *Environmental Science & Technology* **35**(18), 3691–3694.

Stumpf Th., Bauer A., Coppin F., Fanghänel Th., Kim J. I. (2002) Inner-sphere, outer-sphere and ternary surface complexes: A TRLFS study of the sorption process of Eu(III) onto smectite and Kaolinite. *Radiochim. Acta,* **90**, 345–349.

Stumpf T., Hennig C., Bauer A., Denecke M. A., and Fanghänel T. (2004) An EXAFS and TRLFS study of the sorption of trivalent actinides onto smectite and kaolinite. *Radiochimica Acta* **92** (3), 133–138.

Stumpf T., Fernandes M. M., Walther C., Dardenne K., and Fanghänel T. (2006) Structural characterization of Am incorporated into calcite: A TRLFS and EXAFS study. *Journal of Colloid and Interface Science* **302**(1), 240–245.

Suteerapataranon S., Bouby M., Geckeis H., Fanghänel T., and Grudpan K. (2006) Interaction of trace elements in acid mine drainage solution with humic acid. *Water Research* **40**(10), 2044–2054.

Sylwester E. R., Hudson E. A., and Allen P. G. (2000) The structure of uranium(VI) sorption complexes on silica, alumina and montmorillonite. *Geochimica et Cosmochimica Acta* **64**(14), 2431–2438.

Tipping E. (1998) Humic ion-binding model VI: An improved description of the interactions of protons and metal ions with humic substances. *Aquatic Geochemistry* **4**(1), 3–48.

Tombácz E. and Szekeres M. (2004) Colloidal behavior of aqueous montmorillonite suspensions: the specific role of pH in the presence of indifferent electrolytes. *Applied Clay Science* **27**, 75–94.

Tufenkji N. and Elimelech M. (2005) Breakdown of colloid filtration theory: Role of the secondary energy minimum and surface charge heterogeneities. *Langmuir* **21**(3), 841–852.

Tullborg E. L. and Gustafsson E. (1999) C-14 in bicarbonate and dissolved organics – A useful tracer? *Applied Geochemistry* **14**(7), 927–938.

Utsunomiya S., Kersting A. B., and Ewing R. C. (2009) Groundwater nanoparticles in the far-field at the Nevada Test Site: Mechanism for radionuclide transport. *Environmental Science & Technology* **43**(5), 1293–1298.

Van De Weerd H. (1997) Assessment of the effect of kinetics on colloid facilitated radionuclide transport in porous media. *Journal of contaminant Hydrology* **26**, 245–256.

Vilks P. and Bachinski D. B. (1996) Colloid and suspended particle migration experiments in a granite fracture. *Journal of Contaminant Hydrology* **21**(1–4), 269–279.

Vilks P. and Baik M. H. (2001) Laboratory migration experiments with radionuclides and natural colloids in a granite fracture. *Journal of Contaminant Hydrology* **47**(2–4), 197–210.

Walther C., Buchner S., Filella M., and Chanudet V. (2006) Probing particle size distributions in natural surface waters from 15 nm to 2 μm by a combination of LIBD and single-particle counting. *Journal of Colloid and Interface Science* **301**(2), 532–537.

Wang X. K., Rabung T., Geckeis H., Panak P. J., Klenze R., and Fanghänel T. (2004) Effect of humic acid on the sorption of Cm(III) onto gamma-Al_2O_3 studied by the time-resolved laser fluorescence spectroscopy. *Radiochimica Acta* **92**(9–11), 691–695.

Weng L. P., Van Riemsdijk W. H., and Hiemstra T. (2007) Adsorption of humic acids onto goethite: Effects of molar mass, pH and ionic strength. *Journal of Colloid and Interface Science* **314**(1), 107–118.

Wilkinson K. J. and Lead J. R. (2006) Environmental Colloids and Particles Behaviour, Separation and Characterisation. In *IUPAC Series on Analytical and Physical Chemistry of Environmental Systems*, Vol. 10, pp. 702. John Wiley & Sons.

Wimmer H., Kim J. I., and Klenze R. (1992) A direct speciation of Cm(III) in natural aquatic systems by time-resolved laser-induced fluorescence spectroscopy (Trlfs). *Radiochimica Acta* **58–59**, 165–171.

Wold S. and Eriksen T. E. (2003) Diffusion of lignosulfonate colloids in compacted bentonite. *Applied Clay Science* **23**(1–4), 43–50.

Wold S. and Eriksen T. (2007) Diffusion of humic colloids in compacted bentonite. *Physics and Chemistry of the Earth* **32**(1–7), 477–484.

Zachara J. M., Serne J., Freshley M., Mann F., Anderson F., Wood M., Jones T., and Myers D. (2007) Geochemical processes controlling migration of tank wastes in Hanford's vadose zone. *Vadose Zone Journal* **6**(4), 985–1003.

Zeh P., Kim J. I., Marquardt C., and Artinger R. (1999) Reduction of Np(V) in groundwater rich in humic substances. *Radiochimica Acta* **87**, 23–28.

Zhuang J., McCarthy J. F., Tyner J. S., Perfect E., and Flury M. (2007) In situ colloid mobilization in hanford sediments under unsaturated transient flow conditions: Effect of irrigation pattern. *Environmental Science & Technology* **41**(9), 3199–3204.

Section 1
Methods for Actinide Nanoparticle Characterization

Scanning Transmission Electron Microscopy and Related Techniques for Research on Actinide and Radionuclide Nanomaterials

Satoshi Utsunomiya, Masashi Kogawa, Eigo Kamiishi, and Rodney C. Ewing

Abstract The physical and chemical properties of actinide-bearing materials, as well as other radionuclides, such as fission product elements from the nuclear fuel cycle, depend greatly on their electronic configuration, crystal structure, thermochemical parameters, and the amount of impurity elements at the atomic scale. Further, nanoscale particles may have distinctly different properties from the bulk composition. In order to understand the properties of such materials, direct characterization at the nanoscale is essential. This chapter reviews relevant methods for direct analysis of nanoscale materials using a focused electron beam, scanning transmission electron microscopy (STEM), in which the electron probe can be less than an Å size with the current high enough to perform elemental analysis. High-angle annular dark-field STEM (HAADF-STEM) provides an incoherent image by which the intensity correlates with the atomic number. The HAADF-STEM image can be greatly enhanced by a theoretical filtering method, such as the maximum entropy method. Electron energy-loss spectroscopy (EELS) allows the investigation of the chemical state including oxidation state and the electron density of states at the nanoscale. Three dimensional electron tomography with STEM or TEM imaging is another useful method for obtaining morphological and topological information of nanoscale materials. In addition, the recent development of the aberration corrector for spherical aberration (C_S) has achieved a sub-Å probe as small as ~0.5 Å in STEM, which greatly improves the spatial resolution of images and chemical analyses. The application of C_S-corrected STEM has not been explored in actinide research; however, it has great potential in the investigation of the properties of actinide materials at the atomic-level.

S. Utsunomiya (✉), M. Kogawa, and E. Kamiishi
Department of Chemistry, Kyushu University, 6-10-1 Hakozaki, Higashi-ku, Fukuoka 812-8581, Japan
e-mail: utu@chem.rc.kyushu-u.ac.jp

R.C. Ewing
Department of Geological Sciences, University of Michigan, Ann Arbor 48109-1005, MI, USA

S.N. Kalmykov and M.A. Denecke (eds.), *Actinide Nanoparticle Research*,
DOI 10.1007/978-3-642-11432-8_2,

1 Introduction

Detailed characterization of nanoscale materials – their structure, composition, and chemical state – is essential for the evaluation of their physical and chemical properties. One of the most powerful methods for obtaining such information is by the use of transmission electron microscopy (TEM). Conventional TEM (CTEM) including analytical electron microscopy (AEM) and high-resolution TEM have been widely used. Indeed, for actinide research, early TEM and scanning TEM (STEM) studies already provided important knowledge about alteration of nuclear waste glass (e.g., [1–4]) and fission products in spent nuclear fuel (e.g., [5]). The fundamentals of basic CTEM are described in many text books (e.g., [6, 7]); thus, this chapter will not cover CTEM. Rather, this paper reviews various aspects in STEM, because the focused probe is a powerful means for determining both the local composition and the chemical state by using an energy dispersive X-ray spectroscopy (EDS) and an electron energy-loss spectroscopy (EELS) at a spot size that is near the atomic scale. With an HAADF detector, STEM is capable of incoherent imaging that reveals the chemical composition. The combination of HAADF-STEM with EDS or EELS is a very useful method for describing the heterogeneous distribution of radionuclides; and thus, it has been applied to the characterization of nuclear materials, which are summarized in this chapter. In addition, the size of STEM probe has been dramatically improved in the recent several years owing to the application of the spherical aberration (C_S) corrector to STEM. The other capability that is described is tomography, which allows three dimensional visualization of nanoscale materials. These two imaging methods are also reviewed in this chapter.

2 HAADF-STEM

2.1 *Comparison of HAADF-STEM with Conventional TEM*

Although HRTEM and the associated analytical capabilities are generally the most appropriate techniques for investigating the structural and chemical properties of nanoscale particles, the amount of the sample that is actually examined is extremely limited due to the high resolution in TEM. This is an important limitation when the element of interest is sparsely distributed and at low concentrations, particularly because the contrast in CTEM is formed by a coherent imaging process. Even though the atomic scattering factor is greater for the heavy elements, such as actinides, and those atoms appear ideally as darker spots (Fig. 1a, top), the contrast transfer function (CTF) varies as a function of objective lens defocus and the specimen thickness when forming the coherent image in high resolution phase contrast imaging. Hence, it is not the easiest

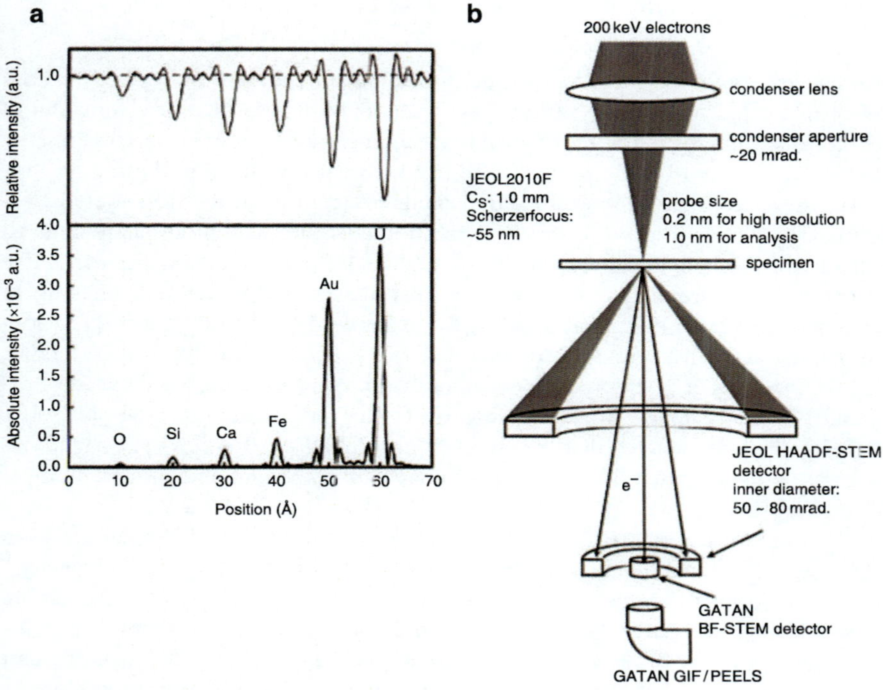

Fig. 1 (**a**) Simulated contrast profile calculated for each atom of each element discussed in the examples of this study. The calculation was completed using the code developed by Kirkland [8]. The atoms were placed in a row at 1-nm intervals. (*top*) BF-CTEM contrast profile. (*bottom*) HAADF-STEM contrast profile. (**b**) Schematic illustration of the electron microscope configuration used in the HAADF-STEM mode (JEOL JEM2010F). Both (**a**) and (**b**) were reproduced from Utsunomiya and Ewing [9]

approach to observe actinides at very low concentrations. Bright-field (BF) images in the STEM mode are formed by a coherent process, essentially the same as BFTEM at high resolution; whereas, STEM turns into a remarkably powerful method when electrons are collected by a high-angle annular detector. Figure 1 shows the configuration of an electron microscope in the HAADF-STEM mode. When the particle consists of relatively heavy elements, as compared to the composition of the matrix material, HAADF-STEM becomes useful for finding the nanoparticles of interest, as the contrast of the image is strongly correlated with atomic number and sample thickness. For example, a contrast profile calculated for each atom with various Z numbers based on the multislice method is illustrated in Fig. 1a (bottom), which clearly reveals a positive correlation of the contrast in HAADF-STEM with Z. The power of the derived Z dependence of the intensity generally ranges 1.5–1.8; that is, the intensity $\propto Z^{1.5\text{–}1.8}$ [10].

2.2 Development in HAADF-STEM

As one can easily imagine, the early studies using annular dark-field (ADF)-STEM were dedicated to imaging various heavy atoms on the surface of a thin film of carbon, and interestingly, the selected targets included U acetate, organic matter with U dimer, and U chloride [11, 12]. It was already realized in those days that, given an appropriate geometry for the annular detector, elastically scattered electrons at large angle (50–100 mrad) form an image with negligible phase-contrast effects. In the late 1980s and the early 1990s, HAADF-STEM was under active development, particularly for obtaining near-atomic scale chemical imaging of inorganic crystals in material science; for example, Pt catalyst clusters [13], ion-implanted Si [14], superconductor material, $YBa_2Cu_3O_{7-\delta}$ [15, 16]. Unlike CTEM, HAADF-STEM is based on the incoherent imaging process, and the contrast of the image is not reversed by defocusing or by the specimen thickness [17, 18]. A theoretical study noted that the intensity in HAADF-STEM can vary slightly as a function of specimen thickness due to contributions from coherent and multiple scattering [19].

Thermal diffuse scattering (TDS), which is a signal used to form the image in HAADF-STEM and which was previously considered as "background intensity," became a powerful source of information by using an HAADF detector. According to Kirkland [8], fundamentally, each atom in the specimen vibrates thermally with a frequency of 10^{12}–10^{13} Hz. The atomic vibration is quantized, and the quantum unit is a "phonon". The electron's velocity in the beam of the electron microscope is approximately 1.5×10^{10} cm/s. Thus, the time required for the electron to pass through the specimen is approximately 0.7×10^{-16} s, which is much shorter than the oscillation period of the atoms. However, the period between successive electrons in an incident electron beam is longer than that of atomic oscillations. Therefore, the image of each configuration of atoms is unique due to the random atomic oscillations; as a result, the average of all the configurations of atoms eventually becomes incoherent. The TDS profile through spatial frequency can be calculated by the "frozen phonon" method, in which the atoms are allowed to deviate from their original position following a Gaussian distribution [20].

2.3 Basic Imaging Theory in HAADF-STEM

The fundamental physics of the HAADF-STEM are described following Kirkland [8]. The wave function of a focused probe at a position, $\mathbf{x}_p$, is expressed as an integration of the aberration wave function over the objective aperture (aperture size is α_{max} in maximum angle):

$$\psi_p(\mathbf{x}, \mathbf{x}_p) = A_p \int_0^{k_{max}} \exp[-i\chi(\mathbf{k}) - 2\pi i \mathbf{k}(\mathbf{x} - \mathbf{x}_p)] d\mathbf{k} \qquad (1)$$

$$\lambda k_{\max} = \alpha_{\max} \tag{2}$$

where $\chi(k)$ is the aberration function, λ is the wavelength of the electron, and A_p is a normalization factor to produce:

$$\int \left| \psi_p(\mathbf{x}, \mathbf{x}_p)^2 \right| d\mathbf{x} = 1 \tag{3}$$

The transmission function of the specimen is:

$$t(\mathbf{x}) = \exp[i\sigma v_z(\mathbf{x})] \tag{4}$$

Here, σ is an interaction parameter expressed as:

$$\sigma = \frac{2\pi}{\lambda V}\left(\frac{m_0c^2 + eV}{2m_0c^2 + eV}\right) \tag{5}$$

where m_0, c, and eV are the rest mass of the electron, the speed of light in vacuum, and the kinetic energy of the electron in vacuum, respectively. The $v_z(\mathbf{x})$ in (4) is the projected atomic potential of the specimen. The resulting transmitted wave function is:

$$\Psi_t(\mathbf{x}) = t(\mathbf{x})\psi_p(\mathbf{x}, \mathbf{x}_p) \tag{6}$$

Finally, the wave function on the diffraction plane can be described as:

$$\Psi_t(\mathbf{k}) = FT(\psi_t(\mathbf{x})) \tag{7}$$

The signal on HAADF detector, $g(\mathbf{x}_p)$, is the integration of intensity on the detector:

$$g(\mathbf{x}_p) = \int D(\mathbf{k})|\Psi_t(\mathbf{k})|^2 d\mathbf{k} \tag{8}$$

$$\begin{aligned} D(k) &= 1 \quad \text{on the detector} \\ &= 0 \quad \text{outside of the detector.} \end{aligned} \tag{9}$$

As expected from (1), (7), and (8), a convergent-beam diffraction pattern is formed on the back-focal plane. Thus, the intensity on the detector is the sum of the pure elastic scattered electrons and the pure TDS electrons. Because the elastic scattering must satisfy Black's condition, that is the reciprocal lattice must cross Ewald's sphere, the intensity of zero order Laue zone (ZOLZ) becomes weak and disappears in the high-frequency region of the diffraction space. As a result, TDS scattering becomes dominant at high angles and is recorded by the HAADF detector. However, the first order or high order Laue zone (FOLZ or HOLZ) is possible, and this signal may also be recorded by the HAADF detector. Eventually, the contrast of the HAADF image is a result of the relative intensities of elastic scattering and TDS on the HAADF detector.

2.4 *Image Simulation and Filtering for HAADF-STEM*

To simulate the high resolution HAADF-STEM image, Bethe's eigenvalue method [18, 21] and the multislice method [22–24] have been employed. An advantage of the multislice method is its ability to account for mixed layers in the calculation of specimen potential, which is performed by a linear superposition of the atomic potential of each atom in the specimen; the atomic potential is related to the scattering factor.

McGibbon et al. [25] demonstrated significant enhancement in the quality of the *Z*-contrast image by using the maximum entropy analysis. The image intensity is generally described by a deconvolution between a point spread function and an object function, and the authors applied the technique of maximum entropy [26, 27] in order to obtain the sharpest "most likely" object function resulting in decrease of the image noise and to locate the atomic column positions with an accuracy of approximately 0.02 nm. This method leads to the much improved atomic-scale *Z*-contrast imaging coupled with C_S-corrected STEM [28], which is described in a later section.

2.5 *Application of HAADF-STEM to Nuclear Materials*

The HAADF-STEM study has not been widely employed in radiological sciences as compared with numerous applications reported in materials science. In this section, we first describe examples of HAADF-STEM applied to uranium nanoparticles in various environments and then discuss topics relevant to nuclear materials.

2.5.1 Uranium in Atmosphere

The occurrence of toxic metals including actinides in the ambient atmosphere is a critical environmental issue. In particular, those metals in fine fraction (PM2.5) can have increased toxicity and most likely penetrate deep into the respiratory system. Utsunomiya et al. [29] performed chemical and structural characterization of individual fine atmospheric particulates collected from Detroit urban area utilizing HAADF-STEM in addition to bulk analysis. Among many heavy metals, the authors found particles with extremely high contrast in HAADF-STEM image, which mainly consist of U and O (Fig. 2a). In this sample, the U concentration was determined to be 0.25 ng/m^3 (9.2 ppm) prior to the TEM investigation by bulk analysis. This work demonstrated that even actinides existing at the very low concentrations can be detected efficiently by means of HAADF-STEM. Interestingly, some of the nanocrystals of uraninite were completely encapsulated by a fulleroid (Fig. 2b). In this "fulleroid shell," the U appears to be partially protected from the ambient oxidizing conditions of the atmosphere, and this reduces its mobility in the environment, as oxidized species of uranium are highly mobile. Rigorous characterization of U in the atmospheric particulates is directly linked to its bioavailability and also provides new insight into the transportation processes of uranium by aerosols.

2.5.2 Uranium Precipitates on the Bacteria

Uranium can be utilized by microbes [30–33], which take advantage of the multiple oxidation states of U(IV and VI) as a source of energy. The use of STEM was not common in this research topic because the experiments were well controlled, and CTEM usually provided the essential data for the characterization of the nanoparticles formed on the surface of bacteria. Fayek et al. [34] demonstrated the value of the coupled analyses of HAADF-STEM and nano-SIMS to characterize U nanoparticles formed on the cell surface. In their study, a *Geobacter sulfurreducens* biofilm was placed in contact with U^{6+}-rich synthetic groundwater. It is obvious that the HAADF-STEM images shown in Fig. 3 generally provide information of U localization both in large and small areas of observation. The associated EDS elemental maps confirm the U distribution that is approximately correlated with the distribution of high contrast spots, although the mapping takes as long as ~1 h, and the X-ray count at low concentrations is often not distinguishable from background counts due to the short dwell time per pixel. Hence, HAADF-STEM imaging is helpful to quickly determine the location of the U at a scale $<$10 nm. These U precipitates were subsequently characterized as uraninite, UO_{2+x}, which were also observed in the previous study [33].

2.5.3 Submicron Scale Phenomena in Uraninite and Coffinite Alteration

Natural U ore deposits mainly consist of uraninite, UO_{2+x}, and this provides a great opportunity for the investigation of the process of uraninite alteration over geologic time scales, specifically reflecting the regional mineralogy, the physicochemical conditions of the fluid, and the age of formation of the U-ore. Because uraninite alteration process in nature can provide an important insight that is applicable to the safe geological disposal of U-bearing nuclear waste, many previous studies have reported bulk and microscopic phenomena in several U deposits all over the world, which have already been summarized in a review article (e.g., [35]). On the other hand, there are limited studies that characterized alteration zone at the submicron scale. Recently, a HAADF-STEM has been successfully applied to

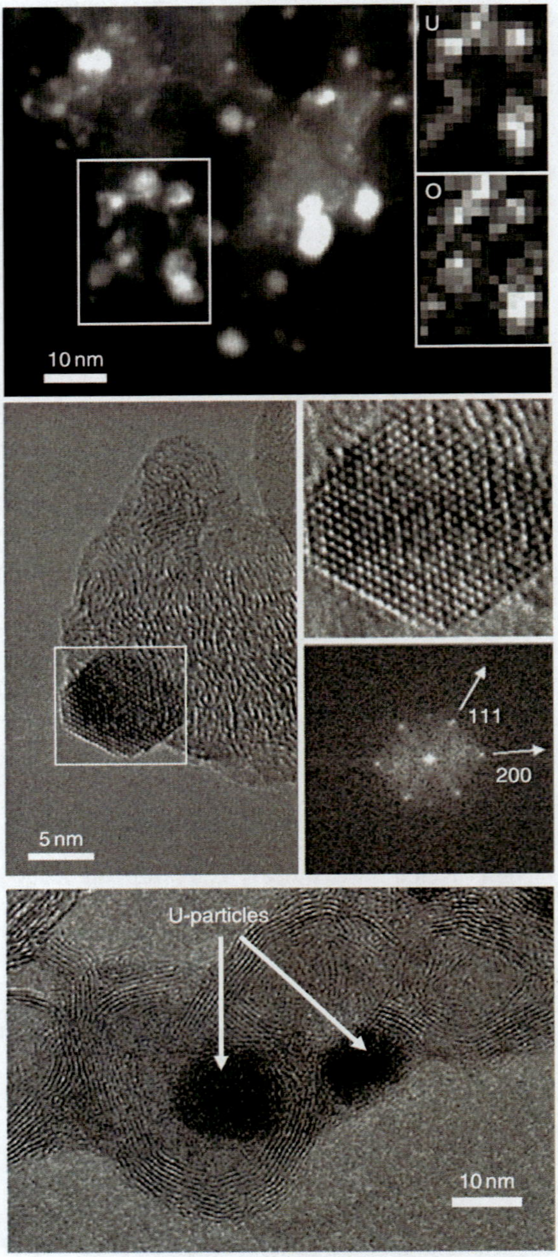

Fig. 2 Application of HAADF-STEM to characterize uraninite nanoparticles in the atmospheric particulates [29] (**a**) BF-TEM image reveals the presence of small particles within the carbonaceous matter of the soot. (**b**) HAADF-STEM image of the area reveals the occurrence of many heavy particles less than 10 nm in size. The *insets* are elemental maps of uranium and oxygen by STEM-EDX. (**c**) HRTEM images show that U-bearing nanoparticles are encapsulated in the "cage" structure of fulleroid

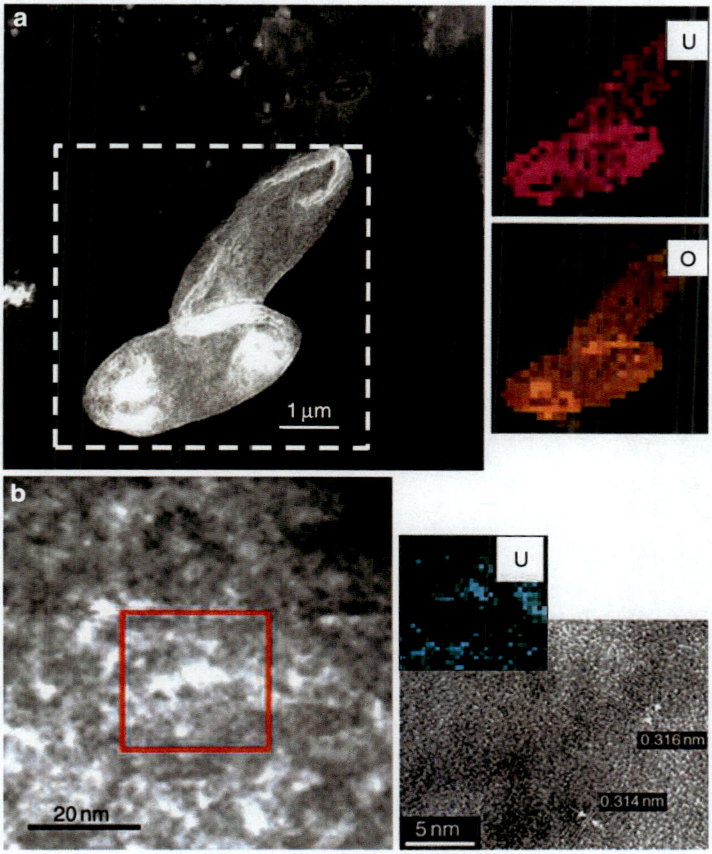

Fig. 3 HAADF-STEM images of Geobacter sulfurreducens after contacting with U solution [34]. (**a**) Magnified image showing the whole bacteria. (**b**) STEM-EDX element-distribution maps of Na, Cl, O, and U showing these elements located on the surface of the bacteria. *Red box* represents area selected for high-resolution (nm) HAADF-STEM analysis shown in (**b**). (**a**) (**b**) HAADF-STEM image of the surface of bacteria (area outlined in *red*, Fig. 2). (**b**) STEM–EDX element-distribution maps of Na, Cl, O, and U from the region outlined in (**a**), showing the congruency between O and U, and Na and Cl. Note that U and Cl are decoupled. (**c**) HRTEM image showing the size of the nanocrystals of uranium oxide on the surface of the bacteria. Numbers are d values, which are indicative of the mineral uraninite (UO_2). Samples were obtained from the 4-day experiments

investigate the alteration and precipitation of coffinite, $USiO_4$ [36, 37], and coffinitization of uraninite [38]. For example, Fig. 4a–c display alteration front of coffinite in the presence of apatite. The sample was from the Woodrow Mine, Grants uranium region, New Mexico. The phase boundary and the dissolution pores were clearly evident even at the thicker parts of the specimen, while they are unambiguously displayed if observed by CTEM. Thus, it is emphasized that HAADF-STEM imaging is suitable for a relatively thick sample. Based on these

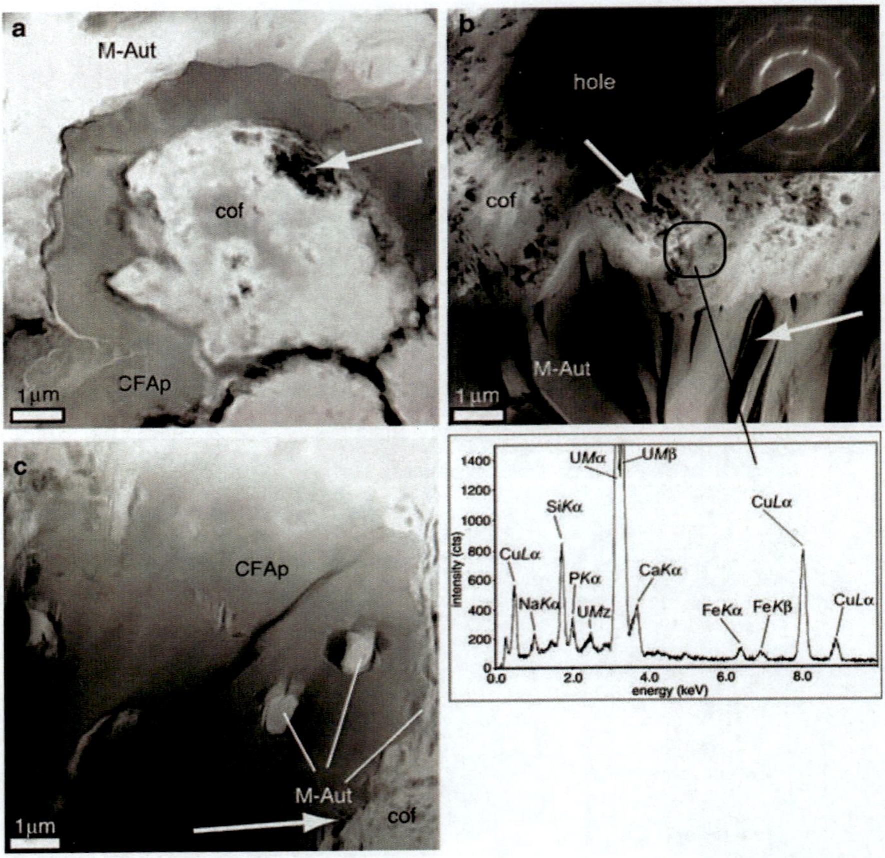

Fig. 4 HAADF-STEM images of the associated coffinite, M-Aut and CFAp [37]. (**a**) Dissolution of CFAp by coffinite and subsequent alteration of coffinite and CFAp to M-Aut; (**b**) Alteration of coffinite to M-Aut with associated SAED pattern of M-Aut and EDS analysis of the porous coffinite enriched with Na (*rimmed area*); the diffused diffraction maxima indicate a low crystallinity of precipitated M-Aut; (**c**) M-Aut precipitating inside CFAp grains and at the interface between CFAp and coffinite; (**a**)–(**c**) Note the extensive formation of porosity

data coupled with the results from electron microprobe, the authors concluded that microscale dissolution of apatite can create conditions conducive to the precipitation of U(IV)- and U(VI)-minerals, leading to the reduced mobility of U-species under both reducing and oxidizing conditions [37].

2.5.4 Groundwater Colloids and Nanoscale Particles in the Contaminated Sites

Nanometer-sized colloids, consisting of inorganic and/or organic compounds, provide an important means of transporting elements with low solubilities, including the actinides [39, 40]. The formation of actinide pseudo-colloids, in which the actinide sorbs onto aquatic colloids, can stabilize actinides in natural waters and increase their concentrations by many orders of magnitude over the values expected from solubility calculations [40, 41]. In spite of the numerous studies of colloids, there is still a lack of information on the speciation of the actinides or the type of colloids with which they are associated, particularly during the transport in the far-field, some kilometers from the source. At the Mayak Production Association, Urals, Russia, Pu was found to be associated with amorphous Fe-hydroxide (Fig. 5a and b) and was transported through the groundwater system [43]. Unfortunately, the HAADF-STEM was not able to capture the adsorbed Pu atoms on those colloids. It was also reported that Pu has migrated 1.3 km in 30 years in groundwater by means of colloids with sizes of 7 nm to 1 mm at Nevada Test Site [44]. In Utsunomiya et al. [42], a variety of radionuclides including U were identified in the same samples of Kersting 0et al. [44] from Nevada Test Site by a HAADF-STEM; cesium uranate, Co in Fe–Ni–Cr alloy, uranyl minerals (Na-boltwoodite and U oxide hydrates) (Fig. 5c–e). The presence of these colloids was not predicted considering the solubility of these phases, which may imply a thin coating layer on the surface of the colloids. Nonetheless, the phase-dependent transport of these elements is of critical importance, as each colloidal phase has characteristic properties in terms of mobility, under the specific physicochemical conditions of groundwater.

2.5.5 Uranium Cluster Formation on Muscovite

Arnold et al. [45] conducted U(VI) adsorption experiments onto muscovite at room temperature, and then the U speciation was examined by HAADF-STEM and time-resolved laser-induced fluorescence spectroscopy (TRLFS). Results from TRLFS indicated the presence of two adsorbed uranium(VI) surface species on edge-surfaces of muscovite with different coordination environments: On the other hand, HAADF-STEM revealed that nanoclusters of an amorphous uranium phase were attached to the edge-surfaces of muscovite. These U-nanoclusters were not observed on {001} cleavage planes of the muscovite. Combined with the TRLFS data, the two surface species were characterized as truly adsorbed bidentate surface complexes, in which the U(VI) binds to aluminol groups of edge-surfaces, and amorphous U(VI) condenses as nanosized clusters of polynuclear uranyl(VI) surface species with a particle diameter of 1–2 nm. Formation of these U nanoclasters on the edge of ferrous mica was also reported in Ilton et al. [46]. This example clearly demonstrates the successful combination of HAADF-STEM with another analytical method, in this case, TRLFS.

2.5.6 Applications to the Nuclear Materials

As a daughter nuclide in the decay chain, Pb is of critical importance in radiochemistry and geosciences. On the other hand, zircon ($ZrSiO_4$) has been proposed as

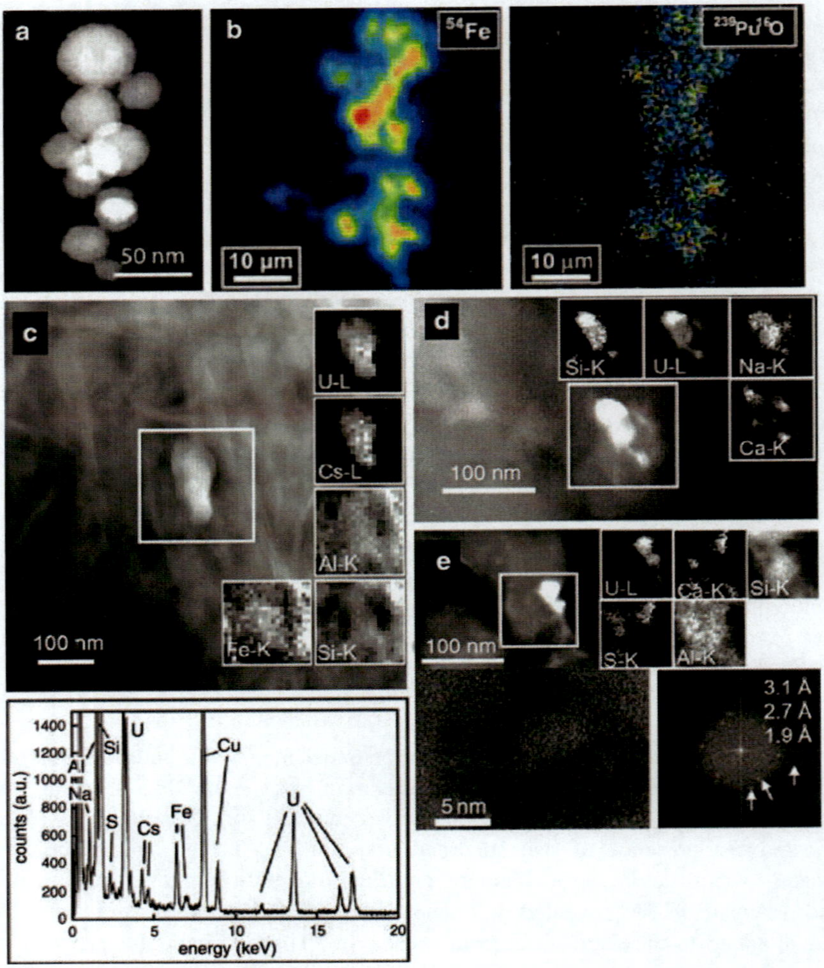

Fig. 5 (**a**) HAADF-STEM image of the spherical colloids in the groundwater collected at the Mayak Production Association, Urals, Russia. (**b**) Nano-SIMS elemental maps of the colloids. (**c**) HAADF-STEM image of Cs–U-phase with the elemental maps. (**d**) EDX spectrum (*bottom*) of this Cs–U-particle. Cu-peak is from Cu-grid. HAADF-STEM image of Na-boltwoodite with elemental maps. (**e**) HAADF-STEM image with the elemental maps of U-phase attached to the zeolite surface. The HRTEM image (*bottom left*) of the U-phase accompanied by the indexed FFT image (*bottom right*) reveals the presence of uraninite nanocrystals after exposure to the electron beam in STEM mode [42]

a waste form for immobilization of Pu from dismantled nuclear weapons [47, 48]. Hence, loss of trace elements, e.g., U, Pb, and Pu, from zircon depends on the form of these elements and is of critical interest. Utsunomiya et al. [49] investigated a high concentration Pb-zone (0.03 wt%) in a natural zircon crystal from meta-sedimentary belt near the Jack Hills [50, 51] using HRTEM and HAADF-STEM.

A HAADF-STEM image revealed some patches of relatively high contrast, ~5 nm in the diameter (Fig. 6a), and the nanoscale elemental map revealed the high concentration of Pb (~3 wt%). The high resolution HAADF-STEM image of a Pb-rich region showed a continuous zircon structure without distortion across the Pb-rich region (Fig. 6b). It was suggested that (1) Pb substitutes directly for Zr in the zircon structure, and that (2) there is no other Pb-rich phase present.

In addition to the patchy occurrence of Pb, numerous fission tracks were observed in <5 nm width, which were created by high-energy (~100 MeV), heavy particles from spontaneous fission events (Fig. 6c). Although spontaneous fission of ^{238}U is infrequent (decay constant $= 10^{-16}$ per year), the old age (>~3.3 Ga) and actinide content of these zircons suggests that fission tracks are to be expected. The majority of the fission tracks showed a dark contrast in the HAADF-STEM images; however, one fission track had a higher concentration of Pb (Fig. 6d) throughout the length of the entire track. The edge of the zircon lattice (Fig. 6e), indicated by a yellow arrow, shows a lattice image with higher contrast than the bulk zircon matrix, indicating the incorporation of Pb atoms into the zircon structure adjacent to the fission track. These results suggest that among many fission tracks and nanoscale Pb-rich domains, one Pb-rich domain was coincidently cut across by a fission track, and Pb preferentially diffused along the amorphous track.

The advantage of HAADF-STEM with EDS mapping was also demonstrated in a study that characterized the metallic epsilon phase nanoparticles in the reactor zone 10 in Oklo natural fission reactor [52]. The epsilon phase (Mo–Ru–Pd–Tc–Rh) is an important host of ^{99}Tc derived from a fission reaction, which has a long half life (2.13×10^5 years). HAADF-STEM images show an Ru–As particle ~300 nm in size surrounded by a Pb-rich region within the uraninite matrix (Fig. 7a). The composition of the particle was determined semiquantitatively as: As, 59.9; Co, 2.5; Ni, 5.2; Ru, 18.6; Th, 8.4; Pd, 3.1; Sb, 2.4 in atomic% (Fig. 7b). The Ru–As phase was not a single particle but an aggregate of 100–200 nm sized particles. The phase was identified as ruthenarsenite, (Ru, Ni)As, based on the HRTEM image and the FFT (fast Fourier transformed) image (Fig. 7c).

Another Ru-phase observed in this sample was 600–700 nm in size. The elemental distribution is rather complicated for this inclusion, as shown in the elemental maps (Fig. 7d). Lead occurs at the core of the particle; whereas, the rim of the inclusion consists of Ni, Co, and As without Ru. The Ru is enriched in the intermediate zone, associated with As and Ni, forming ruthenarsenite. This complexly mixed texture suggests that a process of dissolution and precipitation occurred within this inclusion. Based on the nanoscale characterization by HAADF-STEM combined with EDS and HRTEM, the history that the epsilon phases have experienced may be as follows (1) The original epsilon phase was changed to, in most cases, ruthenarsenite, by As-rich fluids with other trace metals. Dissolution and a simultaneous precipitation may be responsible for the phase change. (2) All Mo and most of the Tc were released from the epsilon phase. Galena precipitated surrounding the 4*d*-metal phases. (3) Once the uraninite matrix has dissolved, the epsilon nanoparticles were released and "captured" within alteration phases that are immediately adjacent to the uraninite.

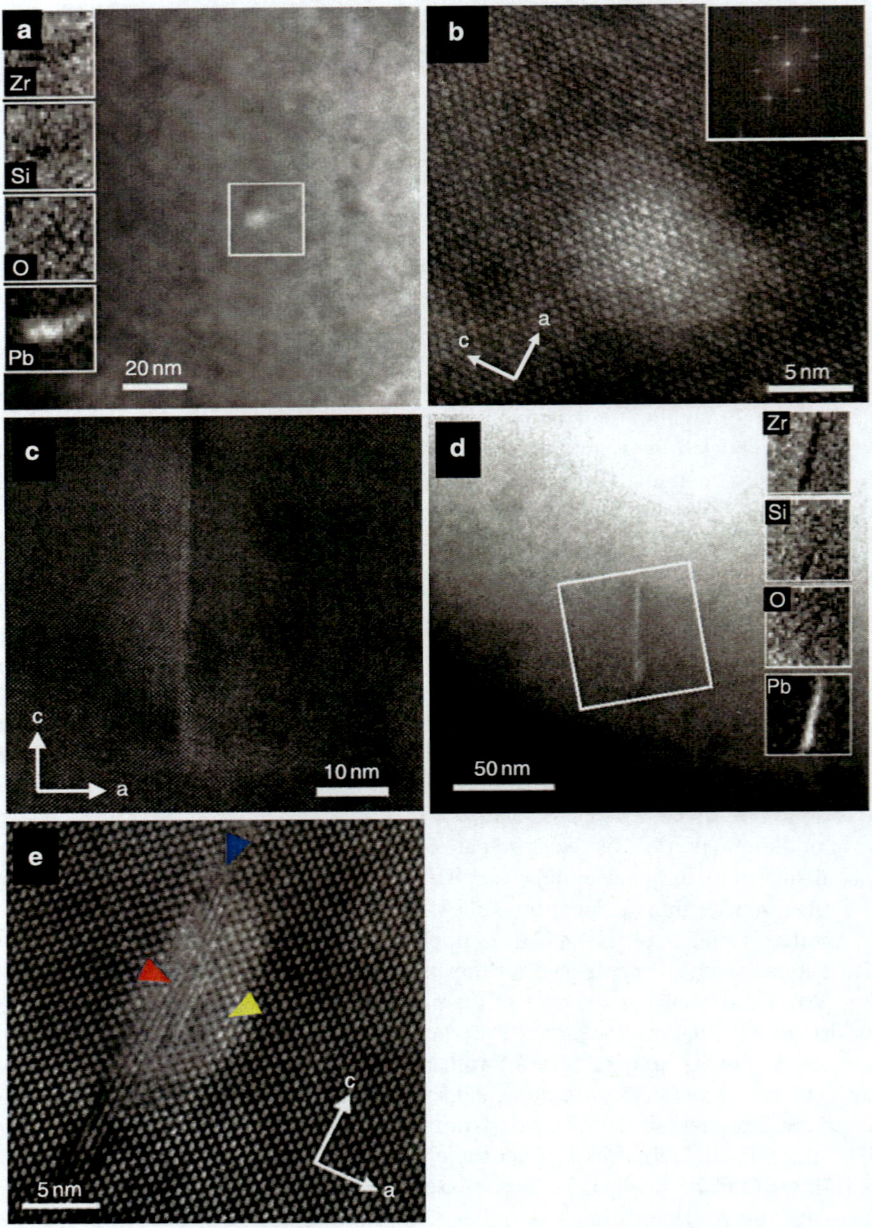

Fig. 6 The occurrence of Pb in the Archean zircon [49]. (**a**) HAADF-STEM image of a patchy concentration of Pb in zircon associated with elemental maps of the *boxed region*. (**b**) High resolution HAADF-STEM image from the view along [010] with the *inset* of the FFT image. (**c**) HRTEM image of the fission track with the view along [010]. The direction of the fission track is parallel to the (010) plane. (**d**) HAADF-STEM image of the fission track with the elemental maps of the *boxed area*. (**e**) High resolution HAADF-STEM image of the fission track from the view along [010]

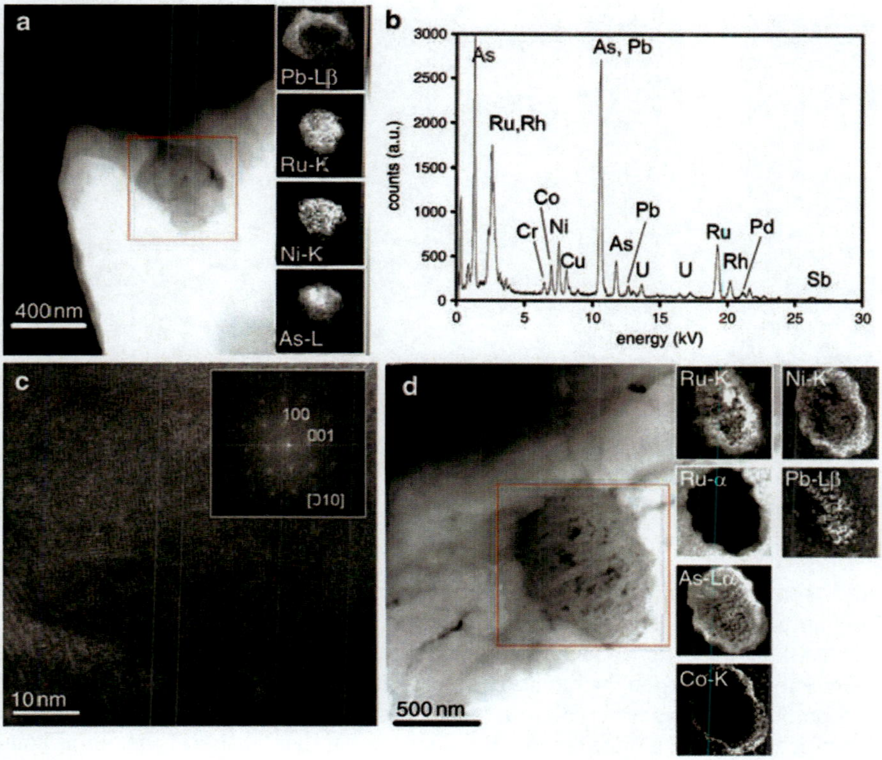

Fig. 7 HAADF-STEM images (**a** and **d**) with EDS maps of the epsilon particles in the uraninite matrix in the reactor zone 10 of Oklo natural fission reactor [52]. (**a**) Ruthenarsenite nanoparticle, which was identified by the EDS spectrum (**b**) and the HRTEM with FFT image (**c**). (**d**) A Ru-phase mixed with the other metals; Ni, Co, As, and Pb

HAADF-STEM imaging was also employed to characterize ion-irradiated synthetic garnet structure that incorporates a variety of potential fission products [53]. Figure 8 shows the high resolution TEM (a) and HAADF-STEM image (b) of the garnet, $(Ca_{2.37}Th_{0.56})^{VIII}(Zr_{2.03})^{VI}(Fe_{2.96})^{IV}O_{12}$, which was irradiated by 1.0 MeV Kr^{2+} at 0.12 dpa (displacement per atom) at room temperature. In the HAADF-STEM image, amorphous domains appear as darker areas. As summarized in the previous section, there are some factors that can make the amorphous matrix have a darker contrast in the HAADF-STEM mode; thickness, density, structural configuration of specimen, and the average mass [9]. In this case, the chemical composition of the amorphous volume does not change significantly as compared to that of the crystalline remnants. Thickness and density in the amorphous domains possibly changed from the original garnet. The previous studies on radiation effects in zircon reported that swelling increases and the density decreases concurrently with the increasing accumulation of dose [54, 55]. The swelling may result in an increase of

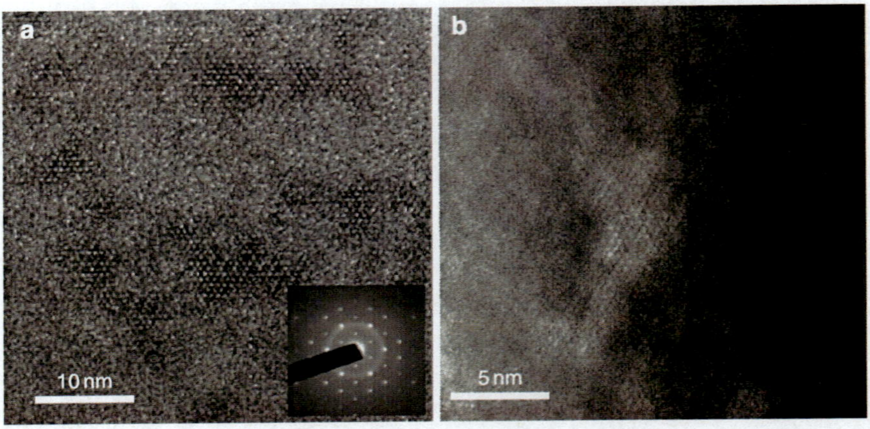

Fig. 8 HRTEM image and HAADF-STEM image of the garnet $(Ca_{2.37}Th_{0.56})^{VIII}(Zr_{2.03})^{VI}(Fe_{2.96})^{IV}O_{12}$, after the irradiation with 1.0 MeV Kr^{2+} at 0.12 dpa (displacement per atom) at room temperature [53]

thickness in the amorphous area, and the thicker amorphous area may have a brighter contrast in HAADF-STEM, but the decrease of density in the amorphous area has an opposite contribution, making the contrast darker. A loss of the structural configuration in radiation-induced amorphous domain can also make the contrast darker as observed in zircon. Thus, the combined effect of the loss on the structural configuration and of the decrease on the density in the amorphous domains contributed to the contrast being greater than the effect of thickness changes.

Xenon is also one of the fissiogenic elements in nuclear reaction. Mitsuishi et al. [56] performed atomic-scale HAADF-STEM imaging of Xe in Al metal combined with multislice image simulation and reconstructed the structure of Xe nanoclusters formed in Al crystalline matrix.

All of these examples illustrate the efficacy and power of *Z*-contrast imaging in HAADF-STEM for characterizing the occurrence of actinides and other heavy trace metals at near the atomic-scale, even at ppm levels of concentration. In particular, these examples provided new insight as to the form of migrating U and the other radionuclides in the environment. Another aspect is the importance of combination of the HAADF-STEM technique with the other state-of-the-art analytical methods: SIMS, TRLFS, and so on. Some of the examples show that multiparameter analysis combined with the STEM capability was still required for a complete understanding of the speciation of the element of interest, partly because of the limited volume of the TEM analysis and the relatively high level (~0.1 wt%) of detection limit in EDS (semi-) quantitative. However, recent developments in aberration correction in STEM has dramatically improved the capability of atomic-scale analysis and overcome some of the previous disadvantages. The advantages of C_S-corrected STEM are summarized in Sect. 5.

3 Electron Energy-Loss Spectroscopy

EELS with TEM is commonly used to determine the chemical state, such as the electronic structure of elements in target materials, in the small volume through which the electron beam passes. Fundamental aspects of EELS have been summarized in [57]. Application of EELS to the actinide research has been recently summarized in extensive reviews by Moore [58] and Moore and van der Laan [59]. As discussed by Moore [58] and Moore et al. [60], the EELS spectral shape is identical to synchrotron-based X-ray absorption spectrum. For actinide materials, several core-loss edges, $O_{4,5}$ edge, $N_{4,5}$ edge, and $M_{4,5}$ edge, can be examined owing to the multiple shells of the electron orbitals. The $O_{4,5}$ edge and $N_{4,5}$ edge are shown in Fig. 9a, b [62, 63]. The $O_{4,5}$ prepeak size and structure depend on the spin–orbit interaction of both the 5*d* and 5*f* states. In the case of the $N_{4,5}$ edge, the ratio of the $N_4(d_{3/2})$ and $N_5(d_{5/2})$ peak intensities can be indicative of the relative occupation of the $5f_{5/2}$ and $5f_{7/2}$ levels.

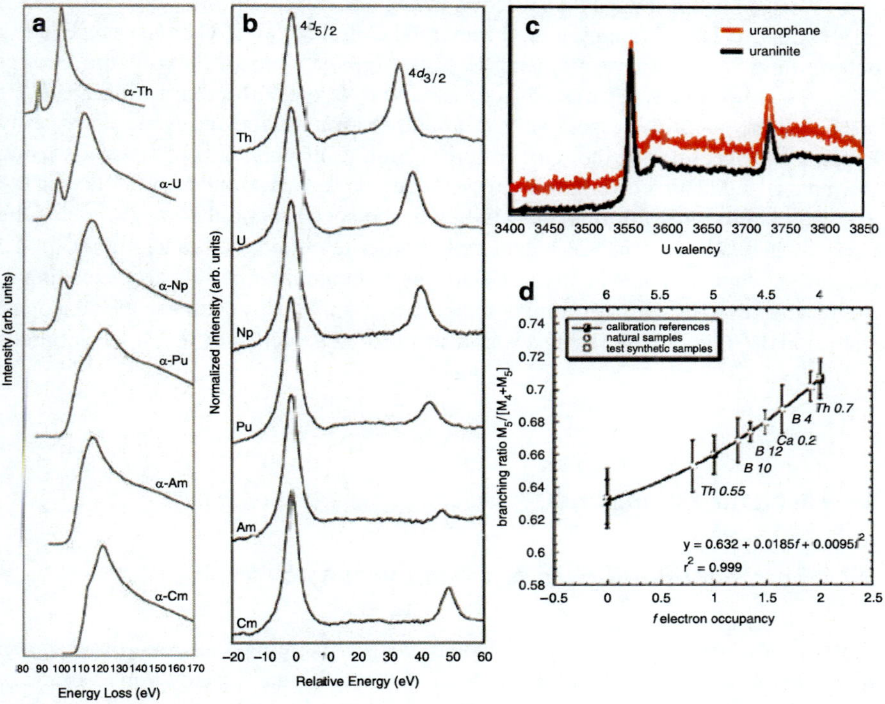

Fig. 9 (**a**) The $O_{4,5}$ (5*d*→5*f*) EELS edges for the ground-state α-phase of Th, U, Np, Pu, Am, and Cm metal. (**b**) The $N_{4,5}$ (4*d*→5*f*) EEL spectra for Th, U, Np, Pu, Am, and Cm metal. Each spectrum is normalized to the N_5 peak height. (**a**) and (**b**) are from Moore and van der Laan [59]. (**c**) The $M_{4,5}$ (3*d*→5*f*) EEL spectra for synthetic uraninite, $U(IV)O_2$, and synthetic uranophane. (**d**) The correlation of the branching ratio $M_5/(M_4+M_5)$ with the number of *f* electrons [61]

As in the EELS studies in nonradioactive elements [64, 65], a procedure to determine the chemical state of radionuclides by using EELS has been explored. A determination of the oxidation state of U at nanoscale is needed because the geochemical mobility of U varies as a function of its oxidation state. On the other hand, U^{5+} tends to disproportionate into U^{6+} and U^{4+}, except that relatively stable U^{5+} has been recognized on mineral surfaces; e.g., on the edge surface of biotite [66]. While X-ray photoelectron spectroscopy (XPS) and synchrotron-based X-ray absorption spectroscopy (XAS) have been widely used to quantify the oxidation state of U in experimentally prepared samples and natural samples, an analytical method at smaller scale than their probe size has been needed because the U frequently occurs localized in submicron domains. Naturally, EELS analysis of U has been attempted due to the high special resolution. Two EEL spectrums of U $M_{4,5}$ edges taken from synthetic uraninite and synthetic uranophane are given in Fig. 9c as representative standards for U^{4+} and U^{6+}, respectively. As shown in the figure, these two spectra have minimum differences. Difficulties in determining U oxidation state are due to the low count in core-loss electrons at >3,000 eV because the signal of EELS spectrum exponentially decreases as a function of energy and from electron beam-induced destruction of the structure with subsequent reduction of U oxidation state in many uranyl minerals. Colella et al. [61] have succeeded in determining U oxidation state using U $M_{4,5}$ edge, of which the two white lines at 3,552 eV (M_5), and 3,728 eV (M_4) correspond to two electronic transitions $3d_{5/2} - 5f_{7/2}$ and $3d_{3/2} - 5f_{5/2}$, respectively. The authors minimized the beam damage by keeping the specimen at the temperature of liquid nitrogen (−178°C) while using the maximized beam current to improve the signal to background ratio. The second derivative numerical filtering was applied to the spectra, and the integration of the peak intensity above zero was calculated in order to obtain the branching ratio, $M_5/(M_4+M_5)$, which polynominally correlates with the number of *f* electrons leading to the U valency (Fig. 9d). A similar method was applied to estimate the oxidation state of Pu [67]. The $M_{4,5}$ edge was also used to detect ppm levels of actinides [68, 69].

4 Electron Tomography

4.1 Introduction of Electron Tomography by TEM

Electron tomography is a powerful method for obtaining the three-dimensional structure of various samples at the nanometer scale. Figure 10 shows an example of TEM three dimensional tomography, the reconstruction of an object viewed from different angles (Kamiishi and Utsunomiya, submitted). The specimen is hydroxyapatite reacted with 0.5 mM Pb nitrate solution for 9 h at room temperature. This three-dimensional reconstruction clearly reveals that Pb-hydroxyapatite precipitates and grows perpendicular to the *c*-axis only on the tip of the hydroxyapatite.

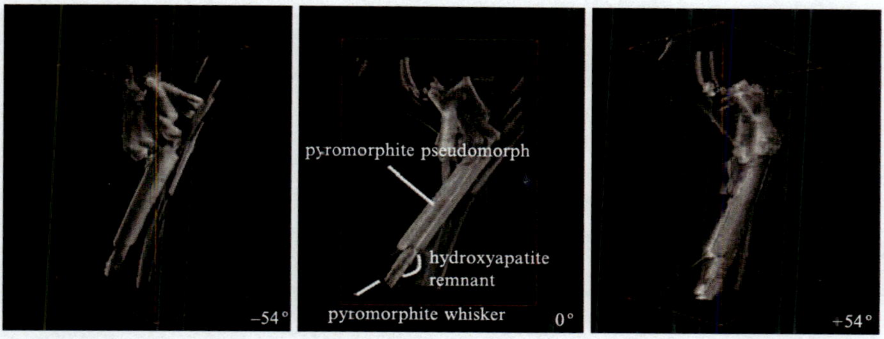

Fig. 10 An example of electron tomography. The images are the ones tilted at 0° and ± 54° of BFTEM tomography. Hydroxypyromorphite nanocrystallite whisker and pseudomorph formed on hydroxyapatite are shown in three dimensions (Kamiishi and Utsunomiya, submitted). Full movie of BFTEM tomography is available on the web site (http://mole.rc.kyushu-u.ac.jp/~ircl/)

Tomography has been utilized in biological and physical sciences as an important technique for the study of the topology of biological materials, such as viruses or macromolecular assemblies. Their shape governs their specific chemical or physical properties. In material science, the use of tomography has been driven by the development of nanoscale devices.

Progress in tomographic reconstruction techniques has been accelerated by medical applications related to the development of the X-ray tomography in the (CT) scanner [70], which required new computational algorithms in order to construct the image. The method of tomography has been applied to many other disciplines including electron microscopy. In the early application of electron tomography, De Rosier and Klug [71] determined the three-dimensional structure of a biological macromolecule by the Fourier reconstruction methods, in which the helical symmetry of the structure allowed for reconstruction from a single projection. Later, Hoppe et al. [72] demonstrated that reconstruction from multiple projections is possible for asymmetrical systems.

4.2 Data Collection and Reconstruction

Data for electron tomography can be obtained by tilting the specimen with respect to the electron beam. In general, a tilt series is taken at 1–2° angular intervals over an angle range of –60° to +60° using specially designed high-tilt specimen holder. This holder is modified by reducing the width of the holder to minimize the effect of shadowing of the specimen by the holder edge. The resolution and quantity of the reconstructed object depend on the angular range and the angular step of the tilt, as well as the resolution of the projections. The acquired projections must be corrected by estimating the sample movement based on the FFT cross-correlation. In most cases, reconstruction in electron tomography is carried out by back-projection procedure, in which a tilt series of the projected BFTEM or HAADF-STEM images were utilized to build the three dimensional structure. The procedure in electron tomography is concisely described in Midgley and Weyland [73]. Briefly, a point in space may be uniquely described by any three rays passing through that point. If the shape of an object is complex, more rays are then required to describe it uniquely. Fundamentally, a projection of an object is an inverse of such a ray, and will describe some of the complexity of that object. Therefore, inverting the projection, "smearing out" the projection into an object space at the angle of projection, generates a ray that will describe uniquely an object in the projection direction.

4.3 Application of HAADF-STEM and Energy-Filtered TEM to Electron Tomography

Most of the previous electron tomography studies are reconstructed from a tilt series of BF-TEM images. In the case of a noncrystalline specimen, the contrast of BF image is correlated with thickness and density of the target. However, for crystalline materials, the contrast depends on the diffraction condition of the crystal, resulting in BF images of this case being less suitable for the tomographic reconstruction [74]. Therefore, application of HAADF-STEM and energy-filtered TEM (EFTEM) to tomography has been recently explored in the physical sciences. As an example, a heterogeneous catalyst composed of Pd_6Ru_6 particles (~1 nm in diameter) within a mesoporous silica was located using this technique [75]. The authors were able to show evidence of a single Pd_6Ru_6 particle anchored to the wall of the mesopore (Fig. 11). In addition, by using a tilt series of EFTEM elemental maps, it is possible to reconstruct a three-dimensional elemental distribution map [76, 77].

4.4 Prospect of Electron Tomography in Actinide Research

Applications of electron tomography continue to expand. A new technique for acquiring a complete tilt series (tilt angular range: ±90°) of a specimen can eliminate the missing tilt angle, which is created in the shadow zone by the holder edge [78]. Aberration-corrected STEM allows the reconstruction of high-resolution atomic-scale tomography by varying the probe focusing point rather than tilting sample [79, 80]. Although the application of electron tomography has not been explored in actinide research, this technique should be a powerful approach to the investigation of the three dimensional distribution, topology, and orientation of actinide nanoparticles, nanodomains, and even single atoms.

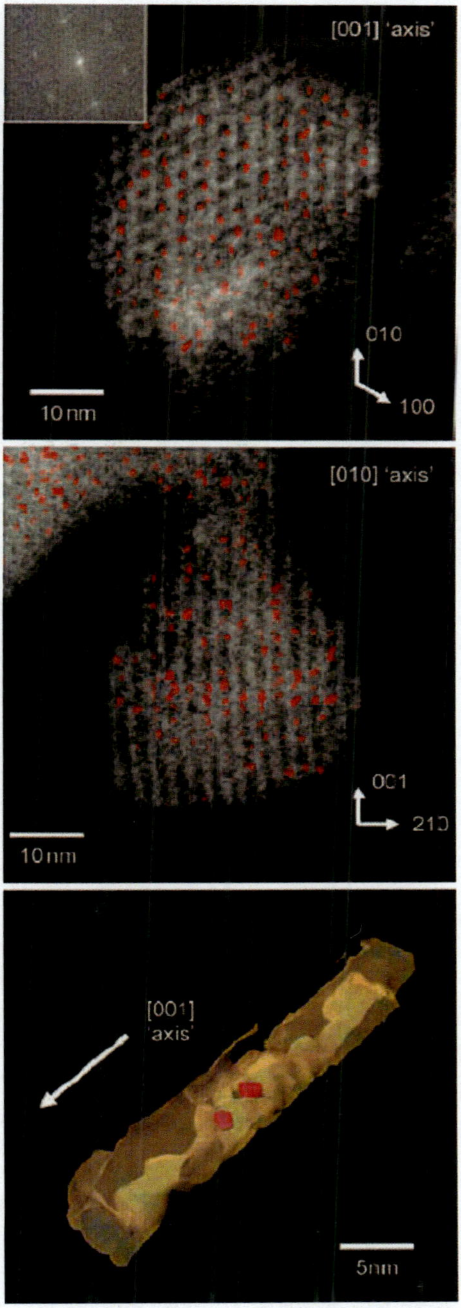

Fig. 11 Three-dimensional structure of an heterogeneous catalyst composed of Pd_6Ru_6 particles, colored *red*, within a mesoporous silica support (MCM-41). (**a**) and (**b**): two perpendicular voxel projections of the reconstruction volume. (**c**) A surface render of a single mesopore extracted from the volume to reveal two nanoparticles within its interior [75]

5 Aberration-Corrected STEM

Recently, the resolution of STEM, as well as TEM, has dramatically improved due to the revolutionary development of aberration correction technology. The achievement of ultra-high resolution is due to the correction of spherical aberration (C_S). Resolution of C_S-corrected STEM now achieved single atomic column or single atom scale. The C_S-corrected STEM and TEM in conjunction with EELS, EDS, and HAADF detector allow us to perform multiparameter analysis with better resolution and quality than the conventional STEM and TEM. This section is intended to concisely summarize several aspects related to C_S-corrected STEM based on its potential for application to actinide research.

5.1 *Basics of C_S Corrector*

The C_S stands for the spherical aberration of rotationally electromagnetic lenses. Spherical aberration was discovered, with chromatic aberration (C_C) as the unavoidable aberration of rotationally electromagnetic lenses [81]; these aberrations cause image blurring and limit the spatial resolution to about 2 Å, which is 50 times the wavelength of electrons accelerated in 100–200 keV. Spherical aberration mainly causes this blurring, and the resolution limit d is expressed by $d = AC_S{}^{1/4}\lambda^{3/4}$ (A = ~0.66) [82, 83], where A is the constant that depends on the condition of microscope, C_S is the spherical aberration coefficient, and λ is the wavelength of the electron. From this equation, it is evident that reducing the value of C_S can lead to an improvement in the resolution. The schematic illustration of spherical aberration and correction by multipole corrector is shown in Fig. 12 [85].

5.2 *The History of C_S Correction*

Since Scherzer [81] pointed out that rotationally symmetric electron lenses of transmission electron microscope have unavoidable aberrations and proposed some alternative ways to improve the resolution by correcting aberrations, many researchers have tried to compensate for the aberration, and C_S-corrected TEM (not STEM) developed with greatly improved resolution in the early 1990s [84, 86].

On the other hand, after the invention of STEM in the late 1960s as a novel alternative technique to the TEM imaging [87], researchers started to correct aberrations in STEM in the early 1970s, and considerable effort to correct the aberration had been made over the last 40 years [88]. This effort led to the development of aberration corrector to compensate for the spherical aberration: hexapole corrector [89–91], quadrupole–octopole corrector, and multipole corrector [92, 93]. A series of articles [94, 95] reported the progress in the development of

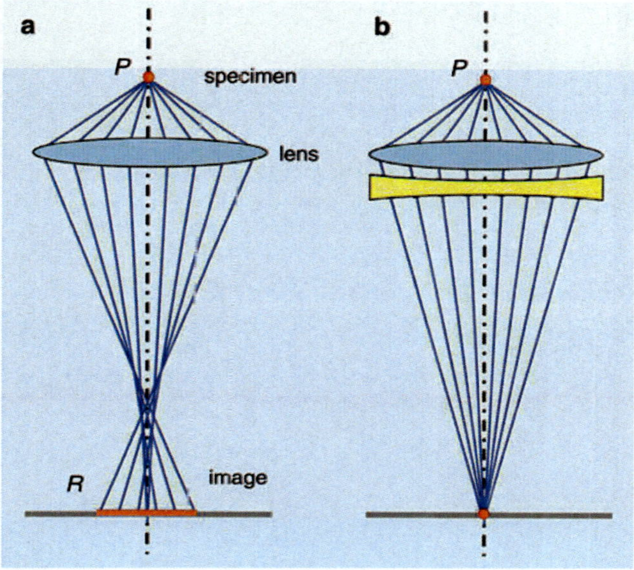

Fig. 12 The schematic illustration of spherical aberration and correction by multipole corrector [84]. (**a**) Spherical aberration of a converging lens. The focal length of the electron beams passing outside of the lens form a focal point distance in front of the image plane that is defined by the paraxial beams. The image of point P is blurred and imaged as disk of radius R on the plane. (**b**) Aberration corrector combined with the lens compensates the spherical aberration

aberration corrector implemented in VG dedicated STEM. By 2003, when the articles were written, a sub-Å sized electron probe and the current of an atom-sized probe by a factor of more than 10 had been achieved. The resolution limit of TEM and STEM is currently achieved at 0.5 Å due to the simultaneous correction of the chromatic aberration and spherical aberration using multipole correctors, and improvement of mechanical stability in the framework of the transmission electron aberration-corrected microscope (TEAM) project that was initiated in the early 2000s [96]. The historical progress in resolution of microscopes is illustrated in Fig. 13 [98].

5.3 *Merit and Application of C_S-Corrected STEM*

At first, one can simply observe and characterize individual atoms with high sensitivity, because the high beam current created by using C_S-corrector allows one to obtain strong signals in imaging and analysis [97, 99]. In addition, C_S-corrected STEM can achieve the sub-Å resolution even at low voltage, thus minimizing damage to the specimen. Hence, some beam-sensitive materials consisting of light elements have been observed recently, such as monolayer BN [10].

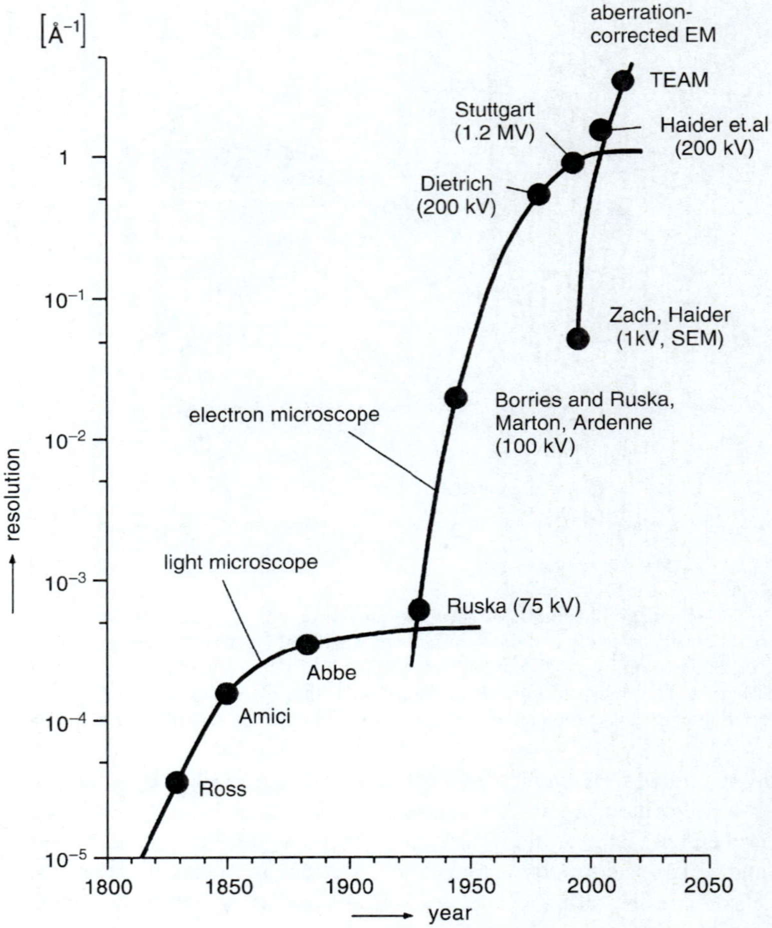

Fig. 13 The historical progress in the resolution of electron microscopes [97]

The focus of current research using C_S-corrected STEM is not only on the light element compounds but also on the analysis of heavy metals such as Au atoms and their behavior on amorphous carbon [97, 100].

5.4 Atomic-Resolution Spectroscopic Imaging

Coinciding with the great improvement in probe size and electron current in STEM, atomic-resolution spectroscopic imaging has greatly advanced in materials research. As a further application, the technique of aberration-corrected STEM in conjunction with EELS has allowed column-by-column resolution analysis of chemical composition, electronic structure, and oxidation state analysis [101–103]. Interestingly, Varela et al. [103] have also demonstrated different methods for obtaining the average oxidation state of Mn in manganite and concluded that the O K edge ΔE (main peak to pre-peak separation) method is the most reliable. In some special cases, chemical analysis at the atom-by-atom level has succeeded [10, 99, 104, 105]. Nevertheless, a caution was raised by Oxley et al. [106] concerning the factors that affect the contrast of atomic-resolution spectroscopic image. Specimen thickness is a particularly important parameter that can potentially modify the EELS image due to a dechanneling phenomenon, in which the heavier elements scatter electrons at higher angle, lowering the number of electrons in the EELS image [106, 107]. As a result, the spectroscopic image simulation is necessary for detailed investigation at the sub Å scale when the specimen thickness becomes thicker than ~100 Å. Further aspects in terms of development in spectroscopic imaging using C_S-STEM have been reviewed in Pennycook et al. [107].

5.5 *Future Developments*

A sub Å electron probe formed by C_S corrector has made it possible to perform multiparameter analysis at the sub Å level. Although the application of C_S-corrected STEM to nuclear science has never been explored, such a capability will have a great impact on our ability to investigate nuclear materials, because the atomic-scale phenomena are closely linked to bulk properties. As mentioned in Pennycook et al. [107], "the intrinsic benefits of STEM for simultaneous imaging of atomic and electronic structure have progressed from a dream limited by noise to a practical reality waiting for application". Still, it is easy to imagine that the resolution limit will be further improved in future with progress in the development of aberration correctors.

Acknowledgments The authors acknowledge the staff of the high voltage electron microscopy (HVEM) laboratory at Kyushu University for their technical support and daily maintenance. SU also thanks the members of the nanogeoscience group in the Department of Chemistry at Kyushu University for their help and constructive discussion. R.C. Ewing acknowledges support from the Energy Frontier Research Center, Materials Science of Actinides, funded by the U.S. Department of Energy, Office of Science, Basic Energy Sciences under Award Number DE-SC0001089.

References

1. Lutze W, Malow G, Ewing RC, Jercinovic MJ, Keil K (1985) Alteration of basalt glasses: Implications for modelling the long-term stability of nuclear waste glasses. Nature 314:252–255
2. Murakami T, Ewing RC, Bunker BC (1988) Analytical electron microscopy of leached layers on synthetic basalt glass. In: Apted MJ and Westerman RE (eds) Scientific Basis for Nuclear Waste Management XI. Mater Res Soc Proc 112:737–748

3. Murakami T, Jercinovic MJ, Ewing RC (1989) Formation and evolution of alteration layers of borosilicate and basaltic glasses I: Initial stage. In: Lutze W, Ewing RC (eds) Scientific Basis for Nuclear Waste Management XII. Mater Res Soc Proc 127:65–72
4. Bates JK, Bradley JP, Teetsov A, Bradley CR, Tenbrink MB (1992) Colloid formation during waste form reaction: implications for nuclear waste disposal. Science 256:649–651
5. Thomas LE, Beyer CE, Charlot LA (1992) Microstructural analysis of LWR spent fuels at high burnup. J Nucl Mater 188:80–89
6. Williams DB, Carter CB (1996) Transmission electron microscopy. A Textbook for Materials Science. Springer, New York
7. Fultz B, Howe JM (2007) Transmission Electron Microscopy and Diffractometry of Materials, 3rd edition. Springer, New York
8. Kirkland EJ (1998) Advanced Computing in Electron Microscopy, 1st editon. Plenum Press, New York
9. Utsunomiya S, Ewing RC (2003) Application of high-angle annular dark field scanning transmission electron microscopy, scanning transmission electron microscopy-energy dispersive X-ray spectrometry, and the energy-filtered transmission electron microscopy to the characterization of nanoparticles in the environment. Environ Sci Technol 37:786–791
10. Krivanek OL, Chisholm MF, Nicolosi V, Pennycook TJ, Corbin GJ, Dellby N, Murfitt MF, Own CS, Szilagyi ZS, Oxley MP, Pantelides ST, Pennycook SJ (2010) Atom-by-atom structural and chemical analysis by annular dark-field electron microscopy. Nature 464:571–574
11. Crewe AV, Wall J, Langmore J (1970) Visibility of single atoms. Science 168:1338–1340
12. Isaacson M, Kopf D, Ohtsuki M, Utlaut M (1979) Atomic imaging using the dark-field annular detector. Ultramicroscopy 4:101–104
13. Liu J, Cowley JM (1990) High-angle ADF and high-resolution SE imaging of supported catalyst clusters. Ultramicroscopy 34:119–128
14. Pennycook SJ (1989) Z-contrast STEM for materials science. Ultramicroscopy 30:58–69
15. Pennycook SJ, Boatner LA (1988) Chemically sensitive structure-imaging with a scanning transmission electron microscope. Nature 336:565–567
16. Chisholm MF, Pennycook SJ (2006) Direct imaging of dislocation core structures by Z-contrast STEM. Philos Mag 86:4699–4725
17. Wang ZL, Cowley JM (1989) Simulating high-angle annular dark-field STEM images including inelastic thermal diffuse scattering. Ultramicroscopy 31:437–454
18. Pennycook SJ, Jesson DE (1991) High-resolution incoherent imaging of crystals. Phys Rev Lett 64:938–941
19. Treacy MMJ, Gibson JM (1993) Coherence and multiple scattering in "Z-contrast" images. Ultramicroscopy 52:31–53
20. Hillyard S, Loane RF, Silcox J (1993) Annular dark-field imaging: resolution and thickness effects. Ultramicroscopy 49:14–25
21. Yamazaki T, Watanabe K, Recnik A, Ceh M, Kawasaki M, Shiojiri M (2000) Simulation of atomic-scale high-angle annular dark field scanning transmission microscopy images. J Electron Microsc 49:753–759
22. Kirkland EJ, Loane RF, Silcox J (1987) Simulation of annular dark field stem images using a modified multislice method. Ultramicroscopy 23:77–96
23. Stadelmann PA (1987) EMS – A software package for electron diffraction analysis and HREM image simulation in materials science. Ultramicroscopy 21:131
24. Ishizuka K (2002) A practical approach for STEM image simulation based on the FFT multislice method. Ultramicroscopy 90:71–83
25. McGibbon AJ, Pennycook SJ, Jesson DE (1999) Crystal structure retrieval by maximum entropy analysis of atomic resolution incoherent images. J Microsc 195:44–57
26. Gull SF, Skilling J (1984) Maximum entropy method in image processing. IEE Proc 131:646–659

27. Gull SF (1989) Developments in maximum entropy data analysis. In Maximum Entropy and Bayesian Methods. Skilling J (ed) 53–71 Kluwer Academic, Boston
28. Chisholm MF, Pennycook SJ (1991) Structural origin of reduced critical currents at $YBa_{2}Cu_{3}O_{7-\delta}$ grain boundaries. Nature 351:47–49
29. Utsunomiya S, Jensen KA, Keeler GJ, Ewing RC (2002) Uraninite and fullerene in atmospheric particulates. Environ Sci Technol 36:4943–4947
30. Lovley DR, Phillips EJP, Gorby YA, Landa ER (1991) Microbial reduction of uranium. Nature 350:413–416
31. Abdelouas A, Lutze W, Nuttall HE (1999) Uranium contamination in the subsurface: characterization and remediation. In: Burns PC and Finch R (eds) Uranium: Mineralogy, Geochemistry and the Environment. Mineral Soc Am 38:433–473
32. Suzuki Y, Banfield JF (1999) Geomicrobiology of uranium. In: Burns PC and Finch RJ (eds) Uranium: Mineralogy, Geochemistry and the Environment. Rev Miner 38:393–432
33. Suzuki Y, Kelly SD, Kemner KM, Banfield JF (2002) Nanometre-size products of uranium bioreduction. Nature 419:134
34. Fayek M, Utsunomiya S, Pfiffner SM, Anovitz L, White DC, Riciputi LR, Ewing RC, Stadermann FJ (2005) Predicting the stability of nano-scale bio-precipitated uranium phases. Can Mineral 43:1631–1641
35. Finch RJ, Murakami T (1999) Systematics and paragenesis of uranium minerals. In: Burns PC and Finch RJ (eds) Uranium: Mineralogy, Geochemistry and the Environment. Rev Miner Geochem 38:91–179
36. Deditius AP, Utsunomiya S, Ewing RC (2008) The chemical stability of coffinite, $USiO_4{\bullet}nH_2O$; $0 < n < 2$, associated with organic matter: A case study from Grants uranium region, New Mexico, USA. Chem Geol 251:33–49
37. Deditius AP, Utsunomiya S, Pointeau V, Ewing RC (2010) Precipitation and alteration of coffinite ($USiO_4{\bullet}nH_2O$) in the presence of apatite. Eur J Mineral 22:75–88
38. Deditius AP, Utsunomiya S, Wall MA, Pointeau V, Ewing RC (2009) Crystal chemistry and radiation-induced amorphization of P-coffite from the natural fission reactor at Bangonbé, Gabon. Am Mineral 94:827–836
39. McCarthy JF, Zachara JM (1989) Subsurface transport of contaminants-mobile colloids in the subsurface environment may alter the transport of contaminants. Environ Sci Technol 23:496–502
40. Kim JI (1994) Actinide colloids in natural aquifer systems. Mater Res Soc Bull 19:47–53
41. Kim JI (1993) The chemical behavior of transuranium elements and barrier functions in natural aquifer systems. Mater Tes Soc Symp Proc 294:3–21
42. Utsunomiya S, Kersting AB, Ewing RC (2009) Groundwater nanoparticles in the far-field at the Nevada test site: mechanism for radionuclide transport. Environ Sci Technol 43:1293–1298
43. Novikov AP, Kalmykov SN, Utsunomiya S, Ewing RC, Horreard F, Merkulov A, Clark SB, Tkachev VV, Myasoedov BF (2006) Colloid transport of plutonium in the far-field of the Mayak Production Association, Russia. Science 314:638–641
44. Kersting AB, Efurd DW, Finnegan DL, Rokop DJ, Smith DK, Thompson JL (1999) Migration of plutonium in groundwater at the Nevada Test Site. Nature 397:56–59
45. Arnold T, Utsunomiya S, Geipel G, Ewing RC, Baumann N, Brendler V (2006) Adsorbed U (VI) surface species on muscovite by TRLFS and HAADF-STEM. Environ Sci Technol 40:4646–4652
46. Ilton ES, Haiduc A, Moses CO, Heald SM, Elbert DC, Veblen DR (2004) Heterogeneous reduction of uranyl by micas: crystal chemical and solution controls. Geochim Cosmochim Acta 68:2417–2435
47. Ewing RC, Lutze W, Weber WJ (1995) Zircon: a host-phase for the disposal of weapons plutonium. J Mater Res 10:243–246
48. Ewing RC (1999) Nuclear waste forms for actinides. Proc Natl Acad Sci USA 96:3432–3439

49. Utsunomiya S, Palenik CS, Valley JW, Cavosie AJ, Wilde SA, Ewing RC (2004) Nanoscale occurrence of Pb in an Archean zircon. Geochim Cosmochim Acta 68:4679–4686
50. Valley JW (2003) Oxygen isotopes in Zircon. In: Hanchar J (ed) Zircon. Mineralogical Society of America, Washington, DC. Rev Mineral Geochem 53:343–385
51. Wilde SA, Valley JW, Peck WH, Graham CM (2001) Evidence from detrital zircons for the existence of continental crust and oceans on the Earth 4.4 Gyr ago. Nature 409:175–178
52. Utsunomiya S, Ewing RC (2006) The fate of the epsilon phase (Mo-Ru-Pd-Tc-Rh) in the UO2 of the Oklo natural fission reactors. Radiochim Acta 94:749–753
53. Utsunomiya S, Yudintsev S, Ewing RC (2005) Radiation effects of ferrate garnet. J Nucl Mater 336:251–260
54. Weber WJ (1991) Self-radiation damage and recovery in Pu-doped zircon. Radiat Effect Defects Solids 115:341–349
55. Weber WJ, Ewing RC, Wang LM (1994) The radiation-induced crystalline-to-amorphous transition in zircon. J Mater Res 9:688–698
56. Mitsuishi K, Kawasaki M, Takeguchi M, Yasuda H, Furuya K (2001) High-angle annular dark-field STEM observation of Xe nanocrystals embedded in Al. Ultramicroscopy 88:25–31
57. Egerton RF (1996) Electron Energy-Loss Spectroscopy in the Electron Microscope, 2nd edition. Plenum, New York
58. Moore KT (2010) X-ray and electron microscopy of actinide materials. Micron 41:336–358
59. Moore KT, van der Laan G (2009) Nature of the 5f states in actinide metals. Rev Mod Phys 81:235–298
60. Moore KT, Chung BW, Morton SA, Schwartz AJ, Tobin JG, Lazar S, Tichelaar FD, Zandbergen HW, Soderlind P, van der Laan G (2004) Changes in the electronic structure of cerium due to variations in close packing. Phys Rev B 69:193104
61. Colella M, Lumpkin GR, Zhang Z, Buck EC, Smith KL (2005) Determination of the uranium valence state in the brannerite structure using EELS, XPS, and EDX. Phys Chem Min 32:52–64
62. Butterfield MT, Moore KT, van der Laan G, Wall MA, Haire RG (2008) Understanding the $O_{4,5}$ edge structure of actinide metals: Electron energy-loss spectroscopy and atomic spectral calculations of Th, U, Np, Pu, Am, and Cm. Phys Rev B 77:113109
63. Moore KT, van der Laan G, Wall MA, Schwartz AJ, Haire RG (2007) Rampant changes in $5f_{5/2}$ and $5f_{7/2}$ filling across the light and middle actinide metals: Electron energy-loss spectroscopy, many-electron atomic spectral calculations, and spin-orbit sum rule. Phys Rev B 76:073105
64. Garvie LAJ, Buseck PR (1998) Ferrous/ferric ratios from nanometer-sized areas in minerals. Nature 396:667–670
65. Garvie LAJ, Buseck PR (1999) Determination of Ce^{4+}/Ce^{3+} in electron-beam-damaged CeO2 by electron energy-loss spectroscopy. J Phys Chem Solids 60:1943–1947
66. Ilton ES, Haiduc A, Cahill CL, Felmy AR (2005) Mica surfaces stabilize pentavalent uranium. Inorg Chem 44:2986–2988
67. Buck EC, Finn PA, Bates JK (2004) Electron energy-loss spectroscopy of anomalous plutonium behavior in nuclear waste materials. Micron 35:235–243
68. Buck EC, Bates JK (1999) Microanalysis of colloids and suspended particles from nuclear waste glass alteration. Appl Geochem 14:635–653
69. Buck EC, Douglas M, Wittman RS (2010) Verifying the presence of low levels of neptunium in a uranium matrix with electron energy-loss spectroscopy. Micron 41:65–70
70. Cormack AM (1963) Representation of a function by its line integrals with some radiological applications. J Appl Phys 34:2722–2727
71. De Rosier DJ, Klug A (1968) Reconstruction of three dimensional structures from electron micrographs. Nature 217:130–134
72. Hoppe W, Langer R, Knesch G, Poppe C (1968) Protein crystal structure analysis with electron rays. Naturwissenschaften 55:333

73. Midgley PA, Weyland M (2003) 3D electron microscopy in the physical science: the development of Z-contrast and EFTEM tomography. Ultramicroscopy 96:413–431
74. Friedrich H, McCartney MR, Buseck PR (2005) Comparison of intensity distributions in tomograms from BFTEM, ADF STEM, HAADF STEM, and calculated tilt series. Ultramicroscopy 106:18–27
75. Midgley PA, Thomas JM, Laffont L, Weyland M, Raja R, Johnson BFG, Khimyak T (2004) High-resolution scanning transmission electron tomography and elemental analysis of zeptogram quantities of heterogeneous catalyst. J Phys Chem B 108:4590–4592
76. Mobus G, Inkson BJ (2001) Three-dimensional reconstruction of buried nanoparticles by element sensitive tomography based on inelastically scattered electrons. Appl Phys Lett 79:1369–1371
77. Weyland M, Midgley PA (2003) Extending energy-filtered transmission electron microscopy (EFTEM) into three dimensions using electron tomography. Microsc Microanal 9:542–555
78. Kawase N, Kato M, Nishioka H, Jinnai H (2007) Transmission electron microtomography without the "missing wedge" for quantitative structural analysis. Ultramicroscopy 107:8–15
79. Sadan MB, Houben L, Wolf SG, Enyashin A, Seifert G, Tenne R, Urban K (2008) Toward atomic-scale bright-field electron tomography for the study of fullerene-like nanostructures. Nano Lett 8:891–896
80. Xin HL, Muller DA (2009) Aberration-corrected ADF-STEM depth sectioning and prospects for reliable 3D imaging in S/TEM. J Electron Microsc 58:157–165
81. Scherzer O (1949) The theoretical resolution limit of the electron microscope. J Appl Phys 20:20–29
82. Cowley JM (1976) Scanning transmission electron microscopy of thin specimens. Ultramicroscopy 2:3–16
83. Smith DJ (2008) Development of aberration-corrected electron microscopy. Microsc Microanal 14:2–15
84. Haider M, Rose H, Uhlemann S, Kabius B, Urban K (1998) Towards 0.1 nm resolution with the first spherically corrected transmission electron microscope. J Electron Microsc 47:395–405
85. Urban K (2008) Studying atomic structures by aberration-corrected transmission electron microscopy. Science 321:506–510
86. Haider M, Uhlemann S, Schwan E, Rose H, Kabius B, Urban K (1998) Electron microscopy image enhanced. Nature 392:768–769
87. Crewe AV, Isaacson M, Johnson D (1969) A simple scanning electron microscope. Rev Sci Instrum 40:241–246
88. Crewe AV (2004) Some Chicago aberrations. Microsc Microanal 10:414–419
89. Beck VD (1979) A hexapole spherical aberration corrector. Optik 53:241–255
90. Crewe AV (1980) Studies on sextupole correctors. Optik 57:313–327
91. Rose H (1981) Correction of aperture aberrations in magnetic systems with threefold symmetry. Nucl Instrum Methods 187:187–199
92. Zach J (1989) Design of a high-resolution low-voltage scanning electron microscope. Optik 83:30–40
93. Zach J, Haider M (1995) Aberration correction in a low-voltage SEM by a multipole corrector. Nucl Instrum Methods A 365:316–325
94. Krivanek OL, Dellby N, Lupini AR (1999) Towards sub-Å electron beams. Ultramicriscopy 78:1–11
95. Krivanek OL, Nellist PD, DellbyN, Murfitt MF, Szilagyi Z (2003) Towards sub-0.5 Å electron beams. Ultramicroscopy 96:229–237
96. Rose H (2005) Prospects for aberration-free electron microscopy. Ultramicroscopy 103:1–6
97. Batson PE, Dellby N, Krivanek OL (2002) Sub-angstrom resolution using aberration corrected electron optics. Nature 418:617–620
98. Rose H (2009) Historical aspects of aberration correction. J Electron Microsc 58:77–85

99. Muller DA, Kourkoutis LF, Murfitt M, Song JH, Hwang HY, Silcox J, Dellby N, Krivanek OL (2008) Atomic-scale chemical imaging of composition and bonding by aberration-corrected microscopy. Science 319:1073–1076
100. Batson PE (2008) Motion of gold atoms on carbon in the aberration-corrected STEM. Microsc Microanal 14:89–97
101. Pennycook SJ, Jesson DE, Browning ND (1995) Atomic-resolution electron energy loss spectroscopy in crystalline solids. Nucl Instrum Methods B 96:575–582
102. Browning ND, Wallis DJ, Nellist PD, Pennycook SJ (1997) EELS in the STEM: determination of materials properties on the atomic scale. Micron 28:333–348
103. Varela M, Oxley MP, Luo W, Tao J, Watanabe M, Lupini AR, Pantelides ST, Pennycook SJ (2009) Atomic-resolution imaging of oxidation states in manganites. Phys Rev B 79:085117
104. Crewe AV, Isaacson M, Johnson D (1971) A high resolution electron spectrometer for use in transmission scanning electron microscopy. Rev Sci Instrum 42:411–420
105. Batson PE (1993) Simultaneous STEM imaging and electron energy-loss spectroscopy with atomic-column sensitivity. Nature 366:727–728
106. Oxley MP, Varela M, Pennycook TJ, van Benthem K, Findlay SD, D'Alfonso AJ, Allen LJ, Pennycook SJ (2007) Interpreting atomic-resolution spectroscopic images. Phys Rev B 76:064303
107. Pennycook SJ, Varela M, Lupini AR, Oxley MP, Chisholm MF (2009) Atomic-resolution spectroscopic imaging: past, present and future. J Electron Microsc 58:87–97

Resonant Inelastic Soft X-Ray Scattering Spectroscopy of Light-Actinide Materials

Sergei M. Butorin

Abstract Resonant inelastic X-ray scattering (RIXS) measurements at the actinide 5*d* threshold provide an opportunity to study elementary excitations in actinide systems in detail . It has turned out that the technique is very sensitive to the valency and the chemical state of actinide in contrast to X-ray absorption spectroscopy, being hampered by the substantial smearing of spectral structures due to a large core-hole lifetime broadening. In this situation, the virtually unlimited resolution (defined by the response function of the instrument) of the RIXS technique and its ability to enhance transitions to low-lying excited states are especially useful. RIXS spectroscopy provides good signatures in terms of new distinct transitions, representing electronic excitations within the 5*f* shell and having a characteristic profile. This helps to distinguish between actinide species with different oxidation states, especially in the case when one of the species has a much lower concentration than another. Experimental data for systems of light actinides, such as U, Np, and Pu, are presented and discussed along with the results of model calculations.

1 Introduction

Actinides are important research subjects from both fundamental (dual localized–delocalized nature of 5*f* states and corresponding properties) and application (issues critical for use, storage, and disposal in industry) points of view. In this chapter, the focus is more on the latter, in particular, on a problem of chemical reduction/oxidation of light-actinide species, with emphasis on Pu, which is addressed by employing the resonant inelastic X-ray scattering (RIXS) technique in the soft X-ray regime.

Many of the fission products and actinides that are produced during irradiation of uranium dioxide fuel in a light water reactor have several valence states. The chemical and physical properties of these elements in water solutions can be very

S.M. Butorin (✉)
Department of Physics and Astronomy, Uppsala University, Box 516, 751 20 Uppsala, Sweden
e-mail: Sergei.Butorin@fysik.uu.se

S.N. Kalmykov and M.A. Denecke (eds.), *Actinide Nanoparticle Research*,
DOI 10.1007/978-3-642-11432-8_3, © Springer-Verlag Berlin Heidelberg 2011

different, depending on the oxidation state. The general trend is that their solubilities increase with oxidation. Uranium, for example, has a solubility of about 10^{-9} mol/kg as U(IV) in carbonate-containing mildly alkaline groundwaters, where $U(OH)_4$ is expected to be the dominant solution species. As U(VI), the corresponding solubility would be 5 orders of magnitude higher with uranyl carbonates as dominating solution species. Obviously, such differences in solubilities will have a large impact on the rate of release of radioactivity from an underground repository to the biosphere.

Deep groundwaters in granitic rock are found to be oxygen-free and reducing. The more reduced forms of actinides are, therefore, expected to be the stable forms in these waters. Because of its radioactivity, however, the spent fuel is capable of decomposing the water by radiation. This radiolysis will produce equivalent amounts of oxidizing and reducing species and should, therefore, not affect the redox potential of the system. The main reducing species produced is hydrogen, and hydrogen is known to be relatively inert at deep ground temperatures. The oxidizing species, on the other hand, are very reactive, and this difference in reactivity of the oxidizing and reducing species is believed to create an oxidizing environment close to the fuel. As a consequence of that, radionuclides may dissolve from the fuel as more oxidized species and, therefore, in larger amounts. Once oxidized, a subsequent reduction when the dissolved species enter a reducing environment may be slow or even hindered. As an example, sulfate ions are stable, in the absence of bacteria, over geological time periods in solutions where sulfide is the thermodynamically stable form of sulfur.

There are a number of fission products and also other actinides that can occur in different oxidation states that have different solubilties in water. One of the most important is plutonium, which has a very low solubility as Pu(IV), while Pu(V) and Pu(III) have much higher solubilities.

PuO_2 had always been thought to be the highest stable oxide of plutonium, but in recent experiments by Haschke et al. [1], plutonium dioxide has been found to react with water to form higher oxides up to $PuO_{2.27}$. A theoretical analysis of this by Korzhavyi et al. [2] on the basis of calculated total energies based on well-established thermodynamic data has shown that the reactions of PuO_2 with either O_2 or H_2O to form PuO_{2+x} are calculated to be endothermic. Their calculations, however, show that PuO_{2+x} can be formed as an intermediate product by reactions with the products of radiolysis of water, such as H_2O_2.

Martin et al. [3] performed X-ray absorption spectroscopy (XAS) experiments to characterize plutonium dioxide oxidation. They showed that their sample preparation method did not enable the preparation of hyperstoichiometric plutonium dioxide. Consequently, they concluded that plutonium dioxide oxidation only occurs under very specific conditions. Neck et al. [4, 5] concluded from the interpretation of solubility data oxidation of hydrated PuO_2(s, hyd) to PuO_{2+x} (s, hyd) and that thermodynamic considerations predict that hydrated PuO_2 (s, hyd) can be oxidized by O_2 whereas dry PuO_2(cr) cannot be oxidized by O_2.

Field and laboratory studies suggest an ability of corroding iron to immobilize solved actinide species, such as U(VI), from aqueous solutions and groundwaters.

Furthermore, O'Loughlin et al. [6] have shown that in the presence of rust, the chemical reduction of U(VI) leads to creation of UO_2 nanoparticles. One can also expect a formation of nanoparticles on corroding iron surfaces for solutions of other actinides.

We have shown that it is possible to determinate the oxidation state of uranium sorbed and reduced onto anaerobically corroding iron from a solution of uranyl carbonate using RIXS technique. It has been also demonstrated that it is possible to monitor in-situ the kinetics of this reduction process using a specially designed reaction cell. In this chapter, the emphasis is on the application of RIXS spectroscopy to studies of the reducibility and reduction kinetics of plutonium. The description includes both model calculations and the obtained experimental results and is mainly limited to experiments at 5*d* edges of actinides (soft X-ray regime).

2 Technique, Models, and Theoretical Approach

To successfully describe the various physical properties of a system in question, it is necessary to obtain knowledge about the ground state and low-energy excited states of this system. For transition element, lanthanide, and actinide compounds with a partly filled *d* or *f* shell, strong correlation effects, when the dispersional part of *d* or *f* bandwidth is smaller than the on-site Coulomb interaction U between localized electrons, break down a single-particle picture, and an atomic-like approach to characterize the electronic structure of these compounds is more appropriate.

In this case, a state of the system without a core hole is described in terms of intra-atomic neutral excitations (a multiplet structure of the ground state electronic configuration due to electrostatic, exchange, crystal field, spin–orbit interactions, etc.) and/or interatomic charge-transfer excitations. The latter are the result of electron hopping from delocalized states to a localized state and are treated by short-range models, such as an Anderson impurity model [7], using a set of parameters. The models are represented by the Hamiltonian

$$H = \sum_{k,\alpha,\sigma} \varepsilon_{k\alpha} n_{k\alpha\sigma} + \sum_{m,\sigma} \varepsilon_m n_{m\sigma} + \sum_{k,\alpha,m,\sigma} \left(V_{k\alpha m} \psi^{\dagger}_{m\sigma} \psi_{k\alpha\sigma} + \text{H.c.} \right) + U \sum_{m,m',\sigma,\sigma'} n_{m\sigma} n_{m'\sigma'} . \quad (1)$$

Important physical quantities included in this Hamiltonian are the delocalized- and localized-state energies $\varepsilon_{k\alpha}$ and ε_m, hopping matrix element $V_{k\alpha m}$, and U. Here, k, α, σ, and m denote a wave vector, an index of the energy level in the valence band, a spin index, and an azimuthal quantum number, respectively. For the description of core spectroscopies, a further term is added to the Hamiltonian to account for coupling between localized electron and a core hole. The values of model parameters are optimized by fitting both high-energy spectroscopic and low-energy transport data and then employed to describe the character of the ground state, different ground-state properties, the nature and size of the band gap in insulators [8], etc.

As the interpretation of transport measurements in these cases is often hampered by the presence of defects and by the importance of electron–lattice interactions, high-energy spectroscopies, which directly probe the electronic degrees of freedom, are often used for preliminary estimations of model parameters. In these estimations, it is important to take into account significant configurational dependence of model parameters, which is predicted by first-principles calculations [9–12]. In particular, removing/adding of a valent d or f electron is expected to result in a decrease/increase in the value of the hybridization strength V, which in turn may lead to renormalization effects for U. These effects are more pronounced for core-level spectroscopies. In the presence of a core hole, V is strongly reduced as the wave functions become more localized. The renormalization of model parameters in the final state can produce a significant uncertainty in estimated values of these parameters in the ground state. In this situation, X-ray scattering techniques become very attractive because the scattering process is charge-neutral.

By now, it has been proven that at some core-thresholds of electron-correlated systems, X-ray emission spectroscopy with monochromatic photon excitation can be considered as an analog of these techniques so that the excitation-radiative deexcitation channel can be treated as RIXS process. Final states probed via such a channel are related to eigenvalues of the ground state Hamiltonian. The core-hole lifetime is not a limit on the resolution in this spectroscopy (see e.g., [13]). It is important to distinguish between the many-body description of RIXS and a single-particle approach, which is usually applied to wide-band materials [14]. The differences between two formalisms are schematically illustrated in Fig. 1.

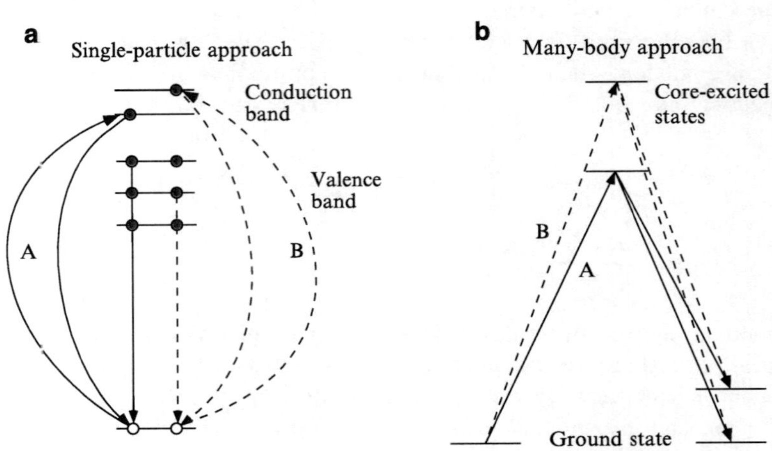

Fig. 1 Schematic representation of the radiative deexcitation process for two different core excitations A and B

According to the many-body picture, an energy of a photon, scattered on a certain low-energy excitation, should change by the same amount as a change in an excitation energy of the primary beam (see a decay route of core excitation *B* versus that of *A*) so that inelastic scattering structures have constant energy losses and follow the elastic peak on the emitted-photon energy scale. In the single-particle view, energy positions of specific inelastic-scattering structures with respect to the elastic peak, which are defined by the momentum conservation rule, may vary only within the energy range covered by the occupied part of the valence band. In Fig. 1a, this is reflected in the situation when, for core excitation *B* with the higher energy, the radiative decay results in a transition with the lower energy than those for *A*, respectively. In spite of simplifications made here, the outlined differences can be used to test the validity of one or another model for a system in question.

As an example, we use data from [15] (see Fig. 2), which were obtained at the U $3d_{5/2}$ edge of UO_3. The inelastic scattering structure with the energy loss of about 5 eV is observed to follow the elastic peak up to 20 eV above the $3d_{5/2}$ threshold,

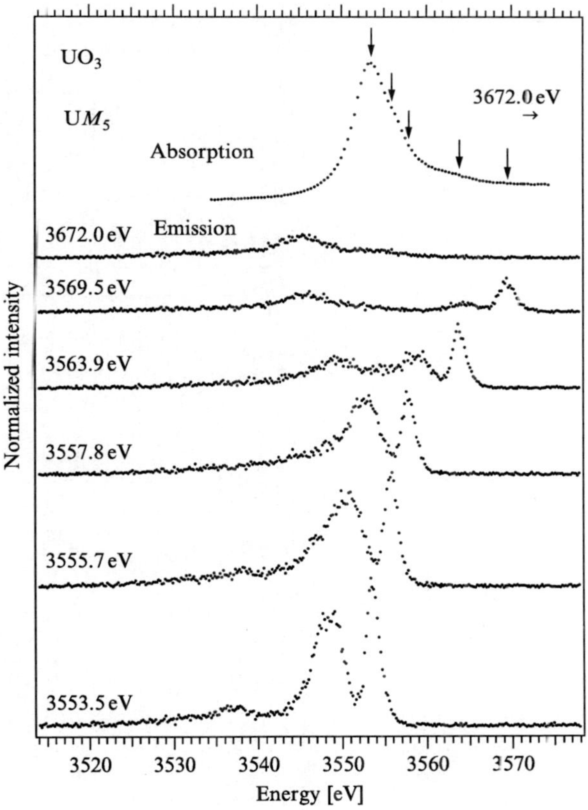

Fig. 2 U M_5 X-ray scattering spectra recorded across the U $3d_{5/2}$ threshold of UO_3 [15]. Excitation energies used in these measurements are indicated by *arrows* on the absorption spectrum

while the width of the occupied part of the valence band is only ~4 eV. This indicates the importance of electron correlation effects in UO_3. Similar results were obtained at the U 5*d* edge of this oxide (Fig. 3) [16, 17].

Although the information provided by the RIXS technique is similar to that obtained from optical absorption or low-energy electron-energy-loss (EELS) spectroscopies, there are some advantages in using this method:

1. The technique is not surface-sensitive, helping to avoid the confusion with a formation of additional states because of surface defects.
2. Its element-specificity enables one to study even very dilute compounds as metal states can be probed separately from ligand states.

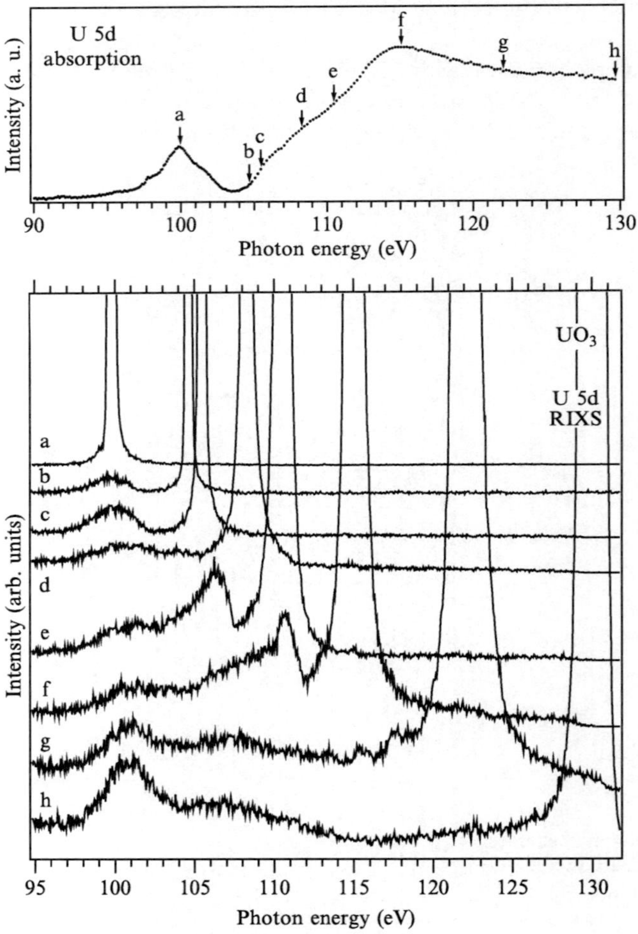

Fig. 3 Resonant inelastic X-ray scattering spectra recorded across the U 5*d* edge of UO_3 [16, 17]. Excitation energies used in these measurements are indicated by *arrows* on the absorption spectrum

3. The cross-section for inelastic X-ray scattering is strongly enhanced on the resonance in contrast to weak dipole-forbidden transitions in optical absorption spectra.
4. The dipole nature of radiative transitions makes it easier to calculate RIXS intensities compared to *d–d* (*f–f*) intensities in optical spectroscopy or in EELS.

In calculations of resonant X-ray scattering as a second order optical process, only a resonant term of the modified Kramers–Heisenberg equation is usually used, where the spectral intensity is given by

$$I_{qq'}(\Omega, \omega) = \sum_f \left| \sum_i \frac{\langle f|T_{q'}|i\rangle\langle i|T_q|g\rangle}{E_g + \Omega - E_i - i\Gamma_i/2} \right|^2 \delta(E_g + \Omega - E_f - \omega). \qquad (2)$$

Here, $|g\rangle$, $|i\rangle$, and $|f\rangle$ are the ground, intermediate, and final states with energies E_g, E_i, and E_f, respectively, while Ω and ω represent energies of incident and scattered photons, respectively. T_q is the dipole operator, Γ stands for the intermediate state lifetime, and q and q' are polarizations of the light with respect to the quantization axis.

For the case of studying the multiplet structure of the ground state configuration, the ability of RIXS spectroscopy to probe low-energy excitations was first discussed by Tanaka and Kotani [18] in the description of RIXS spectra of La_2CuO_4 and CuO at the Cu 2*p* edge. The difference in *d–d* excitation profiles for $3d^9$ multiplets between these two oxides was predicted. However, no experimental data were available with the energy resolution being sufficient enough to support conclusions made by authors. The first experiment, which unambiguously confirmed the ability of the RIXS technique to probe elementary excitations was performed [19] on MnO (2 years later, high-resolution RIXS data at the Cu 3*p* resonance of cuprates were published [13], which are in good agreement with theoretical predictions). Prior to this, probing of *f–f* excitations in rare-earths was discussed in [20]. Figure 4 illustrates how RIXS structures due to *f–f* excitations in UF_4 resonantly enhanced for a number of energies of incident photons tuned to the U 5*d* threshold and how these structures (observed within 4 eV below the elastic peak) follow an elastic peak when sweeping the incident photon energies throughout the U 5*d* X-ray absorption edge.

The efficiency of the RIXS technique in studies of charge-transfer excitations for valence electrons in correlated systems was first demonstrated by Butorin et al. [15] for both soft and intermediate-energy X-ray regions. Various aspects of probing the charge-transfer excitations by this spectroscopy and data interpretation within the Anderson impurity model framework are discussed extensively in the review by Kotani [21]. The reader is also referred to publications by Butorin et al. [15, 22–24], where the latter issue has been addressed.

The intermediate states of the RIXS spectroscopic process [see (2)] are actually the final states of the X-ray absorption spectroscopic process. Therefore, for a better understanding of the excitation-energy dependence of the RIXS profile and

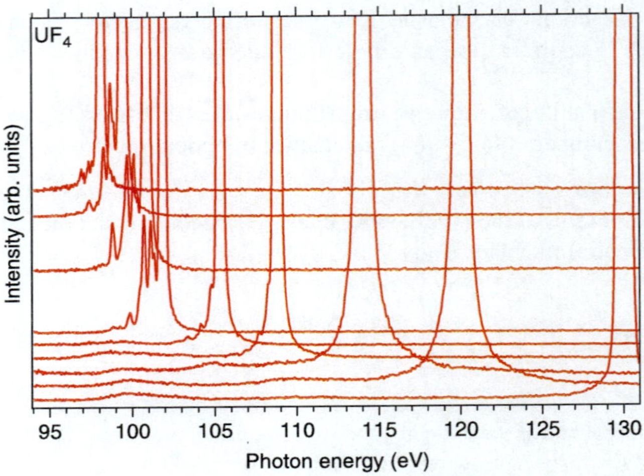

Fig. 4 Resonant inelastic X-ray scattering spectra recorded across the U 5*d* edge of UF_4. Strong elastic peaks indicate/correspond to energies of incident photon beam used to measure RIXS spectra

establishment of the most useful energies of incident photons, it is important to calculate X-ray absorption spectra.

The actinide 5*d* X-ray absorption calculations to account for core excitations such as those represented by path *A* and *B* in Fig. 1b can be performed in a similar fashion as described in [25]. Besides the $5d \to 5f$ excitation process, the calculations include the 5*d*–5*f* 5*f* super-Coster–Kronig decay from the excited state and the interference effect (Fano effect) with the direct $5f \to \varepsilon l$ excitation process.

The Hamiltonian of the system can be expressed as

$$H = H_0 + V_R + V_A, \tag{3}$$

where H_0 includes Coulomb exchange and spin–orbit interactions, V_R represents radiative dipole $5d \to 5f$ and $5f \to \varepsilon l$ transitions, and V_A represents the super-Coster–Kronig $\langle 5d^9 5f^{n+1}|1/r|5d^{10}5f^{n-1}\rangle$ transition, where *n* stands for a number of electrons in the 5*f* shell. The symmetry of continuum *l* can be limited to *g*, which provides the dominant path and the greatest matrix elements. The $5d \to 5f$ transition spectra are obtained by calculating the total yield of the resonant $5d \to 5f$ photoemission process. The intensity of the X-ray absorption spectrum can be written as

$$I_{XA}(\Omega) = \int \mathrm{d}\varepsilon F(\varepsilon, \Omega). \tag{4}$$

$F(\varepsilon, \Omega)$ represents the resonant photoemission spectrum

$$F(\varepsilon, \Omega) = \sum |\langle kj|T|g\rangle|^2 \delta(\Omega + E_g - E_{kj})\delta(\varepsilon - \varepsilon_k), \tag{5}$$

where Ω is the incident photon energy, ε is the kinetic energy of the photoelectron, $|g\rangle$ stands for the ground state of H_0 in the $5f^n$ electronic configuration, $|kj\rangle$ corresponds to the final states of the $5f^{n-1}\varepsilon l$ configuration, and E_g and E_{kj} are the ground state and final state energies, respectively. By neglecting the interaction between the continuum electron and the system left, so that the final states $|kj\rangle$ and their energies E_{kj} can be decoupled as $|kj\rangle = |k\rangle|j\rangle$ and $E_{kj} = E_k + E_j$ ($|k\rangle$ and E_k represent the εl continuum electron state and its energy and $|j\rangle$ and E_j are the states of the $5f^{n-1}$ configuration and their energies, respectively) and by having the T operator as the t-matrix

$$T = V_{\mathrm{R}} + V_{\mathrm{A}} \frac{1}{z - H_0} T \tag{6}$$

with

$$\begin{aligned} z &= \Omega + E_g + i\eta, \\ &(\eta \to +0), \end{aligned} \tag{7}$$

the expression for $F(\varepsilon, \Omega)$ can be rewritten as

$$\begin{aligned} F(\varepsilon,\Omega) = &\sum_{kj} \langle kj|V_{\mathrm{R}}|g\rangle + \sum_{i,i'} \langle kj|V_{\mathrm{A}}|i\rangle\langle i|G|i'\rangle \\ &\times \left(\langle i'|V_{\mathrm{R}}|g\rangle + \sum_{k'j'} \frac{\langle i'|V_{\mathrm{A}}|k'j'\rangle\langle k'j'|V_{\mathrm{R}}|g\rangle}{z - E_{k'j'}} \right) \Bigg|^2 \\ &\times \delta(\Omega + E_g - E_{kj})\delta(\varepsilon - \varepsilon_k), \end{aligned} \tag{8}$$

where $|i\rangle$ represents the excited states of the $5d^9 5f^{n+1}$ configuration, and E_i is their energy. The Green function $\langle i|G|I'\rangle$ is obtained numerically by matrix inversion from

$$\langle i|G|i'\rangle = \frac{1}{z - E_i}\delta_{ii'} + \frac{1}{z - E_i} \sum_{kj,i''} \frac{\langle i|V_{\mathrm{A}}|kj\rangle\langle kj|V_{\mathrm{A}}|i''\rangle}{z - E_{\mathrm{lg}}} \langle i''|G|i'\rangle. \tag{9}$$

To better explain how the RIXS spectroscopic tool can be utilized for addressing the issue of Pu reducibility, examples of similar studies for uranium and neptunium are shown below.

3 Uranium

The case of uranium is maybe the most illustrative for what RIXS technique can do. RIXS measurements at the U $5d$ threshold provide an opportunity to study in detail elementary excitations in U compounds [16, 17] due to the higher resolution of such

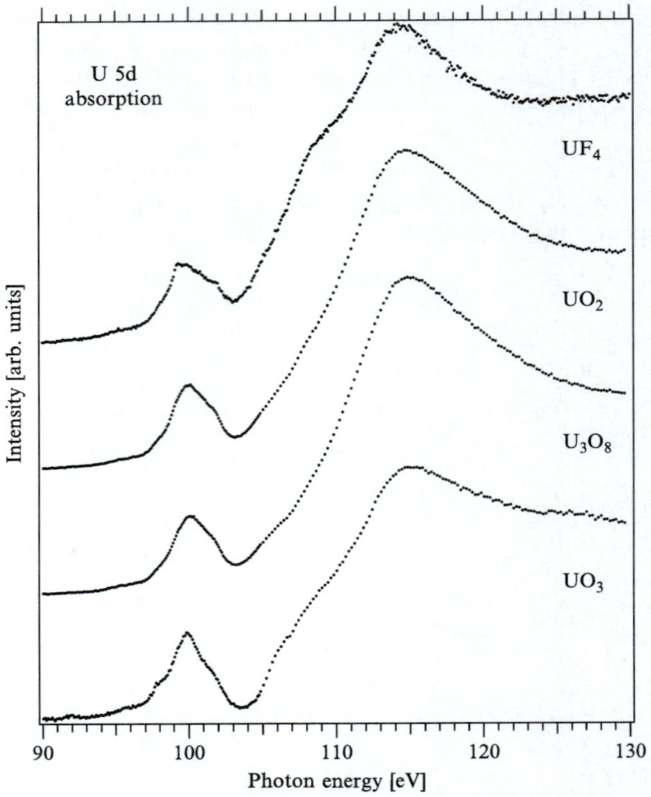

Fig. 5 Total electron yield spectra of UF_4, UO_2, U_3O_8, and UO_3 near the U 5*d* absorption edges [16, 17]

experiments in comparison with those at the U 3*d* and 4*d* thresholds. It has turned out that the technique is very sensitive to the valency and the chemical state of uranium in contrast to XAS. Figure 5 shows X-ray absorption spectra of a number of U compounds recorded across the U 5*d* edge in the total electron yield mode. The 5*d* core-hole lifetime broadening is quite large, thus reducing the utility of XAS. As a result, the U 5*d* absorption spectra in Fig. 5 do not exhibit sharp features. The substantial smearing of spectral structures hampers the analysis of the chemical state and the chemical environment of uranium in various compounds. In particular, it is difficult to distinguish between uranium species with different oxidation states, especially in the case when one of the species has a much lower concentration than another.

In this situation, the virtually unlimited resolution (defined by the response function of the instrument) of the RIXS technique and its ability to enhance transitions to specific excited states are especially useful. The RIXS profile strongly depends on the choice of the excitation energy. The spectral weight corresponding to electronic transitions within the 5*f* shell is enhanced at excitation energies close to 100 eV, while at higher energies of the incident photon beam, set to the main 5*d* absorption edge, interionic excitations of charge-transfer character, such as ligand $2p \rightarrow$ U 5*f* charge-transfer, dominate the RIXS spectra. This has been established by measurements on a set of model U compounds and by model calculations. The spectral pattern of intraionic *f*–*f* excitations is mainly determined by the formal valency of U; in turn, the charge-transfer transitions strongly depend on the chemical environment of U ions.

An example of probing the *f*–*f* excitations in U oxide is shown in Fig. 6 where the RIXS spectra of solid UO_2 [16, 17], recorded for various incident photon energies in the pre-5*d*-threshold region, are displayed. The assignment of sharp inelastic scattering structures to the *f*–*f* transitions is supported by atomic multiplet calculations for the U^{4+} ion. The spectra were calculated using (2). The ground and final states of the spectroscopic process were described by the $5d^{10}5f^2$ electronic configuration, and intermediate or core-excited states were represented by the $5d^95f^3$ configuration. Matrix elements were obtained from Cowan's programs [26] so that Slater integrals $F^k(5f, 5f)$, $F^k(5d, 5f)$, $G^k(5d, 5f)$, and $R^k(5d\varepsilon g, 5f)$ were scaled down to 75%, 75%, 66%, and 80%, respectively, from the Hartree–Fock values. The reduction factors for Slater integrals represent the effect of the intra-atomic configuration interaction; the same factors were used in a number of publications to successfully reproduce 4*d* XA and 4*d* photoemission spectra of various rare-earth systems [27–29].

The calculations reproduce all of the spectral structures very well, especially an enhancement of the peak at about 1.2 eV with increasing excitation energies. The growth of the peak is due to enhanced transitions into the $^1G_4$ state. Changes in relative intensities of inelastic scattering structures corresponding to the *f*–*f* transitions are reproduced on going from spectrum *a* to spectrum *d*.

RIXS profiles, corresponding to the *f*–*f* excitations, are found to be very sensitive to the chemical state of U in different systems [30]. For example, it is a matter of the presence or absence of these excitations when going from U^{4+} to U^{6+} compounds. Therefore, RIXS measurements near the U 5*d* threshold provide good fingerprints for the chemical state of U in different systems in contrast to X-ray absorption spectra, which show only small differences at the U 5*d* edge.

Our findings about the chemical state of U in U_3O_8 and reduced UO_3 are an example. In particular, for U_3O_8, it was discussed in the literature that the chemical state of U is described as either $U^{VI}U_2^{V}O_8$ or $U_2^{VI}U^{IV}O_8$. To our knowledge, there is no clear and convincing answer to this question, and it is still under debate. Thus, core-level photoemission [31–33] and electron spin resonance [34, 35] data have been interpreted in favor of either situation by different groups, and therefore these techniques cannot really provide an unambiguous answer. It turns out that the RIXS technique can. RIXS spectra of *f*–*f* excitations are a good indicator of whether uranium is in the U(IV), U(V) or U(VI) state. The data are easy to interpret and not very difficult to calculate. Establishing the real chemical state of U in oxides is important for applied, environmental, and fundamental sciences (e.g., a development of theory of nonstoichiometry is a fundamental problem).

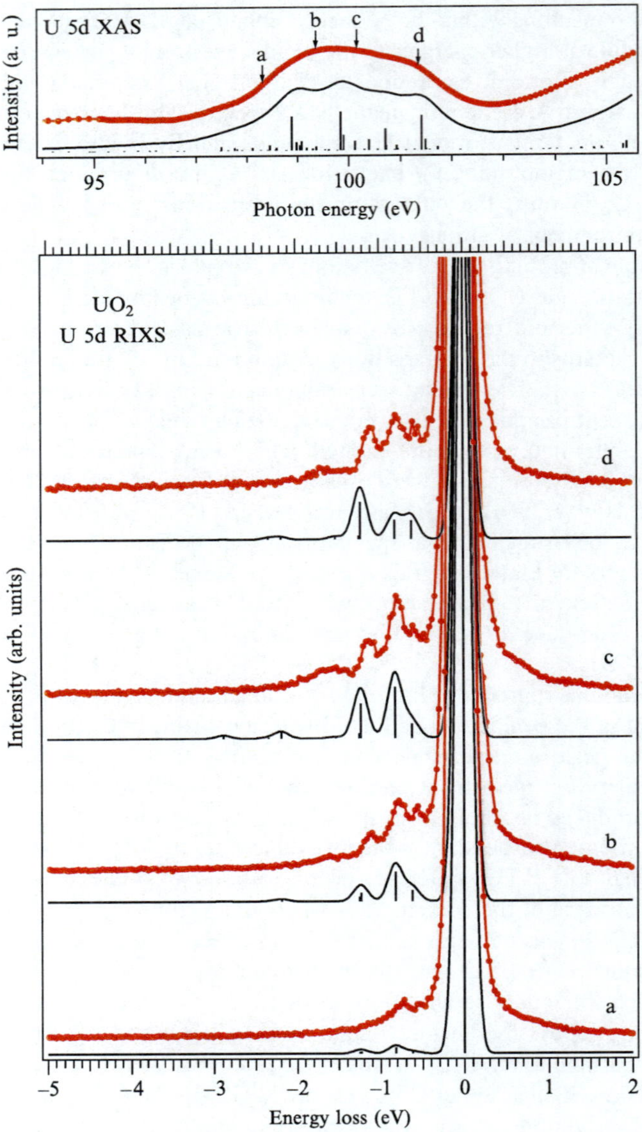

Fig. 6 Resonant X-ray scattering spectra of UO_2 recorded at different excitation energies close to the U 5*d* threshold (*lines with markers*) together with the results of atomic multiplet calculations (*sticks with thin lines*) for the U(IV) ion [16, 17]. Excitation energies are indicated by *arrows* on the total electron yield spectrum at the U 5*d* absorption edge shown in the *top panel*

Figure 7 displays RIXS spectra of *f–f* transitions for a number of U oxides [16, 17]. Close similarity of the *f–f* transitions profile in U_3O_8 to those in UO_2 and UF_4 as well as to that calculated for the U(IV) ion (see Fig. 6) unambiguously

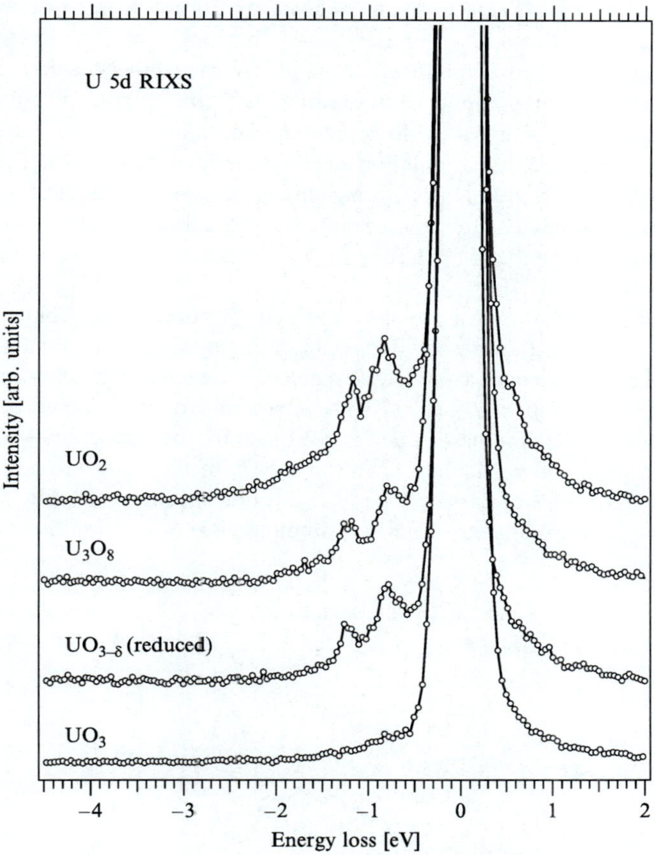

Fig. 7 Resonant inelastic X-ray scattering spectra of U oxides recorded at the incident photon energy of 99.9 eV [16, 17]

indicates the presence of the U(IV) fraction in U_3O_8, thus favoring the $U_2^{VI}U^{IV}O_8$ description of the uranium chemical state in this oxide. Furthermore, similar pattern is observed for reduced UO_3, thus indicating that oxygen deficiency leads to the creation of U(IV) species in the compound.

Charge-transfer effects are expected to be significant in actinide compounds as a result of metal 5*f*–ligand 2*p* hybridization. The analysis of our data [30] obtained at U $3d_{5/2}$ threshold shows that the ligand $2p \rightarrow$ U 5*f* charge-transfer plays an important role in uranium compounds, such as UO_2, $UO_2(NO_3)_2{\cdot}6H_2O$, and even in UF_4.

This is also supported by theoretical studies. Molecular-orbital calculations by several research groups [36–39] gave values for the 5*f* occupancy, which range from 2.3 to 2.9 electrons, while this occupancy was estimated at about 2.3 electrons from the analysis of X-ray absorption and photoemission data within an Anderson impurity model [40–42]. These results indicate significant degree of covalency for U–O chemical bonds in UO_2. For UF_4, a 5*f* contribution of ~0.3 electrons to the bonding orbitals was also predicted from relativistic Dirac–Slater local-density calculations [43]. For compounds containing U^{6+}, the degree of covalency for metal – ligand bonds is expected to be even higher than that for U^{4+} systems. For example, molecular-orbital calculations yielded the 5*f* occupancy of ~2.6 electrons for the uranyl ion UO_2^{2+} [44, 45]. Although the values for the 5*f* occupancy obtained from molecular-orbital calculations seem to be overestimated [46], one cannot rule out the importance of the U 5*f* -ligand 2*p* hybridization even in a compound with "ionic" bonds such as UF_4.

The consequence of high covalency and hybridization in the ground state is an appearance of charge-transfer satellites in high-energy spectroscopic data. Figure 8 illustrates the dependence of the RIXS spectra on the chemical environment of U atoms in various compounds [16, 17]. The spectra were measured under condition when the inelastic RIXS cross-section is enhanced for the ligand 2*p* →U 5*f* charge-transfer transitions, in particular, when the energy of the incident photon beam is tuned to the main edge of the U 5*d* X-ray absorption spectrum (see Fig. 5). At these energies, the charge-transfer excitations dominate RIXS spectra, thus defining the

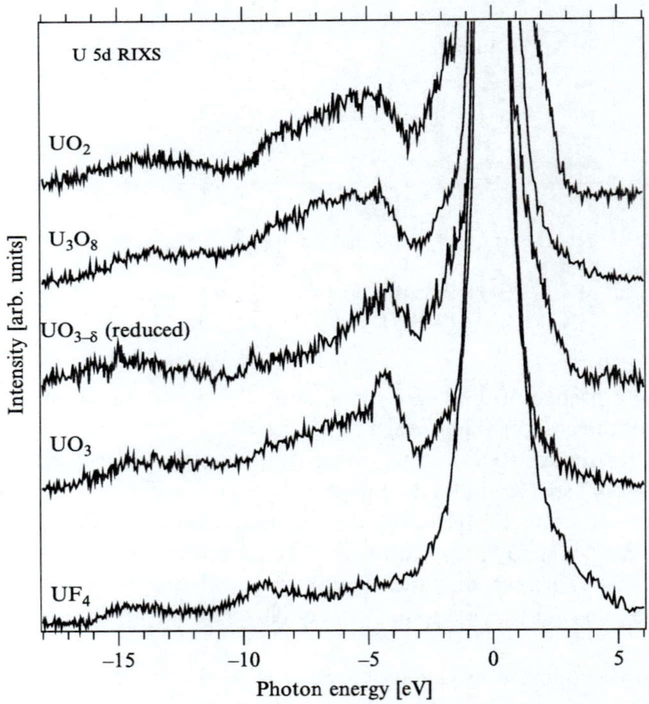

Fig. 8 Resonant inelastic X-ray scattering spectra of a number of U compounds recorded at the incident photon energy of 115.0 eV [16, 17]

RIXS profile. The profiles can be clearly divided into three groups: (1) UF_4, (2) UO_3 and $UO_{3-\delta}$, (3) U_3O_8 and UO_2. The differences between profiles are determined by the character of the bonding and the local geometrical arrangement of ligands, i.e., local crystal structure.

The spectroscopic analysis of samples relevant to the processes of dissolution of uranium dioxide in groundwaters was carried out. The purpose of the work was to determine whether U(VI) can be reduced to U(IV) under conditions relevant to spent fuel disposal, in particular, reduction of U(VI) to U(IV) in the presence of actively corroding canister materials.

Measurements employing RIXS technique for a set of model samples were carried out at beamline 7.0 of the Advanced Light Source, Berkeley, USA. The material studied was a Fe foil polished with diamond spray (1 and 1/4 mm) on one side. The foil was exposed to solution containing U(VI) ions for 17 days. An aliquot of U(VI) was added in the form of uranyl nitrate solution to deaerated Allard groundwater (100 ml) under N_2 atmosphere. The starting concentration of U in solution was 500 ppb. RIXS measurements were performed in an attempt to establish whether the sorption of U(VI) ions onto the Fe surface resulted in some reduction of these ions to U(IV).

Figure 9 displays a set of X-ray scattering data from the Fe sample [47] recorded at the incident photon energy of 100 eV, which corresponds to the energy of the weak prethreshold structure in the U 5*d* X-ray absorption spectrum. A series of spectra were randomly measured from different 1 mm $\times$ 150 μm areas on the Fe sample surface. Six of them are shown in the figure along with spectra of reference oxides UO_2 and UO_3 that contain U(IV) and U(VI), respectively. Setting the energy of the incident beam close to the U 5*d* thresholds enhances the inelastic scattering cross-section and ensures that electronic states of 5*f* symmetry are probed by the spectroscopy due to dipole selection rules. The choice of the excitation energy is also defined by the necessity to selectively look at the intraionic *f–f* transitions of U.

The spectral weight corresponding to these transitions is enhanced at excitation energies close to 100 eV.

An inspection of Fig. 9 shows the presence of two distinct RIXS structures at energy losses of about -0.8 and -1.15 eV in the UO_2 spectrum. These structures represent *f–f* transitions as incident X-rays are inelastically scattered on electronic excitations within 5*f* shell. The structures are very well reproduced by model calculations of RIXS spectra using atomic multiplet theory for the U(IV) ion. Naturally, these structures are absent in the UO_3 spectrum due to the formal $5f^0$ configuration of the U(VI) ions. In this situation, any reduction of U(VI) on the Fe surface should result in an appearance of characteristic RIXS structures. The present measurements indeed reveal the presence of significant inelastic weight in spectra recorded from some areas (spectra 4–6) on the surface of the Fe sample, thus indicating U(VI) reduction in those areas.

A comparison of RIXS spectra [47] of the charge-transfer transitions between the foil and UO_2, made in Fig. 10, reveals that the RIXS profiles of compared spectra are somewhat different. The reason for the observed difference is likely a formation of UO_2 nanoparticles [6] due to a rust presence on the surface of the iron

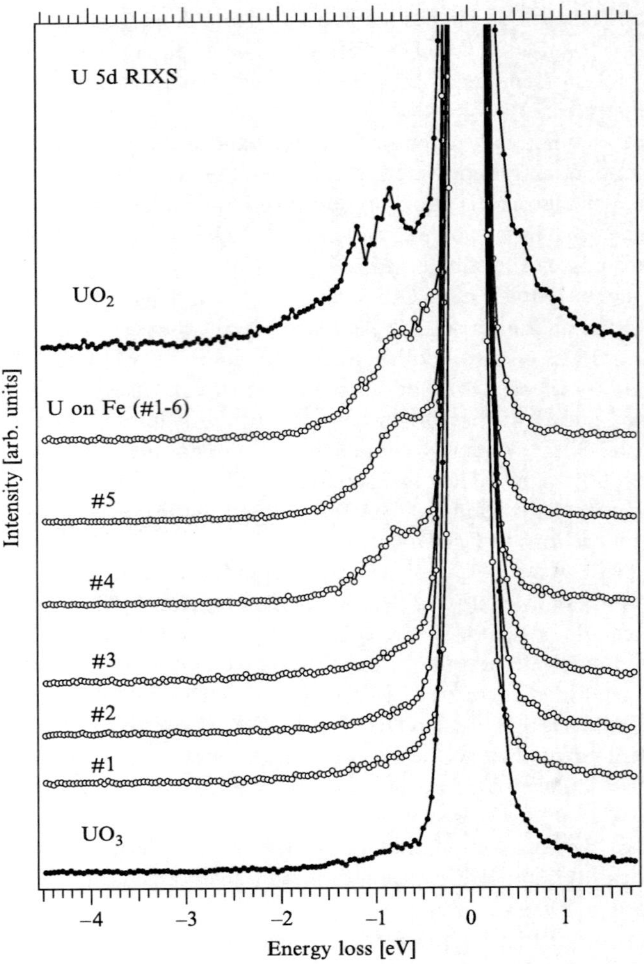

Fig. 9 Enhanced inelastic part of the soft X-ray scattering spectra of UO_2 and UO_3 and UO_2^{2+} adsorbed on the Fe foil (elastic peaks are at 0 eV). The energy of the incident photons was set to 100 eV. The incidence angle of the photon beam to the sample surface was about 15°. Scattered X-rays were detected at the right angle to the direction of the incident beam in the horizontal scattering plane. The six spectra of adsorbed U were randomly measured from different 1 mm × 150 μm areas on the surface of the Fe sample [47]

foil. While the spectrum of *f–f* excitations is not expected to change significantly on going from bulk UO_2 to nanoparticles (except maybe for a narrowing of the spectral lines), the spectrum of charge-transfer excitations can be substantially affected due to changes in the local environment of U atoms.

Similar results were obtained for another Fe foil prepared under the same conditions but with much longer exposure (8.5 month) to the U(VI) solution.

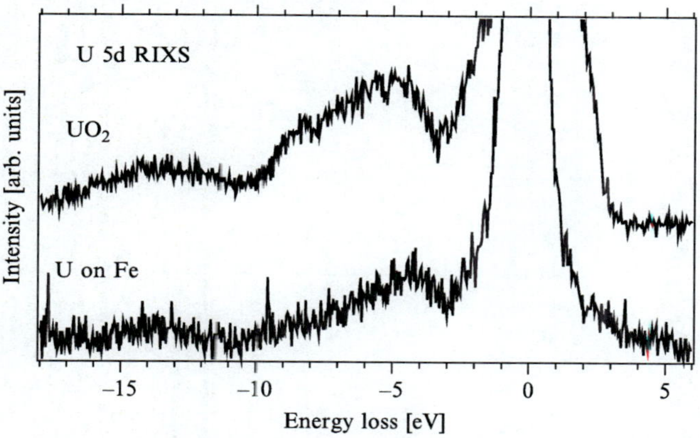

Fig. 10 Enhanced inelastic part of the soft X-ray scattering spectra of UO_2 and UO_2^{2+} adsorbed on the Fe foil recorded at the excitation energy of 115.0 eV [47]

4 Neptunium

The same kind of RIXS measurements were also made for the Np-on-Fe sample prepared at the Chalmers University in a similar manner as for the Fe samples with sorbed UO_2^{2+} discussed above. Figure 11 shows a number of RIXS spectra of NpO_2 used as a reference system with the defined oxidation state for Np (Butorin SM, Shuh DK, Kvashnina K, Ollila K, Albinsson Y, Guo J-H, Werme L, Nordgren J unpublished results). The spectra were recorded at energies of the incident photon beam set to the prethreshold structure in the Np 5*d* X-ray absorption spectrum. At such energies, the X-ray scattering cross-section is enhanced for *f*–*f* transitions. The energy losses of corresponding RIXS structures observed in Fig. 11 are in agreement with optical absorption measurements of *f*–*f* transitions in Np(IV) systems [48, 49].

The spectra in Fig. 11 are compared with the results of the atomic multiplet calculations for the Np(IV) ion. The calculations were performed in a similar manner as described for the case of U(IV) in the previous section; only this time, the $F^k(5f, 5f)$ integrals were scaled down to 70%. The initial and final states of the scattering process were taken to be those of the $5f^3$ configuration, with the intermediate states of $5d^9 5f^4$ character, mainly autoionized to $5f^2 \varepsilon g$ states. The calculated spectra show slight differences in the energy position for some RIXS structures when compared with experimental spectra. This is likely due to the influence of the crystal field interaction and/or the Np $5f - $O $2p$ hybridization, which were not taken into account in the calculations. The extra-structure observed in experimental spectra at an energy loss of around 950 meV may also be a result of the crystal field interaction. Nevertheless, the present calculations account for the overall RIXS profile and behavior and reproduce its dependence on the excitation energy.

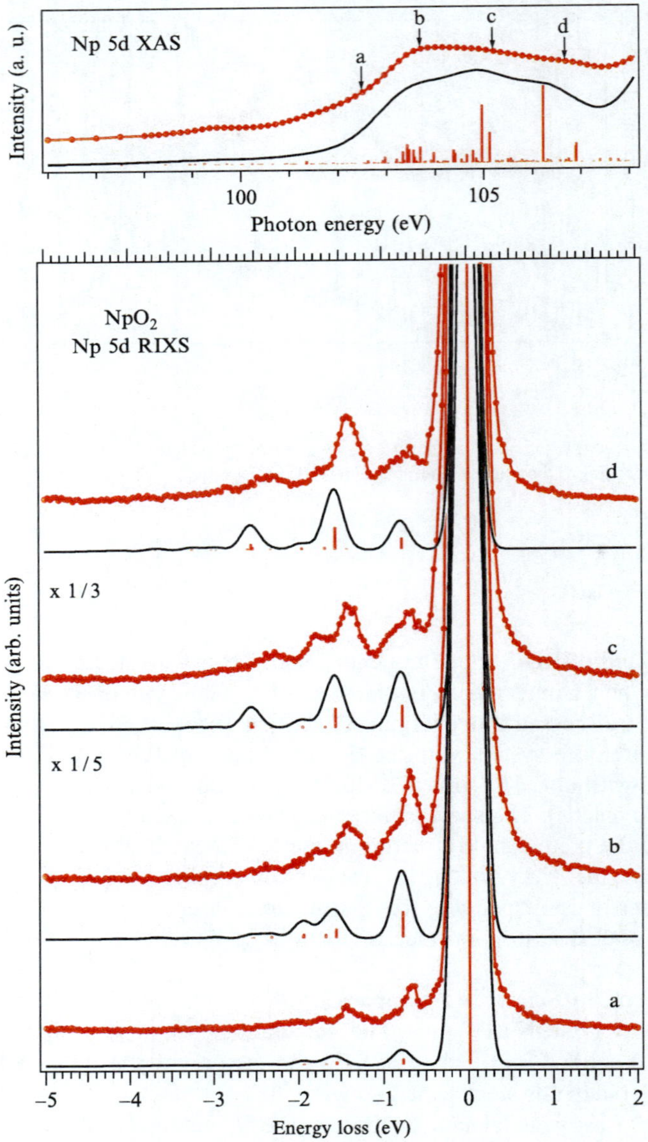

Fig. 11 Resonant X-ray scattering spectra of NpO_2 recorded at different excitation energies close to the Np 5*d* threshold (*lines with markers*) together with the results of atomic multiplet calculations (*sticks with thin lines*) for the Np^{4+} ion (Butorin SM, Shuh DK, Kvashnina K, Ollila K, Albinsson Y, Guo J-H, Werme L, Nordgren J unpublished results). Excitation energies are indicated by *arrows* on the total electron yield spectrum at the Np 5*d* absorption edge shown in the *top panel*

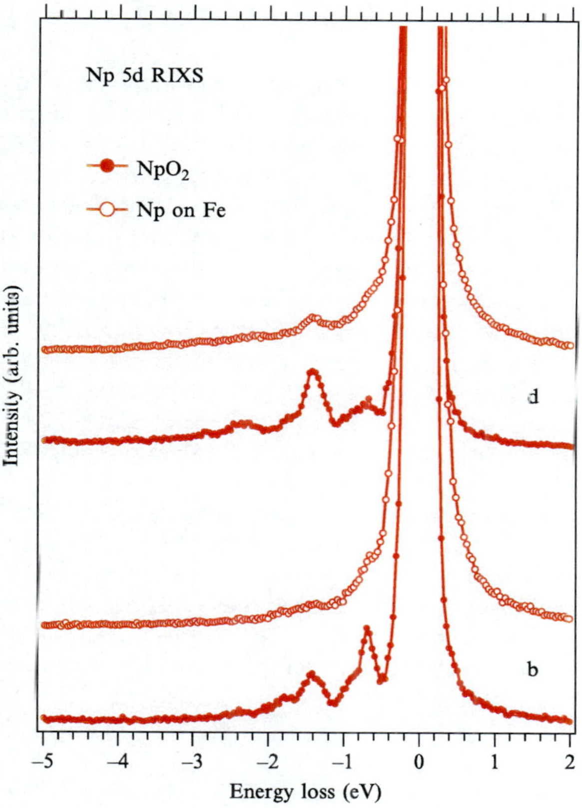

Fig. 12 Comparison of resonant inelastic soft X-ray scattering spectra of Np sorbed on the Fe strip and NpO_2 (Butorin SM, Shuh DK, Kvashnina K, Ollila K, Albinsson Y, Guo J-H, Werme L, Nordgren J unpublished results). *Letters* correspond to the same excitation energies as in Fig. 11

Figure 12 displays RIXS spectra of Np sorbed on the Fe strip along with NpO_2 spectra recorded at the same excitation energies (Butorin SM, Shuh DK, Kvashnina K, Ollila K, Albinsson Y, Guo J-H, Werme L, Nordgren J unpublished results). An inspection of this figure shows that the RIXS structures of both samples closely resemble each other. This resemblance unambiguously indicates the existence of Np in the form of Np(IV) on the studied Fe stripe.

5 Plutonium

The examples described above show that the RIXS technique is quite efficient and successful in studies of the reducibility and reduction kinetics of highly oxidized fission product and light actinide species. In this section, we examine the capabilities of RIXS spectroscopy when applied to plutonium systems.

5.1 Model Calculations for Pu Systems

As shown for U and Np compounds, the atomic multiplet theory is a good first approximation to interpret experimental data and describe RIXS spectral profiles and their changes. In order to understand how efficient the RIXS technique can be in this particular case, we performed model calculations for Pu(III), Pu(IV), and Pu (V) systems. Both X-ray absorption and resonant soft X-ray scattering spectra were calculated for the Pu 5*d* edges in the same manner as for U and Np systems. The ground and final states of the spectroscopic process were described by the $5f^n$ configuration (n = 3, 4, 5), with the intermediate, core-excited states of $5d^9 5f^{n+1}$ character, which are mainly autoionized to $5f^{n-1}\varepsilon g$ states.

The reduced matrix elements required for present calculations were obtained with Cowan's programs [26]. The Slater integrals F^k, G^k, R^k, spin–orbit constants ζ, and radial integrals $(nl|r|n'l')$ were calculated by the Hartree–Fock program with relativistic correction. The values of $F^k(5f, 5f)$, $F^k(5d, 5f)$, $G^k(5d, 5f)$, and $R^k(5d\varepsilon g, 5f^2)$ were reduced to 80%, 75%, 66%, and 80%, respectively. The other parameters were used without reduction. The density of states in the continuum was assumed to be constant so that the R^k and dipole integrals were constant over the whole excitation region. The kinetic energy of the continuum electron was chosen to make the total average energies of $5d^9 5f^{n+1}$ and $5d^{10} 5f^{n-1}\varepsilon g$ configurations identical.

The decay half-width of the core-excited states is expressed as

$$\Gamma_i = \pi |\langle kj | V_A | i \rangle|^2 \tag{10}$$

where $|kj\rangle$ corresponds to the final states of the $5f^{n-1}el$ configuration, $|i\rangle$ represents the core-excited states of the $5d^9 5f^{n+1}$ configuration, and V_A is the super-Coster–Kronig $\langle 5d^9 5f^{n+1} | 1/r | 5d^{10} 5f^{n-1} \rangle$ transition. Γ_i was computed for all the states in present calculations. The other decay channels omitted in the calculations were approximated by including an additional Lorentzian with the width of 1 eV.

The results of calculations of Pu 5*d* X-ray absorption spectra for different chemical states of Pu are displayed in Figs. 13–15 along with computed Γ_i values. Γ_i significantly varies across the 5*d* edge. It has small values for the structure below the ionization threshold but increases substantially above this threshold. The increase causes a large broadening of the states contributing to the main edge so that the significant difference between 5*d* XAS spectra of Pu ions with different valences can be only found in the energy position of the prethreshold structure relative to the main edge. In Fig. 14, the calculated 5*d* absorption spectrum of Pu (IV) is compared with the experimental one (Shuh DK et al. unpublished results) for PuO_2. Relatively good agreement between the two spectra suggests that the atomic multiplet approach is still good approximation for the description of data of Pu compounds.

Figures 16–18 show theoretical X-ray scattering spectra of Pu(III), Pu(IV), and Pu(V) at various excitation energies across the 5*d* edge. The spectra were calculated

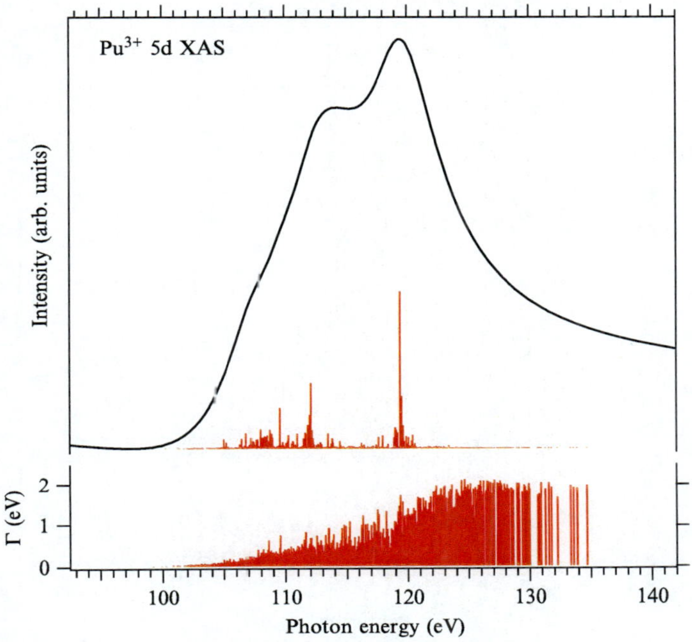

Fig. 13 Calculated 5*d* X-ray absorption of Pu(III) along with decay half-width (*Γ*) values

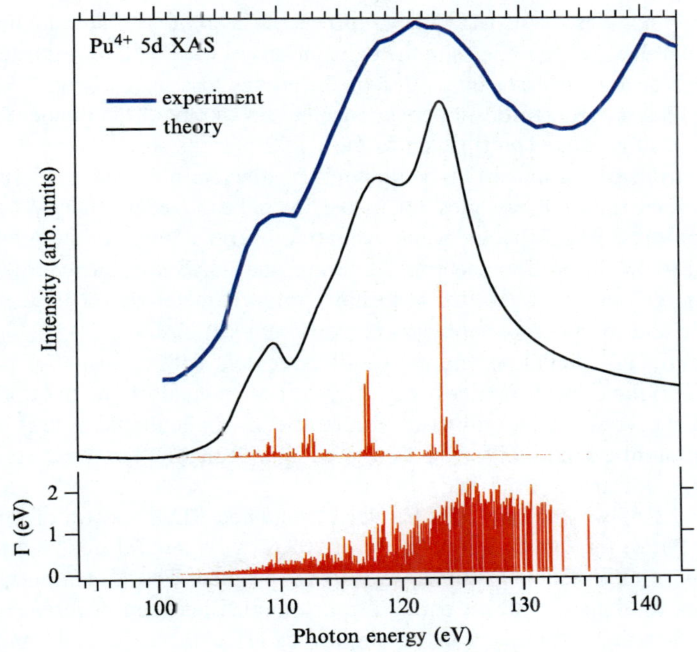

Fig. 14 Calculated Pu(IV) 5*d* X-ray absorption and experimental total-electron-yield spectrum (Shuh DK et al. unpublished results) of PuO_2

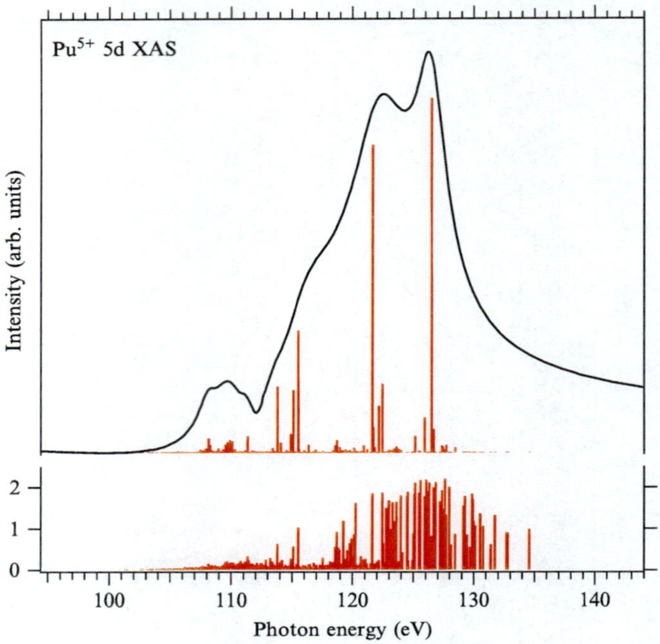

Fig. 15 Calculated *5d* X-ray absorption of Pu(V) along with decay half-width (*Γ*) values

in the same manner as for U(IV) and Np(IV) and using (2). The reduction of the Slater integrals was the same as in the calculations of the Pu *5d* absorption spectra. The same scattering spectra are plotted on the energy loss scale in Figs. 19–21. The RIXS profiles reveal significant dependence on the energy of the incident photons. Although a substantial contribution to the RIXS profiles from ligand *2p* →Pu *5f* charge-transfer excitations in Pu compounds (neglected in the calculations) can be expected for excitation energies set to the main Pu *5d* edge, the *f–f* transitions should dominate RIXS spectra at the excitation energies tuned to the prethreshold structures in *5d* absorption spectra. Therefore, the RIXS spectra recorded at that photon energy range would be the best choice for a characterization of the oxidation state of Pu and for a comparison between different samples.

The results of calculations are in overall agreement with optical absorption data of Pu in various crystal matrices (e.g., [50–56]). The analysis of the optical data showed that crystal field splittings are similar to those in lanthanides, thus justifying the use of atomic multiplet theory as a good approximation for the description of high-energy spectroscopic data.

Figure 22 shows a comparison between calculated RIXS spectra of Pu(III), Pu (IV), and Pu(V) for an excitation energy of 109.6 eV, where the contribution of *f–f* transitions is expected to dominate the RIXS profiles. The three spectra reveal differences that are significant enough to distinguish between different oxidation states so that the technique can be employed, as in the case of U and Np, to

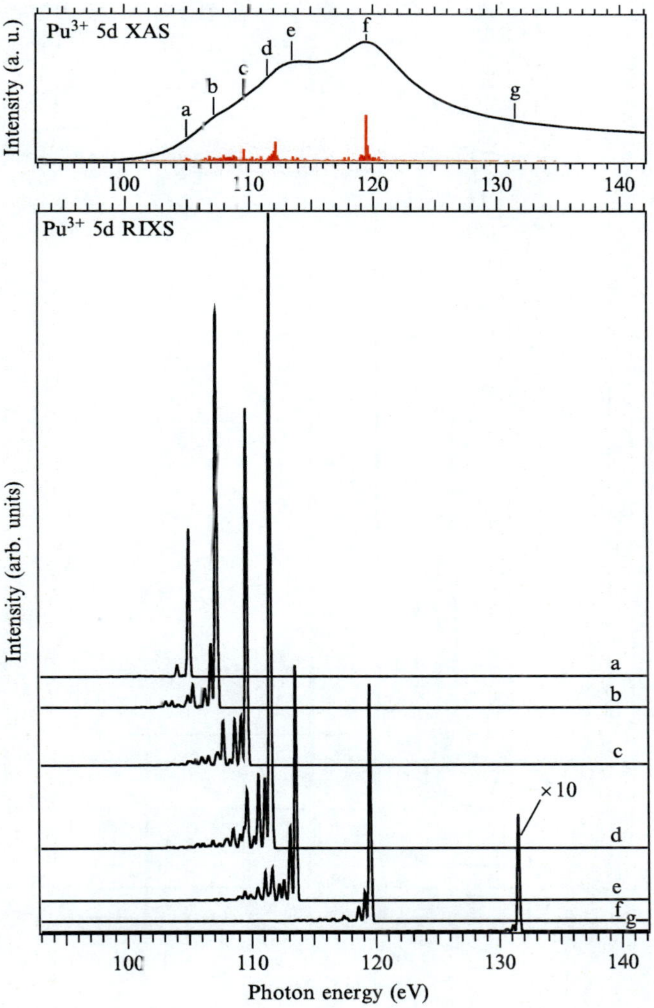

Fig. 16 Calculated resonant soft X-ray scattering spectra of Pu(III) at various excitation energies across the 5*d* edge indicated by *letters* on the corresponding absorption spectrum (*top panel*)

characterize the chemical state of Pu in various systems. In particular, a shift in the energy loss of major RIXS peaks can be used as a corresponding fingerprint.

5.2 Experimental Setup and Safety Issues

The experiments were performed at undulator beamline 7.0 [57] of the Advanced Light Source (ALS), Lawrence Berkeley National laboratory, employing a spherical grating monochromator and at undulator beamline I511-3 [58] of the Swedish

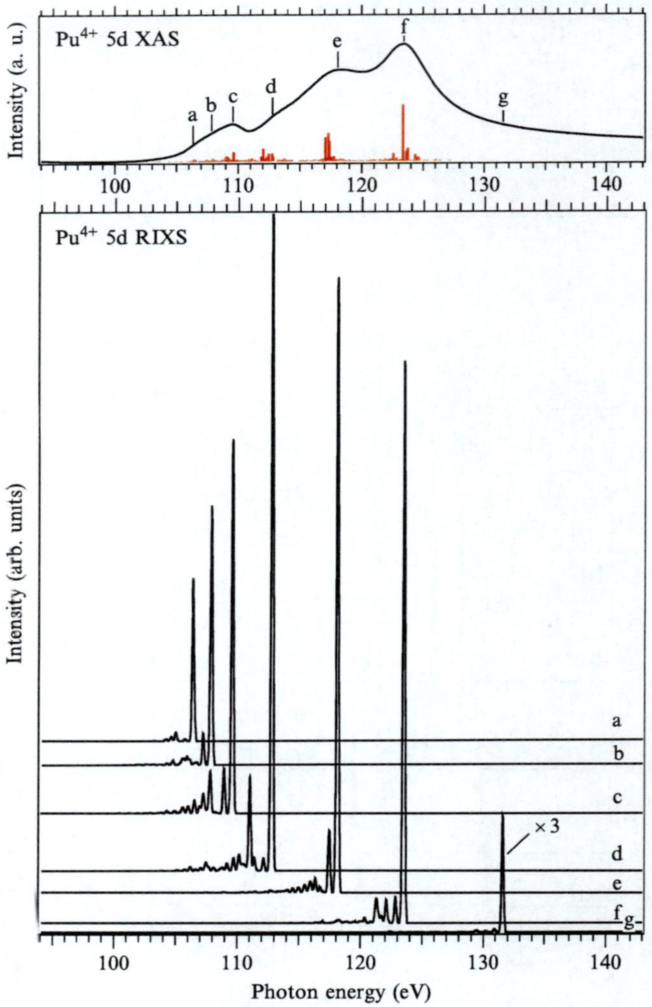

Fig. 17 Calculated resonant soft X-ray scattering spectra of Pu(IV) at various excitation energies across the 5*d* edge indicated by *letters* on the corresponding absorption spectrum (*top panel*)

synchrotron facility MAXLAB, which was equipped with plane-grating monochromator SX-700. Resonant ultrasoft X-ray scattering spectra from the samples were recorded using a grazing-incidence grating spectrometer [59] with a two-dimensional detector. The typical experimental set-up for RIXS measurements (as used at ALS) is shown as an optical layout in Fig. 23 and in the photograph in Fig. 24. The incidence angle of the photon beam was approximately 15° from the sample surface, and the spectrometer was placed in the horizontal plane at an angle of 90° with respect to the incidence beam. The bandwidth of the excitation was about 65 meV. The total energy

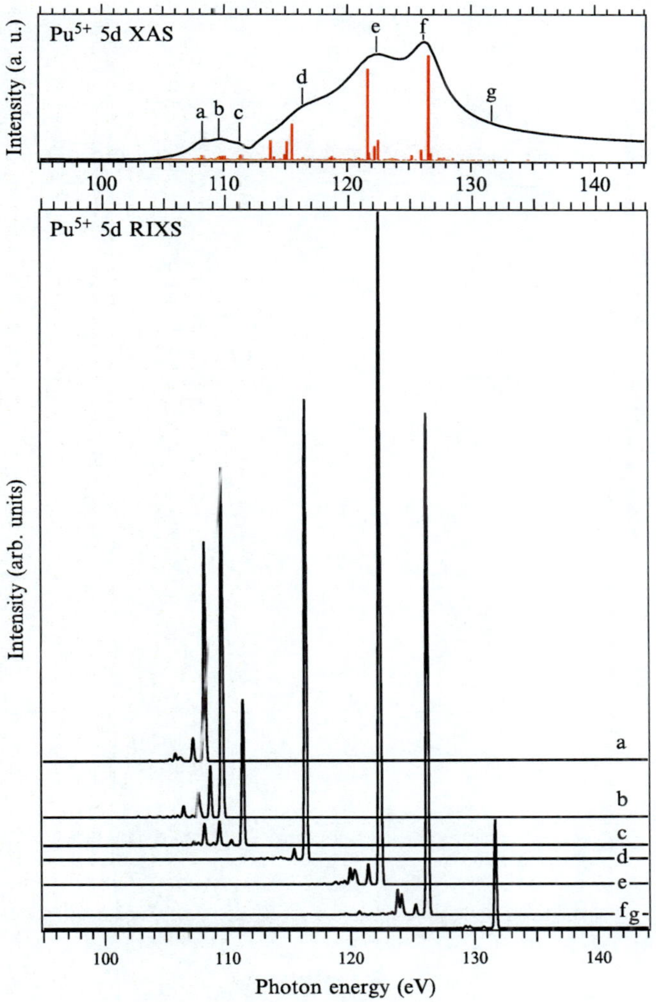

Fig. 18 Calculated resonant soft X-ray scattering spectra of Pu(V) at various excitation energies across the 5*d* edge indicated by *letters* on the corresponding absorption spectrum (*top panel*)

resolution of the RIXS data was estimated from the full width at half maximum of the elastic peak to be 160 meV.

Due to safety regulations at various synchrotron radiation facilities and difficulties to use samples as sealed, closed sources of radioactivity for soft-X-ray spectroscopic measurements, we are forced to greatly minimize quantities of substances in question, in order to reduce the radiological risk to an acceptable level.

One of the measures for preventing radiological contamination of the experimental chamber is the use of specially designed sample holder (see Fig. 25). It is

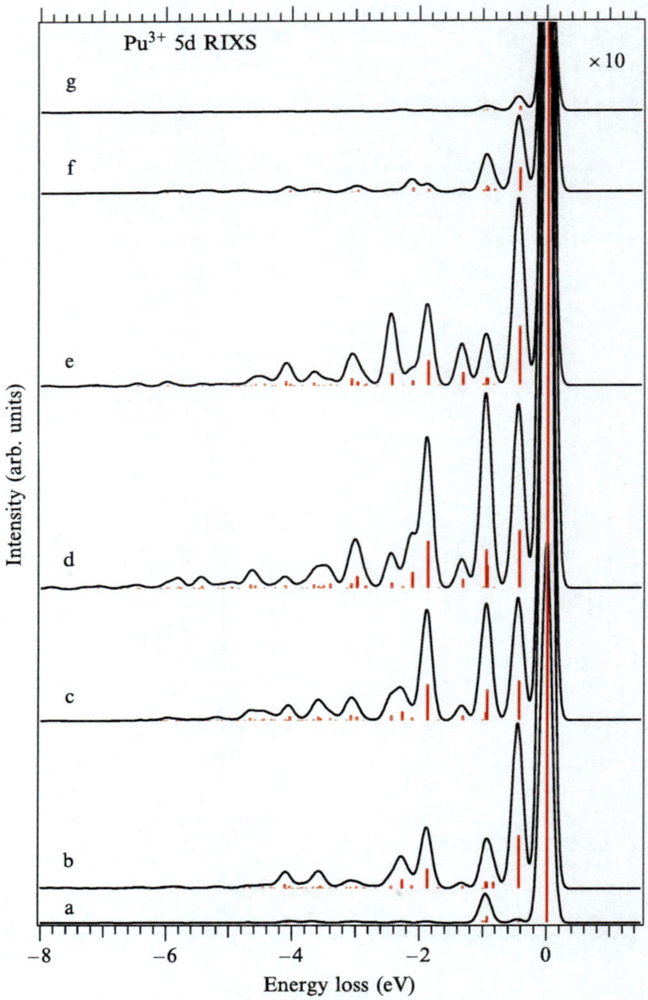

Fig. 19 The same spectra as in Fig. 16 displayed on the energy loss scale

essentially a cylindrical can with slots for incoming and outgoing radiation. The samples are attached to the slab inside the can just behind the slots. Due to this design, the sample holder served as a catch tray for material that might come loose during handling and the measurements, thus ensuring that no residuals will be left in the experimental chamber after the experiment.

Although the sensitivity and selectivity of the RIXS technique is quite high, strict safety limitations on amounts of sample activity/quantity lower the contrast of measurements in some cases as one can see below. Considering that RIXS spectroscopy is a photon-hungry technique (due to low fluorescence yield as compared to electron

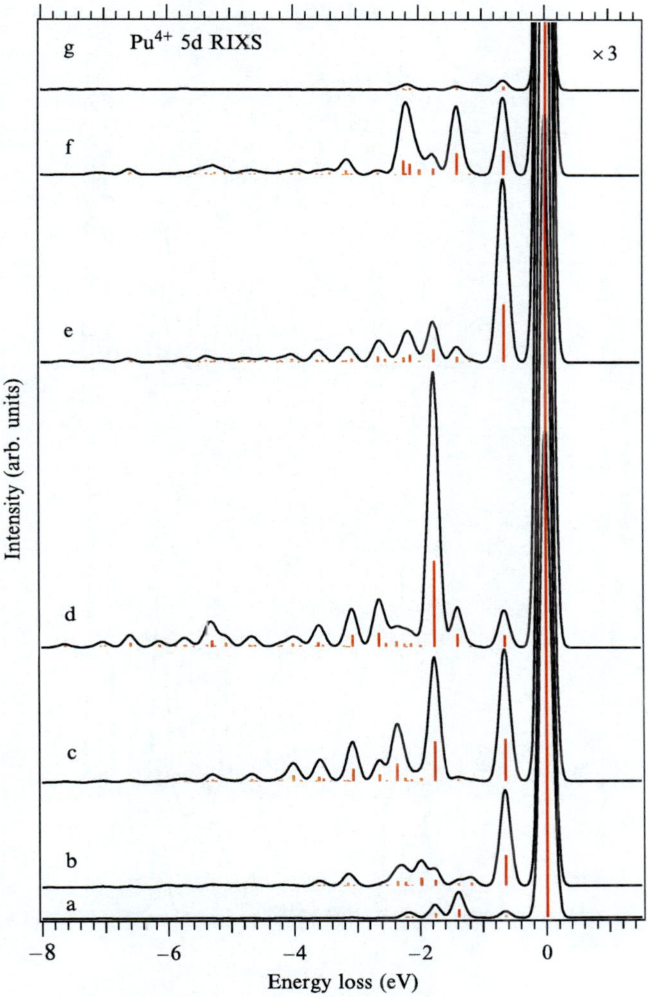

Fig. 20 The same spectra as in Fig. 17 displayed on the energy loss scale

yield per electronic excitation–deexcitation) and generally requires long acquisition times per spectrum, safety restrictions increase acquisition times even more.

One important limitation of the RIXS measurements at the actinide 5*d* edges is the presence of a very strong elastic peak due to enhanced reflectivity at the incident photon energies required for these experiments. The tails of the huge elastic peak reduce signal-to-noise ratio for the signal of the inelastic scattering on low-energy excitations, which are represented by spectral structures close to the elastic peak.

One of the ways to deal with the problem is to significantly improve an energy resolution of instruments used for X-ray scattering experiments (thus to obtain

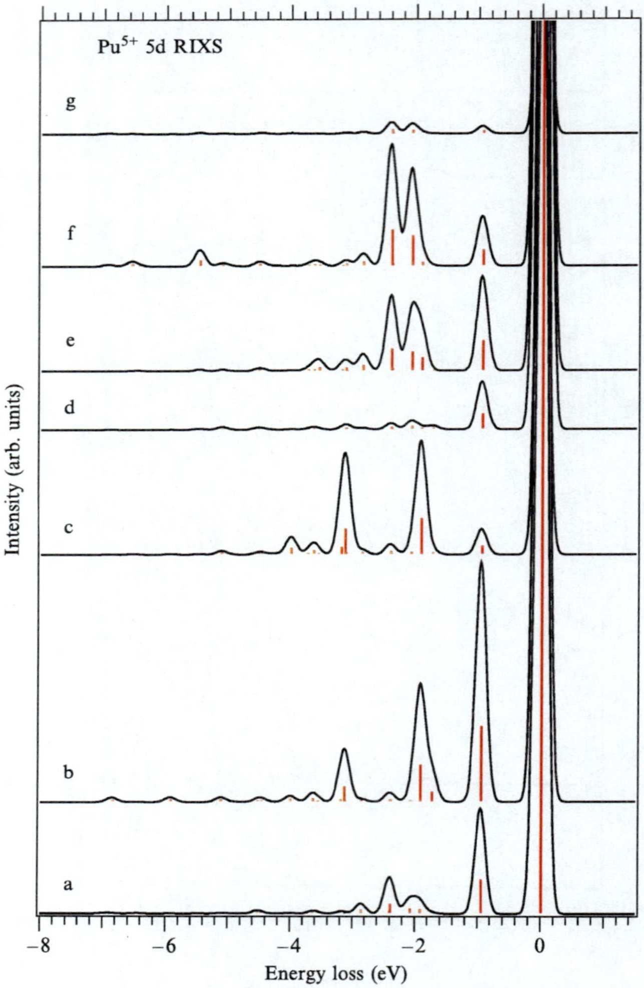

Fig. 21 The same spectra as in Fig. 18 displayed on the energy loss scale

much narrow peaks) while keeping similar efficiency in terms of photon flux throughput. By now, an improved design of the Rowland-circle X-ray fluorescence spectrometer [59] similar to the one used for present measurements is commercially available from Gammadata, Uppsala, Sweden. The spectrometer is equipped with higher groove-density (grooves/mm) diffraction gratings and allows for greater degree of the detector movement, thus providing access to higher diffraction orders and higher energy resolutions. A further step forward would be an employment of the instrument, such as a 2 m variable-line-spacing-grating (VLS) spectrograph [60] installed at beamline 4.0.3 at the Advanced Light Source, Berkeley, USA. The spectrograph is designed to cover the ~30–120 eV range and provides a resolving

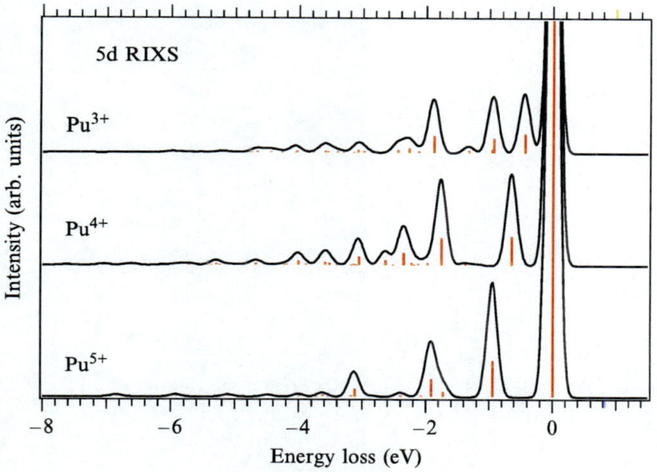

Fig. 22 Calculated resonant soft X-ray scattering spectra of Pu(III), Pu(IV), and Pu(V) at the excitation energy of 109.6 eV

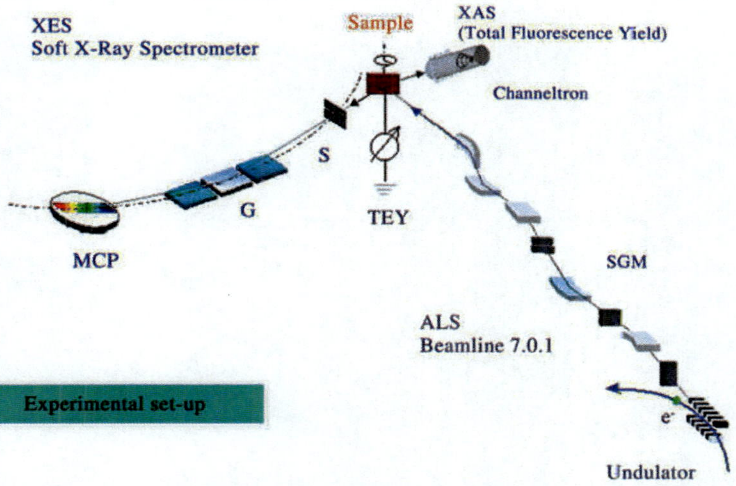

Fig. 23 Typical experimental setup for resonant inelastic soft X-ray scattering measurements as optical layout

power of better than 10,000, although the existence of an additional reflection due to an additional optical element by design puts higher demands on the photon flux from the beam-line at resolutions comparable to that of the spectrograph. The improved resolution might allow for detailed studies of the crystal-field, magnetic, and phonon excitations.

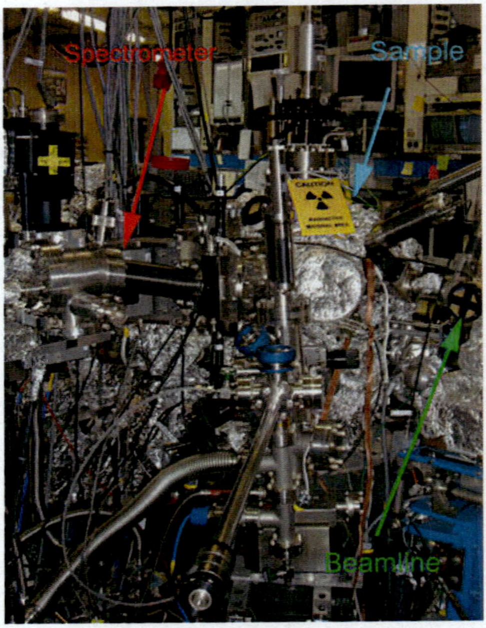

Fig. 24 Photograph of experimental setup for resonant inelastic soft X-ray scattering measurements

Fig. 25 Special sample holder (*middle*), designed for radioactive materials, shown inside the experimental chamber in between the spectrometer slit assembly (*left*) and channeltron (*right*)

Another way would be to perform RIXS measurements at actinide 4*d* edges where the strength of reflectivity and correspondingly the intensity of the elastic peak are greatly reduced while the same final states are probed in the RIXS spectra. The advanced instrumentation, which might allow for highly detailed spectra in the required photon energy range, was recently developed at the ADRESS beamline at the Swiss Light Source. The operational soft X-ray VLS spectrometer [61] there provides resolutions of about 10,000. Furthermore, spectrometers with a resolving power of ~30,000 and ~50,000 are in the process of development at European Synchrotron Radiation Facility (ESRF), Grenoble, France, and MAXLAB, Lund, Sweden. However, the RIXS experiments at the 4*d* edges might not provide as great sensitivity of the RIXS spectra to different types of excitations and to varying excitation energies as those at the 5*d* edges. While the atomic multiplets and significant spectral intensity are spread over the extended energy range at the 5*d* edges with a separation into the prethreshold and main edge regions due to strong 5*d*–5*f* exchange interaction, the $4d_{5/2}$ and $4d_{3/2}$ edges basically consist of a single line with most of spectral intensity and intense atomic multiplets compressed into relatively narrow energy range.

It should be pointed out that the advanced instruments might not be necessarily available for experiments with actinide materials because they are installed as stationary equipment at beamlines where measurements on actinides may not be allowed due to safety restrictions.

5.3 Preparation of Pu Samples

Besides model calculations, it is helpful to have experimental data from reference samples where the oxidation state of actinide element is believed to be well defined.

The plutonium-242 dioxide sample used as a reference in present measurements was fabricated by standard techniques used to prepare radionuclide counting plates. The counting plate was prepared from an aqueous solution of about 0.8 mM plutonium-242 in approximately 0.1 M HCl that was localized onto the surface of high purity platinum substrate (25.4 mm diameter, 0.05 mm thickness) by successive, partial micropipette aliquots into an area of about 2 mm^2. The isotopic composition of the plutonium solution was about 99.9% plutonium-242 and 0.1% in the suite of Pu isotopes (238–241) by mass.

The aqueous solution was allowed to dry and the resulting solid residue was distributed in a ring-shaped manner. This structure was inductively heated to 700°C under atmosphere to oxidize the material and to fix the material onto the substrate to preclude loss in the UHV spectrometer chamber during the measurements. The final amount of Pu-242 on the counting plate was found to be about 1.2 μg.

The portion of the platinum counting plate containing the plutonium was cut into a to 4 mm × 4 mm square and mounted with conductive tape on a rectangular sample holder.

For the Pu-on-Fe sample preparation, 10 ml of the Fe-equilibrated synthetic ground water and 0.7 ml of the Pu stock solution was mixed in a 20 ml glass vial containing a 2.5 mm × 4 mm iron strip. N_2 (g) was bubbled through the solution for a couple of minutes and the vial was closed with a screw cap. The vial was then put in a sealable plastic bag flushed with N_2, which was subsequently put inside a glass jar flushed with N_2. The lid of the glass jar was fitted with a rubber packing effectively sealing the lid. After 14 days, the sample was opened and the pH, E_H, and the Pu-concentration of the liquid mixture were measured (pH 9.7, E_H 335 mV, C_{Pu} 1.7×10^{-7} M). There was some loose material (corrosion products and some staining of the Fe strip), especially on one side. This side had less Pu on it as measured by alpha spectrometry (197 Bq) in comparison with the less stained side, which had a Pu activity of 1,341 Bq. This amounts to about half of the activity used in the experiment and the less stained side was that used for the RIXS measurement.

The plutonium stock solution was prepared by dissolving $^{242}PuO_2$ in a mixture of fairly concentrated $HClO_4$, HNO_3, and HF. From the resulting solution, plutonium hydroxide was precipitated and separated using centrifugation. The precipitate was washed twice with deionized water and then dissolved in 1.5 M $HClO_4$. Undissolved residues were again separated using centrifugation and discarded. Using an electrochemical cell with an Ag/AgCl reference electrode and Pt working and auxiliary electrodes connected to a potentiostat (EG&G Instr. Princeton Applied Research Model 283), the plutonium in the solution was oxidized to Pu (VI). After oxidation, the pH of the solution was adjusted to slightly basic using solutions of NaOH at different concentrations. During this adjustment, the solution was diluted by about one order of magnitude. The solution was left to stand for a couple of days before the first use, and a white fluffy precipitate was observed, which settled nicely to the bottom of the vial. The pH and plutonium concentrations were measured to 9.4 and 0.1 mM, respectively. This is far above the solubility of Pu(IV) at this pH, indicating that the Pu is stable in the pentavalent or hexavalent state. The stock solution was centrifuged a final time just prior to the sample preparation, and the precipitate was discarded.

A synthetic groundwater was prepared according to Vourinen and Snellman [62] and with the approximate composition given in Table 1.

All centrifugation was made using a Fisher Scientific MicroV centrifuge at 10,000 rpm for 10 min. All water was initially Millipore MQ-grade.

Table 1 Composition of ground water with pH 8.8

	(mg/L)	(mmol/L)
Na^+	52.5	2.3
Ca^{2+}	5.1	0.13
Mg^{2+}	0.7	0.03
K^+	3.9	0.10
SiO_2	1.7	0.03
SO_4^{2-}	9.6	0.10
Cl^-	48.8	1.4
HCO_3^-	65.0	1.1

The method described above was also used to prepare another sample of Pu sorbed on the Fe foil with lower Pu content/activity, as low as 2.3 nCi.

5.4 Experimental Results for Pu Systems

Figure 26 displays X-ray absorption and resonant X-ray scattering spectra of the so-called reference sample/counting plate recorded throughout the Pu 5*d* edge. The X-ray absorption spectrum is broad due to autoionization effects and in general agreement with the calculated spectrum for Pu(IV), although when comparing the latter with the calculated spectrum for Pu(V), a large broadening makes difference between the spectra marginal. RIXS spectra, which were measured at an energy of the incident photon beam tuned to various parts of the Pu 5*d* absorption edge, are plotted in Fig. 26 on the energy loss scale. RIXS spectra reveal a number of structures showing dependence on varying energies of incident photons. In particular, structures at energy losses of –0.8, –1.1, –1.6, and –1.8 eV can be identified in some RIXS spectra. Important resonant behavior can be noted for the –1.6 eV and –1.8 eV structures, which are marked by blue- and green-dashed vertical lines, respectively, in Fig. 26. The former is significant in the RIXS spectrum recorded at the excitation energy of 109.5 eV and again becomes pronounced for energies of incident photons being in the range 113–115 eV, while the latter resonates at excitation energies tuned between 110 and 112.5 eV.

On going to the Pu-on-Fe system, first, the sample with lower activity (2.3 nCi) was studied. As a result, a comparison of spectra recorded at two different energies of incident photons with those of the reference sample is made in Figs. 27 and 28. The excitation energies were chosen to represent two regions for which the resonances of RIXS structures at energy losses of –1.6 and –1.8 eV were respectively detected for the reference sample. Although, due to a low content of Pu and an intense contribution of reflectivity in the elastic peak, the signal-to-noise ratio for the useful RIXS signal is reduced, it is still possible to detect RIXS structures.

While the spectra of the reference Pu dioxide and the sample of Pu sorbed on the Fe surface are similar to each other in terms of existing spectral structures and their energies in Fig. 27, they are somewhat different from each other in this respect in Fig. 28. An interesting finding from the measurements is that the resonance of the –1.8 eV structure is missing in the RIXS spectra of the Pu-on-Fe sample. An inspection of RIXS profiles in Fig. 28 shows that only the neighboring –1.6 eV structure is present for the Pu-on-Fe sample under conditions favorable for the –1.8-eV resonance, as found for the spectrum of the reference sample.

Similar results, which are displayed in Figs. 29 and 30, were obtained for the sample with larger quantity of Pu sorbed on the Fe foil. Note that RIXS spectra of the Pu-on-Fe sample in these figures, which were measured at specific excitation energies leading to the above-discussed resonant effects for the reference Pu dioxide, resemble each other quite closely in terms of observed RIXS structures

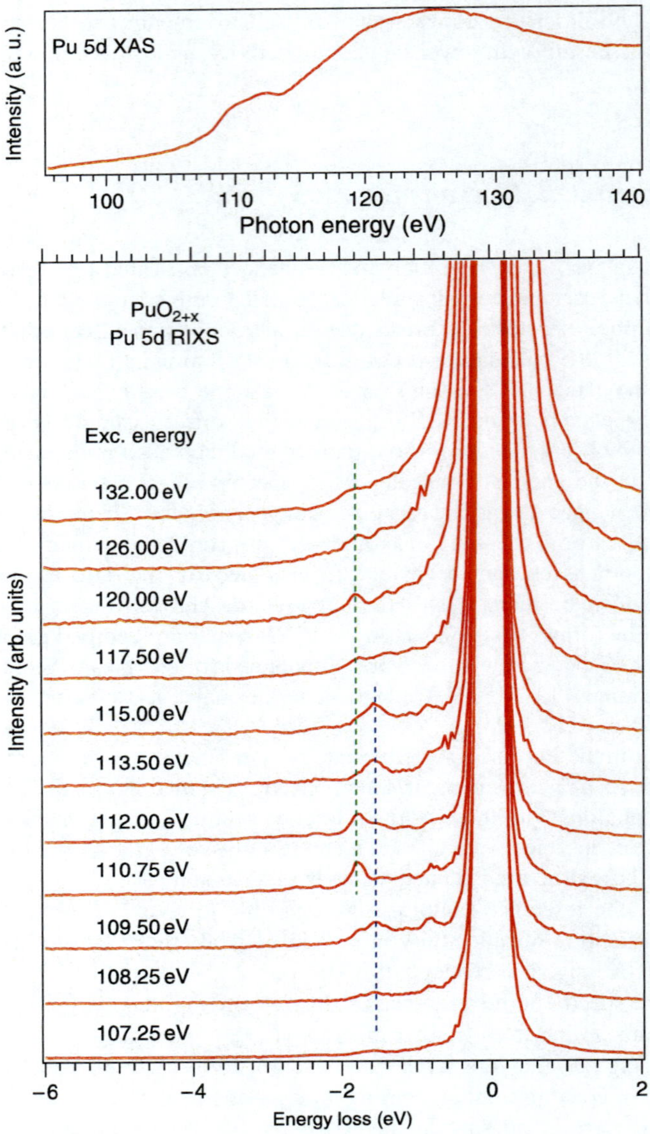

Fig. 26 Resonant X-ray scattering spectra of Pu dioxide recorded at different excitation energies close to the Pu 5*d* threshold. The Pu 5*d* adsorption edge recorded in the total electron yield mode is shown in the *top panel*

and their energies, in contrast to the data for the reference sample. Both spectra of sorbed Pu reveal main RIXS peaks at energy losses of about –0.7 and – 1.6 eV.

The energy scale of the RIXS structures observed in Figs. 26–30 makes it possible to assign them to spectral transitions, which are a result of *f–f* excitations

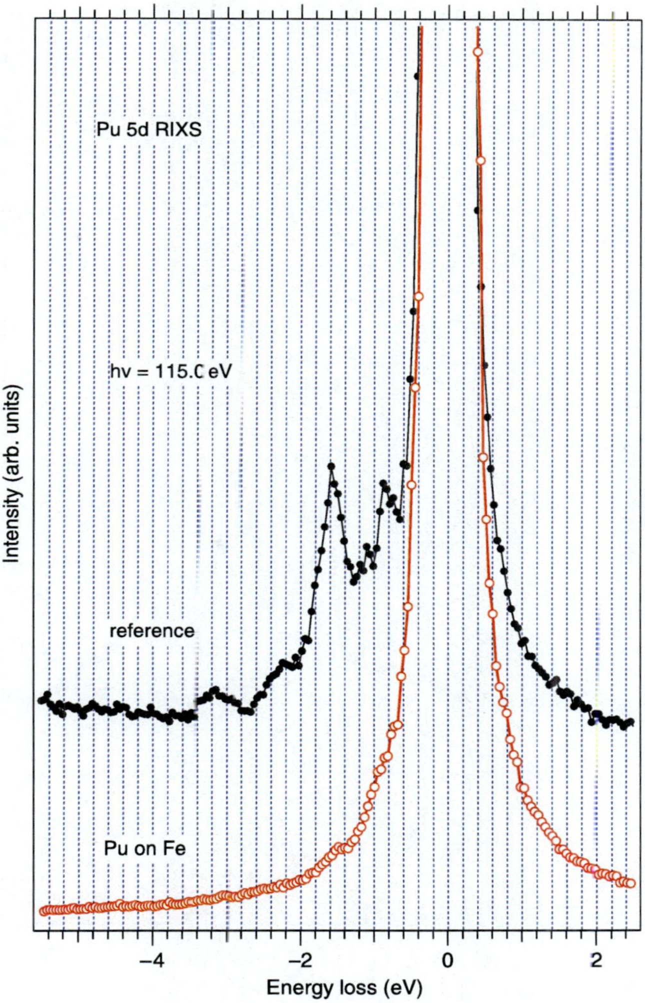

Fig. 27 Comparison of resonant inelastic soft X-ray scattering spectra of Pu sorbed on the Fe strip (sample with low activity) and Pu dioxide at the energy of incident photons tuned to 155.0 eV

within the Pu 5*f* shell. Therefore, the results of the model atomic-multiplet calculations for Pu ions in various oxidation states can be used to aid the interpretation of the experimental data.

When comparing the calculated scattering spectra of Pu(III), Pu(IV), and Pu(V) ions in Fig. 22 (calculated for an excitation energy of 109.6 eV) with the experimental spectrum of the reference sample from Fig. 26, which was recorded at the energy of incident photons of 109.5 eV, we find that the calculated RIXS profile for Pu(IV) is in better agreement with the experimental one. Calculated main RIXS

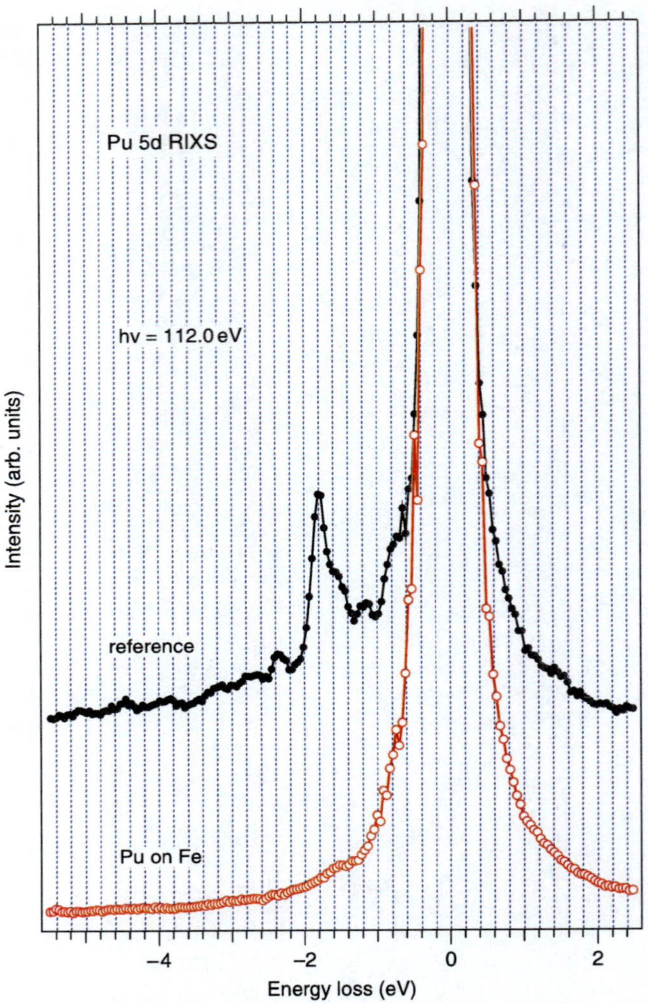

Fig. 28 Comparison of resonant inelastic soft X-ray scattering spectra of Pu sorbed on the Fe strip (sample with low activity) and Pu dioxide at the energy of incident photons tuned to 112.0 eV

peaks at −0.65 eV and −1.75 eV can be related to structures at −0.8 eV and −1.6 eV in the experimental spectrum, respectively. At the same time, the results of calculations for Pu(III) contain too many peaks, while both the calculated main RIXS peaks in the Pu(V) spectrum are shifted by 0.2 eV to higher energy losses as compared to those calculated for Pu(IV).

When analyzing other experimental RIXS data measured on the reference sample, for example, the shape of the spectrum for the excitation energy of 115.0 eV can be reproduced as a superposition of two Pu(IV) RIXS spectra

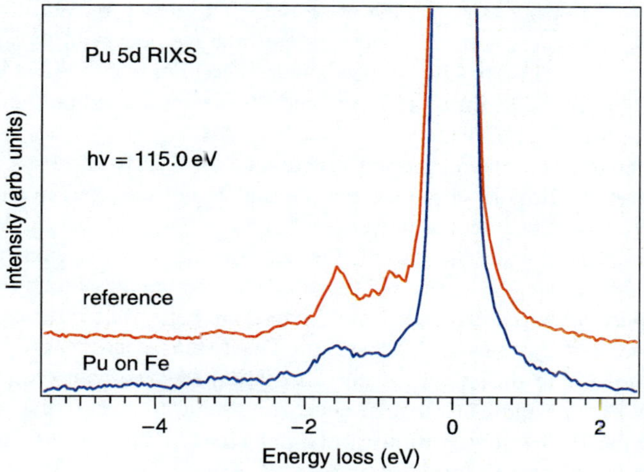

Fig. 29 Comparison of resonant inelastic soft X-ray scattering spectra of Pu sorbed on the Fe strip (sample with high activity) and Pu dioxide at the energy of incident photons tuned to 115.0 eV

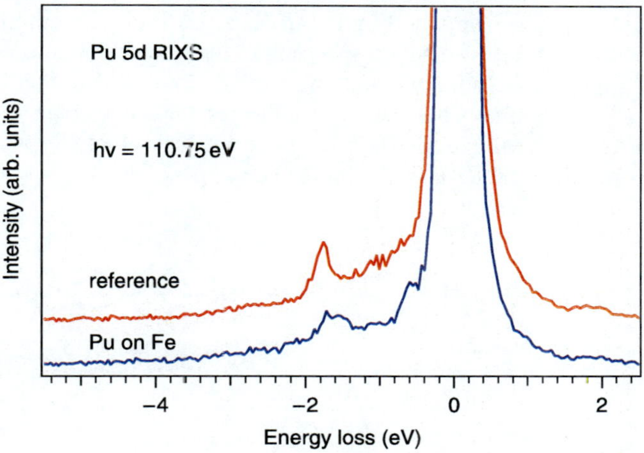

Fig. 30 Comparison of resonant inelastic soft X-ray scattering spectra of Pu sorbed on the Fe strip (sample with high activity) and Pu dioxide at the energy of incident photons tuned to 110.75 eV

calculated for excitation energies of 112.8 eV and 118.1 eV, respectively (since the calculations were performed before the RIXS experiments were carried out, there is no calculated spectrum available for exactly the same excitation energy of 115.0 eV). The RIXS part of the experimental 115.0-eV spectrum is dominated by two structures at energy losses of –0.8 eV and –1.6 eV as in the case of the 109.6-eV spectrum discussed above. Therefore, these structures can be considered intrinsic to Pu(IV) as a result of *f–f* excitations.

Turning to Pu-on-Fe samples, the measured RIXS spectra reveal that the presence of these structures and RIXS profiles are in agreement with those calculated for the Pu(IV) ion. This strongly suggests that plutonium from the Pu(VI) solution sorbed on Fe surfaces is getting reduced, and Pu species sorbed on the Fe foils are mainly in the form of Pu(IV).

What is the origin of the resonating structure that is clearly observed in the RIXS spectra of the Pu dioxide reference sample and is not detected in the Pu-on-Fe samples? We attribute it to the presence of some Pu(V) fraction in the reference sample. According to calculations (see Fig. 22), the main RIXS peaks of Pu(V) are shifted to higher energy losses by 0.2 eV, which matches exactly the energy difference between structures at –1.6 eV, which is believed to belong to Pu(IV), and at –1.8 eV in the measured RIXS spectra. Due to the selectivity properties of the RIXS technique, and differences in the absorption and resonant scattering cross-sections for Pu(IV) and Pu(V), it is quite reasonable to expect that the relative intensity of the Pu(V) contribution can be enhanced versus that of Pu(IV) in the RIXS spectra for specific excitation energies. The presence of the Pu(V) fraction in the reference sample may be due to its preparation procedure or/and specific conditions under which further oxidation of PuO_2 may take place as it has been discussed in scientific literature.

The presence of some fraction of Pu in higher oxidation state than Pu(IV) in the reference sample is also supported by Pu 4*d* X-ray absorption measurements (see Fig. 31). The Pu 4*d* X-ray absorption spectrum of the reference is shifted to higher-energies by approximately 0.3 eV and narrower as compared to that of the high-activity Pu-on-Fe sample. This behavior indicates higher average charge of Pu atoms in the reference sample. On the other hand, analysis of RIXS spectra does not support the presence of Pu(III) in the Pu-on-Fe samples.

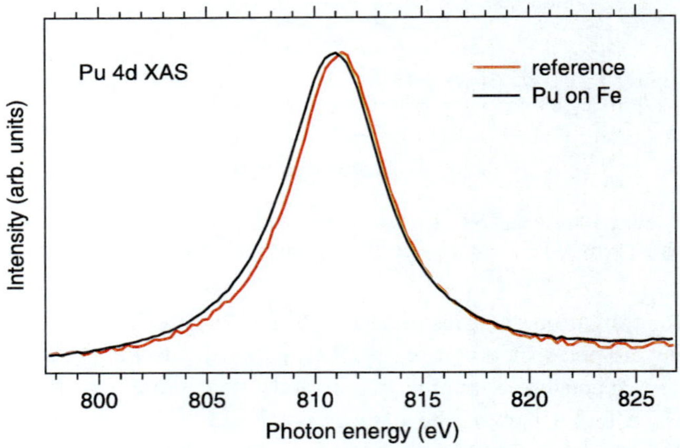

Fig. 31 Total electron yield spectra of Pu dioxide reference and Pu sorbed on Fe foil across the U 4*d* absorption edge

6 Summary

Based on the analysis of spectral shapes, their dependence on the energy of incident photons, and comparison with model calculations and experimental data from the reference sample, it can be concluded that plutonium from the Pu(VI) solution sorbed on Fe surfaces is likely to be reduced and Pu species sorbed on the Fe foils are mainly in the form of Pu (IV). Current results correlate with what was observed Np (V) and U (VI). Furthermore, combined analysis of present data with model atomic multiplet calculations of RIXS and XAS spectra suggests that significant presence of Pu(III) on the Fe foils is unlikely.

Based on experience of the RIXS spectroscopy application to studies of U, Np, and Pu systems in soft X-ray regime, the following deserves to be emphasized:

1. Both localized and delocalized 5*f* components can be selectively probed by RIXS technique, and crystal-field *f*–*f* multiplet and very low-energy excitations of 5*f* nature can be studied in detail when resolution is improved.
2. Physical quantities that enter as parameters in parametric theories, such as Anderson impurity model, can be more accurately determined for the ground state due to charge-neutrality of RIXS technique.
3. Contribution from different species/oxidation states of the same actinide can be enhanced in spectra by tuning the energy of incident photons and studied as in the case of PuO_{2+x}.
4. RIXS is a useful tool for applied problems especially under conditions when few spectroscopic techniques can be employed.

Acknowledgments The experimental work reviewed here was carried out in collaboration with Prof. Y. Albinsson, Dr. J.-H. Guo, Dr. K. Ivanov, Dr. K. O. Kvashnina, Dr. A. Modin, Dr. H. J. Nilsson, Prof. J. Nordgren, Dr. K. Ollila, Dr. D. K. Shuh, Prof. L. Werme, J. Vegelius.
The work was supported by Swedish Research Council, European Comission within 6th EURATOM framework programme, Göran Gustafsson Foundation for Research in Natural Sciences and Medicine, and Svensk Kärnbränslehantering AB.

References

1. Haschke JM, Allen TH, Morales LA (2000) Nature 287:285
2. Korzhavyi PA, Vitos L, Andersson DA, Johansson B (2004) Nat. Mater. 3:225
3. Martin P, Grandjean S, Ripert M, Freyss M, Blanc P, Petit T (2003) J. Nucl. Mater. 320:138
4. Neck V, Altmaier M, Fanghänel Th (2007) J. Alloys Compd. 444–445:464
5. Neck V, Altmaier M, Fanghänel Th (2007) Comptes Rendus Chimie (France) 10:959
6. O'Loughlin EJ, Kelly SD, Cook RE, Csencsits R, Kemner KM (2003) Environ. Sci. Technol. 37:721
7. Anderson PW (1961) Phys. Rev. B 124:41
8. Zaanen J, Sawatzky GA, Allen JW (1985) Phys. Rev. Lett. 55:418
9. Gunnarsson O, Jepsen O (1988) Phys. Rev. B 38:3568
10. Gunnarsson O, Andersen OK, Jepsen O, Zaanen J (1989) Phys. Rev. B 39:1708
11. Gunnarsson O, Schönhammer K (1989) Phys. Rev. B 40:4160

12. Gunnarsson O, Christensen NE (1990) Phys. Rev. B 42:2363
13. Kuiper P, Guo J-H, Såthe C, Duda L-C, Nordgren J, Pothuizen JJM, de Groot FMF, Sawatzky GA (1998) Phys. Rev. Lett. 80:5204
14. Wassdahl N, Skytt P, Guo J, Nordgren J, Johnson PD, Rubensson J-E, Boske T, Eberhardt W, Kevan SD (1992) Phys. Rev. Lett. 69:2598
15. Butorin SM, Mancini DC, Guo J-H, Wassdahl N, Nordgren J, Nakazawa M, Tanaka S, Uozumi T, Kotani A, Ma Y, Myano KE, Karlin BA, Shuh DK (1996) Phys. Rev. Lett. 77:574
16. Butorin SM (2009) In: Proceedings of 5th Workshop on Speciation, Techniques and Facilities for Radioactive Materials at Synchrotron Light Sources, July 15–17, Synchrotron SOLEIL, Saint-Aubin, France, 2008; NEA/NSC/DOC(2009) 15:73
17. Nordgren J, Butorin SM, Duda L-C, Guo J-H (2006) In: Vij DR (ed) Handbook of Applied Solid State Spectroscopy, Springer Science + Business media, New York, p. 595
18. Tanaka S, Kotani A (1993) J. Phys. Soc. Jpn. 62:464
19. Butorin SM, Guo J-H, Magnuson M, Kuiper P, Nordgren J (1996) Phys. Rev. B 54:4405
20. Butorin SM, Mancini DC, Guo J-H, Wassdahl N, Nordgren J (1995) J. Alloys Compd. 225:230
21. Kotani A (2000) J. Electron Spectrosc. 110–111:197
22. Butorin SM, Guo J-H, Magnuson M, Nordgren J (1997) Phys. Rev. B 55:4242
23. Butorin SM, Duda L-C, Guo J-H, Wassdahl N, Nordgren J, Nakazawa M (1997) J. Phys. Condens. Matter 9:8155
24. Butorin SM, Magnuson M, Ivanov K, Shuh DK, Takahashi T, Kunii S, Guo J-H, Nordgren J (1999) J. Electron Spectrosc. 101–103:783
25. Ogasawara H, Kotani A, Thole BT, Ichikawa K, Aita O, Kamada M (1992) Solid State Commun. 81:645
26. Cowan RD (1981) The Theory of Atomic Structure and Spectra, University of California Press, Berkeley
27. Sugar J (1972) Phys. Rev. B 5:1785
28. Kotani A, Ogasawara H (1992) J. Electron Spectrosc. Relat. Phenomena 60:257
29. Kotani A, Ogasawara H (1995) J. Phys. Soc. Jpn. 64:1394
30. Butorin SM (2000) J. Electron Spectrosc. 110–111:233
31. Pireaux JJ, Riga J, Thibaut E, Tenret-Noël C, Caudino R, Verbist JJ (1977) Chem. Phys. 22:113
32. Teterin YuA, Kulakov VM, Baev AS, Nevzorov NB, Melnikov IV, Streltsov VA, Mashirov LG, Suglobov DN, Zelenkov AG (1981) Phys. Chem. Miner. 7:151
33. Madhavaram H, Buchanan P, Idriss H (1997) J. Vac. Sci. Technol. A 15:1685
34. Verbist JJ, Riga J, Tenret-Noël C, Pireaux JJ, Dúrsel G, Caudino R, Derouane EG (1976) In: Blank H, Lindner R (ed) Plutonium and other actinides, North-Holland, Amsterdam, p. 409
35. Allen GC, Griffiths AJ, Suckling CW (1978) Chem. Phys. Lett. 53:309
36. Gubanov VA, Rosén A, Ellis DE (1977) Solid State Commun. 22:219
37. Heera V, Seifert G, Ziesche P (1983) Phys. Stat. Sol. (b) 118:K107
38. Ellis DE, Goodman GL (1984) Int. J. Quantum Chem. 25:185
39. Goodman GL (1992) J. Alloys Compd. 181:33
40. Gunnarsson O, Sarma DD, Hillebrecht FU, Schönhammer K (1988) J. Appl. Phys. 63:3676
41. Gunnarsson O, Li TC (1987) Phys. Rev. B 36:9488
42. Kotani A, Ogasawara H (1993) Physica C 186–188:16
43. Pierloot K, Reinders A, Goodman GL, Devoghel D, Görller-Walrand C, Vanquickenborne LG (1991) J. Chem. Phys. 94:2928
44. Walsh PF, Ellis DE (1976) J. Chem. Phys. 65:2387
45. Wood JH, Boring M, Woodruff SB (1981) J. Chem. Phys. 74:5225 and references therein
46. Cox LE (1982) J. Electron Spectrosc. 26:167
47. Butorin SM, Shuh DK, Kvashnina K, Soroka IL, Ollila K, Roberts KE, Guo J-H, Werme L, Nordgren J (2004) Mater. Res. Soc. Symp. Proc. 807:113
48. Gruber JB, Menzel ER (1969) J. Chem. Phys. 50:3772

49. Jørgensen CK (1955) J. Inorg. Nucl. Chem. 1:301
50. Leontovich AM (1957) Optika i Spektroskopiya 2:695
51. Lämmerman H, Stapleton HP (1961) J. Chem. Phys. 35:1514
52. Lämmerman H, Stapleton HP (1961) J. Chem. Phys. 35:1514
53. McLaughlin R, White R, Edelstein N, Conway JG (1968) J. Chem. Phys. 48:967
54. Edelstein N, Mollet HF, Easley WC, Mehlhorn RJ (1969) J. Chem. Phys. 51:3281
55. Varga LP, Reisfeld MJ, Asprey LB (1970) J. Chem. Phys. 53:250
56. Carnall WT, Fields PR, Pappalardo RG (1970) J. Chem. Phys. 53:2922
57. Warwick T, Heimann P, Mossessian D, McKinney W, Padmore H (1995) Rev. Sci. Instrum. 66:2037
58. Denecke R, Väterlein P, Bässler M, Wassdahl N, Butorin S, Nilsson A, Rubensson J-E, Nordgren J, Mårtensson N, Nyholm R (1999) J. Electron Spectrosc. 101-103:971
59. Nordgren J, Bray G, Cramm S, Nyholm R, Rubensson J-E, Wassdahl N (1989) Rev. Sci. Instrum. 60:1690
60. Chuang Y-D, Pepper J, McKinney W, Hussain Z, Gullikson E, Batson P, Qian D, Hasan MD (2005) J. Phys. Chem. Solids 66:2173
61. Ghiringhelli G, Piazzalunga A, Dallera C, Trezzi G, Braicovich L, Schmitt T, Strocov VN, Betemps R, Patthey L, Wang X, Grioni M (2006) Rev. Sci. Instrum. 77:113108
62. Vuorinen U, Snellman M (1998) Finnish reference waters for solubility, sorption and diffusion studies. Posiva Working Report 98-61, Helsinki

Characterization of Colloid-Borne Actinides by Flow Field-Flow Fractionation (FlFFF) Multidetector Analysis (MDA)

Muriel Bouby and Horst Geckeis

Abstract The flow field-flow fractionation (FlFFF) method coupled with multi detector analytical (MDA) techniques is a powerful method for the characterization of colloid-borne actinides. The present paper summarizes the different techniques available and detection systems currently hyphenated to FlFFF. A review of the FlFFF–MDA applications to the characterization of colloid-borne actinides is presented. It comprises studies on actinide interactions with natural organic matter, bacteria cells, inorganic colloids (iron oxi/hydroxides, clays), and includes actinide speciation in colloid-containing natural systems (deep or surface groundwaters, lake, river, or creek waters) or in sediments from contaminated areas (nuclear material production and (re)processing facilities, nuclear test sites, etc.). The use of FlFFF–MDA allows the identification of the heterogeneities in colloidal matter of unknown composition, the quantification of the actinide associations with diverse colloidal material/nanoparticles and the gaining of mechanistic insight into actinide colloid/nanoparticle-binding modes.

1 Introduction

Appropriate analytical techniques have to be developed or applied to characterize actinide colloidal matter and to investigate their chemical and geochemical properties. Nowadays, various spectroscopic tools are available to gain insight into the binding state of toxic metal ions and radionuclides in colloidal matrices; these methods provide direct insight into the chemical nature of the metal ions. It is thus possible to quantify colloid-bound actinide ion fractions, to gain information on actinide colloid binding modes such as inner-sphere/outer-sphere complex formation and to obtain structural details of colloid-borne actinides. However, most spectroscopic techniques require elevated concentrations, as is the case for X-ray

M. Bouby (✉) and H. Geckeis
Institut für Nukleare Entsorgung, Karlsruhe Institute of Technology (KIT), P.O. Box 3640, 76021, Karlsruhe, Germany
e-mail: muriel.bouby-laliron@kit.edu, horst.geckeis@kit.edu

S.N. Kalmykov and M.A. Denecke (eds.), *Actinide Nanoparticle Research*,
DOI 10.1007/978-3-642-11432-8_4, © Springer-Verlag Berlin Heidelberg 2011

absorption spectroscopy (e.g., EXAFS), or can only be applied to a few selected metal ions with specific properties such as for time-resolved laser fluorescence spectroscopy (TRLFS), which can only be applied to elements exhibiting fluorescence. Recently, mass spectrometry has also been applied to characterize colloidal and notably polymeric actinide species. Applications of those methods are discussed in more detail in other chapters of this issue.

An inherent feature of nanoparticles (NPs) relevant under natural conditions is their heterogeneity in composition and size. The potential impact of nanoparticle size on actinide chemical and geochemical properties can be investigated by size fractionating colloids. In addition to classical ultracentrifugation, ultrafiltration, size exclusion, and hydrodynamic chromatography, the field-flow fractionation (FFF) technique is increasingly being applied as an efficient size fractionation tool. FFF is a separation method first described 40 years ago by Giddings (1966). It offers the possibility to separate colloidal matter ranging from <1 nm up to 100 μm. Unlike other fractionation methods such as size exclusion chromatography, direct contact to, e.g., column surfaces is minimized and, thus, artifacts by colloid interaction with such surfaces avoided. A broad variety of application fields have been documented for the characterization of many types of colloidal particles (inorganic, organic, charged, uncharged, colloids originating from industrial, biological, environmental, or geological samples). By combining FFF with various sensitive detection systems, it is possible to gain a detailed insight into nanoparticle properties and element composition down to the trace concentration range (Fig. 1). The following paper describes the various techniques and applications of FFF-based methods to the characterization of colloidal actinide species.

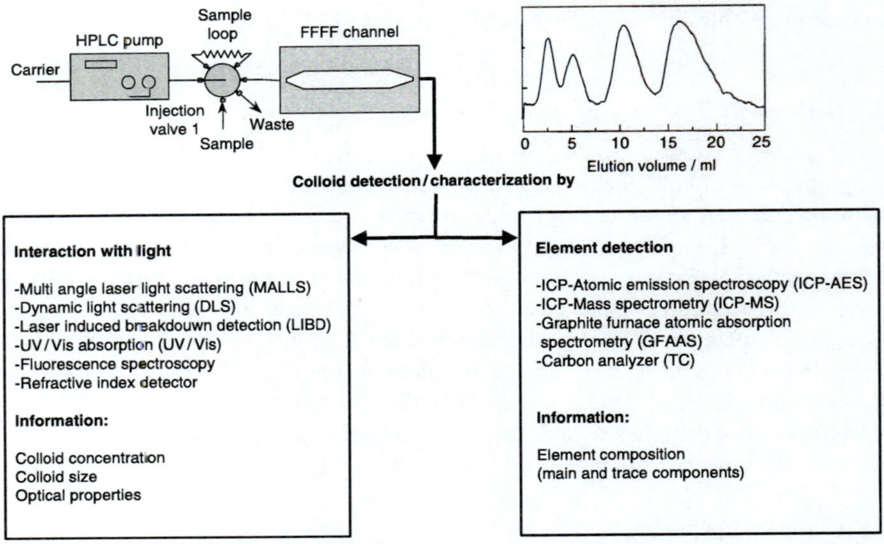

Fig. 1 Schematic flow field-flow fractionation (FlFFF) setup for nanoparticle analysis by combining with multiple detection methods

2 Fundamentals of FlFFF

For detailed information of the technique fundamentals, please refer to Giddings (1993) and Schimpf et al. (2000).

2.1 *Basic Principle*

Separation of NPs by FFF usually takes place in a thin ribbon-like channel containing a carrier solution. The laminar profile inside the channel flow is well defined. Differently sized particles are fractionated due to differing flow velocities across the channel with a force imposed perpendicular to the channel flow. Fractionation can, in principle, be done by applying any kind of field and gradient such as electric, magnetic, or gravitational (gravitational and centrifugal) fields or temperature gradients. The techniques thus allow particle fractionation according to the properties of the particles and the solvents used. They can be selected according to the respective problems and questions at hand, and instrumentation for all techniques is commercially available. In this chapter, we concentrate on the flow field-flow fractionation (FlFFF), where separation is achieved in a solvent flow field acting perpendicular to the channel flow. This field (the cross flow) forces particles, ions, and molecules towards a wall consisting, in general, of a permeable ultrafiltration membrane. While ions and molecules are able to penetrate the filter, particles cannot and they accumulate at the membrane surface. Diffusive forces counteract and equilibrium establishes where an exponential colloid concentration gradient extends a few μm towards the channel center. Separation of NPs by FFF is consequently achieved according to their different diffusion coefficients.

The hydrodynamic colloid diameter is accessible from the Stokes equation without any calibration. For the so-called "symmetric" FlFFF ("SymFlFFF" or "FlFFF") technique (explained below), the relation of the retardation volume V_R to the hydrodynamic diameter and the diffusion coefficient is given by the following simple equation:

$$V_R = \frac{v_c w^2}{6D} = \frac{\pi \eta w^2 v_c}{kT} \times \frac{d}{2}, \tag{1}$$

where η is the medium viscosity; w, the channel width; v_c, the volumetric cross flow; d, the hydrodynamic diameter of the colloidal species; k, the Boltzmann constant; T, the absolute temperature; and D, the particle diffusion coefficient. The hydrodynamic diameter represents the geometrical diameter plus adhering solvent shell for spherical particles. Knowledge of the real dimensions requires information from microscopic methods for particles of other than spherical shapes, e.g., needle- or plate-like particles (see, e.g., Beckett et al. 1997). In this case, FFF only provides "sphere equivalent" size information.

2.2 FlFFF Techniques

SymFlFFF uses a symmetric channel, which is delimited by an overlying porous frit guiding the cross flow uniformly across the channel. The lower channel wall consists of a frit covered by an ultrafiltration membrane, usually consisting of regenerated cellulose or polyethersulphone with variable cutoff. The sample is injected under stopped channel flow and acting cross flow. After a short relaxation time where the diffusion profile evolves, the channel flow is switched on and the particles are eluted towards the detector(s). Flow rates can be programed and varied during fractionation in order to optimize particle fractionation.

Asymmetric FlFFF (AsFlFFF) was first proposed by Wahlund and Giddings (1987); it uses a trapezoidal-shaped channel, and the top wall is impermeable to the solvent. Cross- and channel flow enter the channel from the same inlet. The channel shape aims to maintain constant channel flow rates during fractionation (Williams 1997). AsFlFFF allows particle focusing prior to fractionation by guiding the flow from both ends of the channel during sample injection. In principle, particle Stokes radii can also be determined, although not as easily as in the case of the SymFlFFF [Wahlund in Schimpf et al. (2000)]. AsFlFFF has some advantages with regard to resolution; however, flow conditions are more difficult to control. Size calibration with reference particles is in this case recommended. Technical improvements of the method are taking place continuously. For example, instruments using hollow fibers instead of channels (Wijnhoven et al. 1995) and, in recent days, microfluidic systems (Lenshof and Laurell 2010) are being developed.

2.3 FlFFF Modes

FlFFF can be operated in three different modes [see also Myers (1997)]: (1) In "normal" mode, the distance of the colloid "plume" center of gravity from the accumulation membrane depends on the diffusion coefficient, i.e., on the size of the colloids. The plumes of small colloids with higher diffusion coefficients extend further into the channel than larger particles. In the laminar channel flow, smaller particles elute prior to larger ones due to their average position lying in areas with faster flow gradients. The elution time increases with increasing particle size. (2) As size increases, particles cannot enter the slow velocity streamlines at a distance closer to the accumulation wall than their own diameter. As a result, they protrude into faster flowing regions. Consequently, the elution order is reversed and these larger particles appear in the effluent earlier than predicted by Eq. (1). This separation mode is called "steric" and as a rule of thumb holds for particles with diameters in the range of μm and larger. Smaller particles are usually eluted in the normal mode. Some overlap may occur depending on the cross flow field strength such that at increased flow rates, the steric regime is shifted to smaller particles. (3) By increasing the channel flow even further, the fractionation in the "hyperlayer" mode becomes allowed. Particles in the "steric" mode experience a hydrodynamic lift and are focused in a small flow gradient region relative to their size. "Hyper-layer" FFF allows rapid fractionation of larger sized particles.

2.4 Detection Techniques

A broad variety of detection techniques can be applied to characterize fractionated NPs and colloids. Light interaction with NPs is utilized by light scattering (LS) devices (e.g., DLS, MALLS, ...) and, more recently, by laser-induced breakdown detection (LIBD) (Scherbaum et al. 1996; Bundschuh et al. 2001; Hauser et al. 2002; Walther et al. 2002; Walther 2003; Kim and Walther 2007). Both method types not only detect particle presence but also provide independent size information. According to the Rayleigh relation, the sensitivity of laser light scattering (LLS) techniques is relatively low for samples with low colloid concentration at small colloid sizes (Filella et al. 1997). Notably, for particles with diameters <50 nm, LIBD provides sensitivities that are superior by orders of magnitude compared with those of LS techniques (Fig. 2).

LIBD sensitivity, however, decreases for colloids of certain composition. While it works quite well for polystyrene, clay, silica, ZrO_2, and actinide colloids, signal response decreases significantly for water-containing organic macromolecules of low density such as humic acids and strongly light absorbing particles such as iron oxyhydroxides. For example, for ferrihydrite colloids with an average size of about 10 nm in diameter, detection limits for FlFFF–LIBD lie about two orders of magnitude higher than for polystyrene particles of the same size (Bouby et al. 2004). LIBD has, on the other hand, been helpful for the identification of humic colloid heterogeneities. Apparently, inorganic NPs are often closely linked with humic acid molecules. While the humic colloids are virtually invisible to LIBD, NPs are detected.

UV/Vis detectors are widespread notably for the sensitive analysis of organic colloids and macromolecules. As natural organic matter of different origin has different molar absorbances, analysis of carbon by coupling a dissolved organic carbon (DOC) analyzer to the FFF channel has been suggested (Reszat and Hendry 2005). Refractive index (RI) detectors are commonly used for non-UV/Vis absorbing polymers. Certainly one of the most powerful analytical methods for actinide-containing colloid detection is inductively coupled plasma-mass spectroscopy (ICP-MS). ICP-MS combines the advantages of being fast and offering ultimate sensitivity for inorganic elements and isotopes (Dubascoux et al. 2010). More recently, high resolution ICP-MS (HR-ICP-MS) has also been applied to reduce detection limits and interferences by cluster ion formation in the ICP (Stolpe et al. 2005; Stolpe and Hasselov 2007; Plathe et al. 2010). Limits of detection (LOD) for humic colloid bound actinide (U, Th) and carbon analyses after FFF separation are reported in various papers. This data is plotted in Fig. 2. Values have been normalized to an injection volume of 20 μl to make them comparable to LLS and LIBD data.

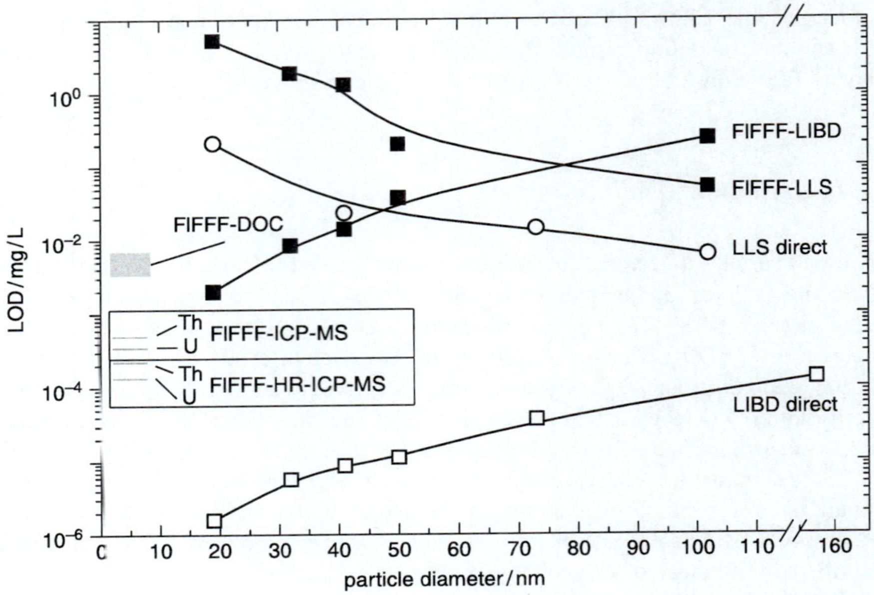

Fig. 2 Limits of detection (LOD) for different detection methods combined to FlFFF; for comparison all data are related to an injection volume of 20 μl; LOD for polystyrene nanoparticles determined by SymFlFFF-LIBD and SymFlFFF-LLS depend on particle size (Ngo Manh et al. 2000); AsFlFFF-DOC data are related to humic colloid characterization (Reszat and Hendry 2005); LODs for Th and U analysis in aquatic humic colloids have been determined by AsFlFFF combined to ICP-MS (Bouby et al. 2008) and HR-ICP-MS (Stolpe et al. 2005)

These element-specific LOD values may vary significantly depending on the selected fractionation conditions. Dilution in the channel will always lead to increasing detection limits. Optimization of the separation conditions by increasing the sample volume (e.g., Stolpe et al. (2005) inject volumes up to 45 ml) and eluting the sample in a narrow peak (small volume) can decrease LOD values by orders of magnitude.

A comprehensive collection of detection techniques combined with Sym- and AsFlFFF for the characterization of NPs is given in Table 1.

2.5 *Quantification*

Quantification of element-specific colloid size distributions requires careful calibration of the overall analytical system. For all types of detection, it is necessary to use reference standards that are closely matched to the sample particles. Notably for heterogeneously composed natural samples, this fact represents a challenge. As mentioned above, SymFlFFF requires no size calibration; diffusion coefficients and hydrodynamic diameters can be directly derived from elution times. For AsFlFFF

Table 1 Detection methods or hyphenations currently reported for the measurement of environmental NPs with (a) SymFlFFF (b) AsFlFFF

Hyphenation	Application	References
(a) SymFlFFF		
UV	Molecular weight distribution of fulvic and humic acids	Beckett et al.(1987)
UV, ICP-MS	Inorganic model compounds: lead chromate, aluminum oxide, iron oxide	Taylor et al.(1992)
RI, MALLS	Dextran, feasibility study	Roessner and Kulicke (1994)
MALLS	Polymer and particles	White (1997)
UV, Fluo, MALLS	Natural colloids in cold water extracts from the Ah-horizon of a gleyic podzol, Neumunster, Germany	Kammer and Förstner (1998)
MALLS	Submicrometer particles sizing	Wyatt (1998)
UV, ICP-MS	Size and trace element distribution of colloidal material in natural water from a small creek (Göteborg, Sweden)	Hassellov et al. (1999)
UV	Molecular weight distribution of natural organic matter	Pelekani et al. (1999)
UV, ICP-MS	Trace metals complexed on soil-derived humic acids	Amarasiriwardena et al. (2000)
MALLS, LIBD	Polystyrene colloids	Ngo Manh et al. (2000)
UV	Characterization of organic colloidal matter	Ngo Manh et al. (2001)
UV, TOC, GFAAS	Molecular weight distribution and binding properties of particulate and colloidal organic matter from the Rio Amazon and Rio Negro	Benedetti et al. (2002)
UV, ICP-MS	Humic colloid borne natural polyvalent metal ions	Geckeis et al. (2002)
UV, ICP-MS	Sediment-bound trace metal characterization	Siripinyanond et al. (2002)
UV, ICP-AES	Iron status in colloidal matter from the Rio Negro, Brazil	Benedetti et al. (2003)
UV, MALLS, ICP-MS	Aquatic colloids relevant to radionuclide migration	Geckeis et al. (2003)
UV, ICP-MS	Iron- and carbon-based colloidal carriers for trace metals in freshwater from a small creek (Göteborg, Sweden)	Lyven et al. (2003)
UV, ICP-MS	Metal–humic substances interaction study	Siripinyanond et al. (2005)
UV, ICP-MS	Colloidal humic-bound Ni and U in the dissolved fraction of contaminated sediment extracts, Savannah River, Aiken, SC, USA	Jackson et al. (2005a)
UV, ICP-MS	U sorption on bacteria cells	Jackson et al. (2005b)
UV, MALLS	Environmental colloids: applicability of the techniques	Kammer et al. (2005)
UV, fluo, HR-ICP-MS	Colloidal size distributions in a natural water sample from a small creek (Göteborg, Sweden)	Stolpe et al. (2005)
UV, fluo, HR-ICP-MS	Colloidal rare earth elements in a boreal river	Andersson et al. (2006)

(*continued*)

Table 1 (continued)

Hyphenation	Application	References
UV, EEMS	Fluorescence variability of marine and terrestrial colloids	Boehme and Wells (2006)
UV, fluo, MALLS, ICP-MS, TEM/ X-EDS	Size-based speciation of the main colloidal carriers of trace elements in the Loire river (France)	Baalousha et al. (2006)
UV, AFM	Natural aquatic colloids from surface lake and river waters from three sites near to Birmingham (UK)	Baalousha and Lead (2007)
UV, Fluo, HR-ICP-MS	Temporal variations of colloidal carrier phases and associated trace elements in a boreal river	Dahlqvist et al. (2007)
ICP-MS	Natural seep water colloidal content from the exploratory studies facility inside the Yucca Mountain, Nevada, USA	Cizdziel et al. (2008)
ICP-MS	pH-dependent uranium (VI) sorption to nanoparticulate hematite	Lesher et al. (2009)
(b) AsFlFFF		
UV, TXRF	Heavy metals associated with colloidal humic substances from seepage water from a municipal waste disposal	Exner et al. (2000)
UV, ICP-MS, LIBD	Aquatic colloids in a natural groundwater from the Gorleben site (Lower Saxony, Germany)	Bouby et al. (2002)
UV, ICP-MS, LIBD	Detection of iron oxi/hydroxide colloids with incorporated elements	Bouby et al. (2004)
ICP-MS	Sorption of radionuclides on bentonite colloids	Schäfer et al. (2004)
UV	Isolated humic and fulvic acid derived from Opalinus clay and Callovo-Oxfordian argillite	Claret et al. (2005)
UV, DOC	On-line dissolved organic carbon measurement	Reszat and Hendry (2005)
UV, ICP-MS	Multi-elements interaction with compost leachate fractions	Bolea et al. (2006)
UV, ICP-MS	Interaction of trace elements in acid mine drainage solution with humic acid	Suteerapataranon et al. (2006)
UV, ICP-MS	Uranium complexation by groundwater dissolved organic carbon	Ranville et al. (2007)
UV, DOC, ICP-MS	Complexation of aqueous elements by DOC in clay aquitard	Reszat and Hendry (2007)
UV, fluo, HR-ICP-MS	Size distribution of fresh water nanoscale colloidal matter and associated elements on mixing with sea water	Stolpe and Hassellöv (2007)
MALLS	Biodegradable nanospheres by different techniques	Augsten et al. (2008)
UV, ICP-MS	Detection of quantum dots and actinide trace metal on clay colloids in the presence of humic acids	Bouby et al. (2008)
UV, MALLS	Metal distribution in soil leachate	Dubascoux et al. (2008)
ICP-MS	Bentonite colloids mixed in Danube river in the presence of radionuclides	Mihai and Georgescu (2008)
UV, SAXS	Maghemite nanorods	Thünemann et al. (2008)

(*continued*)

Table 1 (continued)

Hyphenation	Application	References
UV, MALLS, ICP-MS	Colloidal transport of uranium in soil	Claveranne-Lamolère et al. (2009)
UV, CC-ICP-MS	Nano- to micro-particulates size and elemental distribution in the geochemically stratified Great Salt Lake, Utah, USA	Diaz et al. (2009)
UV	Migration of colloids through nonfractured clay-rich aquitard	Reszat and Hendry (2009)
UV, SAXS	In situ analysis of Resovist suspension (superparamagnetic carboxydextran-coated iron oxide nanoparticles)	Thünemann et al. (2009a)
RI, SAXS, DLS	Polymer standards	Thünemann et al. (2009b)
UV, 7AL	Colloidal organic matter from the Amazonian basin	Alasatoni et al. (2010)
MALLS, HR-ICP-MS	Trace metal–nanoparticles associations in riverbed contaminated sediments of the Clark Fork River, Montana, USA	Plathe et al. (2010)
UV, MALLS, ICP-MS	Colloidal organic matter from wastewater treatment plant effluents	Worms et al. (2010)

as well, this is possible by using a modified retention equation. As flow rates are more difficult to control precisely in AsFlFFF, size determination is done by appropriate size calibration using monodisperse colloid standards of well-known sizes. Such calibration using standards becomes even more important for normal operation, where the ratio of cross- and channel flow is varied during fractionation in programed modes (Botana et al. 1995). Size calibration can be easily done for particles with sizes larger than 10 nm, where particle standards from various materials are available. Appropriate particles with certified diameter <10 nm are more difficult to obtain. The hydrodynamic diameter under the carrier solution composition of interest is often unknown. For example, quantum dots are certified for their core size but, due to organic coating, their actual hydrodynamic diameter in a given solution can be larger (Rameshwar et al. 2006; Bouby et al. 2008). Organic polymers and biopolymers are available with certified molecular masses; there is, however, continuous debate on the actual hydrodynamic diameters of the macromolecules. The relationship between the molecular diffusion coefficient and the sample molecular weight (M_w) is not simple and depends on the molecular conformation. Thus, the difficulty lies in the selection of appropriate standards, e.g., for the size characterization of natural organic matter (Perminova 1999). The use of poly(styrenesulfonate) (PSS) molecular weight standards for calibration has been suggested by Beckett et al. (1987; Beckett and Hart 1993). A direct comparison of hydrodynamic diameters via diffusion coefficient determination under given carrier conditions is more straightforward (e.g., Ngo Manh 2000). The SymFlFFF allows such an analysis. An alternative consists of performing size calibration in the range <10 nm by using PSS standards and converting the molecular mass into hydrodynamic diameters by an appropriate relationship (e.g., Prestel et al. 2005; Bouby et al. 2008).

Some papers on FFF-ICP-MS report procedures on how to quantify metal ions bound to individual colloid fractions (Ranville et al. 1999; Hassellov et al. 1999; Lyven et al. 2003; Reszat and Hendry 2007; Dubascoux et al. 2008). All of these consider the channel flow to be invariant over time. This assumption is certainly justified for most FFF systems. In the absence of flow inconsistencies, ICP-MS data are normalized to an internal standard (In, Rh etc.), which is admixed to the channel outflow. Standard solutions are run immediately before and after an analysis sequence in order to check and correct for variations during a fractionation. Quantification is made by external calibration using reference solutions. Quantification of elements of interest to colloidal species is then performed by peak integration. Flow inconsistencies in AsFlFFF may occur and are reported in Bouby et al. (2008) and Diaz et al. (2009). By relating signals for each element to an internal standard, which is admixed to the outflow of the channel at constant known mass flow and external calibration using the same internal standard, these flow inconsistencies can be corrected for (Bouby et al. 2008).

Chemical speciation provided by FFF-based speciation methods is applicable only for the part that is eluted from the channel. Systematic errors are possible due to analyte losses. Depending on the sample type and the composition of the carrier solution, colloid and colloid-bound element losses are encountered in the channel. Various processes such as colloid attachment to channel surfaces, filtration through the membrane, and desorption of elements can be responsible for this observation. The determination of individual colloid component recoveries is thus mandatory.

3 Application to Actinide Colloids

To date, only 21 publications refer to FFF-studies involving actinide ions (Table 2). These studies describe actinide or actinide homolog ion interaction with natural organic matter (Geckeis et al. 2002; Suteerapataranon et al. 2006; Ranville et al. 2007; Reszat and Hendry 2007; Bouby et al. 2008), bacterial cells (Jackson et al. 2005b) and inorganic colloids (Bouby et al. 2004, 2008; Schäfer et al. 2004; Lesher et al. 2009). Further studies deal with the characterization of trace elements including actinides in colloid containing natural ground or surface water (Bouby et al. 2002; Geckeis et al. 2002, 2003; Lyven et al. 2003; Stolpe et al. 2005; Dahlqvist et al. 2007; Ranville et al. 2007; Reszat and Hendry 2007; Stolpe and Hassellöv 2007; Cizdziel et al. 2008; Mihai and Georgescu 2008; Diaz et al. 2009) or in water samples collected at contaminated sites (Jackson et al. 2005a; Claveranne-Lamolère et al. 2009; Plathe et al. 2010).

Table 2 FlFFF–MDA studies of actinide (or homologs) metal ions nanoparticles/colloids interaction

	Colloids or NPs	Sample origin	Elts.	Application	Results	Refs.
Natural organic matter	Purified Aldrich HA	Brown coal	Eu, Th, U	Lab. versus nat. system: actinide metal ion colloidal distribution	Correlation of dissociation rates with size evolution of the metal-humate complexes established. Metal-induced HA agglomeration process	Geckeis et al. (2002)
	Purified Gorleben HA	Deep groundwater, Gorleben site, Lower Saxony, North Germany	Eu, Th, U, heavy metals	HA is used as a ligand in leaching test solutions to study the release of metal ions from a coal mining tailing area	The size of the metal–humate complexes increase with the metal-ion charge	Suteerapataranon et al. (2006)
			Eu, Th, U	HA is used as a ligand competitor towards the bentonite clay colloids	Complete desorption of Eu in favor of HA, but partial Th desorption	Bouby et al. (2008)
Biocolloids	Bacterial cell	Shewanella oneidensis	U	Metal bacteria sorption	U sorption maximum at pH 5. Identification of a cell exudate of high M_w having a greater affinity for U than cell surfaces	Jackson et al. (2005b)
Inorganic colloids/ nanoparticles (NPs)	Bentonite NPs	Nat. FEBEX bentonite, Spain	Eu, Th, U	Laboratory core colloid migration experiments	U is not colloid borne, different from the Th and Eu, which are associated with colloids	Schäfer et al. (2004)
			Eu, Th, U	Metal ion distribution between HA and clay colloids	Successful separation of different colloidal species present according to their different sizes. Trace metal speciation between both colloidal species	Bouby et al. (2008)

(*continued*)

Table 2 (continued)

	Colloids or NPs	Sample origin	Elts.	Application	Results	Refs.
			Eu, Th, U	Transport of radionuclides by colloids in the Danube river	Bentonite clay colloids play a minor role in the transport of radionuclides in the Danube river due to their instability	Mihai & Georgescu (2008)
	Iron oxi/hydroxide NPs	Synthesized HFO in presence of Lu	Lu	Detection of iron colloids. Interaction and aging in the presence of an actinide homolog	6-Line iron colloids are monomodal. Inhomogeneous metal ion inclusion into the iron colloid over time during the 2-Line transformation in hematite and goethite in favor of hematite	Bouby et al. (2004)
		Synthesized hematite NPs	U	Quantification of colloid-borne U	Quantitative data on the uranium speciation over the pH range of 3–6	Lesher et al. (2009)
Natural systems	Deep GW colloids	Sampling at 128–130 m depth, Gorleben site, Lower Saxony, Germany	Polyvalent metal ions: Ln^{n+}, An^{n+}	Aquatic colloids present and repartition of the polyvalent metal ions	Natural humic colloids are in the size range <3 nm. Inorganic actinide homologs in a bigger size range >10 nm. Different chemical states	Bouby et al. (2002)
				Association nature of colloid-borne elements	Under natural aquatic conditions, metal binding is different from what is observed in the laboratory experiment	Geckeis et al. (2002)

			Similarities between the natural aquatic colloids and those extracted from sediments	Identical location of REE, Th, U and Ca in both samples. Th and REE are mainly in inorganic colloids. Their composition is still under debate	Geckeis et al. (2003)
GW colloids	Pore water sample, clay-rich glacial till aquitard system, western Canada	U	Influence of dissolved organic carbon (DOC) on the U mobility	Powerful method to determine the U-organic matter association. Less than 2% U is associated with the DOC. U transport by DOC may be limited in many groundwater systems where the U (VI) speciation is dominated by carbonate complexes	Ranville et al. (2007)
		U	Same study with additional elements	Under this in situ pH conditions, minor masses of elements are complexed with this DOC. Presence of strongly bound complexes, perhaps in the matrix of the DOC structure	Reszat and Hendry (2007)
Seepage percolated water	Yucca Mountain, Nye County, Nevada	All present	Chemical constituents of percolated water through the mountain	U is eluted in the void volume, i.e., as noncolloidal species. Presence of 10 nm calcite particles	Cizdziel et al. (2008)

(*continued*)

Table 2 (continued)

	Colloids or NPs	Sample origin	Elts.	Application	Results	Refs.
	River water NPs	Kalix River, northern 30 Sweden	Transition, main, Ln, An	Temporal variations of colloidal carrier phases and associated trace elements	Strong seasonal variations of elemental colloidal concentration. U stabilization by colloidal organic matter, higher concentration in spring	Dahlqvist et al. (2007)
	Lake water NPs	Great Salt Lake (GSL), Utah, USA	U in 17 Elts	Size distribution and elemental composition of nanoparticles (NPs) from oxic and anoxic samples	U is associated with small-sized nanoparticles, no change in concentration with depth	Diaz et al. (2009)
	Creek water	Delsjö Creek, Göteborg, Sweden	Th, U in 32 Elts	Identification of trace metals colloidal carriers	Competition between iron-(up to 5 nm) and carbon-(1-2 nm) based colloids	Lyven et al. (2003)
			Th, U in 45 Elts	HR-ICP-MS results compared to previous work (Lyven et al. 2003)	Detection of more elements	Stolpe et al. (2005)
			Th, U on 24 Elts	Size distribution and element composition changes of fresh water NPs during mixing with sea water	The actinide preference to complexation with fulvic-like compounds influences their mobility in estuarine systems	Stolpe & Hasselov (2007)
Samples from contaminated area	Colloids/NPs from leachates of contaminated sediments	Savannah River Site, Aiken, SC, USA	Ni and U	Characterization of colloidal and humic bound U	U presents a bimodal size distribution: association with dissolved organic matter (DOM) and co-elution with Al colloidal phases	Jackson et al. (2005a)

Soil samples (0–30 cm) from a nuclear test site, France	U	Influence of colloids on the migration of uranium	Correlation between the U and the Al and Fe species and/or to small organic matter	Claveranne-Lamolère et al. (2009)
Clark Fork River, Montana, USA	Th among other toxic trace metals	Trace metal–nanoparticle associations in riverbed sediment	Toxic metals are held in the nano-size fraction (<450 nm) of the sediments. Th and other lanthanide species are assigned to Ce-allanite NPs. Presence of numerous Fe and Ti mineral phases	Plathe et al. (2010)

Elts Elements, *HA* humic acid, *GW* groundwater

3.1 Natural Organic Matter

Certainly, a great number of studies have been dedicated to the investigation of metal ions with natural organic matter extracted from sedimentary rock, dissolved in natural water or purified, and well characterized from various sources. The aim of such studies is mainly to quantify metal ion speciation in the presence of natural humic or fulvic acid. Only a few focus their interest on actinide ions or their homologs, mainly rare earth elements (see Table 2). FlFFF studies aimed to quantify actinide association to colloidal humic and fulvic matter. Other investigations also tried to gain mechanistic insight into actinide humic binding modes.

It is well known that natural organic matter can strongly bind actinide ions and either significantly influence their aquatic speciation or contribute to sorption to sediment and rock (Kim et al. 1984; Bertha and Choppin 1978; Kim 1986; Lieser et al. 1990; Choppin 1992; Moulin and Moulin 1995; MacCarthy et al. 1998; Kinniburgh et al. 1999). Mobilization or retention depend on geochemical conditions and the nature of organic matter, which can be stabilized as anionic and colloidal species in solution (DOC) or can be bound as kerogen-like material to mineral phases.

Many experimental studies reveal significant kinetically hindered dissociation of polyvalent actinide ions from their humic/fulvic acid complexes. Dissociation rates usually decrease with increasing contact time (see, e.g., Choppin and Clark 1991; Monsallier et al. 2003). Possible explanations for this behavior are conformational changes in the organic macromolecule or metal-induced agglomeration. By the use of FlFFF–MDA and batch experiments, a correlation of dissociation rates and size evolution of Eu(III)humate species was derived (Geckeis et al. 2002). The observation of decreasing dissociation rates with slight average metal-humate complex size increases suggests a metal-induced agglomeration mechanism. This hypothesis is supported by more studies showing that the size of humate complexes appears to increase with metal ion charge (Suteerapataranon et al. 2006). In this study, the size at fractogram maxima increases from ~1.5 nm for the uncomplexed humic acid to ~1.7 nm for the respective UO_2^{2+} complex, ~2.1 nm for the Ce(III) complex and 2.5 for the Th(IV) complex. The progress of the agglomeration process is currently under investigations. In preliminary studies, polyvalent metal ions (Eu(III), Th(IV), Pu(IV), and U(VI)) were added at trace concentration (~10^{-9} mol L^{-1}) at high pH (9.6) and low ionic strength to a solution containing purified Gorleben humic acid (<1 mg L^{-1}). Even at low occupancy of humic acid carboxylic acid sites (<10%), metal-induced agglomeration was observed to be followed for all cations. Figure 3 reveals the size evolution for the series U(VI)<Eu(III)<Th(IV)<Pu(IV) in agreement with observations made previously (Suteerapataranon et al. 2006). A relatively large sized fraction with a hydrodynamic diameter of about 7 nm evolves within 3 months and is much larger than the average size of the humic acid particles of 2–3 nm. The nature of this large sized fraction is presently not clear and requires further characterization. At least for the tetravalent actinides, formation of $An(OH)_4$ NPs coated and stabilized by humic acid cannot be excluded.

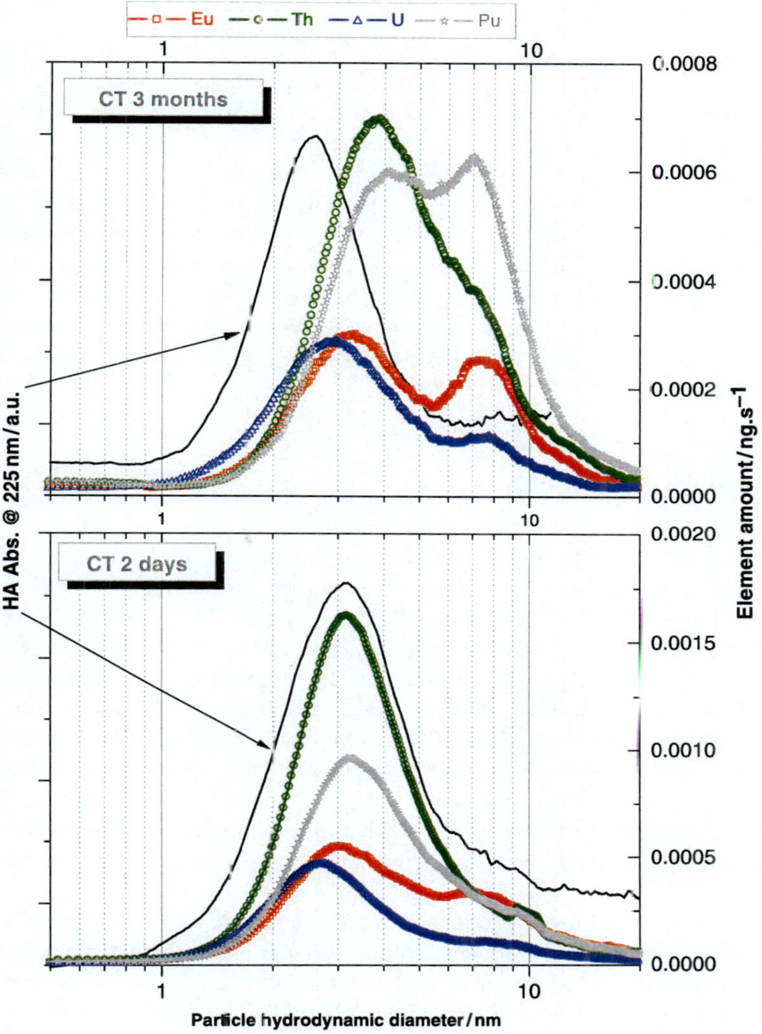

Fig. 3 Metal-induced humic acid (HA) agglomeration at low loading (<10%), high pH, and low ionic strength as a function of contact time (CT). Element-ICP-MS fractograms (*colored lines*) and HA-fractogram (*black line*). Size calibration and elements quantification are performed as in Bouby et al. (2008). Present experimental conditions: regenerated cellulose membrane, 5 kDa; spacer: 0.5 mm; eluent: ultra pure water with NaOH, pH 9.3; outflow rate: 0.2 mL min^{-1}, programed cross-flow: exponential, from 4 mL min^{-1} (i.e., 95% of the total inflow rate) down to 0 mL min^{-1} in 38 min

Interestingly, a larger sized colloid fraction ranging up to hydrodynamic diameters of ~200 nm can also be found in deep groundwater samples rich in humic/fulvic acid (Ngo Manh 2000; Geckeis et al. 2002; Bouby et al. 2002) (see discussion below).

3.2 Bacteria

Bacteria may have an impact on the transport of contaminants as they provide sorption sites or initiate chemical transformations. On the one hand, they may increase actinide mobility if actinide-bearing microbes are transported advectively. On the other hand, they might be filtered out through porous media because their size is in the micrometer range. There is currently only one paper by Ranville et al. (Jackson et al. 2005b) reporting the application of the FlFFF-ICP-MS to study the interaction between U(VI) and bacterial cell suspensions. A maximum of uranium sorption occurred at pH 5. Even more interesting is the presence of a high molecular weight component, presumably a cell exudate. This component complexes U(VI), and at elevated pH, the exudate appears to have even a higher affinity to U(VI) than cell surfaces.

3.3 Inorganic Colloids

The interaction of trace metal elements with inorganic colloids (iron oxi/hydroxide, clays, silica, et) has been the subject of numerous FFF studies (Hassellov et al. 1999; Ranville et al. 1999; Ngo Manh 2000; Bouby et al. 2002, 2004, 2008; Geckeis et al. 2002, 2003; Siripinyanond et al. 2002; Benedetti et al. 2003; Lyven et al. (2003); Schäfer et al. 2004; Jackson et al. 2005a; Stolpe et al. 2005; Baalousha et al. 2006; Suteerapataranon et al. 2006; Stolpe and Hassellöv 2007; Dubascoux et al. 2008; Mihai and Georgescu 2008; Claveranne-Lamolère et al. 2009; Lesher et al. 2009; Diaz et al. 2009; Plathe et al. 2010), but only a few are specifically devoted to actinide (or their chemical homolog) trace metal ions (Ngo Manh 2000; Bouby et al. 2002, 2004, 2008; Geckeis et al. 2002, 2003; Schäfer et al. 2004; Jackson et al. 2005a; Mihai and Georgescu 2008; Claveranne-Lamolère et al. 2009; Lesher et al. 2009).

As in studies on actinide reactions with natural organic matter, the quantification of metal ion sorption onto inorganic colloid surfaces can be investigated by FlFFF and a mechanistic insight gained as well with the use of the FlFFF–MDA approach. In the following paragraph, examples of FFF studies of actinide interaction with inorganic colloid are presented.

Lesher et al. (2009) studied the U(VI) adsorption to hematite colloids (~60 nm) over the pH range 3–6. Iron oxy/hydroxide colloids are generated as secondary phases by weathering of igneous and sedimentary rocks (Jambor and Dutrizac 1998) and also as precipitates from Fe(II)-bearing waters as a consequence of oxygen access (Cornell and Schwertmann 1996). Such particles are considered possible carriers for radionuclide migration from a nuclear waste repository. This is particularly true for nanoparticulate phases with high specific surface area (Hochella et al. 2008; Navrotsky et al. 2008). The results obtained from FlFFF-ICP-MS quantification data were compared with those obtained from centrifugation followed by filtration, and a good agreement was found between the two experimental approaches. A "typical" sorption edge as function of pH was observed, and the authors conclude that the hyphenated flow-based analytical technique is appropriate for the quantification of uranium partitioning between aqueous and colloidal phases.

Surface sorption is not the only conceivable interaction mode with iron oxides forming as secondary phases. Bouby et al. (2004) tested the possibility of actinide ion incorporation into colloidal iron-phases by analyzing colloids resulting from coprecipitation experiments (Dardenne et al. 2001). Aging hydrous ferric oxides or ferrihydrite (HFO) colloids (size range 10–200 nm) results in the transformation into hematite (68%) and goethite (32%). Coprecipitated Lu studied as a chemical homolog to trivalent actinides is found to be preferentially incorporated into hematite colloids rather than larger sized goethite particles. This observation confirmed earlier EXAFS studies (Dardenne et al. 2001).

Work has also been made on the interaction of actinide ions with clay colloids (Geckeis et al. 2003; Schäfer et al. 2004; Bouby et al. 2008, 2010; Mihai and Georgescu 2008), which may be released from natural sedimentary rock and bentonite barriers proposed as hostrock and backfill material in various nuclear waste disposal concepts (Missana et al. 2003). FlFFF–MDA not only allows the characterization of the extent of actinide sorption onto clay colloid, but it also simultaneously provides information on colloid agglomeration reactions (Bouby et al. 2010c). From the experimental findings, one can conclude that even under the low ionic strength and high pH conditions of the investigated groundwater samples, montmorillonite platelets that are initially almost completely delaminated tend to aggregate slowly. This is explained by the impact of mainly Ca^{2+} ions bridging clay entities. After about three years, clay particles partially aggregate to μm sized entities and enter the steric/hyperlayer mode, leading to reduced FFF elution times. The sorption of actinides/lanthanides is obviously not affected by this process. Humic acid can be added as a competing ligand and the actinide/lanthanide dissociation from inorganic colloids to form humate complexes followed by FlFFF–MDA. Results reveal that U(VI) only weakly interacts with montmorillonite colloids under the given conditions, while the main fractions of Eu(III) are colloid bound. Both metal ions rapidly desorb and are subject to complexation with added humic acid. Results are in agreement with speciation calculations. For tetravalent Th(IV), a significant part remained clay colloid bound even though the equilibrium predicted from experiment and calculation should have been on the side of the respective humate complex. This finding points to a hitherto not understood "irreversible" binding mode of Th(IV) to clay particles.

3.4 *Natural Systems*

Natural aquifer systems are usually characterized by heterogeneous composition of rock, minerals, groundwater, and, of course, colloidal and particulate matter. FlFFF investigations have to separate and characterize the various colloidal components of natural water samples. Surface waters are furthermore subject to extensive dynamics – not only due to fast flow rates but also due to chemical gradients. Mixing of water from different sources influences the speciation of trace elements and may induce or eliminate colloidal fractions. FlFFF–MDA is applied to investigate colloidal trace element behavior in general. The natural actinide elements Th and U are often reported without being in the focus of such studies (Lyven et al. 2003; Stolpe et al. 2005; Stolpe and Hassellöv 2007; Dahlqvist et al. 2007).

Hassellöv et al. studied the distribution of trace metals between different colloidal carriers in a creek water sample (Delsjö Creek, Göteborg, Sweden) (Lyven et al. 2003; Stolpe et al. 2005) before and after mixing with seawater (Stolpe and Hasselöv 2007). Only 3–5% of the actinides were found in the inorganic colloid fraction composed mainly of iron particles with an average size of 5 nm. Most of U and Th were apparently bound to DOC being present in a narrow size range of in average 1–1.5 nm (pH 6.5, DOC: 4 mg L^{-1}). Speciation calculations were in agreement with experimental findings. Simulating estuarine conditions by mixing the creek water with saline water removed the Fe-rich colloids >3 nm and a part of the dissolved organic matter (DOM). Obviously low molecular weight fulvic acid-like organic matter (0.5–3 nm) mainly remained in solution, thereby keeping most Th and U in solution. The authors conclude that the actinide preference to complexation with fulvic-like compounds greatly influences their mobility in estuarine systems.

Dahlqvist et al. (2007) studied the temporal variations of colloidal matter and associated trace elements in a boreal river (Kalix River, northern Sweden). Here again, mainly iron and carbon based colloids were identified. The Fe-rich colloid fraction was controlled by seasonal variations: during winter time Fe-oxyhydroxide colloids generated from oxidation of Fe(II) contained in inflowing reducing groundwater were observed to be mostly independent of DOC. However, as the spring flood started, Fe was mainly associated with carbon-rich colloidal matter. Uranium appeared to be stabilized in the river water mainly by association with organic matter, and its concentration was thus highest in spring.

Another study deals with hypersaline stratified lake waters (Great Salt Lake, Utah, USA, Diaz et al. 2009). In this case, NPs were subject to complex generation/agglomeration reactions. However, as the system is not in equilibrium, colloidal precipitates form and at the high density of the brine solutions, aggregates can remain suspended by buoyancy in this hypersaline environment. In the Great Salt Lake (GSL), over 90% of Al, Fe, and Mn mass and over 55% of Co, Cu, Ni, and Zn mass were found associated with particles >450 nm. Only 10% of U was found associated with these particulates >450 nm in reducing deep brine layers, but almost no U was found in this fraction in shallow oxidizing water. Uranium was mainly associated with <20 nm NPs with no change in concentration with depth. The authors discuss potential impact of dilution of the saline solutions into the FFF carrier stream of comparatively low ionic strength. Deagglomeration of colloidal matter cannot be excluded.

Other authors focused their studies on the behavior of actinides in ground or seepage water, especially to gain information on the long-term behavior of the radionuclides after a potential release into the environment. The water samples examined came from areas considered as potential sites for nuclear waste disposal: deep groundwater and sediments from tertiary and quaternary layers at the Gorleben site in northern Germany (Bouby et al. 2002; Geckeis et al. 2002, 2003), groundwater from a clay-rich glacial till aquitard in western Canada (Ranville et al. 2007; Reszat and Hendry 2007), or seepage water percolating from a highly fractured and faulted volcanic area at the exploratory facility at the Yucca Mountain site in USA after high precipitation levels (Cizdziel et al. 2008).

The seepage water samples studied by Cizdziel et al. (2008) are near neutral ($pH = 7.6 \pm 0.2$) and enriched in carbonate ([HCO_3^-]: 123 ± 31 mg L^{-1}) as compared to groundwater collected from the deeper saturated zone. FlFFF-ICP-MS revealed a colloidal fraction consisting mainly of calcite particles with a size around 10 nm. Uranium was, however, eluted in the void volume, i.e., as noncolloidal species. This finding together with speciation calculations suggests the presence of U mainly as the anionic $UO_2(CO_3)_2^{2-}$ complex.

Ranville et al. (2007) studied natural groundwater samples collected from a clay-rich aquitard system in 2–30 m depth, lying close to a tentative repository site for nuclear waste. Colloidal matter under consideration in this case was mainly natural DOM. Naturally abundant dissolved metal ions in this system have been in contact with DOC for long time periods and should, thus, reflect the long-term behavior of actinides in such an environment. Kinetic effects related to variation of colloid size as encountered in many laboratory studies (see discussion above) should be negligible in such a system. A positive correlation was observed between the DOC (4–132 mg L^{-1}) and U concentrations (20–603 µg L^{-1}), suggesting uranium-humic acid complex formation. Only about half of the total DOC was able to pass the asymmetric FlFFF channel and apparently consisted of small-sized entities. Significant fractions of the DOC passed the membrane with 1 kDa cutoff. AsFlFFF-ICP-MS results suggested $\leq 2\%$ of the uranium was associated with the DOC, in agreement with speciation calculations. According to calculations and experiments U(VI), speciation was obviously dominated by carbonate complexes at pH 7–8.1 rather than by DOC bound fractions.

Reszat and Hendry (2007) extended this study to a wider range of elements. In addition, they looked to the effect of pH and carbonate concentration on U(VI) complexation to DOC and the kinetics of the complexation/dissociation of U-DOC in a pH range from around 9.7 to 1.3. U and Zn reacted rapidly upon acidification by dissociating from DOC. This was in clear agreement with previous work from Zeh et al. (1997). Reszat et al., however, observed a small U and Zn fraction that did not dissociate from DOC even under acidic conditions. The authors discuss this finding by postulating the presence of strongly bound complexes, perhaps present in the matrix of the DOC structure. Such interpretation is in agreement with conclusions drawn in Geckeis et al. (2002).

Figure 4 presents AsFlFFF-ICP-MS fractograms of naturally abundant humic colloids without any purification in groundwater taken from the Gorleben site (Northern Germany), which have been already investigated in earlier studies (e.g., Ngo Manh 2000; Geckeis et al. 2002).

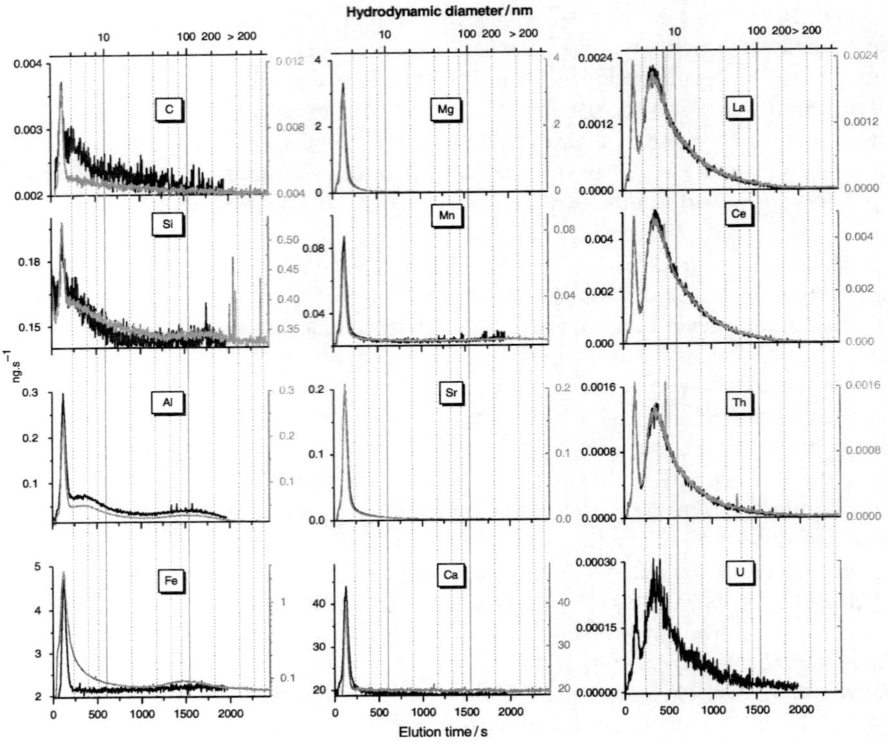

Fig. 4 ICP-MS fractograms obtained by averaging three successive injections of 500 μL aliquots of Gorleben groundwater (borehole 2,227) via crossflow (*black line*) or ultrasonic nebulization (*gray line*); ICP-MS signals for C and Si were normalized to the Rh-signal, but they are not calibrated so that their fractograms do not correspond to ng s^{-1} as for the others. Size calibration is only valid up to 200 nm. Experimental conditions: regenerated cellulose membrane, 5 kDa; spacer: 0.5 mm; eluent: ultra pure water with NaOH, pH 9.3; outflow rate: 0.8 mL min^{-1}, programed cross-flow: exponential, from 3.2 mL min^{-1} (i.e., 75% of the total inflow rate) down to 0 mL min^{-1} in 38 min

A size fraction hosting part of polyvalent trace cations (U, Ln, Th) larger than the average DOC particle size is clearly discernable in Fig. 4. In order to improve detection sensitivity, the ICP-MS fractograms have been recorded using an ultrasonic nebulizer combined with a desolvation device. The ultrasonic nebulizer offers advantages over the standard cross-flow device (1) It enhances the aerosol yield that is transferred to the ICP and (2) It decreases the H_2O content of the aerosol, thus minimizing the tendency of MO^+ cluster ion formation in the plasma by orders of magnitude. A considerable improvement can notably be achieved for the Fe-analysis, where detection limits are decreased by a factor of 25. From the fractograms, we see that Al, Si, and Fe show bigger size fraction at around 100 nm, which does not correlate with the U, Ln, and Th bimodal size distributions. Only ~10 % of Mn, Mg, Sr, and Ca elute with the organic fraction and are thus mainly

(∼90 %) assigned to ionic species. A larger sized colloid fraction, presumably also containing inorganic NPs, has also been identified in this groundwater by LIBD (Bouby et al. 2002). It is still unclear whether actinides originating from a nuclear waste repository can react with natural humic colloids such that they become incorporated. The nature of the strongly DOC bound metal fraction possibly located in larger-sized colloidal phases remains a matter of debate.

3.5 Contaminated Area

A few FFF investigations have been performed on water samples originating from contaminated areas including a former nuclear material production and (re)processing facility (Jackson et al. 2005a), previous nuclear test sites (Claveranne-Lamolère et al. 2009), and a region affected by an accidental release of nuclear or toxic waste (Plathe et al. 2010).

Jackson et al. (2005a) applied FlFFF-ICP-MS to the analysis of aqueous extracts of sediment interstitial water (1:1 sediment: water extracts) from the Tims Branch riparian corridor. This is a U (and Ni) contaminated area located on the Savannah River Site (SRS), Aiken, SC, USA. An estimated amount of 43,500 kg U had been discharged from metallurgical processes directly into wastewater, along with large quantities of Ni, Al, and other metals, as well as chlorinated solvents. Most of the metals were retained in the sediment depositional zones along Tims Branch. Significant erosion, resuspension, and redeposition of sediment-bound contaminants occurred in the meantime. Relatively high concentrations of U (0.2 mg L^{-1}) were found in the interstitial waters, along with variable concentrations of DOC. In Jackson et al., Fl-FFF-ICP-MS results are in agreement with SEC-ICP-MS measurements and indicate a correlation between Al, U, and DOC. U is assumed to be mostly bound to humic and fulvic acids. A bimodal size distribution comparable to Fig. 4 is reported for colloidal U species for some samples. Interestingly, the larger sized fraction of U appears to some extent to be associated both with Al and Fe. Uranium is predominantly present in colloidal form in all water extracts.

Similar to the work of Jackson et al. (2005a), but less clearly documented, is a study performed by Claveranne-Lamolère et al. (2009). They studied the U speciation in leachates (artificial rain water) of a soil sample from a military nuclear site. The soil was a rendosol, i.e., a thin layer of a brown and organic soil on a chalk bed-rock (95% $CaCO_3$). The samples under study were collected at a depth varying from 0 to 30 cm. They are assumed to be rich in humic substances and clays. The supernatants of the leachates were analyzed by a combination of AsFlFFF-UV-MALLS-ICP-MS. Uranium appears to be mainly associated with Al- and Fe-species and/or small sized humic matter. The authors also discuss possible association of uranium to calcite NPs. However, as ICP-MS is not sensitive to Ca-analysis due to interferences (even though equipped with a collision cell), such a conclusion is not clear.

Recently, Plathe et al. (2010) used AsFlFFF-MALLS and TEM to determine trace metal–nanoparticle associations in contaminated riverbed sediments from the Clark Fork River, Montana, USA. A large-scale dam removal project released reservoir sediment contaminated with toxic trace metals, which had accumulated from a century of mining activities upstream. This study, not specifically dedicated to actinides, shows that toxic trace metals, together with Th, are captured in the nanosized fraction of the sediment (<450 nm). The size distribution of Th species derived from the fractograms correlate with those of lanthanides. The Th- and lanthanide-containing species are assigned to Ce-allanite NPs evidenced by a complementary analytical TEM study. Other NPs of similar size distribution contain Fe, Al, and Ti. Consequently, the presence of Fe oxide/hydroxides such as HFO, jarosite, goethite, lepidocrocite commonly found in acid mine drainage solutions and Ti oxide such as brookite were found by TEM analysis.

4 Conclusions

FlFFF combined with multi detector arrays provides an interesting tool to identify heterogeneities in colloidal matter of unknown composition and to study actinide colloid reactions in particular. It can be applied

- To quantify the actinide association to colloidal organic matter (such as humic or the fulvic acid), bacteria cells, or various kinds of inorganic colloids (clays, silica, iron oxy hydroxide, calcite, etc...).
- To gain mechanistic insight into actinide nanoparticle/colloid binding modes (metal-induced humic acid agglomeration process, incorporation into colloid matrices, kinetics of actinide–colloid interaction, etc...).

Important FlFFF–MDA results reported over the last decade give an indication where future efforts should be concentrated:

- Major differences between "laboratory systems", i.e., where actinide colloid complexes are synthesized under well-defined conditions, and the "natural samples" have been shown. Actinide-bearing colloids formed under "real," i.e., close to natural conditions have to be studied in order to understand their chemical nature, stability, and thus their mobility.
- Although actinide elements are observed in fractograms with other elements (e.g., co-elution of Al, Fe, Si with actinides), this does not allow the unequivocal identification of the exact nature of the corresponding NPs and the binding modes of the actinides.

One strength of the FlFFF method is that it allows the collection of size fractions, which can then be further characterized off-line by additional techniques. Numerous other colloid/nanoparticle characterization methods are available (Table 3). Combination of FlFFF with such (mostly microscopic) complementary or ancillary methods should provide more detailed information. A second strength

Table 3 Additional colloid/nanoparticle characterization methods used with or without any fractionation

Methods	Abbreviations	References
Size exclusion chromatography	SEC	Geckeis et al. (2002), Her et al. (2002), Jackson et al. (2005a), Bolea et al. (2006), Yamaguchi et al. (2006), Striegel et al. (2009), Alasatoni et al. (2010), Woods et al. (2010)
Hydrodynamic chromatography	HDCH	Silebi and Dosramos (1989), Stegeman et al. (1991), Thompson et al. (2009), Tiede et al. (2009)
Electrophoretic methods: capillary zone electrophoresis prior to Taylor dispersions analysis	CZE, CTE-TDA	McCormick (1991), Cottet et al. (2007), Surugau and Urban (2009)
Cross-flow (ultra) filtration	CFF	Andersson et al. (2006), Dahlqvist et al. (2007)
Tangential-flow filtration	TFF	Benedetti et al. (2002, 2003), Worms et al. (2010)
Ultra-filtration	UF	Bolea et al. (2006)
Diffusive gradients in thin film	DGT	Andersson et al. (2006), Dahlqvist et al. (2007)
Small angle X-ray scattering	SAXS	Augsten et al. (2008)
Electron paramagnetic resonance spectroscopy	EPR	Benedetti et al. (2003)
Small angle neutron scattering	SANS	Jarvie and King (2007)
Photon correlation spectroscopy	PCS	Filella et al. (1997), Bowen (2002)
Dynamic light scattering	DLF	Bowen (2002), Thünemann et al. (2008, 2009a), Lesher et al. (2009)
Fluorescence correlation spectroscopy	FCS	Lead et al. (2000)
Scanning stripping chronopotentiometry	SSCP	Pinheiro et al. (2007)
Laser-induced breakdown detection	LIBD	Scherbaum et al. (1996), Bundschuh et al. (2001), Hauser et al. (2002), Walther et al. (2002), Walther (2003), Kim and Walther (2007)
Transmission electron microscopy coupled to X-ray energy dispersive spectroscopy	TEM/X-EDS	Bowen (2002), Benedetti et al. (2003), Leppard (2008), Lesher et al. (2009), Plathe et al. (2010)
Scanning electron microscopy	SEM/X-EDS	Diaz et al. (2009), Lesher et al. (2009)
Atomic force microscopy	AFM	Plaschke et al.(1999, 2002), Lead et al. (2005)

of FlFFF-ICP-MS is its high sensitivity for actinide detection (e.g., limit of detection for U: 29 pg mL^{-1}; Claveranne-Lamolère et al. 2009). Ultimately, high sensitivity is notably required for the characterization of colloid samples in real or natural samples, where actinide concentrations are typically low. Finally, fractionation of heterogeneous colloid systems by their size with FlFFF and quantification of individual species allows identification of species predominantly relevant to the transport of individual elements. Such information is indispensable to the understanding of processes, leading to possible mobilization of actinides in natural colloid containing waters.

Nevertheless, FlFFF-based speciation methods are basically invasive techniques. Chemical equilibria are disturbed upon fractionation. Speciation is thus only available for kinetically stable species with insignificant variations during the fractionation time period. Losses of sample components have to be considered and quantified to avoid systematic errors in the analyses. Under certain conditions, colloidal species can be lost due to sorption to equipment surfaces or pass through the accumulation wall membrane. Such typical artifacts of the method have to be carefully quantified for each experimental system under study.

References

Alasonati E, Slaveykova VI, Gallard H, Croue JP, Benedetti MF (2010) Characterization of the colloidal organic matter from the Amazonian basin by asymmetrical flow field-flow fractionation and size exclusion chromatography. Water Res 44:223–231.

Amarasiriwardena D, Siripinyanond A, Barnes RM (2000) Flow field – Flow fractionation – Inductively coupled plasma – Mass spectrometry (flow-FFF-ICP-MS): A versatile approach for characterization of trace metals complexed to soil-derived humic acids. Humic Substances: Versatile Components of Plants, Soil and Water 259:215–226.

Andersson K, Dahlqvist R, Turner D, Stolpe B, Larsson T, Ingri J, Andersson P (2006) Colloidal rare earth elements in a boreal river: Changing sources and distributions during the spring flood. Geochim Cosmochim Acta 70:3261–3274.

Augsten C, Kiselev MA, Gehrke R, Hause G, Mader K (2008) A detailed analysis of biodegradable nanospheres by different techniques – A combined approach to detect particle sizes and size distributions. J Pharm Biomed Anal 47:95–102.

Baalousha M, Lead J (2007) Characterization of natural aquatic colloids (<5 nm) by flow-field flow fractionation and atomic force microscopy. Environ Sci Technol 41:1111–1117.

Baalousha M, Kammer FVD, Motelica-Heino M, Baborowski M, Hofmeister C , Le Coustumer P (2006) Size-based speciation of natural colloidal particles by flow field flow fractionation, inductively coupled plasma-mass spectroscopy, and transmission electron microscopy/X-ray energy dispersive spectroscopy: Colloids-trace element interaction. Environ Sci Technol 40:2156–2162.

Beckett R, Hart B (1993) Use of field-flow fractionation techniques to characterize aquatic particles, colloids and macromolecules. In: Buffle J and van Leeuwen HP (eds.) Environmental Particles, Vol. 2. Lewis Publisher, New York.

Beckett R, Jue Z, Giddings JC (1987) Determination of molecular-weight distributions of fulvic and humic acids using flow field-flow fractionation. Environ Sci Technol 21:289–295.

Beckett R, Murphy D, Tadjiki S, Chittleborough DJ, Giddings JC (1997) Determination of thickness, aspect ratio and size distributions for platey particles using sedimentation field-flow fractionation and electron microscopy. Colloids Surf A 120:17–26

Benedetti M, Ranville JF, Ponthieu M, Pinheiro JP (2002) Field-flow fractionation characterization and binding properties of particulate and colloidal organic matter from the Rio Amazon and Rio Negro. Org Geochem 33:269–279.

Benedetti M, Ranville JF, Allard T, Bednar AJ, Menguy N (2003) The iron status in colloidal matter from the Rio Negro, Brasil. Colloids Surf A Physicochem Eng Asp 217:1–9.

Bertha EL, Choppin GR (1978) Interaction of humic and fulvic acids with Eu(III) and Am(III). J Inorg Nucl Chem 40:655–658.

Boehme J, Wells M (2006) Fluorescence variability of marine and terrestrial colloids: Examining size fractions of chromophoric dissolved organic matter in the Damariscotta River estuary. Mar Chem 101:95–103.

Bolea E, Gorriz MP, Bouby M, Laborda F, Castillo JR, Geckeis H (2006) Multielement characterization of metal-humic substances complexation by size exclusion chromatography, asymmetrical flow field-flow fractionation, ultrafiltration and inductively coupled plasma-mass spectrometry detection: A comparative approach. J Chromatogr A 1129:236–246.

Botana AM, Ratanathanawongs SK, Giddings JC (1995) Field-programmed flow field-flow fractionation. J Microcolumn Sep 7:395–402.

Bouby M, Ngo Manh T, Geckeis H, Scherbaum F, Kim JI (2002) Characterization of aquatic colloids by a combination of LIBD and ICP-MS following the size fractionation. Radiochim Acta 90:727–732.

Bouby M, Geckeis H, Ngo Manh T, Yun JI, Dardenne K, Schafer T, Walther C, Kim JI (2004) Laser-induced breakdown detection combined with asymmetrical flow field-flow fractionation: Application to iron oxi/hydroxide colloid characterization. J Chromatogr A 1040:97–104.

Bouby M, Geckeis H, Geyer FW (2008) Application of asymmetric flow field-flow fractionation (AsFlFFF) coupled to inductively coupled plasma mass spectrometry (ICPMS) to the quantitative characterization of natural colloids and synthetic nanoparticles. Anal Bioanal Chem 392:1447–1457.

Bouby M, Geckeis H, Lützenkirchen J, Mihai S, Schäfer T (2010) Interaction of bentonite colloids with Eu, Th and U in presence of humic acid: a flow field-flow fractionation study. Geochim Cosmochim Acta.

Bowen P (2002) Particle size distribution measurement from millimeters to nanometers, and from rods to platelets. J Dispers Sci Technol 23:631–662.

Bundschuh T, Hauser W, Kim JI, Knopp R, Scherbaum FJ (2001) Determination of colloid size by 2-D optical detection of laser induced plasma. Colloids Surf A Physicochem Eng Asp 180:285–293.

Choppin GR (1992) The role of natural organics in radionuclide migration in natural aquifer systems. Radiochim Acta 58:113–120.

Choppin GR, Clark SB (1991) The kinetic interactions of metal ions with humic acids. Mar Chem 36:27–38.

Cizdziel JV, Guo CX, Steinberg SM, Yu ZB, Johannesson KH (2008) Chemical and colloidal analyses of natural seep water collected from the exploratory studies facility inside Yucca Mountain, Nevada, USA. Environ Geochem Health 30:31–44.

Claret F, Schafer T, Rabung T, Wolf M, Bauer A, Buckau G (2005) Differences in properties and Cm(III) complexation behavior of isolated humic and fulvic acid derived from Opalinus clay and Callovo-Oxfordian argillite. Appl Geochem 20:1158–1168.

Claveranne-Lamolère C, Lespes G, Dubascoux S, Aupiais J, Pointurier F, Potin-Gautier M (2009) Colloidal transport of uranium in soil: Size fractionation and characterization by field-flow fractionation-multi-detection. J Chromatogr A 1216:9113–9119.

Cornell RM, Schwertmann U (1996) The iron oxides: Structure, properties, reactions, occurrence and uses. VCH-Verlag, Weinheim.

Cottet H, Biron JP, Martin M (2007) Taylor dispersion analysis of mixtures. Anal Chem 79:9066–9073.

Dahlqvist R, Andersson K, Ingri J, Larsson T, Stolpe B, Turner D (2007) Temporal variations of colloidal carrier phases and associated trace elements in a boreal river. Geochim Cosmochim Acta 71:5339–5354.

Dardenne K, Schäfer T, Denecke MA, Rothe J, Kim JI (2001) Identification and characterization of sorbed lutetium species on 2-line ferrihydrite by sorption data modeling, TRLFS and EXAFS. Radiochim Acta 89:469–479.

Diaz X, Johnson W P, Fernandez D, Naftz D L (2009) Size and elemental distributions of nano- to micro-particulates in the geochemically-stratified Great Salt Lake. Appl Geochem 24:1653–1665.

Dubascoux S, Le Hécho I, Gautier M P, Lespes G (2008) On-line and off-line quantification of trace elements associated to colloids by As-Fl-FFF and ICP-MS. Talanta 77:60–65.

Dubascoux S, Le Hécho I, Hassellöv M, Von der Kammer F, Potin Gautier M, Lespes G (2010) Field-flow fractionation and inductively coupled plasma mass spectrometer coupling: History, development and applications. J Anal At Spectrom 25:613–623.

Exner A, Theisen M, Panne U, Niessner R (2000) Combination of asymmetric flow field-flow fractionation (AF(4)) and total-reflexion X-ray fluorescence analysis (TXRF) for determination of heavy metals associated with colloidal humic substances. Fresenius J Anal Chem 366:254–259.

Filella M, Zhang J, Newman ME, Buffle J (1997) Analytical applications of photon correlation spectroscopy for size distribution measurements of natural colloidal suspensions: Capabilities and limitations. Colloids Surf A Physicochem Eng Asp 120:27–46.

Geckeis H, Rabung T, Ngo Manh T, Kim JI, Beck HP (2002) Humic colloid-borne natural polyvalent metal ions: Dissociation experiment. Environ Sci Technol 36:2946–2952.

Geckeis H, Ngo Manh T, Bouby M, Kim JI (2003) Aquatic colloids relevant to radionuclide migration: Characterization by size fractionation and ICP-mass spectrometric detection. Colloids Surf A Physicochem Eng Asp 217:101–108.

Giddings JC (1966) A new separation concept based on a coupling of concentration and flow nonuniformities. Sep Sci 1:123–125.

Giddings JC (1993) Field-flow fractionation – Analysis of macromolecular, colloidal, and particulate materials. Science 260:1456–1465.

Hassellov M, Lyven B, Haraldsson C, Sirinawin W (1999) Determination of continuous size and trace element distribution of colloidal material in natural water by on-line coupling of flow field-flow fractionation with ICPMS. Anal Chem 71:3497–3502.

Hauser W, Geckeis H, Kim JI, Fierz Th (2002) A mobile laser-induced breakdown detection system and its application for the in situ-monitoring of colloid migration. Colloids Surf A Physicochem Eng Asp 203:37–45.

Her N, Amy G, Foss D, Cho J, Yoon Y, Kosenka P (2002) Optimization of method for detecting and characterizing NOM by HPLC-size exclusion chromatography with UV and on-line DOC detection. Environ Sci Technol 36:1069–1076.

Hochella MF, Lower SK, Maurice PA, Penn RL, Sahai N, Sparks DL, Twining BS (2008) Nanominerals mineral nanoparticles and earth systems. Science 319:1631–1635.

Jackson BP, Ranville JF, Bertsch PM, Sowder AG (2005a) Characterization of colloidal and humic-bound Ni and U in the "dissolved" fraction of contaminated sediment extracts. Environ Sci Technol 39:2478–2485.

Jackson BP, Ranville JF, Neal AL (2005b) Application of flow field-flow fractionation-ICPMS for the study of uranium binding in bacterial cell suspensions. Anal Chem 77:1393–1397.

Jambor JL, Dutrizac JE (1998) Occurrence and constitution of natural and synthetic ferrihydrite, a widespread iron oxyhydroxide. Chem Rev 98:2549–2585.

Jarvie HP, King SM (2007) Small-angle neutron scattering study of natural aquatic nanocolloids. Environ Sci Technol 41:2868–2873.

Kammer F, Förstner U (1998) Natural colloid characterization using flow field-flow fractionation followed by multi detector analysis. Water Sci Technol 37:173–180.

Kammer F, Baborowski M, Friese K (2005) Field-flow fractionation coupled to multi-angle laser light scattering detectors: Applicability and analytical benefits for the analysis of environmental colloids. Anal Chim Acta 552:166–174.

Kim JI (1986) Chemical behaviour of transuranic elements in natural aquatic systems. In: Freemen AJ and Keller C (eds.) Handbook on the chemistry of the actinides. Elsevier Science Publisher, Amsterdam.

Kim JI, Walther C (2007) Laser-induced breakdown detection. In: Wilkinson KJ and Lead JR (eds.) Environmental colloids and particles. IUPAC.

Kim JI, Buckau G, Baumgärtner F, Moon HC, Lux D (1984) Colloid generation and the actinide migration in Gorleben griundwaters. In: McVay GL (ed.) Scientific basis for nuclear waste management VII. North-Holland, New-York.

Kinniburgh DG, v Riemsdijk WH, Koopal LK, Borkovec M, Benedetti MF, Avena MJ (1999) Ion binding to natural organic matter: Competition, heterogeneity, stoichiometry and thermodynamic consistency. Colloids Surf A Physicochem Eng Asp 151:147–166.

Lead JR, Wilkinson KJ, Starchev K, Canonica S, Buffle J (2000) Determination of diffusion coefficients of humic substances by fluorescence correlation spectroscopy: Role of solution conditions. Environ Sci Technol 34:1365–1369.

Lead JR, Muirhead D, Gibson CT (2005) Characterization of freshwater natural aquatic colloids by atomic force microscopy (AFM). Environ Sci Technol 39:6930–6936.

Lenshof A, Laurell T (2010) Continuous separation of cells and particles in microfluidic systems. Chem Soc Rev 39:1203–1217.

Leppard GG (2008) Nanoparticles in the environment as revealed by transmission electron microscopy: Detection, characterisation and activities. Curr Nanosci 4:278–301.

Lesher EK, Ranville JF, Honeyman BD (2009) Analysis of pH dependent uranium(VI) sorption to nanoparticulate hematite by flow field-flow fractionation – Inductively coupled plasma mass spectrometry. Environ Sci Technol 43:5403–5409.

Lieser KH, Ament A, Hill R, Singh RN, Stingl U, Thybusch B (1990) Colloids in groundwater and their influence on migration of trace elements and radionuclides. Radiochim Acta 49:83–100.

Lyven B, Hassellov M, Turner DR, Haraldsson C, Andersson K (2003) Competition between iron- and carbon-based colloidal carriers for trace metals in a freshwater assessed using flow field-flow fractionation coupled to ICPMS. Geochim Cosmochim Acta 67:3791–3802.

MacCarthy JF, Sanford WE, Stafford PL (1998) Lanthanide field tracers demonstrate enhanced transport of transuranic radionuclides by natural organic matter. Environ Sci Technol 32:3901–3906.

McCormick RM (1991) Characterization of silica sols using capillary zone electrophoresis. J Liq Chromatogr 14:939–952.

Mihai SA, Georgescu II (2008) Contributions to the database concerning the radioactivity of the Danube River and the Black Sea coast Romanian sector. Revis Chim 59:623–627.

Missana T, Alonso U, Turrero MJ (2003) Generation and stability of bentonite colloids at the bentonite/granite interface of a deep geological radioactive waste repository. J Contam Hydrol 61:17–31.

Monsallier JM, Artinger R, Denecke M, Scherbaum FJ, Buckau G, Kim JI (2003) Spectroscopic study (TRLFS and EXAFS) of the kinetics of An(III)/Ln(III) humate interaction. Radiochim Acta 91:567–574.

Moulin V, Moulin C (1995) Fate of actinides in the presence of humic substances under conditions relevant to nuclear waste disposal. Appl Geochem 10:573–580.

Myers MN (1997) Overview of field-flow fractionation. J Microcolumn Sep 9:151–162.

Navrotsky A, Mazeina L, Majzlan J (2008) Size-driven structural and thermodynamic complexity in iron oxides. Science 319:1635–1638.

Ngo Manh T (2000) Charakterisierung aquatischer Kolloide mittels der Größenausschlußchromatographie (GPC) und der Fluß-Feldflußfraktionierung (FfFFF), Dissertation, Universität des Saarlandes, Saarbrücken, Germany.

Ngo Manh T, Knopp R, Geckeis H, Kim JI, Beck HP (2000) Detection of nanocolloids with flow-field flow fractionation and laser-induced breakdown detection. Anal Chem 72:1–5.

Ngo Manh T, Geckeis H, Kim JI, Beck HP (2001) Application of the flow field flow fractionation (FFFF) to the characterization of aquatic humic colloids: Evaluation and optimization of the method. Colloids Surf A Physicochem Eng Asp 181:289–301.

Pelekani C, Newcombe G, Snoeyink VL, Hepplewhite C, Assemi S, Beckett R (1999) Characterization of natural organic matter using high performance size exclusion chromatography. Environ Sci Technol 33:2807–2813.

Perminova IV (1999) Size exclusion chromatography of humic substances: Complexities of data interpretation attributable to non-size exclusion effects. Soil Sci 164:834–840.

Pinheiro JP, Domingos R, Lopez R, Brayner R, Fievet F, Wilkinson K (2007) Determination of diffusion coefficients of nanoparticles and humic substances using scanning stripping chronopotentiometry (SSCP). Colloids Surf A Physicochem Eng Asp 295:200–208.
Plaschke M, Römer J, Klenze R, Kim JI (1999) In situ AFM study of sorbed humic acid colloids at different pH. Colloids Surf A Physicochem Eng Asp 160:269–279.
Plaschke M, Römer J, Kim JI (2002) Characterization of Gorleben groundwater colloids by atomic force microscopy. Environ Sci Technol 36:4483–4488.
Plathe KL, vd Kammer F, Hassellov M, Moore J, Murayama M, Hofmann T, Hochella MF (2010) Using FlFFF and aTEM to determine trace metal-nanoparticle associations in riverbed sediment. Environ Chem 7:82–93.
Prestel H, Schott L, Niessner R, Panne U (2005) Characterization of sewage plant hydrocolloids using asymmetrical flow field-flow fractionation and ICP-mass spectrometry. Water Res 39:3541–3552.
Rameshwar T, Samal S, Lee S, Kim S, Cho J, Kim IS (2006) Determination of the size of water-soluble nanoparticles and quantum dots by field-flow fractionation. J Nanosci Nanotechnol 6:2461–2467.
Ranville JF, Chittleborough DJ, Shanks F, Morrison RJS, Harris T, Doss F, Beckett R (1999) Development of sedimentation field-flow fractionation-inductively coupled plasma mass-spectrometry for the characterization of environmental colloids. Anal Chim Acta 381:315–329.
Ranville JF, Hendry MJ, Reszat TN, Xie QL, Honeyman BD (2007) Quantifying uranium complexation by groundwater dissolved organic carbon using asymmetrical flow field-flow fractionation. J Contam Hydrol 91:233–246.
Reszat TN, Hendry MJ (2005) Characterizing dissolved organic carbon using asymmetrical flow field-flow fractionation with on-line UV and DOC detection. Anal Chem 77:4194–4200.
Reszat TN, Hendry MJ (2007) Complexation of aqueous elements by DOC in a clay aquitard. Groundwater 45:542–553.
Reszat TN, Hendry MJ (2009) Migration of colloids through non fractured clay-rich aquitards. Environ Sci Technol 43:5640–5646.
Roessner D, Kulicke WM (1994) Online coupling of flow field-flow fractionation and multi-angle laser-light scattering. J Chromatogr A 687:249–258.
Schäfer T, Geckeis H, Bouby M, Fanghanel T (2004) U, Th, Eu and colloid mobility in a granite fracture under near-natural flow conditions. Radiochim Acta 92:731–737.
Scherbaum FJ, Knopp R, Kim JI (1996) Counting of particles in aqueous solutions by laser-induced photoacoustic breakdown detection. Appl Phys B 63:299–306.
Schimpf M, Caldwell K, Giddings JC (2000) Field-flow fractionation handbook. Wiley-Interscience, New York.
Silebi CA, Dosramos JG (1989) Separation of submicrometer particles by capillary hydrodynamic fractionation (CHDF). J Colloid Interf Sci 130:14–24.
Siripinyanond A, Barnes RM, Amarasiriwardena D (2002) Flow field-flow fractionation-inductively coupled plasma mass spectrometry for sediment bound trace metal characterization. J Anal At Spectrom 17:1055–1064.
Siripinyanond A, Worapanyanond S, Shiowatana J (2005) Field-flow fractionation-inductively coupled plasma mass spectrometry: An alternative approach to investigate metal-humic substances interaction. Environ Sci Technol 39:3295–3301.
Stegeman G, Kraak JC, Poppe H (1991) Hydrodynamic and size-exclusion chromatography of polymers on porous particles. J Chromatogr 550:721–739.
Stolpe B, Hassellov M (2007) Changes in size distribution of fresh water nanoscale colloidal matter and associated elements on mixing with seawater. Geochim Cosmochim Acta 71:3292–3301.
Stolpe B, Hassellov M, Andersson K, Turner DR (2005) High resolution ICPMS as an on-line detector for flow field-flow fractionation, multi-element determination of colloidal size distributions in a natural water sample. Anal Chim Acta 535:109–121.

Striegel AM, Yau WW, Kirkland JJ, Bly DD (2009) Modern size-exclusion liquid chromatography: Practice of gel permeation and gel filtration chromatography. Wiley, New York.

Surugau N, Urban PL (2009) Electrophoretic methods for separation of nanoparticles. J Sep Sci 32:1889–1906.

Suteerapataranon S, Bouby M, Geckeis H, Fanghanel T, Grudpan K (2006) Interaction of trace elements in acid mine drainage solution with humic acid. Water Res 40:2044–2054.

Taylor HE, Garbarino JR, Murphy DM, Beckett R (1992) Inductively coupled plasma mass-spectrometry as an element-specific detector for field-flow fractionation particle separation. Anal Chem 64:2036–2041.

Thompson JW, Lieberman RA, Jorgenson JW (2009) Hydrodynamic chromatography for the size classification of micron and sub-micron sized packing materials. J Chromatogr A 1216:7732–7738.

Thünemann AF, Kegel J, Polte J, Emmerling F (2008) Superparamagnetic maghemite nanorods: Analysis by coupling field-flow fractionation and small-angle X-ray scattering. Anal Chem 80:5905–5911.

Thünemann AF, Rolf S, Knappe P, Weidner S (2009a) In situ analysis of a bimodal size distribution of superparamagnetic nanoparticles. Anal Chem 81: 296–301.

Thünemann AF, Knappe P, Bienert R, Weidner S (2009b) Online coupling of field-flow fractionation with SAXS and DLS for polymer analysis. Anal Methods 1:177–182.

Tiede K, Boxall ABA, Tiede D, Tear SP, David H, Lewis J (2009) A robust size-characterisation methodology for studying nanoparticle behaviour in 'real' environmental samples, using hydrodynamic chromatography coupled to ICP-MS. J Anal At Spectrom 24:964–972.

Wahlund KG, Giddings JC (1987) Properties of an asymmetrical flow field-flow fractionation channel having one permeable wall. Anal Chem 59:1332–1339.

Walther C (2003) Comparison of colloid investigations by single particle analytical techniques-A case study on thorium-oxyhydroxides. Colloids Surf A Physicochem Eng Asp 217:81–92.

Walther C, Bitea C, Hauser W, Kim JI, Scherbaum FJ (2002) Laser induced breakdown detection for the assessment of colloid mediated radionuclide migration. Nucl Instrum Methods Phys Res B 195:374–388.

White RJ (1997) FFF-MALLS-A new tool for the characterisation of polymers and particles. Polym Int 43:373–379.

Wijnhoven JEGJ, Koorn JP, Poppe H, Kok WT (1995) Hollow-fibre flow field-flow fractionation of polystyrene sulphonates. J Chromatogr A 699:119–129.

Williams PS (1997) Design of an asymmetrical flow field-flow fractionation channel for uniform channel flow velocity. J Microcolumn Sep 9:459–467.

Woods GC, Simpson MJ, Kelleher BP, McCaul M, Kingery WL, Simpson AJ (2010) Online high-herformance size exclusion chromatography-nuclear magnetic resonance for the characterization of dissolved organic matter. Environ Sci Technol 44:624–630.

Worms IAM, Szigeti ZAG, Dubascoux S, Lespes G, Traber J, Sigg L, Slaveykova VI (2010) Colloidal organic matter from wastewater treatment plant effluents: Characterization and role in metal distribution. Water Res 44:340–350.

Wyatt PJ (1998) Submicrometer particle sizing by multiangle light scattering following fractionation. J. Colloid Interface Sci 197:9–20.

Yamaguchi T, Azuma Y, Okuyama K (2006) Development of a photon correlation spectroscopy instrument to measure size distributions of nanoparticles. Part Part Syst Char 23:188–192.

Zeh P, Czerwinski KR, Kim JI (1997) Speciation of uranium in Gorleben groundwaters. Radiochim Acta 76:37–44.

Actinide Nanoparticle Characterization by Mass Spectrometry

Clemens Walther

Abstract This chapter provides an overview of mass spectrometric methods for the investigation of actinide nanoparticles. The great variety of applications requires an equally large number of different mass spectrometric techniques. Trace analysis of bulk fragments on the submicron scale, such as detection of the so-called hot particles from nuclear accidents in the environment, call for ultrasensitive detection with the capability of measuring isotope ratios with high precision. In many applications, imaging mass spectrometric techniques, primarily secondary ion mass spectrometry, are used. Investigations focusing on the formation of nanoparticles in solutions require low invasive ion sources, such as electrospray ionization, which are able to reproduce the species distribution in aqueous solution. The most frequently used ion sources and mass spectrometers are introduced, and progress in characterizing actinide nanoparticles by mass spectrometry is discussed using selected recent examples from the literature. Special emphasis is placed on the observation of nanoparticle formation in solution by electrospray mass spectrometry. Polymerization in solutions containing Th(IV), Pu(IV), and Zr(IV), as a homolog of plutonium, are discussed in greater detail in the last section.

1 Introduction

Mass spectrometry (MS) is a versatile tool for element detection and speciation. The most frequently used application in the nuclear field is the ultrasensitive detection of radionuclides and measurement of isotope distribution without obtaining information on their chemical state. In contrast, the chemical speciation of complexes or molecules and investigation of particulate material is exactly what numerous recent applications focus on. In this chapter, the latter

C. Walther
Karlsruhe Institute of Technology (KIT), Institute for Nuclear Waste Disposal (INE), P.O. Box 3640, 76021 Karlsruhe, Germany
e-mail: walther@ine.fzk.de

S.N. Kalmykov and M.A. Denecke (eds.), *Actinide Nanoparticle Research*,
DOI 10.1007/978-3-642-11432-8_5,

applications are addressed. Section 2 gives a short overview of ion sources and mass spectrometric detection methods. The following discussion of applications is divided into two parts. Section 3 focuses on investigations of nanoparticles formed in the environment including the so-called hot particles emitted from nuclear facilities or nuclear detonations. These particles mostly originate from fragments of solid materials from military sources or nuclear fuels. In contrast, Sect. 4 covers the formation processes and stability of actinide nanoparticles in solution. Due to its unique capability of detecting even minor species in the presence of a large excess of molecules having different mass, MS is suited well to observing formation processes of nanoparticles or polymers in oversaturated solutions. None of the sections can claim to give a complete summary of work done on actinide nanoparticles so far. The aim of this chapter is rather to itemize the capabilities of mass spectrometric techniques and help the reader to get an overview of which method might be most suited to address his or her particular question. Consequently, all the cited works serve merely as examples for the respective application.

2 Mass Spectrometric Techniques

Since its invention in the early twentieth century, the term MS has been used for the separation of atomic or molecular ions by means of their interaction with magnetic and electric fields. In the vast majority of applications, these techniques measure the mass over charge ratio m/q of an ion rather than its mass alone, implying that the charge state must be known. For ion production, there is a choice between many different source types. Only those used for analysis of actinide nanoparticles are discussed in Sect. 2.1. For the performance of the actual MS, three main techniques are used (Sect. 2.2): high frequency devices such as quadrupole (QP) mass analyzers, including radiofrequency (RF) traps, time-of-flight (TOF) spectrometers, and devices combining electric and magnetic fields, the so-called sector field (SF) analyzers.

2.1 The Ion Source

From the multitude of ion sources, only the ones applied to actinide nanoparticle detection are shortly summarized (Table 1, Fig. 1). A straightforward way of probing the element composition of a particle is the introduction in the torch of a conventional inductively coupled plasma mass spectrometer (ICP MS). The particle carried by a well-defined aerosol into the plasma is vaporized (atomized) and a high fraction of the atoms are ionized. The resulting ions are detected by MS. Up to a size of approximately one micron, the particle is completely destroyed and its element composition is measured. However, one cannot distinguish whether the ions result from vaporization of the particle or were present as mononuclear or

Table 1 Characteristics of ion sources used for mass spectrometry on actinide nanoparticles

	Inductively coupled plasma (ICP)	Secondary ion source (SIMS)	Thermal ionization (TIMS)	Resonance ionization (RIMS)	Electrospray (ESI MS)
Mode of operation	Continuous	Pulsed/ continuous	Continuous	Pulsed or continuous	Continuous
Invasiveness	Very harsh	Harsh	Harsh	Harsh	Mild
Species	Atomic ions, very small molecules	Atomic ions, small molecules	Atomic ions, very small molecules	Atomic ions	From atomic ions to large molecules
Charge state	Typically 1+	Typically 1, positive/ negative	Typically 1+	Typically 1+	Typically 1–3 for large organic molecules several 10 charges positive/negative
Sample type	Liquid/surface with laser ablation	Surface with spatial resolution	Surface	Atomic vapor from heated target or laser vaporization	Liquid

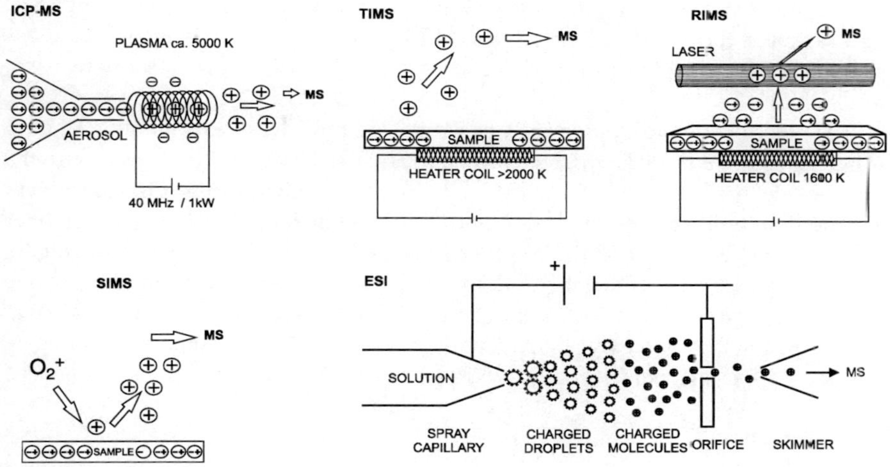

Fig. 1 Schematic representation of ion sources used for mass spectrometry on actinide nanoparticles

small molecular species in solution. Furthermore, most commercial ICP MS instruments are equipped with either RF or SF analyzers (see Sect. 2.2), both not capable of measuring a wide element distribution of a single particle at a time.

The most frequently used MS technique for nanoparticle characterization is secondary ion mass spectrometry (SIMS): The particles are deposited on a surface or a complete bulk sample is analyzed with respect to particulate constituents. Secondary ions are produced by bombarding the surface with fast primary ions (e.g., noble gas ions, oxygen, small molecules such as O_2^+, or metal ions). The secondary ions (from the particle surface) are mass analyzed.

Positive as well as negative ions can be produced, and the source can be operated continuously or in a pulsed mode. SIMS allows the analysis of element composition and isotope distribution at submicrometer spatial resolution and consumes only fg amounts of material. SIMS is hence often considered a nondestructive technique. However, SIMS analysis requires transfer of the sample to ultra high vacuum, bringing with it rather time-consuming preparation and the risk of sample alteration.

In a thermal ionization (TIMS) source, the analyte is heated to approximately 2,000 K in a furnace. At this elevated temperature, the vapor coming off the surface contains not only atoms but also ions, which are introduced into a MS. TIMS has high efficiency and is applied for precise isotope distribution measurements.

Strongly related is resonance ionization (RIMS): In contrast to TIMS, the sample is heated only to approximately 1,300–1,600 K, resulting in atomic vapor with a negligible fraction of ions. Ionization proceeds by resonant absorption of two or more photons from lasers tuned to electronic transitions of the atom to be analyzed. This efficient ionization is highly element and isotope selective. The advantage of RIMS is its superb suppression of isobars or mass neighboring ions, its drawback the time consuming sample preparation. Depending on the mass analyzer (TOF or RF), either pulsed or continuous wave (cw) lasers are used, respectively.

The electrospray ion source (ESI) is a mild and low invasive technique. A liquid is introduced into a small capillary with a fine tip. By applying a high voltage to the tip, the electric field generates charged droplets, which shrink due to solvent evaporation and coulomb explosion until only one molecule with a small number of solvent molecules remains. Developed for the low invasive detection of fragile, large biomolecules, this technique has proven useful for the detection of inorganic complexes and polymers. In contrast to detection of large organic molecules that are charged in the source by attachment of protons, inorganic complexes need to be present as charged species already in solution. Neutral molecules cannot be detected.

2.2 *The Mass Analyzer*

It is beyond the scope of this chapter to describe the manifold of mass spectrometers in detail. Hence, only a short summary of the advantages and shortcomings of the most frequently applied techniques in nanoparticle detection is given here (Table 2, Fig. 2). For a comprehensive overview and a detailed discussion of each technique, the interested reader might refer to one of the many standard text books on MS such as (Gross 2004).

One way of categorizing MS techniques is by means of their detection mode. RF analyzers most often are built in QP geometry, and SF analyzers are operated in the so-called scanning mode. This means that the device is tuned to transmit only one

Table 2 Characteristics of mass analyzers used for investigations on actinide nanoparticles

	RF analyzer (quadrupole)	RF trap (paul trap)	ICR trap (penning trap)	Sector field analyzer	Time-of-flight
Principle	AC and DC electric fields	AC and DC electric fields	DC electric and static magnetic field	DC electric and static magnetic field	DC and pulsed electric fields
Mode of operation	Continuous	Storage up to several minutes	Storage up to several minutes	Continuous	Pulsed
Mass range	Typically 0–3,000 amu up to 10^9 amu possible	Typically 0–3,000 amu up to 10^9amu possible	Typically 0–20,000 amu	Typically 0–3,000 amu	Typically 0–3,000 up to 10^9 amu possible
Mass resolution	Typically <1,000	Typically <1,000	Typically 10^7 up to 10^9	Typically 100,000	Typically 10,000 up to 50,000
Detection mode	Scan	Scan or extraction into TOF	Ejection (destructive) ICR (nondestructive)	Scan (several detectors in parallel possible)	Complete mass range at a time
Detection time	For small mass range ms, for 0–3,000 amu seconds	Scan seconds extraction 10 μs	Small mass range: ms, 0–3,000 amu seconds High resolution: several seconds	Single mass: 10 ms	Typically 10 μs

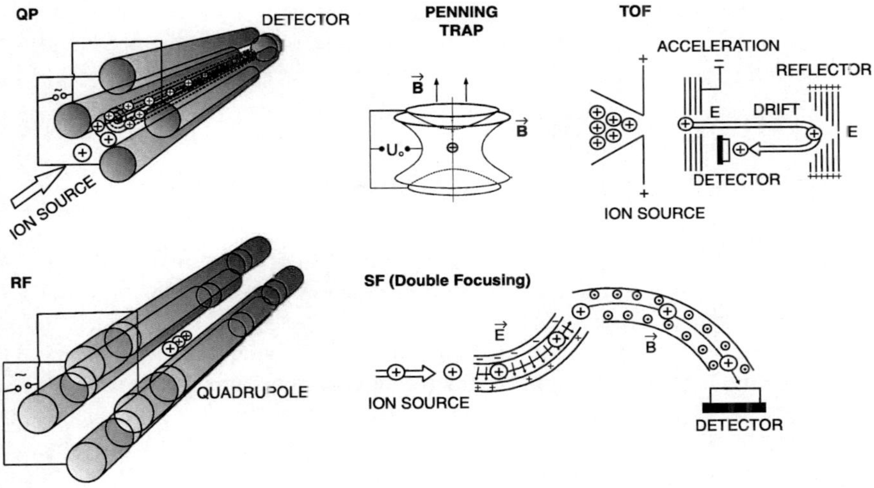

Fig. 2 Schematic representation of mass analyzers used for investigations on actinide nanoparticles

m/q at a time, and in order to obtain a complete m/q spectrum, certain parameters have to be scanned. Detection is performed by faraday cups for high ion currents, and by multichannel plates (MCP) if suited, coupled to photo multiplier tubes for single ion detection. TOF spectrometers, in contrast, are based on pulsed acceleration of the ions by an electric field (U) to kinetic energy $qU = 1/2\ mv^2$ and analysis of their flight-time dispersion due to their different mass. All ions reach the detector typically within a few microseconds, giving a complete spectrum for each single pulse. Consequently, TOF operates in pulsed mode with repetition rates up to several 10 kHz, whereas the first two techniques operate continuously. Somehow in-between are the two trapping techniques. The Paul trap is composed of an electric multipole field providing radial storage and an electric DC potential between two end caps for axial confinement. Hence, this trap is strongly related to the QP and shares most of its properties. The Penning trap, in contrast, stores ions on circular orbits in a homogeneous magnetic field for – in principle – unlimited times. Detection is possible either by resonant ejection of the ions [sequentially by frequency scan or simultaneously by the so-called stored-waveform inverse-Fourier-transform (SWIFT) excitation] or by measurement of the induced current in the storage electrodes, caused by the moving ions' charges (see Fig. 8). The latter technique is nondestructive and allows repeated measurements on the same ion, the so-called MS^n (e.g., after chemical reactions or interaction with light). The principle is called Fourier transform ion-cyclotron-resonance (FT ICR), giving the Penning trap its more commonly used name FT ICR trap.

Which technique is preferred depends on the ion source and on the requirements for resolving power. Pulsed sources, as for instance, laser desorption or ionization sources, harmonize best with pulsed TOF detection. Continuous sources (for instance ICP torches) call for continuous detection in order to achieve maximum sensitivity. However, if – as in the case of single particle element analysis – a complete spectrum is required for each single particle, mass scans are not suited and TOF detection or trapping techniques are preferred.

3 Actinide Particles in the Environment

MS is one of the major tools for the highly sensitive detection of radionuclides in the environment, even at trace concentrations. In particular, ICP-MS and TIMS are the main work horses for radioecological investigations. Chemical speciation is possible by use of soft ionization methods such as ESI (Moulin and Moulin 2001), recently improved for working in the trace concentration regime (Luo et al. 2009). However, most of the studies focus on measuring isotope ratios of radionuclides in order to trace back the origin of the material but do not further classify the chemical (or physical) species. For instance, collecting radioactivity on an air filter followed by ashing and mass spectrometric investigation destroys all traces of the physical form. Only recently has element analysis on single nanoparticles become possible due to the availability of mass spectrometric methods operating in imaging mode, namely nano SIMS, and also laser-ablation (LA) ICP-MS [for a comprehensive overview, see Becker (2003)]. These techniques make it possible to distinguish solid actinide particles – for instance, PuO_2 particles, from nuclear explosives – from pseudocolloids – i.e., natural inorganic particles carrying a small number of actinide atoms. The release of Pu from the Mayak reprocessing plant was investigated by Novikov et al. (2006). Pu sorbed on iron hydroxide colloids (HFO) was imaged, amongst other techniques, by nano SIMS. An example for pseudocolloid characterization by TIMS is given in Dai et al. (2002). In this work, the colloid-mediated transport of plutonium and transplutonium isotopes was investigated at the Savannah River site. From the isotope distribution, global fallout could be excluded as a possible source, and the origin of the Pu was traced back to nuclear reactor production. However, the method required quite extensive pretreatment of the sample. In this case, nanoparticles were separated from ionic species from several kilograms of sample by cross-flow ultrafiltration, subsequently coprecipitated with LaF_3, dissolved in 7.5 M HNO_3, separated several times in exchange columns and eluted onto AG1 $\times$ 4 mesh resin. The latter was placed in a rhenium filament and measured in the TIMS.

While TIMS is well suited for ultratrace detection, it is time consuming and, in addition, gives only integral information. No single particle measurements are possible without pretreatment.

Single particle isotope ratios were obtained by combining surface analytic techniques and ICP MS in Lind et al. (2007). Particles released from the airplane accidents near Palomares, Spain (1966), and Thule, Greenland (1968), were sampled and measured by scanning electron microscopy (SEM), micro X-ray fluorescence (XRF), and micro X-ray absorption near edge spectroscopy (XANES). Using these nondestructive methods, the authors identified the intrinsic actinide particles (UO_2, U_3O_8, PuO_2) down to 1 μm size. Single particles were then selected and digested by microwave boiling in HNO_3; finally, the $^{235}U/^{238}U$, the $^{240}Pu/^{239}Pu$, and the $^{235}U/^{239}Pu$ ratios were determined. Both Palomares and Thule particles contained a mixture of weapon grade plutonium and enriched uranium. A similar study (Shinonaga et al. 2008) combined SEM for identification of uranium particles followed by TIMS and ICP MS measurement on single particles, previously selected and transferred on to a silicon wafer using a manipulator (Esaka et al. 2007). Also using ICP-MS, Zhang et al. measured the uranium isotope ratios of single particles for nuclear safeguard purposes (Zhang et al. 2007).

By far the most frequently used method for single particle analysis, however, is SIMS: Pu particles released during nuclear weapons detonations at the Marshall Islands were analyzed by Jernstrom et al. (2006). The SIMS instrument operated by Jernstrom et al. was equipped with magnetic sector analyzers with a mass resolution of up to $m/\Delta m = 25{,}000$. This high resolution is particularly helpful for discriminating isobaric molecular interferences. The particles from Runit Island were either composed of pure plutonium or embedded in a Si/O-rich matrix. From the rather low $^{239}Pu/^{240}Pu$ ratio, the particles were identified as originating from explosions with low fission yields, which were indeed conducted in the vicinity of Runit Island between 1946 and 1958.

Particles originating from depleted uranium (DU) ammunition in the Kosovo were investigated by Danesi et al. (2003). Samples were suspended in heptane, brought onto a graphite planchet and measured by magnetic SF SIMS (Fig. 3).

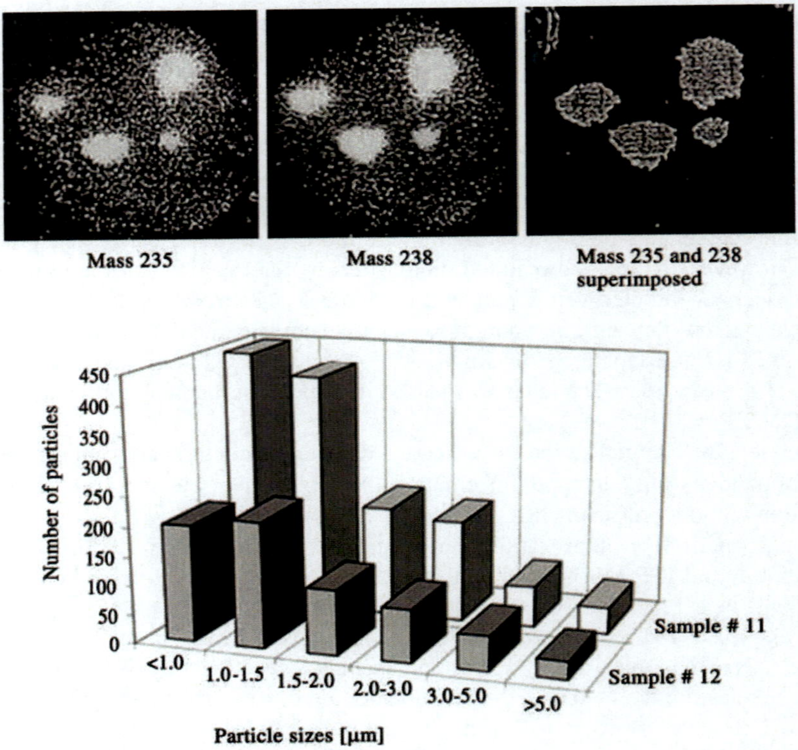

Fig. 3 *Top*: SIMS images for the two uranium isotopes ^{235}U and ^{238}U and for the DU particles. *Bottom*: Particle size distribution measured by SEM. From Danesi et al. (2003)

All particles under investigation had an isotope ratio $^{235}U/^{238}U$ ~0.002 corresponding to DU. More than 50% of the particles were smaller than 1.5 μm, making them potentially harmful as they could be dissolved rather easily or resuspended as an aerosol and inhaled by humans.

Despite its many advantages, SIMS suffers from atomic isobaric interferences. Even the high resolution of SF instruments does not suffice to resolve the isobars $^{238}U/^{238}Pu$ or $^{241}Am/^{241}Pu$. Additional problems may be caused by a high abundance of a mass neighboring element, for instance, many orders of magnitude excess of ^{238}U over ^{239}Pu. In the case of the DU ammunition used in the Kosovo conflict, the question of possible Pu contamination of the uranium ammunition drew considerable attention. However, the high excess of ^{238}U prevented direct determination of the Pu content by SIMS. Bürger et al. (2006) measured a total Pu content of ~15 pg/g uranium in the ammunition by the highly element and isotope selective RIMS (see Table 1) and inferred from the Pu isotope ratios that at least part of the material was reprocessed from uranium previously exposed to a considerable neutron flux. Going one step further, Erdmann et al. used RIMS to post ionize neutrals sputtered by ion bombardment from "hot" micro particles of Thule,

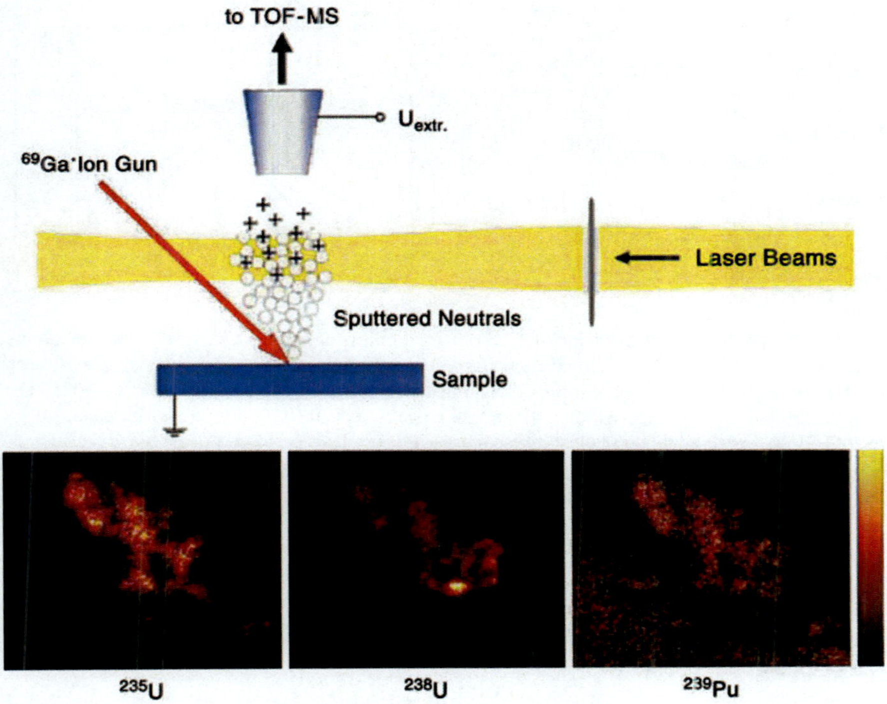

Fig. 4 *Top*: Schematic view of the experimental arrangement for the coupling of ion beam sputtering and resonant laser postionization of sputtered neutrals. Ions are extracted into a TOF mass spectrometer. *Bottom*: Lateral distribution of ^{235}U, ^{238}U, and ^{239}Pu in a "Hot Particle" from Thule, determined by TOF-SIMS. The relative ion intensity in each pixel is visualized by a color scale from *black/red* (low intensity) to *yellow* (high intensity). From Erdmann et al. (2009)

Greenland, in a SIMS machine (Erdmann et al. 2009). This promising approach combined the submicron spatial resolution typical for nano SIMS with the astonishing element and isotope selectivity of RIMS (Fig. 4).

An additional large field of application is nuclear safeguarding. By analyzing single particles, one avoids misinterpretation due to admixtures of radioisotopes from background sources, for instance, from global fallout. While fission track analysis after neutron irradiation provides a fast method to gain a crude estimate of particle size and enrichment (Stetzer et al. 2004), more detailed information is obtained by MS. In Shen et al. (2008), fission track analysis was combined with TIMS for the investigation of single uranium-bearing particles of swipe tests.

Nowadays, SIMS is the most powerful and versatile tool for measuring aerosols from swipe samples to detect undeclared nuclear operations in reprocessing or fuel fabrication plants by means of the ^{235}U/^{238}U ratio (Esaka et al. 2008). Fully automated scanning of a large number of particles on one substrate is possible by use of large geometry SIMS as demonstrated recently on mono disperse uranium oxide particles produced by a vibrating aerosol generator (Ranebo et al. 2009).

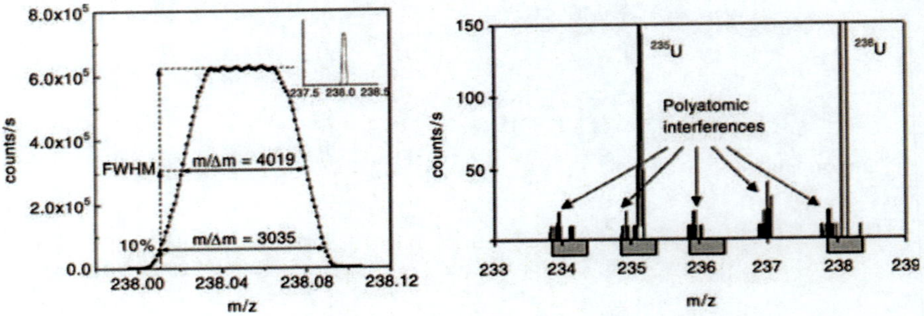

Fig. 5 *Left*: Example of peak shape of ^{238}U obtained on the Cameca 1270 SIMS. Mass resolving power ($m/\Delta m$) at full width half maximum and at 10% of peak height is given. *Right*: A mass scan from a U particle from a swipe sample displaying additional polyatomic peaks at $m/\Delta m$ 3035. The *gray rectangles* represent the width of the uranium peaks at $m/\Delta m$ 450; normal working conditions for SG-SIMS instruments on nuclear swipe samples. From Ranebo et al. (2009)

The latter work also discussed advantages of SIMS versus TIMS after fission track analysis and included a discussion on resolving power needed to resolve most common molecular isobaric interferences (Fig. 5). Fuel production plants were tested for undeclared operations such as fabrication of highly enriched uranium exploiting UF_6 release, which led to the formation of UO_2F_2 nanoparticles (Kips et al. 2009).

By SIMS, the F to U ratio was determined on the 50–100 nm scale making it possible to probe source terms and age (Kips and Kristo 2009) of the particle. This investigation was particularly challenging due to the large mass difference of U and F calling for sensitivity calibration of the mass analyzer over a wide mass-over-charge range.

In a further study, the age of PuO_2 particles was used to validate the cut off treaty for nuclear weapons material. For bulk material, the $^{241}Am/^{241}Pu$ ratio is routinely determined by γ-spectroscopy and used for age determination. However, the tiny amount of γ emitters in a nanoparticle inhibits the use of this method. Using SIMS, the ratios of $^{238}Pu/^{234}U$, $^{239}/Pu^{235}U$, and $^{240}Pu/^{236}U$ were measured and evaluated (Wallenius et al. 2001) to give precise information on the age of particles without the need for chemical separation or pretreatment. The feasibility was demonstrated on well-defined PuO_2 particles from a Round Robin test and applied to a PuO_2 sample seized at Munich airport (Fig. 6).

4 Laboratory Systems

Specific production of actinide nanoparticles can proceed from two directions: For one, particles can be produced from larger entities (bulk material) by breaking and milling, which results in the formation of particles similar to those described in the

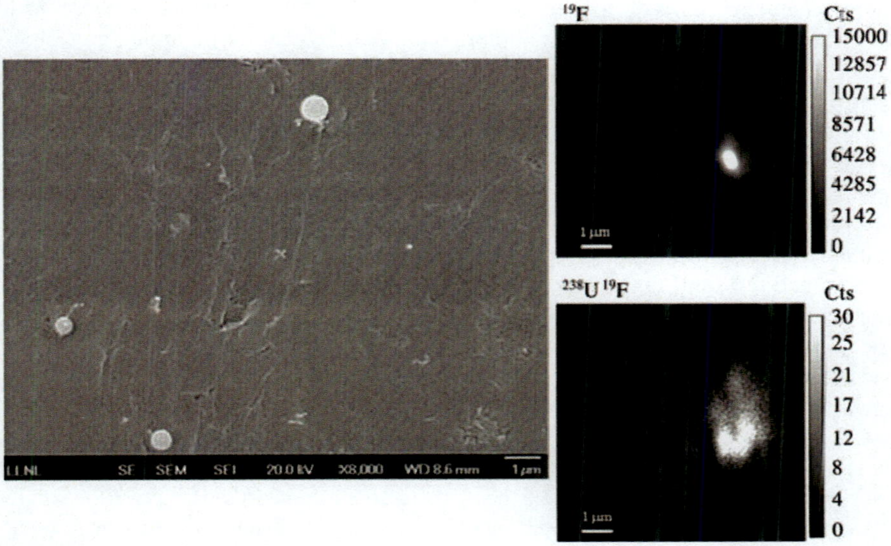

Fig. 6 *Left*: Scanning electron microscope image of a UO_2F_2 particle sample showing the particle distribution on the graphite planchet. *Right*: Ion images for ^{19}F- and $^{238}U^{19}F$-obtained by Nano-SIMS 50 showing a uranium oxyfluoride particle (*bright spot*). From Kips and Kristo (2009)

previous section, i.e., produced by explosions and subsequent weathering of nuclear material. In contrast to environmental samples, particles produced in the lab are well characterized, sometimes even monodisperse, and can serve as reference material. The second approach of particle generation is by growth of nanoparticles from smaller moieties, for example, the formation of nanoparticles in aqueous media by hydrolysis and subsequent polymerization of metal ions. Analysis of nanoparticles >10 nm is performed best by (LA) ICP MS or SIMS, techniques which are also applied frequently for environmental particles as described in Sect. 3. The formation from molecules, however, is best observed by soft ionization techniques such as matrix-assisted laser-desorption ionization (MALDI) or ESI. So far, only the latter was applied for investigation of actinide polymers.

4.1 *Investigation of Preformed Nanoparticles*

Varga produced nanoparticles from uranium-oxide powder and measured isotope distributions of single uranium particles by LA ICP-MS (Varga 2008). The particles were deposited on double-sided tape on a cardboard holder and vaporized by a 213 nm YAG laser. The ablated material was carried into the plasma by argon carrier gas. Depleted, natural, lowly enriched and highly enriched uranium particles were characterized with respect to their ^{234}U, ^{235}U, ^{236}U, and ^{238}U content with typical precisions of 1–5%. The authors point out the need for reducing molecular interference by operating the instrument at high mass resolution (SF MS), which is particularly important for investigating small particles and increasing the accuracy of the abundance of minor isotopes. Actinide nanoparticles suspended in solution (colloids) were investigated by Degueldre et al.: A stock solution was prepared by milling commercially available thoria powder with an initial grain size of some 10 μm until nanoparticles between 50 nm and several hundred nm in diameter were obtained (Degueldre and Favarger 2004). These particles were suspended in milliQ water, and the resulting sample was diluted until the nanoparticle number density ranged between 10^5 and 10^6 ml^{-1}; subsequently, it was introduced slowly into the nebulizer of the ICP-MS machine using a syringe pump. The authors stressed the necessity for introducing the nanoparticles slowly at low concentration in order to guarantee that at most one nanoparticle entered the plasma plume at all times. Mass analysis was performed in a QP instrument. In several time scans, the signal intensities at mass 232 amu ($^{232}Th^+$) and in separate scans at mass 248 amu ($^{232}Th^{16}O^+$) were observed. Clear peak structures allowed the detection of single nanoparticles and also measurement of their size by evaluating the peak area. Correlating the results with SEM measurements indicated a lower detection limit of 80 nm diameter of the nanoparticles. A similar study was performed on uranium particles obtained by milling of UO_2 powder (Degueldre et al. 2006). In this case, the signals at masses 235 amu, 238 amu, and 254 amu ($^{238}U^{16}O^+$) were used for nanoparticle size analysis. Besides the lower detection limit around 100 nm, the authors discuss also an upper limit around 6,000 nm caused by the finite residence time of the particles in the plasma torch. Larger particles cannot be fully vaporized, which leads to signal decrease and underestimation of their size.

It is obvious that the precision up to which isotope ratios can be measured depends on particle size. 100 nm sized particles contain only some 10^8 atoms, making precise determination of minor isotopes a real challenge. Esaka et al. (2008) proved in a study on certified uranium reference material (CRM) that precisions of <10% can be achieved for SIMS investigations on single submicron particles even for the low abundant isotopes ^{234}U and ^{236}U ($^{234}U/^{238}U = 3 \times 10^{-4}$, $^{236}U/^{238}U = 5 \times 10^{-4}$, corresponding to only ~1 fg). Erdmann et al. produced monodisperse uranium oxide particles using a commercial vibrating orifice aerosol generator. Certified isotope ratios were reproduced within 1–8% statistical error by measuring single particles using a double focusing SF SIMS optimized for fast switching between masses to be analyzed (Fig. 7).

4.2 Chemical Speciation

For investigation of complex formation in solution, ESI becomes increasingly popular. The information obtained by MS is much more direct than results obtained by titration, requiring model assumptions for interpretation. A great

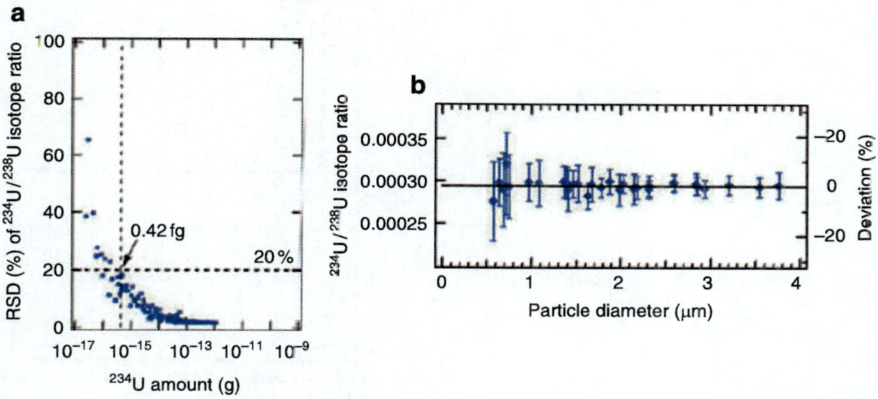

Fig. 7 *Right*: The $^{234}U/^{238}U$, isotope ratios measured for the CRM U050 particles with the diameter between 0.58 and 3.76 μm. The horizontal lines represent the certified value. *Left*: The RSD (%) values of $^{234}U/^{238}U$ isotope ratios measured for the CRM U010, U050, U350 and U500 particles, plotted against the amount of ^{234}U in the particles. From Esaka et al. (2008)

number of studies aim at finding the composition of metal ligand complexes in solution recently reviewed in Di Marco and Bombi (2006), some of which address actinide chemistry: Keith-Roach et al. investigated the formation of thorium hydroxamate siderophore (desferrioxamine and ferrichrome) complexes and exchange reactions with Fe(III) (Keith-Roach et al. 2005) in the range pH 5–9. They used an RF trap spectrometer equipped with a flow injection ESI, i.e., small amounts of sample are added to a continuous stream of liquid supplying the source.

Complexation of actinide ions by organic ligands is particularly relevant for ultimate disposal of nuclear waste in deep geological repositories. The large inventory of citric acid and its anhydrites in the Waste Isolation Pilot Plant (WIPP) (Carlsbad NM, USA) requires detailed investigation of the relevant actinide chemistry. Several studies were performed using ESI: Francis et al. focused at the biotransformation of Pu(IV) citric acid complexes (Francis et al. 2006). In addition to mononuclear species, the formation of dimeric citrate complexes, $\{Pu_2(Hcit)(cit)(H_2O)_4(OH)_2\}^+$, was observed (cit denotes $C_6H_5O_7{}^{3-}$). Furthermore, biodegradation of Pu citric acid complexes by Pseudomonas fluorescens bacteria induced polymerization of Pu(IV), i.e., formation of small plutonium nanoparticles between pH 0 and pH 4. In contrast to plutonium, polymerization of hexavalent uranium proceeds without bacterial digestion of citrate. Somogyi et al. investigated complexation of uranyl ions with citric acid (Somogyi et al. 2007) and detected polynuclear complexes of the general formula $\{(UO_2)_n(Hcit)_2(cit)(HNO_3)_p(X)_q\ (H_2O)_r\}^+$, $X = \{Na^+;K^+\}$. This study was performed using two different ESI ion trap instruments. One is a so-called 3D trap (RF, or Paul trap), the second one a high resolution FT ICR trap (Penning trap). The paper includes a detailed investigation of ion gas phase reactions, which

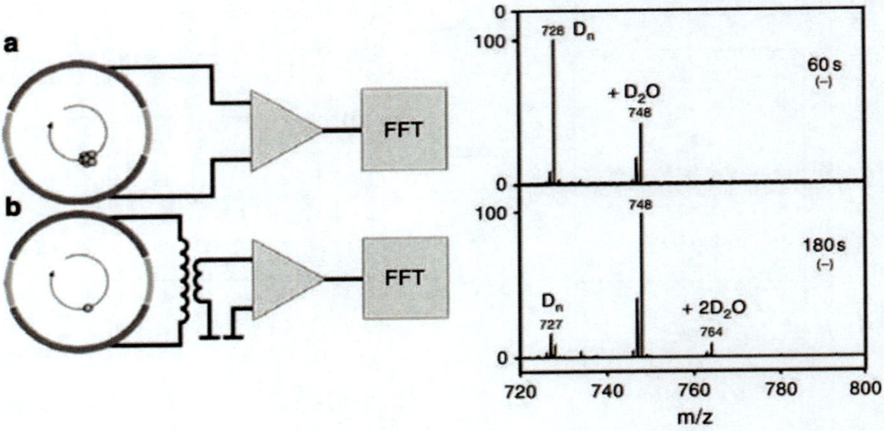

Fig. 8 *Left*: Detection of ions in a Penning trap. The moving ions induce mirror charges in the segmented trap electrodes. The Fourier transform of this temporal signal gives a frequency spectrum, which is converted into mass over charge. Building a superconducting resonant circuit allows single ion detection. From Ketelaer et al. (2009). *Right*: FT ICR mass spectrum showing ion-molecule reactions between D_2O and monoisotopically selected uranyl-monocitrate 2:1 complexes. Reaction times applied for $[(UO_2)_2(C_6H_4O_7)]^-$ (*m/z* 728) at 60 s (*top*), and 180 s (*bottom*) From Somogyi et al. (2007)

proceed particularly fast during storage in the 3D trap due to elevated residual gas pressure. By successive association steps, $\{(UO_2)_nO_m\}^-$ nanoparticles formed (Fig. 8).

Uranyl polymerizes in the absence of organic ligands as well. In combined ESI laser-induced-fluorescence (TRLIF) studies, mono- and polynuclear uranyl hydroxide complexes were observed (Moulin et al. 2000; Moulin 2003). The authors pointed out that combining the complementary techniques TRLIF and ESI-MS is particularly helpful. TRLIF is highly sensitive but limited to investigation of fluorescing metals, whereas ESI-MS is less sensitive but can be applied to any element and provides direct information on the stoichiometry of the complexes (Moulin et al. 2006). Tsierkezos et al. (2009) reported hydrolysis of uranyl monomers and formation of polymers $\{U_mO_{2m}(X, Y)_{2m-1}(H_2O)_n\}^+$; ($X, Y$ = OH, NO_3) m <5, in acidic solutions over a wide concentration range of $[U]_{tot}=10^{-8}$–10^{-2} M. Though the unambiguous identification leaves no doubt about the polynuclear species forming in solution, the authors did not measure pH, which precludes the use of these findings for calculating equilibrium constants. However, the authors discuss the risk of potential artifacts caused by up-concentration effects using high flux electrospray sources (typically >5 μl min^{-1}). The primary droplets exceed micrometer size and contain many uranyl molecules. In this case, solvent evaporation takes several 100 ms. During this time, the concentration of nonvolatile species in the droplet increases by several orders of magnitude, which might initiate fast complexation processes.

4.3 Direct Observation of Nanoparticle Formation

Up-concentration effects are strongly reduced by generating smaller primary droplets in the case of the so-called nanoESI, a particular soft variation of electrospray invented by Wilm and Mann (1996). By substituting the metal capillary by a glass capillary of only approximately 2 μm tip diameter, droplets of less than 50 nm radius are generated. These droplets contain only a small number of nonvolatile compounds, making polymerization during droplet shrinkage less likely. Second, evaporation is much faster than for conventional ESI. Operative parameters of the source can be chosen to be mild such that a solvent shell is maintained surrounding the molecules during all stages of detection, minimizing also the risk of fragmentation or dehydration.

Results of the polymerization in solutions of three metals are compared: zirconium (Walther et al. 2007) being a frequently used homolog of tetravalent plutonium, thorium (Walther et al. 2008), and plutonium (Walther et al. 2009b), which is, of course, the most relevant actinide with regard to long-term safety considerations of spent nuclear fuel.

All experiments were performed using an electrospray TOF mass-spectrometer (Bergmann et al. 1990) comprising a home built nanospray ion source equipped with spray capillaries of borosilicate glass. Primary droplets shrink within a few microseconds to tiny droplets of some 50 water molecules, each containing only one complex, for instance, $\{M_xL_y{}^{4x-y}{\cdot}mCl^{-}nH_2O\}^{4x-y-m}$. The chloride ions, presumably located in an outer shell, compensate part of the charge of the complex. The TOF mass-spectrometer ($m/\Delta m = 26{,}000$) provides molecular isobaric resolution and allows one to identify, as a function of pH, polymerization x, number of hydroxide groups y, and via $z = 4x{-}y$, the charge of each oligomer (Walther et al. 2007). Often, several hydroxide complexes with the same degree of polymerization x, but different number of hydroxide groups y are present simultaneously. The natural isotope abundances of $^{35,37}Cl$ cause isotopic patterns unique for each complex in the mass spectrum ($\{M_xL_y{\cdot}{}^{35}Cl_p{}^{37}Cl_q\ n(H_2O)\}^{4x-y-p-q}$) which facilitates unequivocal identification.

Samples of [Zr]=1.5–10 mM at $0 < pH_C < 3$ were prepared by dissolving appropriate amounts of zirconyl chloride hydrate in hydrochloric acid of predefined pH. The frequently used approach of working at rather high, but constant, ionic strength, e.g., at I = 1 M, is not possible here, as the amount of electrolyte present in solution is limited by the ESI technique: Salt deposition clogs the tip of the needle or the small orifice where the spray enters the vacuum chamber. Hence, no electrolyte was added to the samples, and the ionic strength of the solution was determined by the molality of H^+, Cl^-, and the zirconium hydroxide complexes.

One typical example of a TOF mass-spectrum is shown in Fig. 9 (top) in a logarithmic representation ([Zr] = 2.5 mM, pH_C 0.2). Three major species are distinguished: the hydrolyzed monomers (1,*y*), tetramers (4,9), and pentamers (5,11). Due to the varying number of water molecules, the peaks group into clusters (Fig. 9, middle). Further magnification (bottom) reveals a substructure due to the

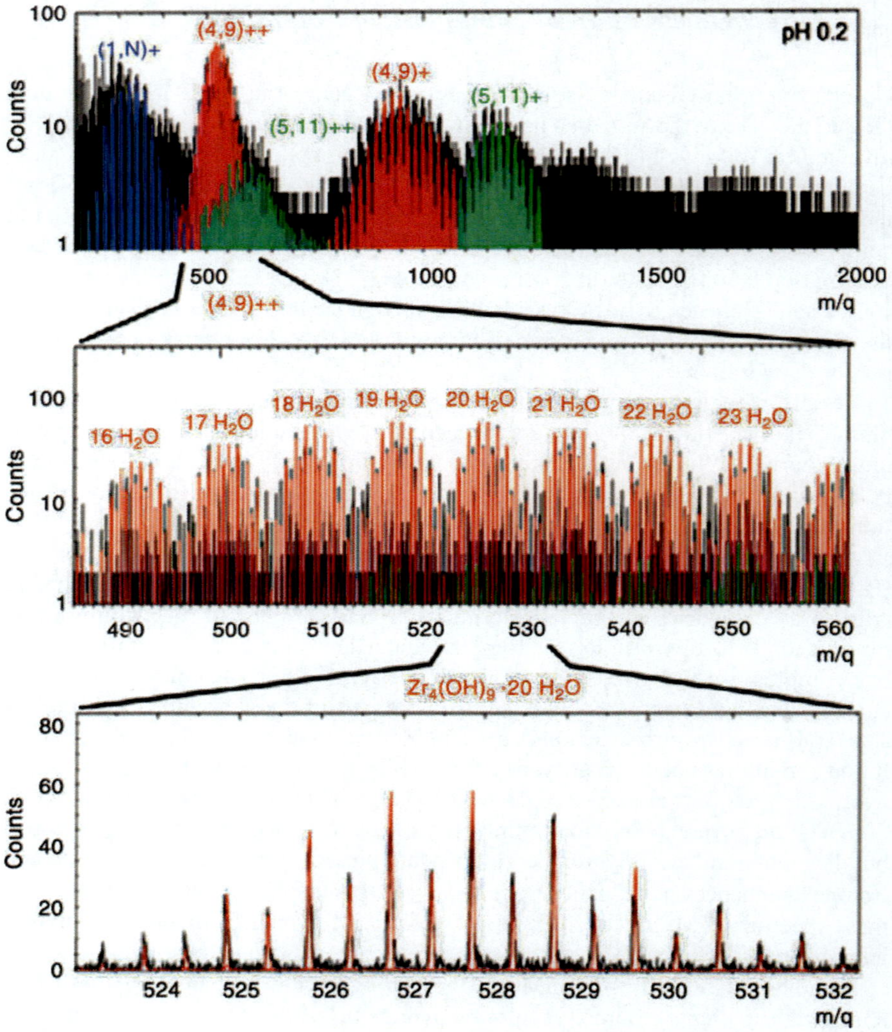

Fig. 9 Mass spectra of a solution with [Zr] = 2.5 mM, pH 0.2. *Top*: Monomers (1,*N*), tetramers (4,9), and pentamers (5,11) are identified. *Middle*: A magnification reveals that the cluster of doubly charged tetramers is in turn composed of peak clusters representing different numbers of water molecules in the solvation shell. *Bottom*: The isotope distribution (Cl and Zr isotopes) of each group is shown by further magnification

natural isotope distributions of Zr and Cl, which agrees well with the theoretical distribution, allowing an unambiguous identification of the complexes $[Zr_x(OH)_y{}^{4x-y}mCl^- \, nH_2O]^{q+}$.

The measured species distributions ([Zr] = 1.5 mM 0 < pH <3, Fig. 10, right, symbols) are dominated by polymers. With increasing pH, the fraction of

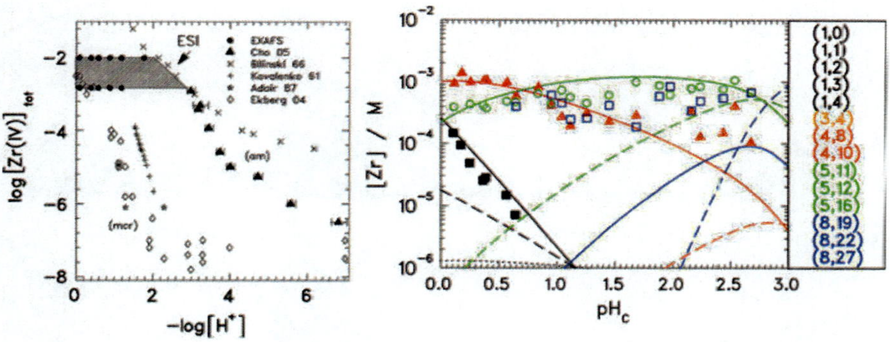

Fig. 10 *Left*: Solubility data for microcrystalline $ZrO_2 \times H_2O$ (mcr) and amorphous zirconium hydroxide [$Zr(OH)_4$(am)] [from Walther et al. (2007)]. *Right*: Species distribution at [Zr(IV)] = 1.5 mM in units of molar Zr concentration within each size class for the ESI TOF data [monomers (*black squares*), tetramers (*red triangles*), pentamers (*green circles*), and octamers (*blue open squares*)]. *Lines*: speciation model. (*x*, *y*) denotes the complex $Zr_x(OH)_y^{4x-y}$

Zr(IV)-monomers decreases even more. Above pH 1, pentamers and larger complexes form in addition to the well-known tetramers ($Zr_4(OH)_8^{8+}$), resulting in a complex mixture of many solution species. In the course of pH increase, the number of hydroxide groups *y* increases and, consequently, the charge decreases.

Under conditions (H^+ and Zr concentrations) close to the solubility of $Zr(OH)_4$(am) (Fig. 10, left), the most abundant species (pentamers as well as octamers) approach charge *z* ~2. This is in agreement with the slope analysis of previous work (Cho et al. 2005) and with calculations using equilibrium constants of formation from Brown et al. (2005). An unpredicted finding, however, was the strong abundance of pentamers.

Solutions of thorium [Th(IV)] salts in the concentration range ([Th(IV)] $>10^{-4}$ M) (Rand et al. 2008) show similar polymerization behavior as does Zr(IV). However, due to the weaker hydrolysis, polymerization starts only at higher pH. Stoichiometries for dimers ($x = 2$) (Johannson 1968; Wilson et al. 2007), tetramers, and hexamers have been suggested and considerable efforts have been made to obtain equilibrium formation constants. In Walther et al. (2008, 2009a), samples between $[Th(IV)]_{tot} = 6 \times 10^{-6}$ M and $[Th(IV)]_{tot} = 1 \times 10^{-1}$ M at $1 < pH_C < 4$ up to the solubility of amorphous Th(IV) hydroxide (Ekberg et al. 2000; Rai et al. 2000) were investigated. Samples were prepared by dilution of a stock solution of $[Th(IV)]_{tot} = 0.3$ M in HCl at pH_C 3.1 with appropriate HCl (pH 0.3–4), followed by slow titration with MilliQ water to avoid nanoparticle formation due to local pH gradients. All samples were undersaturated with respect to the formation of amorphous thorium hydroxide.

At $[Th(IV)]_{tot} = 6 \times 10^{-6}$ M (Fig. 11), no polymers were detected, in accordance with Ekberg et al. (2000). With increasing pH, 3.0–4.3, the dihydroxide complex $Th(OH)_2^{2+}$ formed at the expense of both hydrated ion Th^{4+} and first

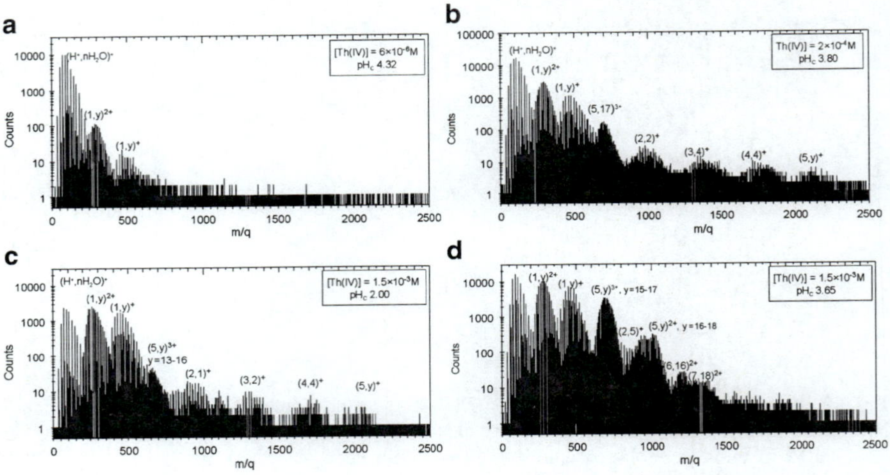

Fig. 11 (**a**) No significant amount of polymers is detected for the dilute solutions at ($[Th(IV)]_{tot}$ $= 6 \times 10^{-6}$ M) between pH 3.25 and 4.32. (**b**) At $[Th(IV)]_{tot} = 2 \times 10^{-4}$ M, initially polymer-free solutions form polymers upon pH increase. (**c**) For increasing Th(IV) concentration ($[Th(IV)]_{tot} = 1.5 \times 10^{-3}$ M), polynuclear hydroxide complexes already form at pH = 2.0 and (**d**) dominate the spectrum close to the solubility limit of $Th(OH)_4$(am) at pH = 3.65

complex $Th(OH)^{3+}$, which was in quantitative agreement with the hydrolysis model of Rand et al. (2008). The most abundant polymer was expected to contribute only $[Th_4(OH)_{12}{}^{4+}] = 1 \times 10^{-8}$ M, which was below the detection limit of the instrument. At $[Th(IV)]_{tot} = 2 \times 10^{-4}$ M and pH 2.5, Th^{4+} and $Th(OH)^{3+}$ contributed approximately 40% of the total Th(IV) concentration each. The fraction of dimers remained below the detection level. At pH 3.3, dimer, tetramer, and pentamer formed in approximately equal amounts, though they were clearly minor to the mononuclear complexes Th^{4+} and $Th(OH)_3{}^+$. With increasing pH, the relative abundance of the pentamer increased strongly and eventually accounted for almost 50% of the $[Th(IV)]_{tot}$ concentration at pH >3.6 (Fig. 11). In this pH range, the fraction of Th(IV) pentamers agreed well with the polymer fraction predicted by Rand et al. (2008). At pH >3.5, the pentamer coordinated 16–17 hydroxide ligands: $Th_5(OH)_{17}{}^{3+}$. The unexpected presence of pentamers $Th_5(OH)_y{}^{z+}$ was confirmed by the use of X-ray absorption fine structure spectroscopy (EXAFS) and high energy X-ray scattering measurements combined with quantum chemical calculations. Upon aging, Th polymers undergo an Ostwald ripening-like process and form nanoparticles (Walther et al. 2009a).

Of the three metals discussed here, tetravalent plutonium shows the highest tendency to form polymeric complexes and nanoparticles (Johnson and Toth 1978a, b; Lloyd and Haire 1978; Toth et al. 1981, 1983; Thiyagarajan et al. 1990). From early work on Pu separation chemistry (Clark et al. 2006) to today's assessment of nuclear waste repositories (Kim 1994), understanding Pu polymer chemistry has remained a challenge. Hence, the subject of hydrolysis and polymerization of tetravalent actinides is an active area of research.

From EXAFS data (Rothe et al. 2004), polymerization of tetravalent Pu was shown to proceed via aggregation of cubic Pu-dihydroxy building blocks ($Pu(OH)_2^{2+}{\cdot}6H_2O$) followed by partial condensation $2OH^- \rightarrow H_2O + O^{2-}$, leading to the formation of larger moieties such as dimers, trimers, and so on. Further indication of the formation of small polymers was obtained by optical absorption spectroscopy of strongly undersaturated Pu(IV) solutions (Walther et al. 2009b), but their exact stoichiometry was not known. This piece of information was gained from ESI-TOF investigations (Walther et al. 2009b): At $[Pu]_{tot} = 3 \times 10^{-5}$ M and pH_C 0.65 (undersaturated by a factor of >50), only mononuclear ions (Pu^{3+}, Pu^{4+}, $Pu(OH)^{3+}$, PuO_2^+, and PuO_2^{2+}) were detected (Fig. 12). Their relative abundances were in good accordance with absorption spectroscopy. A slight decrease in the acidity down to pH 0.82 resulted in spontaneous formation of tetramers, and at pH 0.94 (undersaturation ~10), trimers and tetramers account for >75% of $[Pu]_{tot}$. In contrast to the experiments on Th(IV) and Zr(IV), however, the spectra could not be fitted by including $M_x^{(IV)}(OH)_y^z$ polymers alone. Instead, mixed valence polymers are identified, where Pu(IV) is partly substituted by Pu(V), such as the tetramer $Pu_3^{(IV)}Pu^{(V)}O_2(OH)_y^z$.

There are numerous parallels of polymer and colloid chemistry of Pu(IV), Zr(IV), and Th(IV): Polymer formation is the first step towards nanoparticle formation and subsequent precipitation. The ESI-TOF measurements on Zr(IV) and Th(IV) supply direct evidence that both monomers and polymers hydrolyze

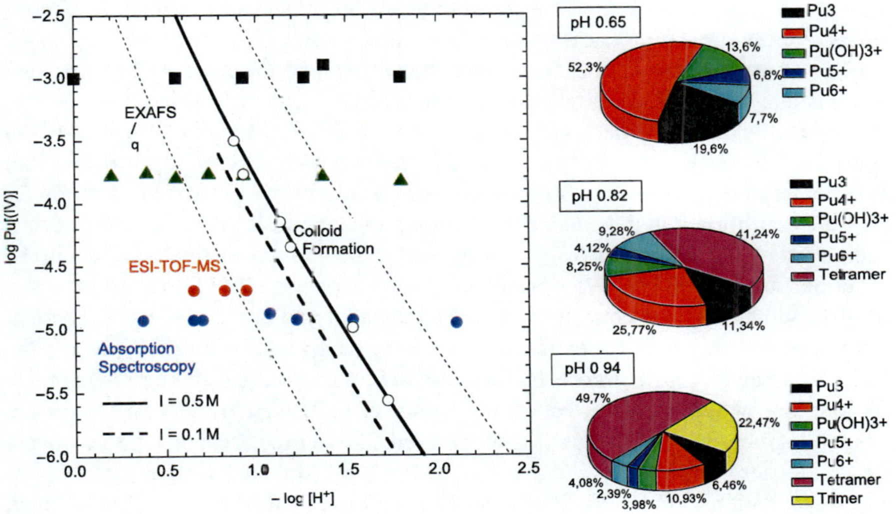

Fig. 12 *Left*: Solubility of amorphous hydrated plutonium hydroxide calculated according to (Guillaumont et al. 2003) ($\log K^\circ_{sp}(Pu(OH)_4(am, hyd) = -58.3$). Samples examined by ESI MS are marked in *red*. *Right*: Species distributions at $[Pu]_{tot}] = 3 \times 10^{-5}$ M

continuously, leading to charge compensation with decreasing acidity, not only in the case of monomeric species, $M(OH)_y^{4-y}$, but also in the case of polymers $M_x(OH)_y^{4x-y}$. The simultaneous presence of three or four different complexes with the same degree of polymerization x [for instance (5,13), (5,14), (5,15), (5,16)] is frequently observed.

However, it is just as clear that there exist pronounced differences between Th(IV) and Zr(IV), both of which play important roles as model substances for Pu(IV). For [Zr(IV)] ≥1 mM at pH >0.1, mononuclear complexes are minor species. The expected stability range of the tetramer $Zr_4(OH)_8^{8+}$ is confirmed followed by formation of pentamers and octamers, the latter being in agreement with previous investigations by X-ray scattering (Singhal et al. 1996). The degree of polymerization gradually increases as the conditions (Zr(IV) and H^+ concentration) approach the solubility limit of $Zr(OH)_4$(am). The polymerization process culminates in the formation of nanoparticles and finally precipitation. In contrast to Zr(IV) solutions, Th(IV) in 1 M HCl is dominated by monomeric species. The fraction of polynuclear hydroxide complexes increases with increasing pH and increasing $[Th(IV)]_{tot}$ concentration. In solutions of millimolar Th(IV) concentration, the stability field of polymers extends from pH 2 up to the limit of solubility of $Th(OH)_4$(am). Pentamers ($Th_5(OH)_y^{20-y}$) dominate the species distribution of acidic Th(IV) solutions over a wide range of concentrations. Tetramers, hexamers and larger polymers ($Th_{10}(OH)_{32}^{8+}$, $Th_{11}(OH)_{38}^{6+}$ and $Th_{12}(OH)_{42}^{6+}$) are observed only as minor species or in freshly prepared nonequilibrated solutions. Formation of larger polymers eventually culminates in the generation of amorphous $Th(OH)_4$ – like nanoparticles. Polymerization in Pu(IV) solutions proceeds much more similar to Zr(IV) than to its actinide analog Th(IV): Pu(IV) hydrolyzes as strongly as Zr(IV), i.e., at pH >0.5, $Pu(IV)_{aq}$ is dominated by hydroxide complexes. Consequently, polymerization starts at about the same range as in the case of Zr(IV). However, in contrast to the continuous transition from monomers via growing polynuclear complexes to a well-defined onset of nanoparticle formation close to the solubility limit of the amorphous hydroxide for Zr(IV) and Th(IV), Pu(IV) behaves differently. In strongly undersaturated solutions, observation of a small number of nanoparticles sized some tenths of a micron was reported. It might be that the equilibrium and, hence, coexistence of nanoparticles and solution species, which was proven for Th(IV), spans a much wider range in the case of Pu(IV) solutions and that monomeric ions coexist with nanoparticles (Thiyagarajan et al. 1990) and polymers of considerable size (Ockenden and Welch 1956). This coexistence would underline the extraordinary role of nanoparticles and polymers in the Pu(IV) chemistry. Furthermore, Pu forms up to four oxidation states simultaneously in solution, which has an effect on the nature of polymers: trimers and tetramers containing both $Pu^{(IV)}$ and PuO_2^+ are observed. These mixed valence polymers are of great interest since they might present the long looked for transition species during oxidation from Pu(III)/Pu(IV) to the trans dioxo ions PuO_2^+ and PuO_2^{2+} and, vice versa, reduction of PuO_2^+ and PuO_2^{2+} analogously to the model proposed by Neck et al. (2007) for redox equilibria in near neutral nanoparticle containing Pu solutions.

5 Future Perspectives

MS is well suited to answer a wide range of questions concerning nanoparticle formation. Particles from solid, liquid, and also gaseous (aerosols) samples can be detected and analyzed. Formation processes can be monitored by observing aggregation of single atoms or small molecules to polymers and subsequent growth to nanoparticles almost "in-situ".

Recent developments of MS aim at ever increasing sensitivity. Mass analyzers themselves, e.g., the TOF machines are mature with transmissions close to 100% and low mass dependency. Ion detectors reach quantum efficiencies close to one over a wide mass range so do ion traps allowing one to detect single ions by FT ICR (see above). For this reason, research and development focus on improving ion sources. Recently, extractive electrospray was improved to open up speciation of actinide ions in environment-relevant trace concentrations down to <1 ng/l (Luo et al. 2009), a concentration range where so far only element characterization and isotope distribution measurements were possible (TIMS, ICP MS). Single particle element analysis from the liquid phase should soon be applied to actinide nanoparticles as it is routinely performed in aerosol science (Reents et al. 2000; Trimborn et al. 2000).

As the second current trend, strong efforts are being made to improve spatial resolution. Ion production by interaction of the sample with tightly focused beams of lasers (MALDI) leads to resolutions in the μm range, and interaction with energetic ions allows the resolution of structures on the nanoscale. Real imaging of isotope distributions of nanoparticles is now possible down to a few tens of nm and will certainly improve in the near future. The amount of material desorbed from the surface of the particle is so minute that even small particles are not consumed by the detection allowing one to perform repeated measurements on the same particle.

Examples for combining sensitivity, selectivity, and spatial resolution are, for instance, the coupling of RIMS with MALDI (Maul et al. 2004) or with SIMS for postionization of neutrals (Erdmann et al. 2009). The ultimate aim of single ion detection on atomic scales might become reality not too far in the future.

References

Becker JS (2003) Mass spectrometry of long-lived radionuclides. Spectrochimica Acta. Part B: Atomic Spectroscopy 58: 1757–1784

Bergmann T, Martin TP, Schaber H (1990) High-resolution time-of-flight mass spectrometers. 2. Reflector design. The Review of Scientific Instruments 61: 2592–2600

Brown PL, Curti E, Grambow B, Ekberg C, Mompean FJ, Perrone J, Illemassene M (2005) Chemical thermodynamics of zirconium. Elsevier, Amsterdam

Bürger S, Buda RA, Geckeis H, Huber G, Kratz JV, Kunz P, Gostomski CLv, Passler G, Remmert A, Trautmann N (2006) Isotope selective ultratrace analysis of plutonium by resonance ionisation mass spectrometry. In: Povinec PP, Sanchez-Cabeza JA (eds) Radioactivity in the environment. Elsevier, Amsterdam, pp 581–591

Cho HR, Walther C, Rothe J, Neck V, Denecke MA, Dardenne K, Fanghänel T (2005) Combined LIBD and EXAFS investigation of the formation and structure of Zr(IV) colloids. Analytical and Bioanalytical Chemistry 383: 28–40

Clark DL, Hecker SS, Jarvinen GD, Neu MP (2006) Chapter 7: Plutonium. In: Morss LR, Edelstein NM, Fuger J, Katz JJ (eds) The chemistry of actinide and transactinide elements. Springer, Heidelberg, pp 813–1203

Dai M, Kelley JM, Buesseler KO (2002) Sources and migration of plutonium in groundwater at the Savannah River Site. Environmental Science & Technology 36: 3690–3699

Danesi PR, Markowicz A, Chinea-Cano E, Burkart W, Salbu B, Donohue D, Ruedenauer F, Hedberg M, Vogt S, Zahradnik P, Ciurapinski A (2003) Depleted uranium particles in selected Kosovo samples. Journal of Environmental Radioactivity 64: 143–154

Degueldre C, Favarger PY (2004) Thorium colloid analysis by single particle inductively coupled plasma-mass spectrometry. Talanta 62: 1051–1054

Degueldre C, Favarger PY, Rosse R, Wold S (2006) Uranium colloid analysis by single particle inductively coupled plasma-mass spectrometry. Talanta 68: 623–628

Di Marco VB, Bombi GG (2006) Electrospray mass spectrometry (ESI-MS) in the study of metal-ligand solution equilibria. Mass Spectrometry Reviews 25: 347–379

Ekberg C, Albinsson Y, Comarmond MJ, Brown PL (2000) Study on the complexation behaviour of thorium(IV). 1. Hydrolysis equilibria. Journal of Solution Chemistry 29: 63–86

Erdmann N, Kratz JV, Trautmann N, Passler G (2009) Resonance ionization mass spectrometry of ion beam sputtered neutrals for element- and isotope-selective analysis of plutonium in micro-particles. Analytical and Bioanalytical Chemistry 395: 1911–1918

Esaka F, Esaka KT, Lee CG, Magara M, Sakurai S, Usuda S, Watanabe K (2007) Particle isolation for analysis of uranium minor isotopes in individual particles by secondary ion mass spectrometry. Talanta 71: 1011–1015

Esaka F, Watanabe K, Onodera T, Lee CG, Magara M, Sakurai S, Usuda S (2008) Dependence of the precision of uranium isotope ratio on particle diameter in individual particle analysis with SIMS. Applied Surface Science 255: 1512–1515

Francis AJ, Dodge CJ, Gillow JB (2006) Biotransformation of plutonium complexed with citric acid. Radiochimica Acta 94: 731–737

Gross JH (2004) Electrospray ionization mass spectrometry. Springer, Berlin

Guillaumont R, Fanghänel T, Fuger J, Grenthe I, Neck V, Palmer DA, Rand MH (2003) Update on the chemical thermodynamics of uranium, neptunium, plutonium, americium and technecium. Elsevier, Amsterdam

Jernstrom J, Eriksson M, Simon R, Tamborini G, Bildstein O, Marquez RC, Kehl SR, Hamilton TF, Ranebo Y, Betti M (2006) Characterization and source term assessments of radioactive particles from Marshall Islands using non-destructive analytical techniques. Spectrochimica Acta. Part B: Atomic Spectroscopy 61: 971–979

Johannson G (1968) The structure of the dinuclear complex of thorium. Acta Chemica Scandinavica 22: 389–398

Johnson GL, Toth LM (1978a) Plutonium (IV) and thorium(IV) hydrous polymer chemistry. Report 3: 43386

Johnson GL, Toth LM (1978b) Plutonium(IV) and thorium(IV) hydrous polymer chemistry. Oak Ridge National Laboratory, TM-6365, Tennessee

Keith-Roach MJ, Buratti MV, Worsfold PJ (2005) Thorium complexation by hydroxamate siderophores in perturbed multicomponent systems using flow injection electrospray ionization mass spectrometry. Analytical Chemistry 77: 7335–7341

Ketelaer J, Blaum K, Block M, Eberhardt K, Eibach M, Ferrer R, George S, Herfurth F, Ketter J, Nagy S, Repp J, Schweikhard L, Smorra C, Sturm S, Ulmer S (2009) Recent developments in ion detection techniques for Penning trap mass spectrometry at TRIGA-TRAP. The European Physical Journal. A: Hadrons and Nuclei 42: 311–317

Kim JI (1994) Actinide colloids in natural aquifer systems. MRS Bulletin 19: 47–53

Kips RS, Kristo MJ (2009) Investigation of chemical changes in uranium oxyfluoride particles using secondary ion mass spectrometry. Journal of Radioanalytical and Nuclear Chemistry 282: 1031–1035

Kips R, Pidduck AJ, Houlton MR, Leenaers A, Mace JD, Marie O, Pointurier F, Stefaniak EA, Taylor PDP, Van den Berghe S, Van Espen P, Van Grieken R, Wellum R (2009) Determination of fluorine in uranium oxyfluoride particles as an indicator of particle age. Spectrochimica Acta. Part B: Atomic Spectroscopy 64: 199–207

Lind OC, Salbu B, Janssens K, Proost K, Garcia-Leon M, Garcia-Tenorio R (2007) Characterization of U/Pu particles originating from the nuclear weapon accidents at Palomares, Spain, 1966 and Thule, Greenland, 1968. The Science of the Total Environment 376: 294–305

Lloyd MH, Haire RG (1978) The chemistry of plutonium in sol-gel processes. Radiochimica Acta 25: 139–148

Luo MB, Hu B, Zhang X, Peng DF, Chen HW, Zhang LL, Huan YF (2009) Extractive electrospray ionization mass spectrometry for sensitive detection of uranyl species in natural water samples. Analytical Chemistry 82: 282–289

Maul J, Berg T, Eberhardt K, Hoog I, Huber G, Karpuk S, Passler G, Strachnov I, Trautmann N, Wendt K (2004) A laser desorption/resonance enhanced photoionisation TOF-system for the spatially resolved trace analysis of elements. Nuclear Instruments & Methods in Physics Research. Section B, Beam Interactions with Materials and Atoms 226: 644–650

Moulin C (2003) On the use of time-resolved laser-induced fluorescence (TRLIF) and electrospray mass spectrometry (ES-MS) for speciation studies. Radiochimica Acta 91: 651–657

Moulin V, Moulin C (2001) Radionuclide speciation in the environment: A review. Radiochimica Acta 89: 773–778

Moulin C, Charron N, Plancque G, Virelizier H (2000) Speciation of uranium by electrospray ionization mass spectrometry: Comparison with time-resolved laser-induced fluorescence. Applied Spectroscopy 54: 843–848

Moulin C, Amekraz B, Colette S, Doizi D, Jacopin C, Lamouroux C, Plancque G (2006) Electrospray mass spectrometry for actinides and lanthanide speciation. Journal of Alloys and Compounds 408: 1242–1245

Neck V, Altmaier M, Seibert A, Yun JI, Marquardt CM, Fanghänel T (2007) Solubility and redox reactions of Pu(IV) hydrous oxide: Evidence for the formation of PuO$_{2+x}$(s,hyd). Radiochimica Acta 95: 193–207

Novikov AP, Kalmykov SN, Utsunomiya S, Ewing RC, Horreard F, Merkulov A, Clark SB, Tkachev VV, Myasoedov BF (2006) Colloid transport of plutonium in the far-field of the Mayak Production Association, Russia. Science 314: 638–641

Ockenden DW, Welch GA (1956) The preparation and properties of some plutonium compounds. V. Colloidal quadrivalent plutonium. Journal of the Chemical Society 1: 3358–3363

Rai D, Moore DA, Oakes CS, Yui M (2000) Thermodynamic model for the solubility of thorium dioxide in the Na+–Cl–OH–H2O system at 23 degrees C and 90 degrees C. Radiochimica Acta 88: 297–306

Rand MH, Fuger J, Grenthe I, Neck V, Rai D (2008) Chemical thermodynamics of thorium. Elsevier, Amsterdam

Ranebo Y, Hedberg PML, Whitehouse MJ, Ingeneri K, Littmann S (2009) Improved isotopic SIMS measurements of uranium particles for nuclear safeguard purposes. Journal of Analytical Atomic Spectrometry 24: 277–287

Reents WD, GE ZZ (2000) Simultaneous elemental composition and size distributions of submicron particles in real time using laser atomization/ionization mass spectrometry. Aerosol Science and Technology 33: 122

Rothe J, Walther C, Denecke MA, Fanghänel T (2004) XAFS and LIBD investigation of the formation and structure of colloidal Pu(IV) hydrolysis products. Inorganic Chemistry 43: 4708–4718

Shen Y, Zhao YG, Guo SL, Cui JY, Liu Y, Li JH, Xu J, Zhang HZ (2008) Study on analysis of isotopic ratio of uranium-bearing particle in swipe samples by FT-TIMS. Radiation Measurements 43: S299–S302

Shinonaga T, Esaka F, Magara M, Klose D, Donohue D (2008) Isotopic analysis of single uranium and plutonium particles by chemical treatment and mass spectrometry. Spectrochimica Acta. Part B: Atomic Spectroscopy 63: 1324–1328

Singhal A, Toth LM, Lin JS, Affholter K (1996) Zirconium(IV) tetramer/octamer hydrolysis equilibrium in aqueous hydrochloric acid solution. Journal of the American Chemical Society 118: 11529–11534

Somogyi A, Pasilis SP, Pemberton JE (2007) Electrospray ionization of uranyl-citrate complexes: Adduct formation and ion-molecule reactions in 3D ion trap and ion cyclotron resonance trapping instruments. International Journal of Mass Spectrometry 265: 281–294

Stetzer O, Betti M, van Geel J, Erdmann N, Kratz JV, Schenkel R, Trautmann N (2004) Determination of the U-235 content in uranium oxide particles by fission track analysis. Nuclear Instruments & Methods in Physics Research Section A, Accelerators Spectrometers Detectors and Associated Equipment 525: 582–592

Thiyagarajan P, Diamond H, Soderholm L, Horwitz EP, Toth LM, Felker LK (1990) Plutonium (IV) polymers in aqueous and organic media. Inorganic Chemistry 29: 1902–1907

Toth LM, Friedman HA, Osborne MM (1981) Polymerization of plutonium (IV) in aqueous nitric acid solutions. Journal of Inorganic and Nuclear Chemistry 43: 2929–2934

Toth LM, Friedman HA, Osborne MM (1983) Aspects of plutonium (IV) hydrous polymer chemistry. ACS Symposium Series 216: 231–240

Trimborn A, Hinz KP, Spengler B (2000) Online analysis of atmospheric particles with a transportable laser mass spectrometer. Aerosol Sci Techn 33: 191

Tsierkezos NG, Roithova J, Schroder D, Oncak M, Slavicek P (2009) Can electrospray mass spectrometry quantitatively probe speciation? Hydrolysis of uranyl nitrate studied by gas-phase methods. Inorganic Chemistry 48: 6287–6296

Varga Z (2008) Application of laser ablation inductively coupled plasma mass spectrometry for the isotopic analysis of single uranium particles. Analytica Chimica Acta 625: 1–7

Wallenius M, Tamborini G, Koch L (2001) The "age" of plutonium particles. Radiochimica Acta 89: 55–58

Walther C, Rothe J, M. Fuss, S. Büchner, S. Koltsov, Bergmann T (2007) Investigation of polynuclear Zr-hydroxide complexes by nano-electrospray mass-spectrometry combined with XAFS. Analytical and Bioanalytical Chemistry 388: 409–431

Walther C, Fuss M, Büchner S (2008) Formation and hydrolysis of polynuclear Th(IV) complexes – a nano-electrospray mass-spectrometry study. Radiochimica Acta 96: 411–425

Walther C, Fuss M, Buchner S, Geckeis H (2009a) Stability of Th(IV) polymers measured by electrospray mass spectrometry and laser-induced breakdown detection. Journal of Radioanalytical and Nuclear Chemistry 282: 1003–1008

Walther C, Rothe J, Brendebach B, Fuss M, Altmaier M, Marquardt CM, Büchner S, Cho H-R, Yun J-I (2009b) New insights in the formation processes of Pu(IV) colloids. Radiochimica Acta 97: 199–207

Wilm M, Mann M (1996) Analytical properties of the nanoelectrospray ion source. Analytical Chemistry 68: 1–8

Wilson RE, Skanthakumar S, Sigmon G, Burns PC, Soderholm L (2007) Structures of dimeric hydrolysis products of thorium. Inorganic Chemistry 46: 2368–2372

Zhang XZ, Esaka F, Esaka KT, Magara M, Sakurai S, Usuda S, Watanabe K (2007) Application of inductively coupled plasma mass spectrometry to the determination of uranium isotope ratios in individual particles for nuclear safeguards. Spectrochimica Acta Part B: Atomic Spectroscopy 62: 1130–1134

Synchrotron-Based X-Ray Spectromicroscopy of Organic Nanoparticles Complexing Actinides

Markus Plaschke, Jörg Rothe, and Melissa A. Denecke

Abstract Humic acids (HA) have a high binding capacity toward traces of toxic metal cations, including actinides, and have a propensity for forming nano- to micron-sized colloids. They can therefore act as transport vehicles for actinides in aquatic systems, which can have detrimental consequences for the safety of nuclear waste disposal. Modern scanning transmission X-ray microscopy (STXM) at the carbon K(1*s*)-edge is an excellent tool for investigations of such and other processes. STXM studies, combined with other sophisticated techniques and quantum chemical calculations, provide the necessary spatial resolution in the sub-μm range to resolve characteristic aggregate morphologies and identify molecular processes involved in HA–actinide interaction. Systematic experimental and theoretical investigations of reference systems allow successful interpretation of C 1*s* near edge X-ray absorption fine structure (NEXAFS) extracted from STXM image stacks of HA–metal aggregates. The results show that different HA domains (of likely structurally supramolecular HA associations) exhibit different interactions with metal cations. Metal cations bound to HA are enriched in a minority fraction containing higher densities of complexing carboxylic sites. This fraction is likely to play a dominant role in HA colloid-mediated transport of actinides and other toxic trace metals in the hydrosphere.

1 Introduction

The performance and safety assessment of any concept of nuclear waste repository envisaged for the medium- or long-term future will rely on quantified modeling of the individual mobilities of different radionuclide species in the various compartments constituting the geological pathway between the disposal site and the earth's surface. Due to the long half-lives of transuranium elements generated during

M. Plaschke, J. Rothe, and M.A. Denecke (✉)
Karlsruhe Institute of Technology (KIT), Institute for Nuclear Waste Disposal,
P.O. Box 3640, 76021 Karlsruhe, Germany
e-mail: melissa.denecke@kit.edu

S.N. Kalmykov and M.A. Denecke (eds.), *Actinide Nanoparticle Research*,
DOI 10.1007/978-3-642-11432-8_6, © Springer-Verlag Berlin Heidelberg 2011

uranium oxide fuel burn-up in present nuclear power plants, a detailed understanding of their geochemical behavior is crucial. Most repository failure scenarios involve water influx and subsequent container/waste matrix corrosion as initial steps preceding the release of radionuclides/actinides, which can potentially become mobile and reach the environment. Although the aqueous solubilities of actinide elements at typical (neutral to slightly basic) groundwater pH conditions are generally low (e.g., $\log K^{\circ}_{sp} = -14$ for PuO_{2+x}(s, hyd) [1]), actinide interaction with or uptake by nano-sized groundwater particles may enhance their mobility by several orders of magnitude. Colloid-borne actinide transport is a major topic in this compendium. The reader is also referred to the chapter in this book, "Actinide–Nanoparticle Interaction: Generation, Stability and Mobility", for a discussion of actinide–HA interaction and its impact on actinide transport under natural conditions. The present chapter deals with the special case of actinide interaction with water-soluble humic acids (HA). HA are polyfunctional organic macromolecules, which can form strong complexes with metal cations including actinide or chemically homologue lanthanide cations (cf. Sect. 4.1). Recent improvements in the understanding of molecular processes of metal–HA complexation reactions and associated changes in HA morphology attained by combining soft X-ray spectromicroscopy with other advanced techniques and computational chemistry are presented. In section 2, the technique of soft X-ray spectromicroscopy is introduced. The next section briefly summarizes the information content of this technique on hand application examples from different scientific fields. Section 4 describes results from studies of reference systems, which establish spectral peak assignments as basis for the successful interpretation of HA spectral and results from studies of HA interaction with actinide and chemically homologue cations.

2 Synchrotron-Based Soft X-Ray Spectromicroscopy

2.1 Basics and Definitions

X-ray absorption spectroscopy (XAS) has advanced rapidly in the last decades, particularly due to the increasing availability of synchrotron radiation (SR) sources around the world (cf. the compilation in [2]). The increasing source brightness has driven the development of X-ray microscopy techniques which enable XAS studies at high spatial resolution. Experimental stations have been developed for spectromicroscopy (i.e., imaging with spectral sensitivity at tunable wavelength/energy), microspectroscopy (i.e., recording spectra from small sample areas or volumes) or a combination of both. Generally these novel analytical tools are summarized as X-ray spectromicroscopy techniques, emphasizing that the combination of spectroscopy and microscopy is highly beneficial for chemical speciation in heterogeneous materials. The reader is referred to [3] for an up-to-date compilation of technical developments and applications.

There are many problems in numerous fields including material and environmental sciences, which require detailed chemical analysis at a sub-micron spatial scale, e.g., contaminant speciation, morphology determination in polymer blends, interface chemistry in soils and sediments, biomineralization and biofilm formation, to name just a few examples. Traditional chemically sensitive spectroscopy techniques such as infrared, Raman and nuclear magnetic resonance can differentiate chemical species, but they do not have the necessary sub-micron spatial resolution. Electron (SEM, TEM) and scanning probe (AFM) microscopy techniques have excellent spatial resolution to visualize structures down to the atomic scale, but they typically have limited chemical sensitivity for chemical state mapping. Compared to the radiation damage with the use of electron microscopy techniques, the radiation damage using X-ray excitation is reduced by about two orders of magnitude, making it ideally suited for imaging radiation-sensitive samples such as organic polymers or biofilms in their native state.

2.2 Zone Plate Based X-Ray Microscopy Techniques

First attempts to image specimens with synchrotron radiation in the soft X-ray regime (i.e., the electromagnetic spectrum spanning the vacuum ultraviolet to the shorter wavelength, hard X-ray region, where Bragg crystal monochromators become available) relied on simple pinholes [4]. Several focusing devices for soft and hard X-rays have been envisaged and successfully applied in the past decades, including Kirkpatrick–Baez (KB) mirrors [5], polycapillary optics [6], Fresnel zone plates (FZP) [7], and – more recently – compound refractive lenses (CRL) [8]. All these techniques strongly benefit from the small source size, low divergence, and/or partial coherence of X-ray light generated by second- and third-generation synchrotron sources. Nowadays, soft X-ray spectromicroscopy is dominated by FZP focusing optics. FZPs can be conceived of as circular diffraction gratings. Illuminated by a plane wave, an FZP generates a focus at the wavelength-dependent distance $f = 4N(\Delta r)^2/\lambda$ (with the number of zones N, the width of the outermost zone Δr, and the radiation wavelength λ). The diffraction-limited resolution or minimum spot size is given by the Rayleigh criterion as $1.22 \times \Delta r$ – typically a few tens of nm at the carbon K-edge energy (~284 eV, corresponding to λ ~4 nm). FZPs can also be used as near-optical-axis imaging devices, where the point-by-point reconstruction of the object at distance p and the (magnified) image at distance q obeys the thin lens formula: $1/p + 1/q = 1/f$ [7].

Two types of soft X-ray microscopes (XRM) based on FZP lenses have been developed and are routinely used at undulator or bending magnet beamlines (cf. Sect. 2.5). In a full-field transmission X-ray microscope (TXM), a partially transparent specimen is illuminated by a condenser zone plate (CZP). A pinhole at variable distance from the CZP selects the energy to image the object. The radiation transmitted by the specimen is projected by an objective FZP onto a CCD detector, which records a magnified transmission image of the specimen. In the second type,

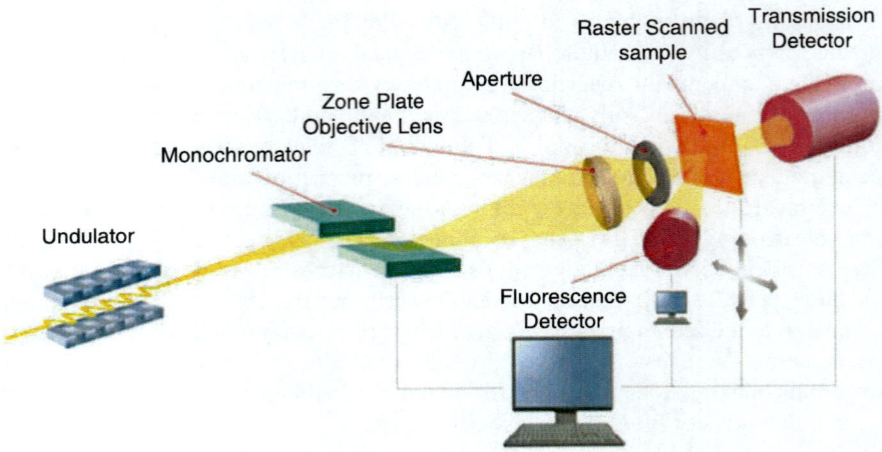

Fig. 1 STXM setup scheme at an undulator synchrotron radiation source (image taken from [9] with kind permission from Xradia company)

scanning transmission X-ray microscope (STXM), a single FZP is used to focus monochromatic X-rays onto the specimen. Images are generated by monitoring the X-ray signal transmitted through a transparent section of the specimen as it is raster-scanned in the FZP focal plane. The principle is shown in Fig. 1. Spectra are either measured by holding the beam at the spot of interest on the sample while scanning the photon energy (micro/nanospectroscopy mode) or by registering stacks of images at consecutive energy positions around the absorption edge of interest (spectromicroscopy mode) and later extracting the spectra from the stack of images. In both microscope concepts, the image contrast is simply generated by the varying sample absorbance, possibly enhanced by intense spectral features, or – more sophisticated – by phase shifts modulating the wave front as it passes through the irradiated material. The latter contrast mechanism requires partially coherent sample illumination and special segmented detectors [10].

For quantitative analysis of XRM transmission data, the intensity behind the sample is converted into the optical density $OD = \ln(I_0/I)$, where I_0 is the incident X-ray flux at a given X-ray energy E, I is the transmitted flux through the sample, and ln denotes the natural logarithm. OD is related to the sample properties by $OD = \mu(E)t$, where $\mu(E)$ is the energy-dependent linear absorption coefficient and t the sample thickness.

Samples investigated by soft X-ray microscopy can be mounted as microtome thin sections on TEM grids (preferentially in vacuum or under helium atmosphere to reduce background absorption) or enclosed between X-ray transparent silicon nitride (Si_3N_4) membrane windows. The latter approach is used to study samples embedded in a thin water film when drying artifacts need to be avoided, e.g., for biological or environmental specimens. Samples must be partly transparent at the chosen X-ray energy, which means a sample thickness of a few tens to a few hundreds of nm is required for carbon studies, and it can reach 1–2 μm for higher energy edges (e.g., transition metal L-edges) or low-density samples. TXM setups make it possible to record images with typically 10–20 μm field of view at a spatial resolution down to 20 nm. The spectral monochromaticity of the CZP/pinhole assembly is typically limited to $\lambda/\Delta\lambda = 500$ [11]. High-energy resolution XAS spectra recorded around or near energies of the atom of interest with $\Delta E \leq 0.1$ eV are obtained by STXM spectromicroscopy/microspectroscopy endstations, where high spectral monochromaticity is achieved by employing spherical grating monochromators in the beamline optics.

2.3 NEXAFS: Near Edge X-Ray Absorption Fine Structure

Interaction of soft X-rays with matter comprises (elastic and inelastic) scattering and direct (photo) absorption. The impinging radiation is attenuated by the sample to an extent which depends on the nature of the specimen, its thickness and density and the photon energy. The absorbed photons can excite inner shell electrons of the atoms in the specimen. Depending on the photon energy, these electrons are promoted to unoccupied energy levels of the absorbing atom or molecule, or they are removed completely to leave behind an ionized state. These short-lived excited states decay under emission of secondary electrons or (fluorescence) photons. Experimentally, one observes element-specific absorption edges at the onset of inner-shell ionization. Fig. 2a depicts the consecutive series of absorption edges obtained for the organic polymer Kapton® (calculated based on data from [12]). The K-edges are associated with excitation of 1*s* inner shell energy levels of C, N, and O between ~284 eV and ~543 eV. The different "edge-heights" reflect the atomic composition and element-specific photo-absorption cross-section. Therefore, the amount of a particular element in the sample can be determined quantitatively from the difference in the absorption above and below the edge. While this aspect of X-ray absorption provides the basis for elemental analysis, the chemical and electronic structure of the excited element can also be identified from its NEXAFS (cf. the comprehensive introduction to NEXAFS spectroscopy and its

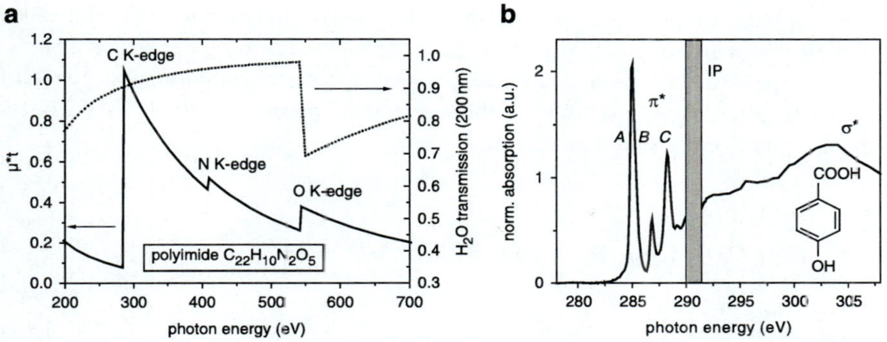

Fig. 2 (**a**) *Solid line*: carbon, nitrogen and oxygen X-ray absorption K-edges for polyimide (Kapton®), $t = 200$ nm; *dotted line*: transmission of a 200-nm-thick water layer. (**b**) C 1*s*-NEXAFS of 4-hydroxy benzoic acid (*inset*): *A* aromatic, *B* phenolic and *C* carboxylic π^* transitions

many applications by J. Stöhr [13]). NEXAFS spectra exhibit a series of peaks or resonances in the vicinity of the rising edge, which may strongly modulate the step function like behavior of the X-ray attenuation cross-section. These features generally reflect transitions of the excited photoelectron to unoccupied electronic states associated with the absorbing atom species in a specific bonding environment. As for all spectroscopy techniques based on photon-induced electron transitions, promotion of core electrons into (quasi) bound excited states is governed by (dipole) selection rules. They are the basis for NEXAFS sensitivity to molecular symmetry. As the incident X-ray energy approaches the ionization potential (IP, Fig. 2b), the first NEXAFS spectral features are associated with excitation into the lowest unoccupied molecular orbitals (LUMO), e.g., π^* orbitals for unsaturated carbon bonds. These are followed by structures associated with higher energy molecular states, typically of σ^* character associated with saturated (single) chemical bonds. The IP position for inner-shell ionization is generally masked between the π^* and σ^* transition features. Its direct determination requires additional techniques such as photoelectron spectroscopy (XPS).

NEXAFS spectra of unknown species are often analyzed based on a simple "fingerprint" approach or "functional group fingerprinting" by comparison with spectra obtained for pure, well characterized standards [14]. In many cases, enough is known about the relation between chemical structures or molecular (sub-)units and NEXAFS spectral features to allow identification of unknown species from measured NEXAFS spectra – an approach designated as "building block principle", where the NEXAFS of an unknown is considered a linear combination of spectra derived from simpler subunits (e.g., benzene and formic acid as building blocks for benzoic acid). These approaches have been especially successful in the analysis of NEXAFS data obtained for polymer systems (e.g., [15, 16]). Attempts to establish a theoretical foundation for the "building block principle" by "molecular modeling" calculations have been reported for selected systems such as hydroxyl-, carboxyl-, or halogen-substituted benzenes [17–19]. Individual spectral features, particularly the low energy π^* transitions, are often sufficient for qualitative identification in reasonably well-characterized systems. However, some experimentally observed peak positions, peak shapes, or line widths cannot be unambiguously interpreted based on these simple methods. This limitation is especially obvious for molecules with electronically interacting functional groups or for metal complexing ligands. The interpretation of such experimental spectra can be supported by quantum chemical calculations. Examples of these different approaches for NEXAFS interpretation will be presented in Sect. 4.

2.4 *Wet Cells and the Water Window*

Many problems where chemical state imaging by soft X-ray spectromicroscopy is applied demand the investigation of hydrated specimens. For instance, spectral analysis of environmental or biological samples may require the presence of a thin water layer or at least high humidity to keep the specimen in its native state. XRM endstations operated at soft X-ray energies generally allow to retain samples in a relatively thick water layer. As illustrated in Fig. 2a, the transparency of water is high at energies just below the oxygen K-edge near 543 eV and drops off above this value. At photon energies needed for spectroscopy of organic, carbonaceous material, water is sufficiently transparent to tolerate thicknesses of up to ~0.5 μm. Accordingly, the spectral region from ~284 eV (C 1*s*) to ~543 eV (O 1*s*) is often referred to as the "water window". A special setup is used for experiments with aqueous solutions and suspensions both for full-field and scanning XRMs. The "wet cell" consists of two silicon nitride membranes of 100–200 nm thickness and about 1–3 mm width, in between which a microliter volume droplet of sample solution is sandwiched. A water layer of optimum thickness allowing carbon K-edge spectroscopy is formed due to capillary forces spreading the droplet between the membranes. The cells are either directly sealed by glue or grease, or they are mounted into an O-ring sealed sample holder [20]. Although brittle, the Si_3N_4 membranes trap water for hours and prevent the specimen from drying during spectral image stack data acquisition.

2.5 *Sources and Instrumentation*

A comprehensive overview on the technical development of (soft) X-ray spectromicroscopy was recently presented by Kirz and Jacobsen [21]. Schmahl et al. from Göttingen University built the first TXMs employing FZPs which were used both with laboratory X-ray sources and SR at BESSY-I [22]. The STXM instrumentation was originally developed by Kirz and Jacobsen from SUNY-Stony Brook, New York [23] and there are three STXMs at the X1A undulator beamline at the NSLS in Brookhaven, New York [24]. Further soft X-ray TXM and STXM instruments are available at the Advanced Light Source (ALS) at Berkeley, California [25], the Swiss Light Source at PSI, Villigen [26], the Canadian Light Source [27] at Saskatoon, British Columbia, and BESSY-II at HZB in Berlin [28]. ELETTRA at Turino, Italy, operates TWINMIC, a novel XRM aiming at integrating both TXM and STXM microscope types in a single instrument with easy switching capabilities between the two modes [29]. The increasing demand for soft X-ray microscopy instrumentation has meanwhile encouraged commercial suppliers such as Xradia [9] and Accel/Bruker ASC [30] to develop, manufacture, and install TXM and STXM stations for laboratory and SR applications.

3 STXM/NEXAFS at the Carbon K-Edge: Selected Applications

Numerous C 1*s*-NEXAFS/STXM investigations of carbonaceous materials have been reported in the past two decades. These studies focused on various aspects of material, environmental and life sciences, including (multicomponent) polymer films and fibers, organic geochemical materials such as coals and soils, biological materials like proteins and biofilms and even extraterrestrial materials collected from meteorites and comets [31]. The following section is intended to exemplify typical carbon K-edge XRM applications – it is not intended to be a comprehensive review on this active field; this would go far beyond the scope of this chapter.

3.1 Polymer and Material Science

Materials of high commercial significance in the polymer industry are usually multicomponent systems composed of one or several elastomers and various fillers or modifiers. The complex morphology that can arise from blending organic materials into such multicomponent systems affects their physical and chemical properties. In many applications, the properties of polymer systems can be specifically tailored by altering the chemical and morphological structure. Since the groundbreaking work by Hitchcock, Ade and Urquhart (e.g., [15, 32–34]), STXM in combination with C 1*s*-NEXAFS spectroscopy has been a well-established tool for the characterization of organic polymer materials that may be difficult or impossible to elucidate with other techniques. In addition to merely imaging or quantifying carbon species distribution in polymer blends, X-ray linear dichroism measurements (i.e., determination of varying absorbance of the monomers as a function of their orientation relative to the polarized SR field) can be exploited to determine the orientation of polymer moieties at high spatial resolution [35].

A recent example from the Ade group deals with polymer solar cells, which are the subject of intense research due to their promise as energy and cost-efficient alternatives to classical silicon technology [36]. Currently, the most advanced polymer solar cell designs are based on bulk structures of poly(3-hexylthiophene) (P3HT) and (6,6)-phenyl C_{61} butyric acid methyl ester (PCBM) [37]. The electron donor material P3HT and the electron acceptor material PCBM are blended together in order to optimize exciton separation and charge transport. Thermal annealing increases phase segregation of both components, which improves the hole transport within the polymer film and thus the device efficiency. STXM was applied to gain a detailed understanding of the diffusion and aggregation properties of PCBM molecules within the P3HT matrix. The depletion of PCBM in the vicinity of large PCBM crystals that form during P3HT:PCBM thermal annealing was measured. C 1*s*-NEXAFS spectra of pure P3HT and PCBM films exhibit strong resonances with the highest absorbance for P3HT at 285.1 eV (thiophene ring C 1*s* (C–H) $\rightarrow \pi^*_{C=C}$ transitions) and for PCBM at 288 eV (C 1*s* (COOR) $\rightarrow \pi^*_{C=O}$ transitions). Thus, STXM images taken at these photon energies provided the chemical contrast to derive a quantitative distribution of the two components. The PCBM concentration profiles around the crystals were fitted to Fick's second law of diffusion and the corresponding diffusion constant and the temperature-dependent PCBM equilibrium concentration in the P3HT matrix could be determined.

3.2 Microbiology

Antimicrobial agents can contaminate aquatic ecosystems if they are not efficiently removed by wastewater treatment [38]. Many of these compounds retain their biological activity after release into the environment and disrupt the balance of ecosystems such as biofilms. Understanding the ecological impact of an antimicrobial agent on natural biofilms requires quantitation of accumulated amounts of agent and its spatial distribution, identification of the biochemistry of active sites responsible for its sorption and characterization of its impact on the biofilm structure. The Hitchcock group has recently shown how soft X-ray STXM spectromicroscopy can provide chemical species mapping needed to address these questions [39].

Chlorhexidine (1,1′-hexamethylene-bis[5-(*p*-chlorophenyl) biguanide]) is one of the most widely used antimicrobial agents for, e.g., disinfection of skin and surgical instruments [40]. The spatial distribution of chlorhexidine following exposure of a river biofilm to chlorhexidine-contaminated river water was studied by a combination of STXM and confocal laser scanning microscopy (CLSM, cf. Sect. 4.4). The distribution of specific fluorescence markers imaged by CLSM was used to characterize microbial species and extracellular polymer substances (EPS) in the biofilm. STXM image sequences between 280 and 320 eV were used to quantitatively map organic substances (chlorhexidine, proteins, lipids, polysaccharides) and inorganic species (K^+, CO_3^{2+}, SiO_2) in diatom algae and bacterial colonies in the biofilm. C 1*s*-NEXAFS spectra of biofilm organic species reference substances and the disinfection agent are depicted in Fig. 3a. Fig. 3b shows an STXM image of a hydrated biofilm imaged in a wet cell. The image was recorded at 288.2 eV, an energy where microbial organisms are preferentially visualized due to absorption by the C 1*s* → $\pi^*_{C=O}$ peptide band of proteins. A diatom is visible adjacent to the left of the bacterial colony. It appears bright due to the strong (but featureless) absorption of SiO_2 at this photon energy. The spectrum extracted from a central region of the bacterial colony (indicated with a white dot in Fig. 3b and labeled "chlorhexidine-internal" in Fig. 3a) shows excellent agreement with the spectra of chlorhexidine references in the energy range of the characteristic C 1*s* → π* transitions (284–289 eV). Based on these features, chlorhexidine is readily distinguishable from the biological macromolecules present in the specimen. The chemically specific imaging capabilities of STXM showed explicitly that chlorhexidine bioaccumulates in both intra- and extracellular structures of the diatom algae and in intracellular components of the bacteria, preferentially associated with lipids present in the microorganisms.

3.3 Environmental Science

Environmental samples are typically heterogeneous in their elemental and chemical composition on any length scale down to the molecular level. The intimate mixture of species, structures and compartments, which determine speciation, transport and

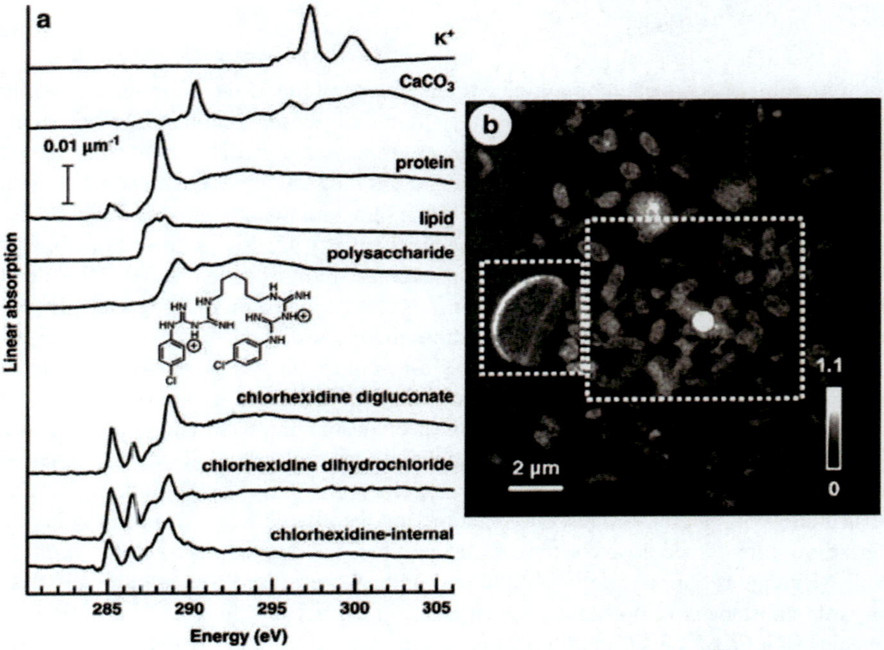

Fig. 3 (**a**) NEXAFS spectra of pure K^+ (difference spectrum of K_2CO_3 and $CaCO_3$), $CaCO_3$, protein, lipid, polysaccharide, chlorhexidine digluconate and chlorhexidine dihydrochloride. The spectrum "chlorhexidine-internal" was extracted from a 500 nm region (*white dot*) in Fig. 3b. (**b**) STXM image recorded at 288.2 eV of a diatom (*small rectangle*) and adjacent bacterial colony (*large rectangle*) in the chlorhexidine exposed biofilm (reprinted from [39] with kind permission from Elsevier)

bioavailability of nutrients, pollutants or contaminants, is highly demanding for any method providing molecular-level process understanding of environmental phenomena. Molecular environmental science (MES) has been established as an interdisciplinary tool to study microbiological, chemical and geological processes at surfaces and interfaces. From the onset of this new scientific approach, SR-based methods – especially soft X-ray spectromicroscopy – have played a crucial role in the attempt to correlate chemical and morphological information obtained for environmentally relevant specimens (cf. the recent reviews by Yoon [41] and Thieme et al. [42]).

X-ray microscopy images are usually two-dimensional projections of complex three-dimensional structures. Soil or sediment samples represent typical specimens studied by environmental scientists. They are composed of a large variety of organic and inorganic colloidal constituents. A two-dimensional projection image lacks the ability to clarify the exact spatial arrangement of such colloidal structures. Therefore, soft X-ray tomography reconstructed from TXM images can be applied to visualize the complete three-dimensional morphology of soil samples. As an

Fig. 4 The reconstruction of a tomographic data set of 50 TXM images taken above the carbon K-edge spanning 180° shows the network-like structure of microbially reduced humic substances (*left*) and the change in appearance to isolated flocks after reoxidation (*right*). The spatial resolution is ~50 nm (images reproduced from [43] with kind permission from http://pubs.acs.org)

example, Fig. 4 depicts images taken from a study by Thieme et al. [43] where the ALS soft X-ray TXM was used to investigate morphological changes of originally anaerobic humic substances (HS) upon reoxidation in air. Projection images from HS flocks enclosed in glass capillaries were recorded for each sample over an angular range of 180° to reconstruct the spatial arrangement of the organic material. Starting from a network or sponge-like appearance at anaerobic conditions (left image), the HS flocks form larger isolated particles when contacted with oxygen (right image).

4 Organic Nanoparticles as Actinide Carriers

4.1 Impact of Humic Substances on the Migration of Actinides in the Geosphere

Humic substances (HS) can be described as refractory, chemically heterogeneous and structurally ill-defined, dark-colored organic compounds left over from biodegradation of living organisms. This assembly of decomposition products belonging to the most widely distributed organic material on earth contributes to various processes in the geosphere, including sequestration from soil, redox chemical transformation reactions, mobilization and transport of organic pollutants or trace metal contaminants in natural aquifers. In general, HS can be characterized by their origin, composition, properties or by the method for their isolation. The latter characterizes HS based on the solubility of different fractions under acidic or alkaline pH conditions (a) humins, the insoluble fraction under all conditions; (b) humic acid (HA), isolated and separated by flocculation in acid medium (generally pH 2 or lower); and (c) fulvic acid (FA), the fraction soluble under all pH conditions. Even though differences in solubility of HS fractions are determined by their chemical composition and physical properties, this classification remains operational and does not allow any conclusions about their molecular structure.

In the search for a general structural model, HA appear in some respects "chameleon-like" by adapting their "molecular" structure and physicochemical behavior to the chemical environment. One early concept adapted from polymer science – the "polymer model" – describes HA as randomly coiled macromolecules with elongated shapes under basic or coiled shapes under acidic pH conditions. This model is in accordance with the colloidal properties of HA suggesting hydrodynamic radii in the range of a few nm and average molecular masses in the range of some 10 kDa. However, neither the coiling/uncoiling process could be directly observed nor any kind of "humic molecule" with a unique sequence of structural subunits identified until now. Recent microscopic and spectroscopic investigations point to a new model – that of the "supramolecular association" – where relatively small and chemically diverse organic molecules (including biomolecules) associate to clusters, held together by various types of chemical bonds, e.g., van der Waals, H-bonds or metal bridges. This new concept allows for varying structures on a molecular level depending on the chemical milieu (e.g., pH, ionic strength, metal cations) by association or disassociation of smaller entities. In addition, this model emphasizes that key properties of HA arise mainly through intermolecular interactions, not necessarily molecular composition. In a corollated concept – that of the "micellar structure" – HS are thought to be amphiphilic organic molecules, which arrange in water to form HA micelles with hydrophilic "head" regions in contact with surrounding water and the hydrophobic tail regions orientated toward the micelle center.

HA themselves clearly demonstrate colloidal properties and also undergo important colloidal interactions with other organic or inorganic substances and at solid–liquid interfaces. Specifically, they not only interact with other naturally occurring materials (such as clays or iron oxide/hydroxides) or mineral surfaces, but also demonstrate binding to man-released toxic components such as pesticides or heavy metal ions. As a consequence, HA colloid-associated pollutants can be mobilized and transported potentially over long distances.

A repository design for nuclear waste disposal in deep geological formations includes redundant engineered and natural barriers that ensure long-term containment in waste packages, limited groundwater access to the radioactive substances, and retardation of radionuclide transport to the geosphere. A key criterion for an appropriate disposal strategy is the capability for long-term isolation of long-lived radioelements, e.g., actinide elements. Actinide mobility strongly depends on the chemical state of species in groundwater and the geochemical characteristics of related surfaces, e.g., host rock minerals and engineered barrier materials. Key factors are charge differences between surfaces and solution species and specific chemical interaction affinities. In contact with groundwater, most mineral surfaces are negatively charged and actinide cation mobility is highest if its solution species are negatively charged as well. Therefore, typical mobile species are complex ions or negatively charged colloids, including HA. In contrast, positively charged species tend to adsorb on mineral surfaces. The geochemical stability of positively charged species decreases with increasing charge. Elements in higher oxidation states can form oxo-ions or oxo-/hydroxo "eigencolloids" thereby transforming them to lower charged species. Positive ions can also decrease their charge and increase their mobility by complexation with inorganic ligands (e.g., carbonate) or adsorption to colloids, including HA colloids.

The importance of understanding actinide–HA interaction in nuclear repository safety is obvious. Despite the fact that a consistent HA structural model on a molecular level is currently not available, considerable progress has been made in understanding and modeling HA behavior in both lab-scale experiments and the environment. The role of proton-exchanging functional groups such as carboxylic or phenolic groups has been investigated in order to describe HA (or FA) actinide complexation behavior. Spectroscopic studies reveal carboxylic groups as the primary HA metal ion binding sites [44–46]. Characterization of the number and density of these proton-exchanging binding sites is accessible by titration experiments. These values are correlated to the metal loading capacity, a key feature in the description of HA metal cation interaction according to the so-called charge neutralization model. Experimental observations such as varying complexation strength (described in terms of stability or complexation constants) or quasi-irreversible reaction kinetics observed for certain HA fractions (sometimes associated with empirical "strong and weak" binding sites) are typically attributed to the high structural and chemical diversity of HA. Note that bulk characterization techniques merely provide averaged data, thereby neglecting the inherent HA heterogeneity. Due to this heterogeneity and adaptive behavior, a scientific strategy utilizing XRM, combined with other sophisticated analytical, spectroscopy, and microscopy techniques, is a promising avenue for advancing our understanding of HA and actinide–HA interaction.

In the following, such an experimental approach for HA investigations is presented, which provides structural and chemical information on a (sub-) micrometer scale. Integral to this approach are the application of a combination of complementary experimental techniques, comparative investigations of well-defined reference systems, and accompanying quantum chemical calculations to provide a foundation for sound interpretation of spectral features of spectroscopic results obtained in HA/metal cation interaction experiments.

4.2 Simple Organic Molecules as Building Blocks

In Sect. 2.3, concepts for the analysis of C 1*s*-NEXAFS spectra were introduced: "functional group fingerprinting", the "building block principle", and "molecular modeling" [13]. In this section, the NEXAFS spectra of a set of model substances of carboxyl- and hydroxyl-substituted benzenes are presented. Carboxyl and phenol groups are important candidate HA functional groups for actinide binding. These model substances help identify peak assignments (fingerprinting) and the applicability of the building block principle in NEXAFS spectroscopy of HA.

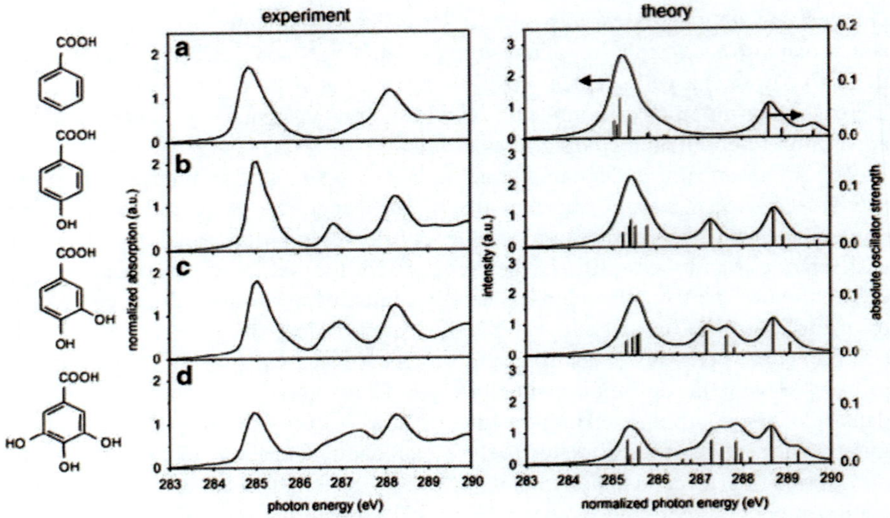

Fig. 5 Normalized experimental (*left-hand side*) and calculated (*right-hand side*) C 1*s*-NEXAFS spectra of benzoic (**a**), 4-hydroxybenzoic (**b**), protocatechuic (**c**), and gallic acid (**d**) (image taken from [19] with kind permission from Elsevier)

Experimental and theoretically calculated C 1*s*-excitation spectra of benzoic acid and hydroxylated benzoic acids are shown in Fig. 5. Details of the ab initio computations are discussed in the literature [19]. Theoretically calculated energy positions and oscillator strengths for the discrete transitions are plotted as bars in Fig. 5. The spectral envelope is generated by convoluting the discrete transition values with Lorentz functions of 0.5 eV width (full width at half maximum, FWHM) [19]. The two prominent features in the experimental benzoic acid spectrum (Fig. 5a, left) are the "aromatic" peak (C 1*s*(ring) → π*) at 284.9 eV and the "carboxylic" peak (C 1*s*(COOH) → π*) at 288.2 eV. The computational results distinguish for the aromatic peak excitations of all six ring C 1*s* electrons into the lowest π* orbital, all in a narrow energy region between 284.9 and 285.5 eV. The carboxylic excitation energy is calculated to be 288.6 eV. The difference between the carboxylic and the aromatic peak positions of the calculated spectral envelope (3.4 eV) is in excellent agreement with the experimental finding (3.3 eV). Above the carboxylic peak, photon absorption is mainly due to excitations of C 1*s*(ring) electrons to higher π* orbitals.

The computations reproduce one, three, or five distinct hydroxyl C 1*s*(OH) → π* transitions for 4-hydroxybenzoic, protocatechuic and gallic acid spectra, respectively, lying between the aromatic and carboxylic C 1*s* excitations in the range of 287.2–288.2 eV. While the energy splitting of the C 1*s*(OH) excitations in protocatechuic and gallic acid is not completely resolved in the experimental spectra, a corresponding broadening is clearly visible in Figs. 5c, and d (left). In the 4-hydroxybenzoic acid NEXAFS, the theoretical and experimental energy shift between the

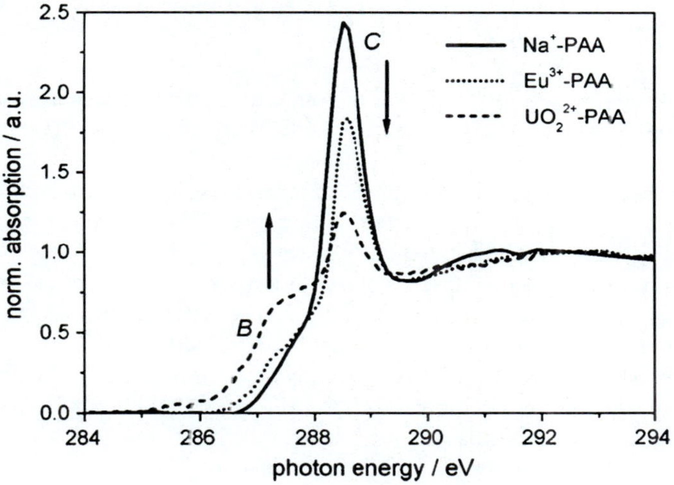

Fig. 6 Normalized experimental C 1*s*-NEXAFS spectra of Na–PAA and PAA complexed with Eu^{3+} and UO_2^{2+} (figure taken from [47] with kind permission from Elsevier)

relatively sharp hydroxy C 1*s*(OH) → π* and the aromatic C 1*s*(ring) → π* peak are in perfect agreement (both 1.8 eV).

The experimental NEXAFS peak intensities change with increasing OH-substitution of the hydroxylated benzoic acids in Fig. 5. The aromatic peak decreases, the phenolic peak increases,and carboxylic peak intensity remains essentially constant. These trends are validated by ab initio calculations, similar to the previous calculations for multiple fluorine-substituted benzenes [17]. However, the most important result is the reproduced splitting of transitions associated with phenolic C in the presence of neighboring OH-groups. Obviously electronic interaction of neighboring OH-groups influence the corresponding C 1*s*(OH) → π* transitions leading to a superposition of a number of peaks, as confirmed here by theory (Fig. 5d, right). Such spectral complexity for relatively simple molecules with a limited number of functional groups and structural subunits suggests that mere building block interpretation of C 1*s*-NEXAFS for complex macromolecules with unknown chemical structures such as HA, i.e., considering them as combination of spectra of smaller organic units, may be insufficient more often than not.

4.3 Polyacrylic Acid as Model Ligand

Polyacrylic acid (PAA) is a water-soluble polymer with a high content of carboxyl functional groups attached to an aliphatic chain and has found widespread application to mimic HA metal ion complexation behavior [48]. Figure 6 shows the normalized C 1*s*-NEXAFS spectra of Na^{+}-, Eu^{3+}-, and UO_2^{2+}-PAA complexes

(Eu^{3+} is chosen as a chemical homologue for trivalent actinide ions). The spectra of PAA are dominated by the C 1*s* (COO^-) → $\pi^*_{C=O}$ transition at ~288.4 eV (peak *C* in Fig. 6). Complexation of polyvalent metal ions (M^{n+}) with these polymers induces a decrease of peak *C*, while at the same time a shoulder appears at slightly lower energy (~0.5–1 eV), adjacent to the carboxylic transition (shoulder *B* in Fig. 6) [49].

Similar spectral behavior is observed for several HA/M^{n+}–HA samples (cf. next section) [62]. Moreover, the extent of the decrease of the carboxyl resonance and corresponding increase of the shoulder at lower energy observed for M^{n+}-PAA has been shown to depend on the type of metal cation and on the chemical conditions. Figure 7 depicts the C 1*s* NEXAFS of Zr(IV)-PAA at different pH. Zr(IV) serves as a chemical homologue for tetravalent actinides, e.g., Pu(IV) [50].

The C 1*s*-NEXAFS of Zr(IV)-PAA strongly depends on the solution pH (Fig. 7). At pH 5, the spectrum of uncomplexed PAA is observed, with a NEXAFS comparable to the Na-form. Similar to the spectral changes observed for metal–HA complexation in Fig. 6, the carboxyl peak intensity decreases and a shoulder adjacent to the carboxyl peak appears at pH 4.3 (see arrows in Fig. 7), indicating complex formation at this pH. This shoulder becomes more pronounced at pH 3. These trends are interpreted qualitatively as follows: at pH 5, Zr(IV) hydrolysis and oxo/hydroxide formation is dominant. PAA merely coats and coprecipitates [51] with these inorganic particles; no specific metal organic interaction is discernible from the C 1*s*-NEXAFS. This interpretation is corroborated by Zr K-EXAFS measurements (unpublished results), giving evidence for Zr(IV) colloid formation under these conditions. Decreasing the pH leads to an increased fraction of ionic Zr(IV), which can be complexed by PAA carboxyl groups. This reaction becomes

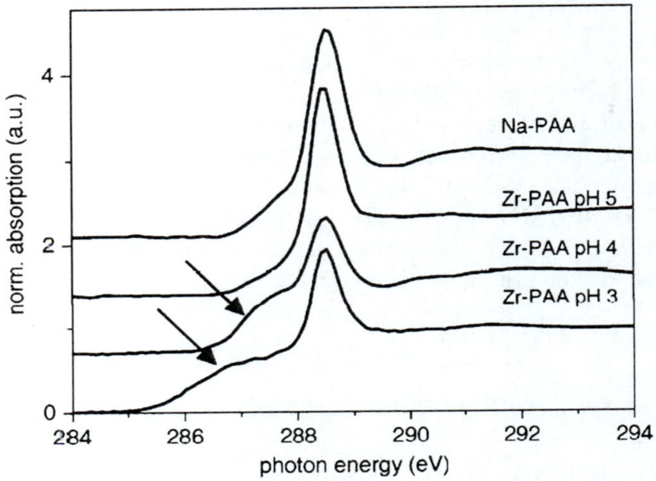

Fig. 7 Normalized C 1*s*-NEXAFS for Na-PAA and Zr^{4+}-PAA at different pH (figure taken from [49], with permission from the American Institute of Physics; Copyright 2007). *Arrows* indicate the shoulder forming upon metal–PAA complex formation (cf. Fig. 6)

more favorable at pH 3, indicated by the stronger shoulder associated with complexation in the C 1*s*-NEXAFS.

To correctly interpret the observed metal ion complexation effects in the C 1*s*-NEXAFS, i.e., C 1*s* (COO^-) → $\pi^*_{C=O}$ intensity decrease and new peak formation as a shoulder at its low energy flank, ab initio calculations of the core excitation spectra were performed. Quantum chemical calculations of geometrically optimized metal acetate complexes, which serve as the simplest structural metal–HA models, and the energies of their C 1*s* (COO^-) → $\pi^*_{C=O}$ transitions are performed using the RI-ADC(2) approach implemented in the TURBOMOLE program package [47]. Results show that the calculated spectra of acetate complexes with various metal cations exhibit a shift of the C 1*s* (COO^-) → $\pi^*_{C=O}$ transition to lower energies compared to the free acetate anion (Fig. 8). This is defined as the complexation shift Δf, the value of which is found to correspond well with the energy difference between the peak *C* and the feature *B* in Fig. 6.

The ab initio calculations reveal that structural modifications of the ligand following metal–cation complexation (i.e., the O–C–O bond angle) and the type of binding mode of the carboxylate groups to the metal centers (bidentate versus monodentate) determine the magnitude of Δf, with only small contributions affected by metal center electronic structure or hydration. Although the calculation

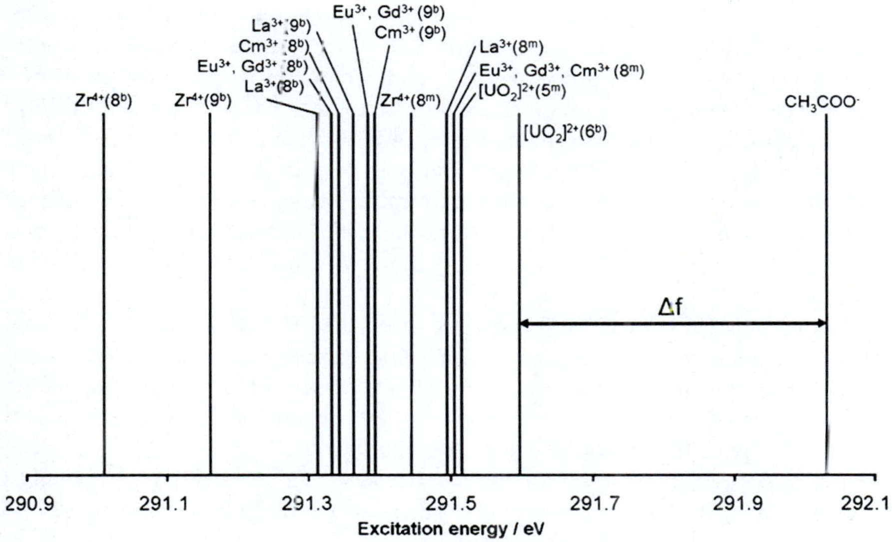

Fig. 8 Calculated C 1*s* (COO^-)→$\pi_{C=O}$* transition energies in eV for the free acetate ion (CH_3COO^-) and for the metal cation–acetate complexes indicated each having different coordination structures. The *numbers in parentheses* after the element symbols represent the metal coordination number; subscripts *b* and *m* indicate bi- and monodentate acetate bonding, respectively. The absolute values of the theoretically calculated excitation energies are not the same as experimental values, but the complexation shift, Δf, compares well with experiment (figure taken from [47], with permission from Elsevier)

of C 1*s* core excitation spectra of complex organic macromolecules (PAA or HA) is not yet possible with currently available quantum chemical methods, selected spectral features can be reproduced for structurally well-defined subunits. As calculated Δf values for carboxyl transitions in metal acetates at the ab initio level (in the range of 0.44–1.02 eV) correspond well to the experimental values obtained for PAA/M^{n+}–PAA (0.5–1 eV), these calculations allow assignment of the experimentally observed shoulder on the carboxyl resonance in C 1*s* NEXAFS of M^{n+}-PAA samples to an energy shifted C 1$s \rightarrow \pi^*_{C=O}$ transition (increase in C electron density) associated with M^{n+} complexation to carboxylic acid functionals. This assignment indirectly implies that the observed decrease in organic acid carboxyl peak intensity upon complexation with M^{n+} occurs because a portion of its original intensity becomes associated with the new lower energy transition (feature *B* in Fig. 6).

4.4 In Situ Studies of Purified Humic Acid

As part of the overall assessment of the impact of HA on actinide transport phenomena, characterization of both the nature of HA/actinide ion complexes and HA hetero-aggregates with actinide oxo/hydroxide colloids are mandatory. STXM/NEXAS studies have great potential to provide insight into HA interactions with trace metal ions (Me^{n+}) and inorganic colloids [52–55]. In the following, commercial HA, as well-characterized model HA, its interaction with Me^{n+} and its formation of hetero-aggregates with other colloids is studied. The utility of NEXAFS to identify and characterize the primary nature of HA–metal cation interaction gained from investigations of model systems and presented in the previous subsections will assist the interpretation of the results from these studies of real, heterogeneous HA. The commercial HA (obtained from Aldrich, Deisenhofen, Germany) was purified from inorganic constituents according to a procedure described in the literature [56].

The normalized C 1*s*-NEXAFS spectra of purified HA reacted with U(VI) and Th(IV) oxo/hydroxide [57] are depicted in Fig. 9. For comparison, the HA spectrum before metal loading is plotted in Fig. 9 as dashed line. The corresponding morphologies of the aggregates forming upon HA–metal interaction are compared in Fig. 10. Figure 10a shows the overall morphology of U(VI)–HA aggregates to be very similar to those reported for Eu(III)–HA (see Fig. 11b below) [55]. Patches with higher optical density are embedded in an optically less dense matrix of carbonaceous material. The spatial distribution of carboxyl functional groups in the U(VI)–HA aggregate is better visualized by calculating ratio images ($-\log [I(\Delta E)/I(\Delta E0)]$, $\Delta E0 = 280$–283 eV, the spectral energy range for the background and $\Delta E = 287.8$–289.2 eV, the range for the C 1$s \rightarrow \pi^*_{C=O}$ transition; Fig. 10b, c). The characteristic white spots in the ratio images represent dense HA material of high carboxylic functionality. At least two different spectral signatures characterize the C 1*s*-NEXAFS of U(VI)–HA: signature I (Fig. 9a), exhibiting a strong

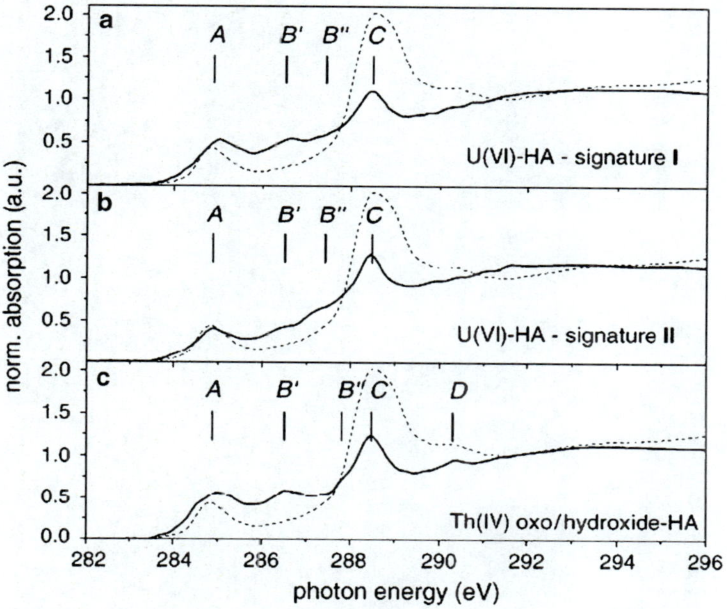

Fig. 9 Normalized C 1*s*-NEXAFS spectra of (**a**) U(VI)–HA extracted from aggregate in Fig. 10b and (**b**) from aggregate in Fig. 10c, and (**c**) from HA-coated Th(IV) oxo/hydroxide depicted in Fig. 10d. The spectrum of HA before metal loading (*dashed line*) is plotted for comparison (figure taken from [57], with permission from Elsevier)

resonance peak B' and signature II (Fig. 9b), with features marked B' and B''. The peak position B' (286.6 eV) coincides with the phenolic C 1*s* (C–OH) $\rightarrow \pi^*_{C=C}$ transition obtained for hydroxybenzoic acid (Fig. 5). In analogy to the feature near 287.5 eV appearing in Me^{n+}-PAA NEXAFS (Figs. 6 and 7), feature B'' is attributed to C 1*s*$\rightarrow \pi^*_{C=O}$ transitions following Me^{n+}-complexation. Signature II also exhibits a decrease of the carboxyl resonance (marked C in Fig. 9), as one would expect for a shift of a portion of its original intensity to the complexation feature B''.

The NEXAFS of HA reacted with Th(IV) oxo/hydroxide present at pH 5.3 (Figs. 10c and 11d) exhibits strong similarities to the NEXAFS previously recorded for HA coprecipitated with Eu(III) oxo/hydroxide particles [58]. The faint structures observed to surround the hetero-aggregate in Fig. 10d are at the STXM resolution limit of around 100 nm. Only a single NEXAFS signature is observed, irrespective of the location in the image stack and the material density within the hetero-aggregates where the NEXAFS is extracted. The additional weak resonance at 290.4 eV in the spectrum (position *D* in Fig. 9c) indicates the presence of carbonate in the sample [14]. A carbonate peak was also observed in C 1*s* NEXAFS for Eu(III)-oxo/hydroxide particles [58]. The B'' position (287.8 eV) for Th(IV) oxo/hydroxide reacted with HA is apparently higher than the energy of B'' for U(VI)–HA (287.4 eV). Further investigation of reference systems and comparison

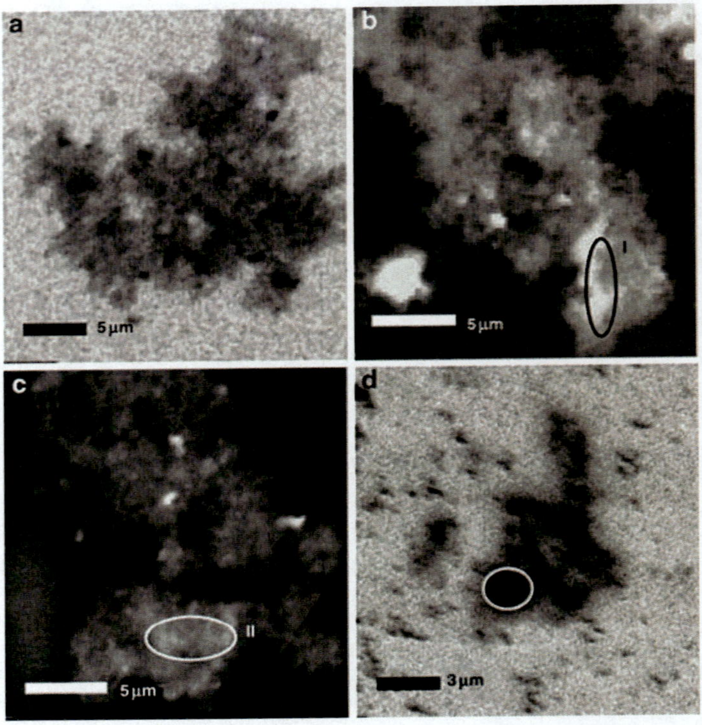

Fig. 10 (**a**) STXM-image of U(VI)–HA recorded at 288.5 eV; (**b, c**) STXM ratio images of U (VI)–HA aggregates in the spectral region of the carboxyl resonance; (**d**) STXM-image of HA coated Th(IV) oxo/hydroxide at 288.5 eV; C 1*s*-NEXAFS signatures depicted in Fig. 9 are extracted from encircled areas (figure taken from [57], with permission from Elsevier)

to theoretical models are necessary before this difference can be attributed to any specific HA interaction with the Th(IV) oxo/hydroxide species. When secured, such information can be potentially used to support interpretation of other HA-(eigen-) colloid systems.

Aggregates formed upon reaction of Eu(III) with purified HA are investigated in situ by laser scanning luminescence microscopy (LSLM) combined with STXM/NEXAFS. In Fig. 11, LSLM- and STXM-micrographs of Eu(III)–HA aggregates are compared. Although similar at first glance, the contrast mechanism in these images is based on different physical processes, i.e., optical Eu(III) luminescence emission (LSLM) and specific C 1*s* level X-ray absorption (STXM) [59]. The STXM image in Fig. 11b exhibits dark areas of carbonaceous material with high optical density, similar to the U(VI)–HA aggregates in Fig. 10a. Dark areas in the LSLM image correlate with areas showing relative high Eu(III) luminescence intensity ($^5D_0 \rightarrow {}^7F_{1,2}$ transitions). Both microscopy techniques reveal similar aggregate morphologies: confined dark patches with diameters ranging from 200 nm to 2 µm embedded in a lighter matrix. Although STXM and LSLM

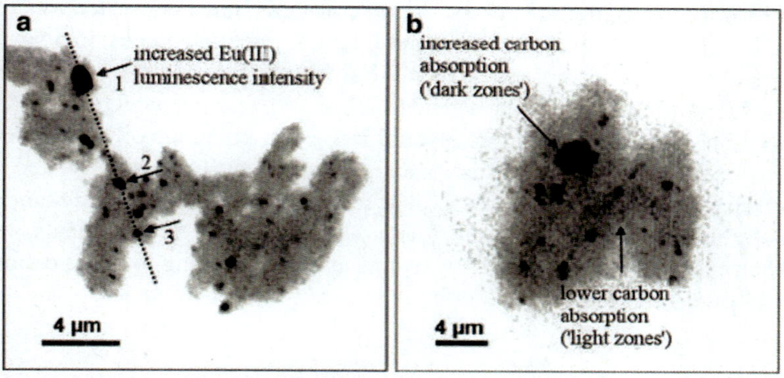

Fig. 11 Comparison of (**a**) Eu(III) luminescence intensity distribution in Eu(III)–HA by LSLM (excitation wavelength 393.5 nm) and (**b**) specific carbon absorption by STXM (recorded at 288.4 eV) of Eu(III)–HA aggregates; confined patches with increased Eu(III) luminescence intensity or carbon absorption are marked with *arrows* (figure reprinted from [59], with permission from Elsevier)

cannot be applied at the same time to the same sample, we may safely assume that zones with high optical density in the STXM image are correlated with zones with high Eu(III) luminescence intensities in the LSLM images. Therefore, one can conclude that the metal cations are enriched in these zones. This conclusion is supported by the relative intensities of light and dark zones in the images. The intensity line profile of the Eu(III) luminescence extracted along the dashed line in Fig. 11a (not shown) indicates that most of these patches exhibit a three- to fourfold higher Eu(III) luminescence intensity compared with that of the surrounding matrix. This corresponds well to the three to four times higher carbon absorption detected for the dark zones compared with the light zones in the STXM micrographs.

The combination of two spectromicroscopy methods probing the organic ligand (STXM/C 1*s*-NEXAFS) and the metal center (LSLM/luminescence spectroscopy) reveals the existence of chemically different fractions in HA, which exhibit different morphologies and spectral features upon metal complexation. Note that the compiled results indirectly favor an HA structural model of a "supramolecular association". The STXM results allowed identification of HA fractions having different interaction with metal cations by their optical densities and spectral fingerprints. However, the metal cations associated with these fractions could only be localized from ancillary LSLM information. Note that comparative investigations of well-defined reference systems and quantum chemical calculations were necessary to provide a foundation for sound interpretation of spectral features obtained in HA/metal cation interaction experiments. Our experiments have shown that M^{n+} bound to HA are enriched in a minority fraction containing higher densities of complexing carboxylic sites. This fraction is likely to play a dominant role in HA colloid-mediated transport of actinides and other toxic trace metals in the hydrosphere. Macroscopic properties attributed to HA, including cation loading capacities and complexation constants, must be treated as averaged empirical values. These values are not sufficient for a detailed understanding of the underlying molecular processes on metal–HA interaction. We still need to explore how such molecular processes alter HA size and charge, which determine its stability and mobility and ultimately influence the potential of HA to serve as vehicles for actinide transport. There is an obvious necessity to establish and further develop spatially resolved spectroscopic techniques to investigate these (and other) inherently inhomogeneous natural samples. Through such developments, we will refine our understanding of the HA fractions and their binding to metal cations to a degree allowing long-term prediction.

Reference

1. V. Neck, M. Altmaier, T. Fanghaenel, in: R. Alvarez, N.D. Bryan, I May (Eds.), Ad-vances in Actinide Science, Royal Society of Chemistry Publishing, Cambridge 2006, 479–484
2. http://www.lightsources.org
3. Ch. David, F. Nolting, Ch. Quitmann, M. Stampanoni, F. Pfeiffer (Eds.), Proceedings of the 9th International Conference on X-Ray Microscopy, Journal of Physics: Conference Series, IOP Publishing, Bristol 2009
4. P. Horowitz, J.A. Howell, Science 178 (1972) 608–611
5. P. Kirkpatrick, A.V. Baez, J. Optical Soc. Am. 38 (1948) 766–774
6. Y. Yan, W. M. Gibson, JCPDS-International Centre for Diffraction Data, Adv. X-Ray Anal. 45 (2002) 298–305
7. A.G. Michette, Optical Systems for Soft X-Rays, Plenum Press, London 1986
8. A. Snigirev, V. Kohn, I. Snigireva, B. Lengeler, Nature 384 (1996) 49–51
9. http://www.xradia.com/technology/microscopes/types-of-xray-microscopes.php
10. B. Hornberger, M.D. de Jonge, M. Feser, P. Holl, C. Holzner, C. Jacobsen, D. Legnini, D. Paterson, P. Rehak, L. Strüder, S. Vogt, J. Synchrotron Rad. 15 (2008) 355
11. W. Chao, B.D. Harteneck, J. Alexander Liddle, E.H. Anderson, D.T. Attwood, Nature 435 (2005) 1210–1213
12. W.H. McMaster, N.K. Del Grande, J.H. Mallett, J.H. Hubbell, Compilation of X-ray cross sections, Lawrence Livermore National Laboratory Report UCRL-50174 (sec-tion I 1970, section II – section IV 1969)
13. J. Stöhr, NEXAFS Spectroscopy, Springer Series in Surface Sciences 25, Springer-Verlag, Berlin 1992
14. A.P. Hitchcock, D.C.J. Mancini, J. Electron Spectrosc. Relat. Phenom. 67 (1994) 1–132
15. H. Ade, S.G. Urquhart, in: T.K. Sham (Ed.), Chemical applications of synchrotron radiation, World Scientific Publishing, Singapore 2002, 285–355
16. O. Dhez, H. Ade, S.G.J. Urquhart, J. Electron Spectrosc. Relat. Phenom. 128 (2003) 85
17. O. Plashkevych, L. Yang, O. Vahtras, H. Ågren, L.G.M. Petterson, Chem. Phys. 222 (1997) 125
18. V. Carravetta, O. Plashkevych, H. Ågren, Chem. Phys. 263 (2001) 231
19. I. Bâldea, B. Schimmelpfennig, M. Plaschke, J. Rothe, J. Schirmer, A.B. Trofimov, Th. Fanghänel, J. Electron Spectrosc. Relat. Phenom. 154 (2007) 109
20. U. Neuhäusler, C. Jacobsen, D. Schulze, D. Stott, S. Abend, J. Synchrotron Rad. 7 (2000) 110–112
21. J. Kirz, C. Jacobsen, in: Ch. David et al. (Eds.), Proceedings of the 9th International Conference on X-Ray Microscopy, Journal of Physics: Conference Series, IOP Pub-lishing, Bristol 2009, 012001

22. B. Niemann, D. Rudolph, G. Schmahl, Appl. Opt. 15 (1976) 1883–1884
23. H. Rarback, D. Shu, S.C. Feng, H. Ade, C. Jacobsen, J. Kirz, I. McNulty, Y. Vladimirsky, D.P. Kern, T.H.P.Chang, in: Sayre et al. (Eds.), X-ray Microscopy II, Springer Series in Optical Sciences 67, Springer-Verlag, Berlin 1988
24. http://xray1.physics.sunysb.edu
25. http://www.als.lbl.gov/als/microscopes/
26. http://sls.web.psi.ch/view.php/beamlines/pollux/index.html
27. http://www.lightsource.ca/experimental/sm.php
28. http://www.bessy.de/cms.php?idcat=166&changelang=5
29. http://www.elettra.trieste.it/beamlines/TWINMIC/
30. http://www.bruker-est.com/synchrotron-inst.html
31. G.J. Flynn, S. Wirick, L.P. Keller, C. Jacobsen, in: Ch. David et al. (Eds.), Proceed-ings of the 9th International Conference on X-Ray Microscopy, Journal of Physics: Conference Series, IOP Publishing, Bristol 2009, 012085
32. S.G. Urquhart, A.P. Hitchcock, A.P. Smith, H.W. Ade, W. Lidy, E.G. Rightor, G.E. Mitchell, J. Electron Spectrosc. Relat. Phenom. 100 (1999) 119–135
33. S.G. Urquhart, A.P. Smith, H.W. Ade, A.P. Hitchcock, E.G. Rightor, W. Lidy, J. Phys. Chem. B 103 (1999) 4603–4610
34. A.P. Hitchcock, I. Koprinarov, T. Tyliszczak, E.G. Rightor, G.E. Mitchell, M.T. Dineen, F. Hayes, W. Lidy, R.D. Priester, S.G. Urquhart, A.P. Smith, H. Ade, Ultramicroscopy 88 (2001) 33–49
35. H. Ade, B. Hsiao, Science 262 (1993) 1427–1429
36. B. Watts, W.J. Belcher, L. Thomsen, H. Ade, P.C. Dastoor, Macromolecules 42(21) (2009) 8392–8397
37. J.J.M. Halls, C.A.Walsh, N.C. Greenham, E.A. Marseglia, R.H. Friend, S.C. Moratti, A.B. Holmes, Nature 376 (1995) 498–500
38. B.A. Wilson, V.H. Smith, F. Denoyelles Jr., C.K. Larive, Environ. Sci. Technol. 37 (2003) 1713–19
39. J.J. Dynes, J.R. Lawrence, D.R. Korber, G.D.W. Swerhone, G.G. Leppard, A.P. Hitchcock, Sci. Total Environ. 369 (2006) 369–383
40. N.S Ranganathan, in: J.M Ascenzi (Ed.), Handbook of disinfection and antiseptics, Marcel Dekker Inc., New York 1996, 235–264
41. T.H. Yoon, Appl. Spectrosc. Rev. 44 (2009) 91–122
42. J. Thieme, J. Sedlmair, S.-C. Gleber, J. Prietzel, J. Coates, K. Eusterhues, G. Abbt-Braun, M. Salome, J. Synchrotron Rad. 17 (2010) 149–157
43. J. Thieme, I. McNulty, S. Vogt, D. Paterson, Environ. Sci. Technol. 41 (2007) 6885–6889
44. M.A. Denecke, C.M. Marquardt, J. Rothe, K. Dardenne, M.P. Jensen, J. Nucl. Sci. Technol. 3 (2002) 410–413
45. M.A. Denecke, D. Bublitz, J.I. Kim, H. Moll, I. Farkes, J. Synchrotron Rad. 6 (1999) 394–96
46. M.A. Denecke, T. Reich, S. Pompe, M. Bubner, K. H. Heise, H. Nitsche, P. G. Allen, J. J. Bucher, N. M. Edelstein, D. K. Shuh, K. R. Czerwinski, Radiochim. Acta 82 (1998) 103–108
47. M.K. Armbruster, B. Schimmelpfennig, M. Plaschke, J. Rothe, M.A. Denecke, R. Klenze, J. Electron Spectrosc. Relat. Phenom. 169 (2009) 51–56
48. J. Rothe, M. Plaschke, B. Schimmelpfennig, M.A. Denecke, in: Speciation Techniques and Facilities for Radioactive Materials and Synchrotron Light Sources, Nuclear Energy Agency (NEA), OECD-NEA, Issy-les-Moulineaux, France, No. 6288 (2007) 283
49. J. Rothe, M. Plaschke, M.A. Denecke, in: B. Hedmann and P. Pianeta (Eds.), X-ray Absorption Fine Structure – XAFS13, American Institute of Physics, CP882, Melville, New York 2007, 193
50. C. Walther, J. Rothe, M. Fuss, S. Büchner, S. Koltsov, T. Bergmann, Anal. Bioanal. Chem. 388 (2007) 409

51. H.-R. Cho, C. Walther, J. Rothe, V. Neck, M.A. Denecke, K. Dardenne, Th. Fanghänel, Anal. Bioanal. Chem. 383 (2005) 28–40
52. S.C.B. Myneni, Rev. Min. Geochem. 49 (2002) 485
53. J. Thieme, C. Schmidt, G. Abbt-Braun, C. Specht, F.H. Frimmel, in: F.H. Frimmel, G. Abbt-Braun (Eds.), Refractory Organic Substances in the Environment, Wiley, New York 2000
54. A.C. Scheinost, R. Kretzschmar, I. Christl, C. Jacobsen, Spec. Publ. Roy. Soc. Chem. 273 (2002) 39
55. M. Plaschke, J. Rothe, T. Schäfer, M.A. Denecke, K. Dardenne, S. Pompe, K.-H. Heise, Coll. Surf. A 197 (2002) 245
56. J.I. Kim, G. Buckau, G.H. Li, H. Duschner, N. Psarros, Fresenius J. Anal. Chem. 338 (1990) 245
57. M. Plaschke, J. Rothe, M. Altmaier, M.A. Denecke, T. Fanghänel, J. Electron Spectrosc. Relat. Phenom. 148 (2005) 151–57
58. J. Rothe, M. Plaschke, M.A. Denecke, Radiochim. Acta 92 (2004) 711
59. A. Naber, M. Plaschke, J. Rothe, H. Hofmann, T. Fanghänel, J. Electron Spectrosc. Relat. Phenom. 153 (2006) 71–74

Section 2
Modeling of Actinide Nanoparticle Structure and Behavior

From Molecules to Nanoparticles: A Computational Chemist's Point of View

Bernd Schimmelpfennig

Abstract The computational investigation of nanoparticles and especially those of actinides is still a challenge. Due to the sheer number of atoms to be treated, approximations have to be employed in electronic structure calculations. The approximations required in the special case of actinide(oxide) nanoparticles, the motivation/reasoning behind the approximations, and the accuracy of results are discussed. An example of a structure of a $(Th_{147}O_{292})^{4+}$ model cluster calculation reveals trends in Th–O bond distances, which compare well to recent experiments.

1 Introduction

Computational chemistry has over the last decades developed into a predictive tool to investigate the chemistry of materials from atoms to solids. It is nowadays used routinely to verify or predict various types of data as different as structure, thermodynamic properties, as well as the response of material to external perturbation such as strong laser or magnetic fields. The progress in this research area can be partly attributed not only to the still ongoing development of computer hardware (Moore's Law) but also to the new algorithmic approaches, which take advantage of ingenious approximations; smart parallelization schemes also play a role. When it comes to actinide compounds though, computational chemistry is still in its teenage years; approaches suitable for light-element compounds cannot be applied with comparable accuracy for the actinide elements. Application of quantum chemical calculations of actinide particles with nanoscale dimensions pushes computational chemistry approaches to their present limits. In the following section, a point of view is presented (certainly with some bias) as to what challenges are faced in computational chemistry when dealing with actinides. The emphasis is on the approaches and approximations that have to be chosen, the effects that can be

B. Schimmelpfennig
Karlsruhe Institute of Technology (KIT), Institute for Nuclear Waste Disposal,
P.O. Box 3640, 76021 Karlsruhe, Germany
e-mail: Bernd.Schimmelpfennig@kit.edu

S.N. Kalmykov and M.A. Denecke (eds.), *Actinide Nanoparticle Research*,
DOI 10.1007/978-3-642-11432-8_7, © Springer-Verlag Berlin Heidelberg 2011

covered with such approaches, and the accuracy that can be expected for actinide (nano-particle) systems. After this overview, some preliminary results for a model system reaching the nanoscale are presented. From the experimental side, much experience exists for nanoparticles of lighter elements. As an example, numerous C 1*s*-NEXAFS/STXM investigations of carbonaceous materials have been reported in the past two decades. These studies focused on various aspects of material, environmental and life sciences, including (multicomponent) polymer films and fibers, organic geochemical materials such as coals and soils, biological materials such as proteins and biofilms, and even extraterrestrial materials collected from meteorites and comets (Westphal et al. 2010).

2 Quantum Chemistry

One of the obvious questions to be raised in computational chemistry exercises dealing with actinide nanoparticles is why one should go for such an obviously demanding approach as quantum chemistry and not restrict oneself to empirically adjusted potentials, e.g., based on the well-characterized structure of thorium dioxide, or to employing methods from computational solid-state physics. The answers are that any empirically adjusted potential derived to describe the solid cannot necessarily be expected to accurately describe the surface structure of a nanoparticle, and a computational solid-state physics approach relies on a comparably small unit-cell, extended by employing techniques of periodic boundary conditions. Quantum chemistry, treating the nanoparticle as a large molecule, appears then to be the method of choice in such exercises.

The quantum chemistry toolbox has enlarged over the last decades and there are tools available for various purposes. At least for small molecules, it is often possible to reach accuracies below experimental error bars. Unfortunately, all these approaches have the unpleasant feature that they scale uncomfortably poorly as the size of the system under investigation increases. Despite much excellent work to improve this behavior, e.g., localization and multipole expansion techniques, for nanoparticles with a few hundred atoms, the only computationally feasible approach is the density functional theory (DFT) and only in case the number of explicitly treated electrons can be reduced drastically compared to an all-electron approach. A limiting factor is the choice of basis sets to be used for describing the molecular orbitals and, ultimately, the overall electron density. In principle, it is possible to choose basis sets large enough to reach the basis set limit, which is equivalent to a numerically exact solution. However, this is only practically possible for small molecules.

Having discussed the limitations to be faced for a single-point calculation (i.e., with fixed nuclei), the next point is making an educated guess of the molecular structure, from which the optimum structure of minimized total energy can be found and compared to experimental data. In a DFT-approach, this is a comparably simple task as gradients with respect to nuclear coordinates can be easily obtained and used afterwards in iterative schemes to reach a (local) minimum starting from an initial guess. Considering the vast number of local minima possible in a system of a hundred or more atoms, it is obvious that a simple algorithmic search for minima is most likely doomed to failure; chemical intuition or experience is needed for selecting reasonable initial guesses of structure. A technical detail can be of assistance: just as symmetry can be helpful in analyzing/interpreting experimental data, symmetry can also offer an essential way to simplify computation by blocking the equations (reducing the number) to be solved.

3 Relativistic Effects

A pedagogical overview by Pyykkö on relativistic effects in chemistry may be found in Pyykkö (2002). A common way to define relativistic effects, according to Einstein's special theory of relativity, is to include all differences between a world with infinite speed-of-light and (for the human mind hard to grasp) one with a finite speed-of-light of 300,000 km/s. As long as one deals with lighter elements, up to around argon, one can fairly well ignore relativistic effects in chemistry. However, one should keep in mind that many interesting effects such as phosphorescence or, as a prominent and beautiful example, the aurora borealis visible near or above the polar circles, are due to relativistic effects in light elements. Some standard experimental tools such as NMR would also not exist in a nonrelativistic world. When dealing with heavy elements (including the actinide elements) though, even a rough, realistic description of molecular and electronic structure requires including relativistic effects. A common way of explaining this is the change (decrease) of the modified Bohr radius for larger nuclear charges when the "relativistic mass-increase" is taken into account. Typically, s- and p-orbitals contract as a direct relativistic effect. On the other hand, contracted s- and p-orbitals lead to a better screening of the nuclear charge for d- and f-shells, which expand as a result. This is referred to as an indirect relativistic effect. Two often sited examples of relativistic effects are the color of gold and the low melting point of mercury. Contraction of the 6*s* orbital in Hg leads to a lowering of bonding interaction between Hg atoms, resulting in its remaining a liquid down to low temperatures (to $-39°$C) and a monomer in gaseous form. The colored luster of gold can be attributed to the contraction of 6*s*-orbitals and associated expansion of 5*d* orbitals, which shifts absorption for the 5*d* to the 6*s* transition to relatively smaller energies up to yellow visible light. These scalar effects are simply modifications of the ordinary chemical picture. A nonscalar relativistic effect is the spin–orbit coupling due to magnetic interaction between an electron spin and the moving charges (thus current generating) of nuclei and other electrons. This leads to a further splitting of the atomic and molecular orbital set. Multiplets obtained in calculations neglecting this effect may get mixed up beyond recognition in the final spectrum of ground and excited states such that comparison to experiment is futile.

These are general relativistic effects, which can be found mentioned in mainstream chemistry textbooks. How can these effects be included in quantum chemical calculations? For the purist, the Dirac equation is the obvious choice. However, this puts demands on computers that are too big for applications on larger systems. The main purpose of the Dirac approach is to provide nonexperimental benchmark data, which can be used to judge the performance of approximate schemes to treat relativity. As the Dirac approach describes positron states, which show up only in case of electron–positron pair creation at high energies, in addition to the electrons, an obvious approach is to decouple electrons and positrons and to explicitly treat the electrons only. There is no unique way to pursue this decoupling, and several approaches have been developed and implemented in available software. As mentioned previously, for large nanoparticle approaches, dealing with all the electrons is presently too demanding, and further approximations are mandatory. From a chemical point of view, inner-shell or core orbitals are chemically inert. A suitable approach is therefore to parameterize core orbital interaction with valence. Such an approach has been developed over the last decades, and it is even possible to include relativistic effects in such parameterizations. These are referred to as pseudopotentials (PP) or effective core potentials (ECP).

A number of careful studies have been carried out to determine which electron shells can safely be parameterized by a potential. For the actinides, the majority of researchers agree that everything up to 4*f* can be included in the potential, and orbitals 5*s* and upwards must be treated explicitly. Such parameterized potentials are commonly called small-core type. One should realize though, for an ion such as Th(4+), 26 electrons still need to be treated explicitly! Over the last years, Dolg et al. developed oxidation-state specific PPs, which parameterize everything up to the 5*d*-shell and include the open-shell 5*f*-electrons characteristic for the oxidation state under investigation (Moritz et al. 2007). This f-in-core approach was first developed for the lanthanides. In this case, just eight electrons (5*s*/5*p* for the lanthanides and 6*s*/6*p* for the actinides) have to be treated explicitly. For lanthanides, this approximation works extremely well, the main reason being the compact -and therefore chemically inert- 4*f* electrons. The studies by Dolg et al., as well as our own studies of trivalent actinides coordinated with soft- and hard-donor ligands (Banik et al. 2010; Apostolidis et al. 2010), show that this approach has to be applied with some skepticism for the lighter actinides. However, it is still the only approach feasible for nanoparticles involving several actinides. It should be emphasized that this is not only due to the number of electrons explicitly treated but also due to the fact that it is the only sensible solution to avoid getting lost in the maze of possible electronic states involving hundreds of unpaired, but nevertheless interacting, 5*f*-electrons.

4 Expected Accuracy

Taking all the approximations discussed above into account, not to the mention limitations of the currently available approximate functionals available for DFT, it is obvious that such an approach can give qualitative results at best. From general experience, we can expect bond-distances to be overestimated by as much as 10 pm. This overestimation, however, is a systematic effect, and errors for relative bond distances are expected to be necessarily much smaller. For any chosen model system, it should thus be possible to deduce relative changes in bond length between atoms in the solid-like embedded part of a nano-particle and those comprising the surface region.

5 Results

Several attempts to calculate a thorium oxide cluster of nanometer-dimensions were performed and results checked for structural artifacts, which might arise from the initial structure used or the imposed symmetry and charge. The cluster shown in Fig. 1 consists of 147 Th^{4+} atoms and 292 O^{2-} atoms with a total charge of +4; the O_h point-group was imposed. The geometry was optimized with the BP86 functional using RI-DFT as implemented in the TURBOMOLE-package (2010) with an *f*-in-core pseudo potential for Th^{4+} and corresponding basis sets of double-zeta quality (Moritz et al. 2007). For oxygen, a split-valence plus polarization (SVP) basis set was employed. With this number of atoms, the cluster approaches a diameter of 3 nm and can be considered a small actinide oxide nanoparticle

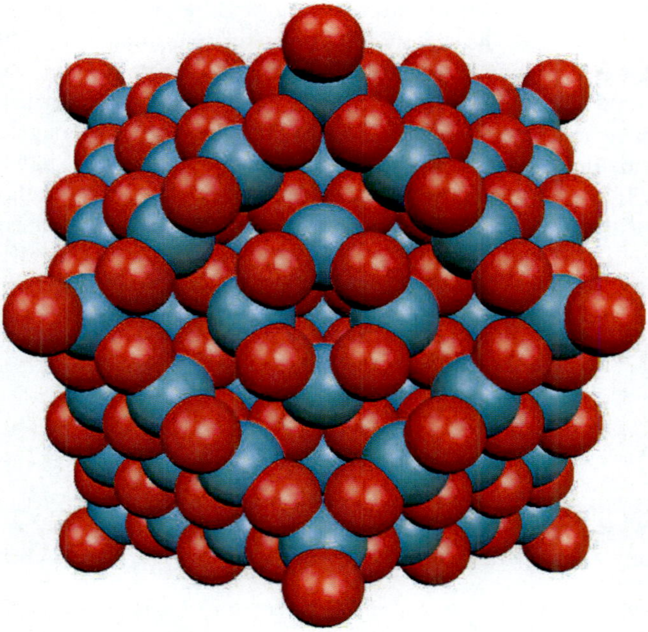

Fig. 1 Optimized structure of a $(Th_{147}O_{292})^{4+}$ cluster with O_h symmetry

model. The thorium at the center of the particle is symmetrically surrounded by eight oxygen atoms at a distance of 247 pm. As discussed above, this is larger than the experimental (XRD) Th–O bond distance of 242 pm in ThO_2(cr) (Magini et al. 1976) but within the deviation range to be expected using the approach employed. Eight symmetry-distinct thorium atoms are located at distances of 405, 558, 708, 812, 900, 984, 1,085, and 1,249 pm from the center. The local symmetry at each of these thorium centers varies, and we give just the average Th–O bond distance at these centers: 248, 245, 249, 246, 244, 243, 237, and 234 pm. It should be noted that for the last two Th-centers, which are located at a corner or at the edge of the cluster, the Th–O distances scatter over a range of 40 pm; especially oxygens with just one bond to thorium are much shorter. If one eliminates these especially short bonds from the average Th–O bond distances, the average values for the last two Th-centers become 242 and 241 pm. We find a decrease in Th–O bond distances from the center of the particle to the surface. This can be rationalized by stronger and shorter Th–O bonds at the surface due to fewer number of bound oxygen atoms and, therefore, also less repulsion between them. We compare this to experimentally determined interplanar distances in synthetic ThO_2 nano-particles (Batuk et al. 2010). The average lattice parameter observed for 2 nm ThO_2 nanoparticles is 5% smaller than the bulk value, and there is a trend towards longer lattice parameters with increasing particle size. The observed decrease in average lattice parameter corresponds to a decrease in average Th–O bond distances. This is in accord with the calculated 439 atom $(Th_{147}O_{292})^{4+}$ cluster structure. The experimental Th–O distance increases for the larger nanosized particles, which can be considered a result of the decrease in surface-to-bulk ratio and the associated growing dominance of longer bonds in the center (decrease in relative amounts of shorter surface Th–O bonds), as was found in the calculated thorium oxide nanoparticle model.

These results should be considered more of a feasibility study than a definite answer to the understanding of actinide oxide nanoparticles. Although the calculations are at the limits of present possibilities, the general progress in hardware and software will soon allow tackling even larger clusters and to relax the tight symmetry restrictions in the computations. Having reached timescales of an hour or shorter per single point calculation, it will be feasible to dynamically relax such clusters and thereby get closer to the global minimum for a given set of atoms and charge.

References

Apostolidis Ch, Schimmelpfennig B, Magnani N, Lindqvist-Reis P, Walter O, Sykora R, Morgenstern A, Colineau E, Caciuffo R, Klenze R, Haire RG, Rebizant J, Bruchertseifer F, Fanghänel Th (2010) Angew Chem Int Ed Engl, 122:6487–6491, http://dx.doi.org/10.1002/anie.201001077

Banik NL, Schimmelpfennig B, Marquardt CM, Brendebach B, Geist A, Denecke MA (2010) Dalton Trans, 39:5117–5122

Batuk ON, Szabó DV, Denecke MA, Vitova N, Kalmykov SN (2010) Radiochim Acta (Submitted)

Magini M, Cabrini A, Scibona G, Johansson G, Sandström M (1976) Acta Chem Scand A, 30:437–447

Moritz A, Cao X, Dolg M (2007) Theor Chem Acc, 118:845

Pyykkö P (2002) Chem Rev, 88:563–594

TURBOMOLE V6.2 2010, a development of University of Karlsruhe and Forschungszentrum Karlsruhe GmbH, 1989-2007, TURBOMOLE GmbH, since 2007; available from http://www.turbomole.com

Westphal AJ, et al. (2010) Non-destructive search for interstellular dust using synchrotron microprobes. In: Denecke MA, Walker CT (eds.) X-ray Optics and Microanalysis. American Institute of Physics: Melville, NY

Theoretical Analysis of Colloid-facilitated Transport of Radionuclides by Groundwater

Viktor Malkovsky

Abstract Mechanisms of colloid-facilitated transport of radionuclides by groundwater are considered for three types of radioactive colloids: intrinsic colloids, primary colloids and pseudocolloids. Effect of elevated velocity of colloid facilitated migration is analyzed. The general characteristic is introduced for the estimation of the role of colloidal species of radionuclide migration velocity. A new probabilistic model of colloid retardation by the host rocks is developed taking into account heterogeneity of the colloid particles.

1 Introduction and Background

The only mechanism of radioactive contamination propagation in underground systems is the transport of radionuclides by groundwater which flows through hydraulically connected systems of pores and fracture voids that are called flow channels. The contaminant is carried in a flow channel by some volume of surrounding water. As a result of Brownian motion of water molecules, the contaminant may come close to the channel walls. Due to sorption, which we assume here to be reversible, it could be attached to the walls. The water volume that transported the contaminant moves farther on. After a period of time, the contaminant could be detached from the surface by a new volume of the groundwater and removed along the channel. However, the travel path of the contaminant during this period is shorter than the travel path of the water volume that carried the contaminant initially (Fig. 1). As a result, the average velocity of the contaminant is less than the groundwater velocity. The higher the sorption properties of the host rocks (which form the walls of the channels) towards the contaminant, the less is the ratio of the

V. Malkovsky
Institute of Geology of Ore Deposits, Petrography, Mineralogy and Geochemistry, Russian Academy of Sciences (IGEM RAS), Staromonetnyi per. 35, Moscow 119017, Russia
and
D.Mendeleyev University of Chemical Technology of Russia, Miusskaya sq. 9, Moscow, Russia
e-mail: malk@igem.ru

S.N. Kalmykov and M.A. Denecke (eds.), *Actinide Nanoparticle Research*,
DOI 10.1007/978-3-642-11432-8_8,

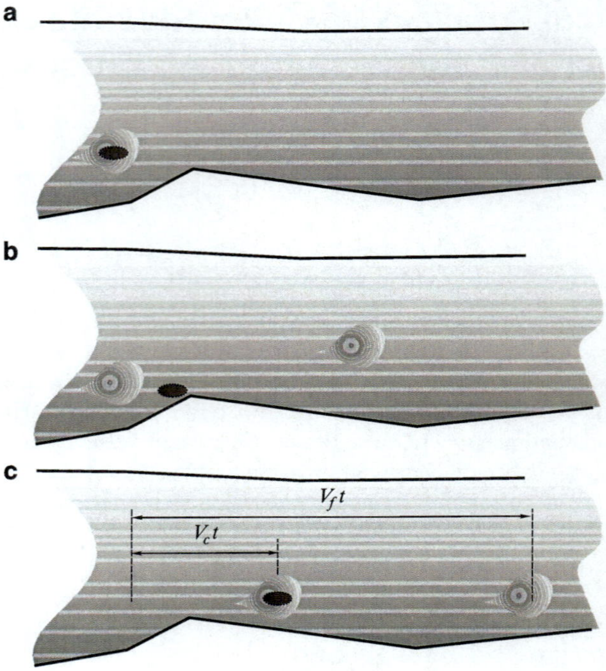

Fig. 1 Migration of colloid particle in the flow channel. (**a**) The particle is carried by the surrounding volume of the groundwater; (**b**) the particle precipitates on the channel wall, while the water volume that carried the particle moves farther; (**c**) another volume of water mobilizes the particle and carries it along the channel. V_f is the groundwater flow velocity; V_c is average velocity of the contaminant; t is time

contaminant migration velocity to the groundwater velocity. The sorption properties of rocks to many radionuclides are very high. Therefore, their migration velocity should be low. However, several studies at the contaminated sites have shown that the actual velocities of the contaminants were much higher than the values predicted on the basis of the radionuclide retardation caused by their sorption by the host rocks (Nyhan et al. 1985; Penrose et al. 1990; McCarthy et al. 1998a, b; Zachara et al. 2002; Smith et al. 2003). The possible reason is that these radionuclides were transported by the groundwater not only as a solute, but also with colloid particles.

The colloid particles in some transport medium by definition should be large enough that their boundary with the medium can be considered as an interphase, but they should be so small that surface effects exert influence on the state of the substance in the whole volume of the particle (Frolov 1988). As a result of significant variety of the shape and composition of the particles and properties of the transport medium, the limits of the size range of the particles which are determined on the basis of this definition are conventional. For this reason, different authors establish different limits of the size range of the colloid particles in the groundwater: 1–10^4 nm (Mironenko and Rumynin 1998), 1–10^3 nm (Klyachko and Apel'tsyn 1971), 1–450 nm (Vilks and Degueldre 1991).

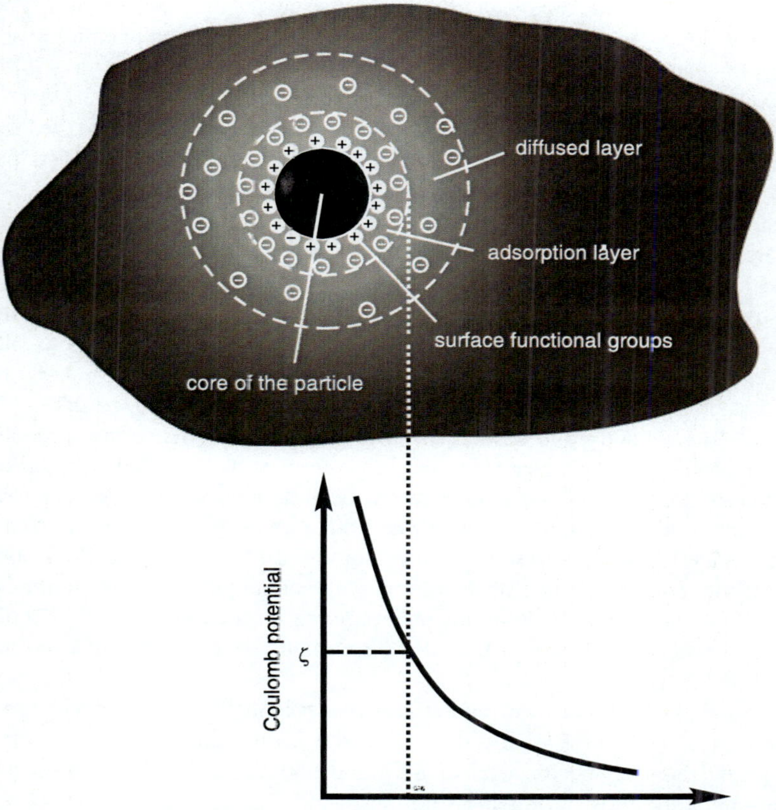

Fig. 2 The structure of colloid interface

The colloids in the groundwater are represented by a micellar structure. The micelle consists of a hard core and two surrounding layers of ions and molecules of the water: adsorption layer and diffused layer (Fig. 2). As colloids move, they carry only the adsorption layer; the coulomb potential of the outer boundary of the adsorption layer is called ζ potential (Klyachko and Apel'tsyn 1971). The surface of the rock has its own ζ potential. Because the core of the colloid and the rock can be of the same composition, their ζ potentials can be of the same sign. As a result, the Coulomb interaction of the colloid particles and the walls of the flow channel forms a repulsion barrier. From these considerations, the assumption is made that precipitation/sorption of radiocolloid on the host rock is weak.

As the sorption properties of the host rock to many radionuclides that are carried by the groundwater in ionic form are high, one would expect that the fraction of the radionuclides that are carried by the colloid particles can be substantial. According to the results of groundwater sampled at the Nevada test site, the colloid particles bind up to 95% of Cs, 91% of Co, and 99% of Pu+Eu in the groundwater (Kersting et al. 1999). Analysis of the groundwater from the site where liquid radioactive wastes (RW) from the central reprocessing plant of the Los Alamos National

Laboratory had been discharged since the 1940s showed that 85% of Pu was bound to particles of the size of 25–450 nm and 31% of Am was bound to colloid particles the size of which exceeded 5 nm (Penrose et al. 1990). The lake Trawsfynydd (Wales, UK) was used as a cooling pond for the nuclear reactor. In addition to that, some volume of solutions contaminated by radionuclides was discharged into the lake. As much as 85–90% of Am and Pu in the water samples from the lake were detained by filters with a pore diameter of 15 nm (Orlandini et al. 1990).

Uranium ore deposits are often considered analogs of underground repositories of high-level RW and spent nuclear fuel (SNF). In this connection, results of studies of colloid-facilitated transport of U and Th by groundwater are of interest for safety assessment of potential underground repository of RW and SNF. Data of such studies obtained at different deposits differ from each other. According to (Airey 1986), particles of the size of 5–10^3 nm contain 0.01–2% of U and 0.3–39% of Th in the groundwater. According to (Short and Lowson 1988), particles of the size of 20–10^3 nm bind 5–30% of U and up to 98% of Th. The significant difference in the amount of radionuclides carried by the groundwater in the form of the radiocolloid is caused by variations of Eh, pH, and differences in the chemical composition of the groundwater. In spite of these differences, the results of the studies of the groundwater samples from uranium mines showed that a substantial part of U and Th is bound to colloid particles.

It follows from published observations at the contaminated sites, prediction of radionuclide migration by groundwater requires consideration of their colloid-facilitated transport. Failure to take this into account leads to a substantial under-estimation of migration rates.

Sustainable development of the nuclear power is inevitably associated with the problem of disposal of RW and SNF (Laverov 2006). Consideration of different approaches showed that disposal of SNF and high-level RW in deep geological formations is the most promising solution of the problem from the viewpoint of ecological safety and cost efficiency (Krauskopf 1988; Hara et al. 1996).

Colloids that carry radionuclides from the underground repositories of SNF and high-level RW could be classified into three groups according to their origin: intrinsic colloids, primary colloids, and pseudocolloids (Inagaki et al. 1998; Buck and Bates 1999). The intrinsic colloids consist of particles composed of two oxy-hydroxides of highly hydrolyzable cations, e.g., Pu(IV), Th(IV), Zr(IV), Tc(IV). The primary colloid represents colloid particles composed of leaching products of solidified RW and SNF at their contact with groundwater. Pseudocolloid consists of colloid particles that existed in the groundwater before the contact with RW and SNF. As a result of such interaction, particles sorb radionuclides.

The formation of intrinsic colloids in aqueous solutions containing Am, Pu, and Np is noted in (Olofsson et al. 1982). Probably, intrinsic colloids are formed in this case as a result of the dynamic equilibrium between the processes of formation and the destruction of aggregates from actinide hydroxides. Average size of the particles in the system and number of the particles per unit volume of the solution are most likely monotone increasing functions of the total concentration of the actinides in the solution.

It is assumed that geologic disposal of HLW will be accompanied by the creation of engineering barriers between the HLW and the groundwater to decrease the ingress of the radionuclides into the groundwater (Laverov et al. 1991, 1994). In

Russia, the main volume of HLW is liquid waste from radiochemical enterprises. These wastes are presumed to be solidified before the disposal. In this case, radionuclides are immobilized into glassy or crystalline waste forms that are resistant to leaching when in contact with the groundwater (Ringwood 1980; Harrington and Apps 1982; Tait et al. 1989; Glagolenko et al. 1996). The solidified wastes are presumed to be placed into steel canisters, preventing the contact of the waste with the groundwater (Laverov et al. 1994). A layer of buffer filling is characterized by low permeability and low diffusion coefficient of the radionuclides can be put between the canister and walls of the repository cells. From the viewpoint of the requirements to the characteristics and relatively low cost, bentonite clay is the most promising material (Laverov et al. 1994). Processes of primary colloid formation as a result of leaching of the waste forms from borosilicate glass are studied in (Bates et al. 1992; Ebert and Bates 1993; Inagaki et al. 1998; Buck and Bates 1999). It was established that chemical and mineralogical composition of the colloid particles is not characteristic to natural colloid in the groundwater. Therefore, these particles are the primary colloid (Buck and Bates 1999). The experiments with the waste forms containing Am, Pu, and Np showed that the primary colloid particles larger than 1 nm contain more than 79% of Np, and 99% of Pu and Am in the groundwater after its interaction with the waste form (Bates et al. 1992). The leaching of Pu and Np bearing borosilicate waste forms was studied by Inagaki et al. (1998). It was shown that the presence of colloid particles in the initial groundwater due to corrosion of the steel canister can exert influence on the content of Pu and Np in colloidal form in the groundwater as a result of leaching of the glassy waste forms. The buffer layer can also be a source of colloid particles. Properties of the bentonite colloid were studied in (Missana et al. 2000).

It is presumed in many concepts of HLW and SNF management that SNF can be disposed in an underground repository without preliminary chemical processing (Laverov et al. 2003). As a result, the colloid particles formed upon SNF leaching by the groundwater can also be ascribed to the primary colloid. Properties of the colloids formed upon SNF leaching by the groundwater were studied in (Finn et al. 1994) where it was shown that the colloid particles larger than 3.6 nm contain up to 90% of actinides in the groundwater after its interaction with SNF. Besides, the process of SNF leaching depends on the presence of colloid particles in the initial groundwater (Amme et al. 2004).

Pseudocolloids are formed due to sorption or coprecipitation of radionuclides onto colloid particles of natural origin present in the groundwater. Pseudocolloids represent particles of clayey minerals, iron, aluminum and manganese oxy-hydroxides, carbonates, silicates, humic and fulvic acids, bacteria, and viruses (Klyachko and Apel'tsyn 1971; Keswick and Gerba 1980; Morel and Gschwend 1987; Vilks et al. 1991; Kersting et al. 1999; Fortner et al. 2003; Kretzschmar and Schäfer 2005).

The role of colloids in the transport of radionuclides by groundwater in the mathematical models is characterized by the radionuclide flow rate with colloid $\vec{j}_c$, which can be expressed as:

$$\vec{j}_c = n_c \gamma_c \vec{V}_c, \tag{1}$$

where n_c is the amount of colloid per unit volume of the groundwater, γ_c is the amount of radionuclides associated with each colloid particle, and $\vec{V}_c$ is the velocity of colloid migration.

Further we will consider from this viewpoint the role of each group of radiocolloid (intrinsic and primary colloid and pseudocolloid) in models of radionuclides transport by groundwater.

2 Modeling of Radionuclides Transport by Intrinsic Colloids

2.1 Formation of the Intrinsic Colloids

Intrinsic colloids consist of hydroxides of highly hydrolyzable elements, we consider their formation using the example of formation of polymeric Pu(IV) hydroxide. The theoretical model of the colloid formation is based on the results of the experiments carried out by Dr. Kulyako and his colleagues in the Vernadsky Institute of Geochemistry and Analytical Chemistry of RAS. Simulated groundwater solution corresponded to Japanese standard of the groundwater PNC TN 1412 98-013 that is slightly oxic carbonate groundwater. Concentration of $Pu(OH)_4$ was measured by radiometry. Fractions of polymeric $Pu(OH)_4$ were determined by successive filtration of the solution through membranes with MWCO of 100, 50, 30, 10, and 3 kDa, respectively. We use the data of the filtration through the membrane with MWCO of 100 kDa. The fraction of retained particles was measured in the initial solution. Then the solution was diluted with the same simulated groundwater and the fraction of retained particles was measured again, and so on. By so doing, the concentration of colloidal Pu was measured as a function of the total concentration of Pu. The data are presented in Table 1.

One could see that equilibrium concentration of polymeric Pu(IV) is not proportional to the total concentration of Pu in the solution. Filtration through the membranes with different pore sizes showed that the particles were heterogeneous in size.

We consider the formation of polymeric particles from the monomers. We designate the particle of the mass of km_0 as k polymer (where m_0 is the mass of the monomer), denote the number of k polymeric particles per unit volume of the solution as n_k, and assume that movement of monomers and polymers in the

Table 1 Dependence between Pu fraction retained by the membrane (100 kDa) and the total concentration of Pu in the solution

Dilution factor	Total concentration of Pu, M	Fraction of colloidal Pu
1	2.4×10^{-5}	~1
10^2	3.2×10^{-7}	0.999
10^4	4.6×10^{-9}	0.99
10^6	4.3×10^{-11}	0.09

solution is caused by the mechanism of Brownian motion. It is natural to suppose that the polymeric particles are formed as a result of interaction of two smaller particles (polymer + polymer, polymer + monomer, or monomer + monomer). We assume that the polymer could be disintegrated due to breakup of the bond between two of its neighboring monomers. These events of breakup between neighboring monomers are independent and equiprobable, and the probability of breakup of a bond for the short time interval δt is equal to $\lambda\ \delta t$, where λ is a constant coefficient. Then the probability of breakup of the bond in the k polymer is:

$$1 - (1 - \lambda\delta t)^k = k\lambda\delta t + o(\delta t). \tag{2}$$

By virtue of the above assumptions, the functions $n_k(t)$ satisfy the modified Smoluchowski–Einstein equation (Voloshuk 1984)

$$\begin{aligned}\frac{\partial n_k}{\partial t} = \frac{1}{2}\sum_{i=1}^{k-1}\beta[(k-i)m_0, im_0]n_{k-i}n_i - \sum_{i=1}^{\infty}\beta(km_0, im_0)n_k n_i - \lambda(k-1)n_k \\ + \lambda\sum_{i=k+1}^{\infty}(2-\delta_{k,i-k})n_i.\end{aligned} \tag{3}$$

Here, $\delta i, k$ is Kronecker delta and $\beta(m_1, m_2)$ is a function that characterizes the frequency of collision of the particles with masses m_1 and m_2. In the case of Brownian motion, the function takes the form

$$\beta(m_1, m_2) = \frac{2}{3}\frac{k_B T}{\mu}\left(m_1^{1/3} + m_2^{1/3}\right)\left(m_1^{-1/3} + m_2^{-1/3}\right), \tag{4}$$

where k_B is the Boltzmann constant, T is temperature of the medium, and μ is the dynamic viscosity of the solution.

For practical computability, it is convenient to consider the functions f_k, instead of n_k, where f_k is the fraction by weight of k polymers in the total mass of the polymer and monomer particles in the unit mass of the solutions:

$$f_k = \frac{n_k k m_0}{\sum_{i=1}^{\infty} n_i m_0 i}. \tag{5}$$

Equation (3) in the functions f_k takes the form:

$$\begin{aligned}\frac{df_k}{dt} = \frac{1}{2\tau}\sum_{i=1}^{k-1} b_{k-i,i}\frac{k}{i(k-i)}f_{k-i}f_i \\ -\frac{f_k}{\tau}\sum_{i=1}^{\infty}\frac{b_{k,i}}{i}f_i - \lambda(k-1)f_k + \lambda k\sum_{i=k+1}^{\infty}(2-\delta_{k,i-k})\frac{f_i}{i^2}\end{aligned} \tag{6}$$

where

$$\tau = \frac{3m_0\mu}{2\rho C k_B T}, \quad b_{k,i} = \left(k^{1/3} + i^{1/3}\right)\left(k^{-1/3} + i^{-1/3}\right), \tag{7}$$

C is the total concentration of the substance that constitutes the monomer and polymer particles in the solution and ρ is the solution density.

The boundary conditions could be written as:

$$t = 0, \quad f_k = \delta_{1,k}, \ k \geq 1 \tag{8}$$

We denote Km_0 as the mass of the particle which corresponds to the MWCO of the membrane. Then the fraction by weight of the particles of the considered substance which are retained by the membrane (in the total mass of the substance before filtration) is:

$$F(C,\lambda) = \sum_{k=K}^{\infty} \lim_{t\to\infty} f_k(t;C,\lambda) \tag{9}$$

where $f_k(t;C,\lambda)$ is the solution of the equations set (6) at boundary conditions (8) and certain specified values of C and λ.

We consider the function

$$\Phi(\lambda) = \sum_{n=1}^{N} \left[F(C_n,\lambda) - F_n^{(\mathrm{exp})}\right]^2, \tag{10}$$

where C_n is the concentration of Pu in solution and $F_n^{(\mathrm{exp})}$ is the measured fraction of colloidal Pu at the total concentrations C_n.

The unknown value of λ is determined from the condition of the minimum of the function $\Phi(\lambda)$.

The Cauchy problem (6)–(8) was solved numerically by the semi-implicit method. Equation (6) was integrated with the use of the finite difference approximation of the first order of accuracy.

$$\begin{aligned} f_k^{n+1}\left(1 + \frac{\Delta t}{2\tau}\sum_{i=1}^{\infty}\frac{b_{k,i}}{i}f_k^n + \frac{\lambda(k-1)\Delta t}{2}\right) &= \frac{\Delta t}{4\tau}\sum_{i=1}^{k-1}\frac{b_{k-i,i}k}{i(k-i)}\left(f_{k-i}^n f_i^n + f_{k-i}^{n+1}f_i^{n+1}\right) \\ &\quad - \frac{f_k^n}{2\tau}\Delta t\sum_{i=1}^{\infty}\frac{b_{k,i}}{i}f_k^n - \frac{\lambda(k-1)\Delta t}{2}f_k^n \\ &\quad + \lambda\sum_{i=k+1}^{\infty}\left(2 - \delta_{k,i-k}\right)\frac{kf_k^n}{i}, \end{aligned} \tag{11}$$

where Δt is the time step of integration, $f_k^n = f_k(n\Delta t; C, \lambda)$.

Since the presented finite difference approximation is not conservative, the solution is corrected at each time step taking into account the condition

$$\sum_{k=1}^{\infty} f_k^{n+1} \equiv 1, \quad n = 0, 1, ... \tag{12}$$

The integration process is over when the time changes of f_k become insignificant.

The dynamic viscosity and density of the solution are determined from the tabular data for water at the pressure of 1 bar and temperature of 293 K. The value of m_0 is determined from the mass of Pu ion without taking into account the hydroxyl groups. The value of K (in expression (9)) for the membrane, which retains particles with a mass of more than 100 kDa, is approximately equal to 420.

The minimum value of the function $\Phi(\lambda)$ is determined by the modified gradient method with the control of the convergence.

It is shown from the calculations that the function $\Phi(\lambda)$ reaches its maximum at $\lambda = 1.343 \times 10^{-5}\ s^{-1}$. One can see in Fig. 3 that the measured and calculated

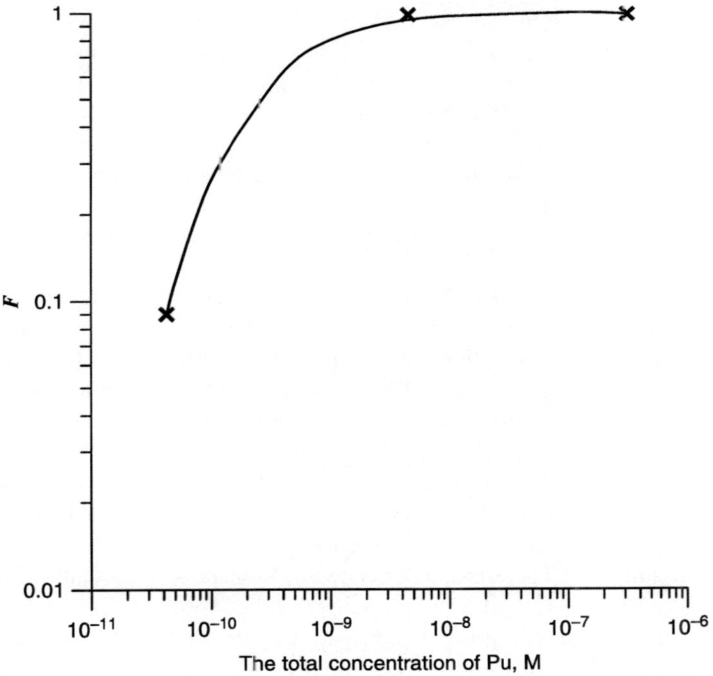

Fig. 3 Dependence between the fraction of the polymeric Pu retained by the filter (F) and the total concentration of Pu in the solution. The solid line is the calculated value and the crosses are experimental data

values of F are in good agreement at this value of λ. The corresponding equilibrium size distributions of the particles at different total concentrations of Pu are shown in Fig. 4a, b.

Attention should be paid to possible sources of inaccuracy of measurement in the theoretical model.

We consider one of the reasons for the possible errors. Use of the modified Smoluchowski equation in the form (3) is based on the assumption that each collision of the particles leads to their coagulation. However, the validity of this assumption is not unconditional. We consider to what extent its fallacy exerts influence on the results of the modeling. We denote the fraction of the collision that leads to the coagulation as α ($0 \leq \alpha \leq 1$). Then it follows from (3) that the value 1.343×10^{-5} s^{-1} obtained from the calculations is the value of λ/α, not λ. The equilibrium size distributions of the particles remain the same in this case, as shown in Fig. 4.

Another reason of possible inaccuracy could be associated with the fact that the value of m_0 is accepted to be equal to the mass of the cation Pu without taking into account the possible hydroxyl groups. However, this approximation seems to be legitimate because the mass of the cation is rather high, so the simplified estimation of m_0 cannot lead to substantial errors in the calculation of F.

At last, the error can be caused by the measurement method. The membranes with the pore size that corresponds to the particle mass of 100 kDa are used in the experiments. It is possible that some fraction of the particles with less masses is retained by the membrane (Fane et al. 1981; Salbu and Bjørnstaad 1990; Buesseler et al. 1996; Pham and Garnier 1998). This can lead to the inaccuracy in measurement of the fraction of the particles mass of which exceeds 100 kDa (i.e., in measurements of F). It is obvious that such an error is of random nature and should change from experiment to experiment. However, the results of the measurements showed a good repetition. This permits the expectation that the considered error is not significant.

One should emphasize that all calculations of the parameter λ are carried out at $C \leq 3.2 \times 10^{-7}$ M (see Table 1). This is caused by two reasons. At first, the fraction of the particles with the mass lower than 100 kDa is less than 0.1% (i.e., $1 - F = 0.001$) even at $C = 3.2 \times 10^{-7}$ M. This follows from either measurement data or the calculation results. It is obvious that we obtain $1 - F < 0.001$ from the calculations at $C > 3.2 \times 10^{-7}$ M. Therefore, it follows from (9) that the data obtained from the measurements at $C > 3.2 \times 10^{-7}$ M cannot exert a significant influence on the value of the function Φ. Therefore, these data cannot exert influence on the value of the parameter λ which is determined from the condition of the minimum of Φ. Therefore, there is no need to take into account these data in calculation of λ.

As the total concentration of Pu increases, the size distribution of the particles shifts to the right along the abscissa axis (see Fig. 4). As it follows from the results of the experiments at lower concentrations, the distribution curve shape remains the same. The distribution curve remains smooth with the only maximum. Taking into account the fact that measured values of F are close to each other at filtration

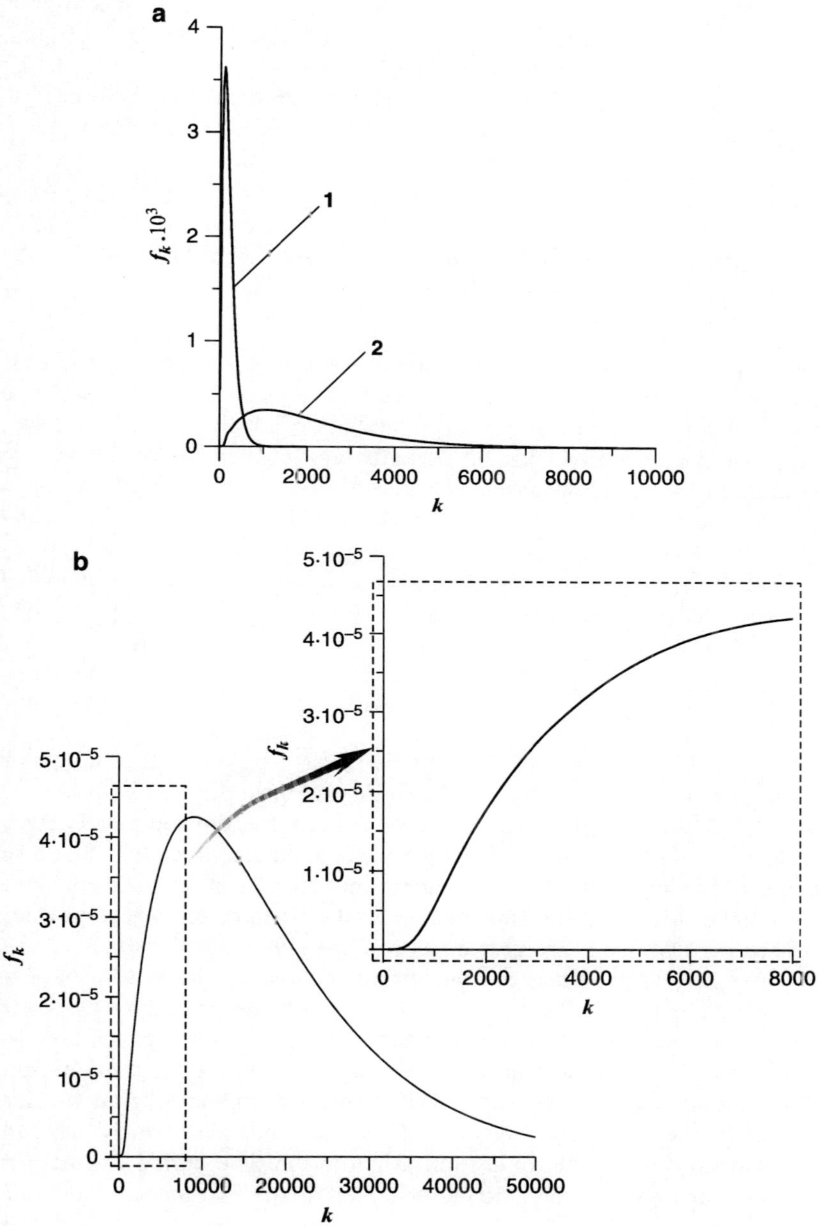

Fig. 4 Equilibrium size distributions of the polymeric Pu particles at different total concentrations of Pu in the solution. (**a**) (1) $C = 4.3 \times 10^{-11}$ M, (2) 4.6×10^{-9} M; (**b**) 3.2×10^{-7} M

through the membranes with the pore size corresponding to particle mass from 10 to 100 kDa, one may expect occurrence of the inflection in the left parts of the distribution curves at $k > K$ and total concentrations of Pu of the order of 10^{-5} M and at higher values. This is in good agreement with the curve shape in Fig. 4b, where this inflection exists even at the concentration of 3.2×10^{-7} M.

2.2 *Influence of the Intrinsic Colloids on Velocity of Radioactive Plume Movement*

As it is shown in this analysis, establishing of a new equilibrium between the formation and disintegration of intrinsic colloids at variation of concentration of the radionuclides in ionic form in the solution is relatively fast. We estimate the travel path of the plume, which is sufficient for breakup of the polymeric particles at a significant decrease of the total concentration of the radionuclide in the groundwater. We denote this travel path as L_b, and the velocity of the contaminant plume movement is denoted as V_m. The dimensionless Damköhler number Da which characterizes how fast is the reaction relative to the velocity of the contaminant plume movement is introduced:

$$Da = \frac{L_b \lambda}{V_m}. \tag{13}$$

Let us take V_m equal to 100 m per year $= 31.75 \times 10^{-7}$ m s^{-1}. As it follows from the modeling, $\lambda = 1.343 \times 10^{-5}$ s^{-1}. Then Da is of the order of unity if L_b is of the order of 0.24 m. Therefore if the plume reaches the domain where the total concentration of radionuclides is negligibly small, polymeric particles are disintegrated. As it follows from this estimation, if the plume reaches the domain where the total concentration of the radionuclides in the groundwater is negligibly small, polymeric particles are disintegrated fast with the plume travel distance.

Let us consider the influence of the intrinsic colloids on the movement of some contaminant plume which is sketched in areal coordinates in Fig. 5.

The movement of the contaminant plume is caused by the regional flow of the groundwater, and the direction of its velocity is shown by the arrow in Fig. 5. Let us assume that the velocity of the intrinsic colloid migration exceeds significantly the migration velocity of the radionuclides which are carried by the groundwater in the ionic form. Then in the time period Δt, the part of the contaminant plume where the radionuclides are present predominantly in the colloid form passes ahead of the plume part where the radionuclides are in the ionic form and comes to the domain of the underground medium where the concentration of the radionuclides in ionic form is near zero (Fig. 6). In consequence, the new equilibrium between the processes of formation and destruction of the intrinsic colloids establishes relatively fast. As a result, the number of the

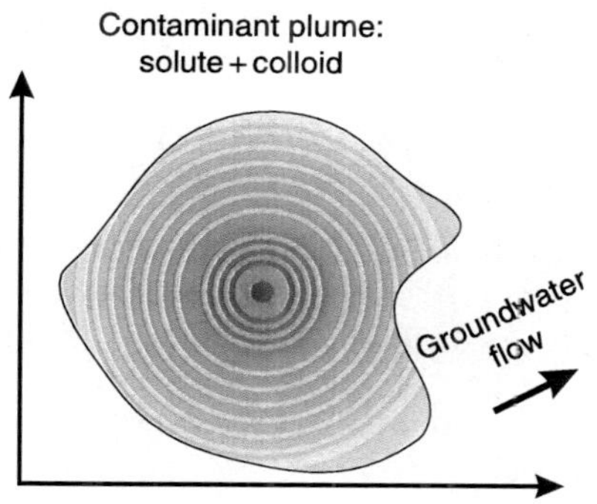

Fig. 5 Initial position of the radioactive contaminant plume in areal coordinates

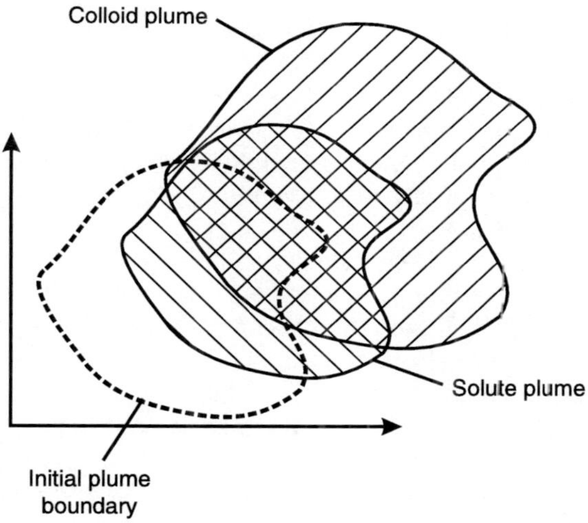

Fig. 6 Position of the radioactive contaminant plume in the time Δt

intrinsic colloid particles and their average size decrease significantly, and radionuclides in this part of the plume are predominantly in ionic form. Therefore, the movement velocity of this part of the contaminant plume decreases fast to the value which corresponds to migration velocity of the radionuclides in ionic form. Therefore, the influence of the intrinsic colloid does not lead to a significant increase in the velocity of radioactive contamination propagation in underground systems.

Expression (1) characterizes the role of the radiocolloid of a certain type on the velocity of radionuclide migration. As it is shown, two of three efficients in the right-hand part of the expression tend to zero very fast with time. The third coefficient remains limited. Therefore, their product tends to be zero, which implies that the role of the intrinsic colloid in radioactive plume migration in underground medium is not substantial.

3 Modeling of Radionuclides Transport by Primary Colloid

3.1 *Sorption of the Primary Colloid on the Host Rocks*

The radionuclides are approximately uniformly distributed in each particle of the primary colloid; the radionuclides are often constituents of the lattice of the particle substance. For this reason, one can expect that the radionuclides are strongly bound to the particles of the primary colloid.

The initial values of n_c in the plume can be determined from analysis of the samples of the polluted groundwater. Since the radionuclides are strongly bound to the particles of the primary colloid, one can expect that γ_c remains practically constant in time. As it follows from the section, the ratio of the colloid velocity to the groundwater velocity depends on sorption properties of the host rocks to the primary colloid. Therefore, determination of the ratio of the colloid velocity to the groundwater flow velocity calls for development of a model of colloid sorption/precipitation on the surface of the flow channels.

Composition of the primary colloid particles differs due to their origin from the composition of the host rocks. No such particles are bound initially to the surfaces of the flow channels. Therefore, all potential places of the particles sorption/precipitation are free.

The colloid particles differ in size. This should be taken into account in the model because particle sorption rate can depend on the particle size.

Let us consider the theoretical model of the precipitation of the radiocolloid, which heterogeneous in size, on the potential places of sorption. We subdivide the system of the colloid particles into *N* types so that the particles of the same type are approximately identical in size whereas the particles of different types are different. The process of sorption/desorption in this system of the particles is shown as the state diagram of a potential sorption place in Fig. 7. Absence of the precipitated particles corresponds to zero state (i.e., the sorption place is free); the state *j* corresponds to presence of a particle of the type *j* on the sorption place. It is evident that the state "0" can be changed for any state "*j*", whereas any nonzero state "*j*" can be changed only for the state "0".

We denote $P(n, t_1 | j, t_0)$ the probability that the sorption place at the time moment t_1 is occupied by a particle of the type *n* under condition that this place was occupied by the particle of the type *j* at the time moment t_0 ($n = 0$ implies that the sorption place is free).

Let us denote conditional probability that the system is in the state *n* at the moment *t* if it was in the state *m* at the moment t_0 as $P(n,t|j,t_0)$. Let us denote

$$W(n|m,t) = \lim_{\Delta t \to 0} \frac{P(n, t + \Delta t|m, t)}{\Delta t}. \tag{14}$$

The process of sorption/desorption shown in Fig. 7 satisfies to the set of Kolmogorov–Feller equations of the following type (Gardiner 1983):

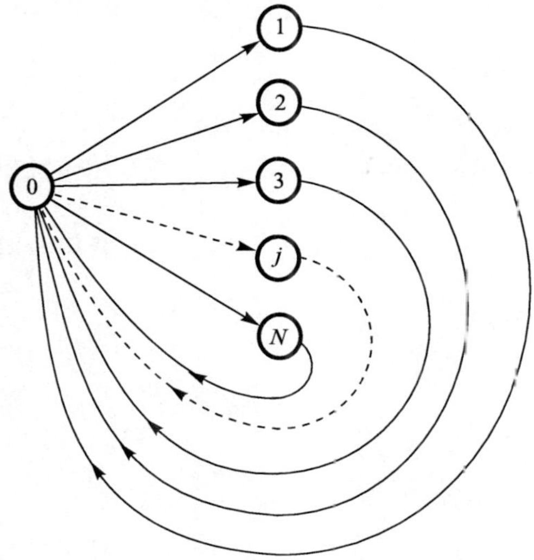

Fig. 7 The state diagram of a potential sorption place of the colloid particles on the host rocks

$$
\begin{aligned}
\frac{\partial P(1,t|j,t_0)}{\partial t} &= \sum_{m=0}^{N} [W(1|m,t)P(m,t|j,t_0) - W(m|1,t)P(1,t|j,t_0)], \\
&\dots \\
\frac{\partial P(N,t|j,t_0)}{\partial t} &= \sum_{m=0}^{N} [W(N|m,t)P(m,t|j,t_0) - W(m|N,t)P(N,t|j,t_0)], \\
& j = 1,\dots,N.
\end{aligned}
\tag{15}
$$

It follows from the scheme in Fig. 7 that $W(n|m,t) = 0$, if neither n nor m are not equal to zero. As the probability of transition from the state 0 to the state n depends on the volume concentration of particles of the nth type and velocity of the chaotic momentum of these particles, $W(n|0,t)$ must depend on n. At the same time, the desorption of particles is governed mainly by the character of chaotic eddy currents in the liquid in the vicinity of the potential sorption place; therefore, one can suppose that the quantity $W(0|m,t)$ does not depend on m. Then we obtain

$$
W(n|m,t) = \begin{cases} k_n & \text{if} \quad m = 0; \\ \kappa & \text{if} \quad n = 0,\ m \neq 0; \\ 0 & \text{if} \quad mn \neq 0; \end{cases}
\tag{16}
$$

Substitution of the explicit expressions for $W(n|m,t)$ into Kolmogorov–Feller equations gives:

$$
\frac{\partial P(n,t|j,0)}{\partial t} = k_n P(0,t|j,0) - \kappa P(n,t|j,0), \quad n = 1,\dots,N; \quad j = 0,\dots,N. \tag{17}
$$

The initial conditions for system (17) take the form at $n \neq 0$.

$$P(n,0|j,0) = \delta_{n,j}, \tag{18}$$

where $\delta_{n,j}$ is Kronecker delta.

We sum all equations from (17) from $n = 1$ to $n = N$. We obtain:

$$\frac{\partial P_\Sigma(t|j,0)}{\partial t} = k_\Sigma P(0,t|j,0) - \kappa P_\Sigma(t|j,0); \quad j = 0, ..., N \tag{19}$$

where

$$P_\Sigma(t|j,0) = \sum_{n=1}^{N} P(n,t|j,0), \quad k_\Sigma = \sum_{n=1}^{N} k_n. \tag{20}$$

Since

$$P(0,t|j,0) = 1 - P_\Sigma(t|j,0), \tag{21}$$

Equation (19) can be written as

$$\frac{\partial P_\Sigma}{\partial t} = k_\Sigma(1 - P_\Sigma) - \kappa P_\Sigma. \tag{22}$$

As the sorption place at the initial time moment is free, $j = 0$, we obtain the initial condition for (22) at $j = 0$ from the definition of P_Σ and the initial conditions (18):

$$P_\Sigma(0|0,0) = 0 \tag{23}$$

The solution of the Cauchy problem (3.5)–(3.6) takes the form

$$P_\Sigma(t|0,0) = \frac{k_\Sigma}{k_\Sigma + \kappa}\{1 - \exp[-(k_\Sigma + \kappa)t]\}. \tag{24}$$

Substitution of (21) and (24) into equations of the system (17) gives

$$\frac{\partial P(n,t|0,0)}{\partial t} = \frac{k_n}{k_\Sigma + \kappa}\{\kappa - k_\Sigma \exp[-(k_\Sigma + \kappa)t]\} - \kappa P(n,t|j,0), \; n = 1, \ldots, N. \tag{25}$$

The solution of the Cauchy problem (25)–(18) is written in the following form

$$P(n,t|0,0) = \frac{k_n}{k_\Sigma + \kappa}\{1 - \exp[-t(k_\Sigma + \kappa)]\}. \tag{26}$$

It follows from (26) that

$$P(n,t|0,0) \underset{t\to\infty}{\to} \frac{k_n}{k_\Sigma + \kappa}. \tag{27}$$

In accordance with the law of large numbers, $P(n,t|0,0)$ can be considered as a fraction of the potential places of sorption occupied by particles of the nth type.

If the number of the particles in the liquid is low, and their mutual influence can be neglected, quantities k_i can be represented as

$$k_i = k_0 \langle v_n \rangle_i n_i, \tag{28}$$

where $\langle v_n \rangle_i$ is the average component of velocity of the particles of the ith type normal to sorption surface and n_i is the number of the particles of the ith group per unit volume of the liquid.

Let us assume that particles of all types are in thermodynamic equilibrium with liquid molecules, and that the particles execute a continuous chaotic motion. Taking into account these assumptions, one can expect that particle momentum satisfies the Maxwell distribution. Then, the distribution function of the momentum vector for particles of the ith type in any system of spherical coordinates takes the form

$$\phi(I,\psi,\vartheta) = \frac{1}{(2\pi M_i k_B T)^{3/2}} \exp\left\{-\frac{I^2}{2M_i k_B T}\right\} I^2 \sin\psi \tag{29}$$

where $\{I, \psi, \vartheta\}$ are spherical coordinates of the vector of the particle momentum (I is the absolute value of momentum, ψ is the latitude, ϑ is the longitude), M_i is the mass of a particle of the ith type, k_B is the Boltzmann's constant, and T is the absolute temperature.

Let us select such spherical coordinates in a point of sorption surface that the plane $\psi = 0$ coincides with a tangent plane to the sorption surface in this point. Then an average velocity of particles of the type i toward the sorption surface in this point is determined by the expression

$$\langle v_n \rangle_i = \frac{1}{M_i} \int_0^\infty dI \int_0^{2\pi} d\vartheta \int_0^{\pi/2} I \cos\psi \; \phi(I,\psi,\vartheta) d\psi. \tag{30}$$

Substitution of the expression for the Maxwell distribution function into this formula gives

$$\begin{aligned} \langle v_n \rangle_i &= \frac{1}{M_i} \frac{1}{(2\pi M_i k_B T)^{3/2}} \int_0^\infty dI \int_0^{2\pi} d\vartheta \int_0^{\pi/2} I^3 \sin\psi \cos\psi \exp\left\{-\frac{I^2}{M_i k_B T}\right\} d\psi \\ &= \sqrt{\frac{k_B T}{2\pi M_i}}. \end{aligned} \tag{31}$$

This implies that

$$k_i = k_0 n_i \sqrt{\frac{k_B T}{2\pi M_i}}. \tag{32}$$

Let us assume that particles are of the same density ρ_s. Then

$$M_i = \frac{\pi d_i^3 \rho_s}{6}. \tag{33}$$

Hence,

$$k_i = \frac{k_0}{\pi}\sqrt{\frac{3k_B T}{\rho_s}} \frac{n_i}{d_i^{3/2}}. \tag{34}$$

As a scale for d_i, we select a quantity d_0 that is characteristic of the considered size distribution of the particles (e.g., average size of particles) and introduce a dimensionless variable

$$\delta_i = \left(\frac{d_i}{d_0}\right)^{3/2}. \tag{35}$$

Then the quantity k_i can be represented as

$$k_i = \frac{k_0}{\pi}\sqrt{\frac{3k_B T}{\rho_s d_0^3}} \frac{n_i}{\delta_i}. \tag{36}$$

The quantity $P(n,t|0,0)$ can be considered as a fraction of all the potential places of sorption (i.e., of the total sorption capacity of the host rocks relative to the colloid particles) occupied by the colloid of the type n. This implies that expression (27) can be represented in the form

$$\frac{w_n}{w_{\lim}} = \frac{k_n}{\kappa + \sum_{i=1}^{N} k_i}, \quad n = 1, \ldots, N, \tag{37}$$

where w_n is the number of particles of the nth type adsorbed in a unit volume and $w_{\lim}$ is the number of potential sorption places in a unit volume.

Substitution of the expression for k_i into this formula gives

$$\frac{w_m}{w_{\lim}} = \frac{n_m}{\delta_m \left(K + \sum_{i=1}^{N} n_i/\delta_i\right)} \tag{38}$$

where

$$K = \frac{\kappa\pi}{k_0}\sqrt{\frac{\rho_s d_0^3}{3k_B T}} \tag{39}$$

(i.e., the constant K is the same for particles of all types).

If $N = 1$, (38) takes the form of the Langmuir' isotherm of sorption (Gregg and Sing 1967).

The unknown parameters k_0, κ, and the total sorption capacity of the host rocks relative to the colloid particles w_0 can be determined from an experiment.

3.2 *Convective Transport of the Colloidal Particles Under Equilibrium Sorption*

As an example of one-dimensional flow of a liquid with colloidal particles through a porous medium, we consider a flow in a cylindrical tube filled with rock (Fig. 8).

In this case, the transport of particles of the mth type is described by the equation of transient convective mass transfer

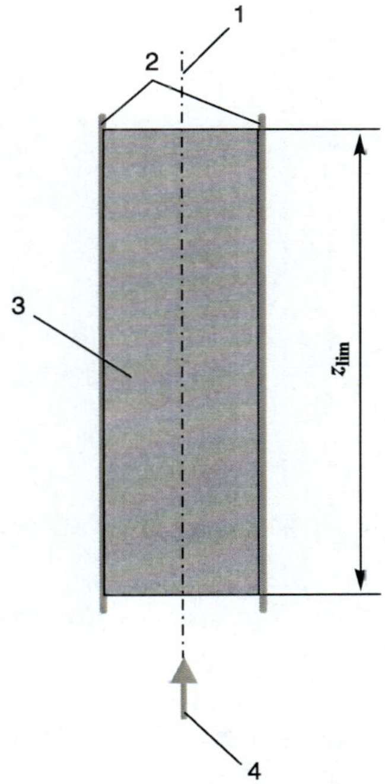

Fig. 8 Flow of a liquid in a tube filled with a granular adsorbent. (1) Tube axis, (2) tube walls, (3) the granular layer, and (4) the direction of the liquid flow

$$\frac{\partial n_m}{\partial t} + \frac{V}{\varphi}\frac{\partial n_m}{\partial z} = -\frac{Q_m}{\varphi}, \quad m = 1, \ldots, N, \tag{40}$$

where V is velocity of the liquid flow in the granular layer, φ is porosity of the layer, Q_m is the number of the particles of the mth type sorbed per unit time from unit volume of the liquid on the rock in unit volume of the layer, z is the coordinate directed downstream along the tube axis, and t is the time.

We assume here that Darcy velocity of the liquid flow through the porous medium is so small that sorption of colloidal particles can be considered as equilibrium sorption. This implies that interaction of the particles with the potential sorption place can be described with a sufficient accuracy by the steady-state solution (27). Then the number of the colloidal particles sorbed on the solid phase is determined by (38). Let us substitute the expression of Q_m from (38) into (40). We obtain

$$\varphi\frac{\partial n_m}{\partial t} + V\frac{\partial n_m}{\partial z} = -\frac{\partial w_m}{\partial t}, \quad m = 1, \ldots, N \tag{41}$$

Since this is an equilibrium process, we can substitute the explicit expression of the quantity wi from (38) into (41). Then (41) after simple formal transformations is rewritten in the form

$$\frac{\partial n_m}{\partial t}\left[\varphi\delta_m\xi^2 + w_0\xi\right] - w_0 n_m \sum_{i=1}^{N}\frac{1}{\delta_i}\frac{\partial n_i}{\partial t} + V\delta_m\xi^2\frac{\partial n_m}{\partial z} = 0, \quad m = 1, \ldots, N \tag{42}$$

where

$$\xi = K + \sum_{i=1}^{N}\frac{n_i}{\delta_i}. \tag{43}$$

This set of equations describes in equilibrium approximation one-dimensional convective transport of colloidal particles at flow of liquid through porous medium.

4 Modeling of Radionuclides Transport by Pseudocolloid

4.1 *Sorption of the Pseudocolloid on the Host Rock*

The particles of pseudocolloid represent the main part of the radiocolloid. The groundwater contained colloid particles of natural origin before the radioactive pollution. These particles are termed protocolloid. Pseudocolloid is formed due to pollution of protocolloid by radionuclides from the contaminated groundwater. A substantial part of the composition of protocolloid (and hence, pseudocolloid) particles is close to that of the host rocks. Therefore, ζ potentials of the particles and the surface of the immovable solid phase are of the same sign. As a result, the particles are sorbed weakly on the host rocks due to the repulsion barrier, and mobility of the radionuclides bound to these particles is higher than the mobility of the radionuclides carried by the groundwater in ionic form (as the solute) (Smith and Degueldre 1993). This is the simplified image of the pseudocolloid migration. One should take into account that the host rocks are heterogeneous. The role of this heterogeneity was especially clearly shown in the experiments in (Elimelech et al. 2000). The heterogeneous rock was represented in these experiments as a layer of solid grains with different signs of ζ potential. The colloid precipitated on the layer even when the average measured ζ potential of the layer was of the same sign as the ζ potential of the colloid. However, the actual role of the rock heterogeneity (and, hence, the role of the colloid composition heterogeneity) is damped by the formation of a cover from the humic and fulvic acids on the surfaces of the particles. The assumption that inorganic particles are covered by humic and fulvic acid was made in (Davis and Gloor 1981; O'Melia 1987; Ledin et al. 1994; Fairhurst et al. 1995). It was confirmed by laboratory experiments in (Tipping 1981; Davis 1982). The presence of the organic cover on inorganic pseudocolloid from the groundwater sampled at the site of significant radioactive pollution was shown by the method of ESCA in (Malkovsky et al. 2009). The results of the study demonstrated that the cover can be noncontinuous. One can expect that the surfaces of the flow channels are also covered by the organic coat. The ζ potential of this cover can be of the same sign as the ζ potential of the particles covered by organics, which prevents sorption of the particles on the host rock. Possible discontinuities of this cover can be potential places of the pseudocolloid sorption. However, in the assessment of sorption properties of the host rock to the pseudocolloid, it should be taken into account that the groundwater with the protocolloid particles was in interaction with the host rocks for a long time before the radioactive pollution initiated Therefore, the distribution of the protocolloid particles between the groundwater and the host rocks is already in equilibrium by the time of contaminant plume approaching. As a result, the sorption of the pseudocolloid from the polluted groundwater is not the sorption on free places, but the substitution of the previously sorbed nonradioactive colloid of practically the same composition (Fig. 9).

Expression (26) describes, in particular, the process of protocolloid sorption. Since the protocolloid in the groundwater interacts with the host rocks for a long time, the probability that the potential sorption place is free by the time of the radioactive contaminant plume approaching is determined from solution (27) at $t \to \infty$:

$$p_0 = 1 - \lim_{t\to\infty} \sum_{n=1}^{N} P(n,t|0,0) = \frac{\kappa}{\kappa + k_\Sigma}. \tag{44}$$

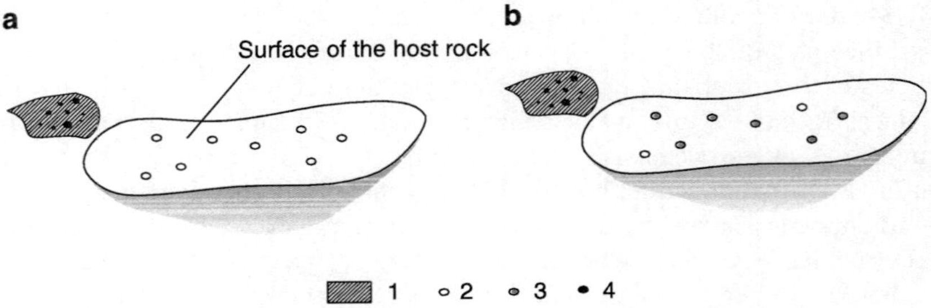

Fig. 9 Sorption of the pseudocolloid on the wall of the flow channel. (**a**) Sorption of pseudocolloid on the surface of the channel wall if the all potential sorption places are free; (**b**) sorption of pseudocolloid as substitution of previously sorbed protocolloid. 1. Contaminated groundwater; 2. sorption site; 3. protocolloid; 4. radiocolloid

Sorption of a radioactive colloid particle of the type n can take place as a result of one of two antithetical events:

- The potential sorption place is free at the plume approaching, and the particle of the type n sorbs in it.
- The sorption place is occupied by the particle of the protocolloid; then the sorption place becomes empty, and only after that it is occupied by the particle of the pseudocolloid of the type n.

Let us denote the time moment of the contaminant plume approaching to the sorption place as t_f. We introduce the new variable $t' = t - t_\mathrm{f}$ which is the time from the time moment of the plume approaching. Let us consider the probability that the protocolloid particle desorbs from the sorption place for a time τ $(0 < \tau < t')$. We subdivide the interval $(0, \tau)$ into small segments of the length $\Delta\tau$. The probability that the protocolloid particle desorbs from the sorption place is $\kappa\Delta\tau$. The probability that the particle does not desorb from the sorption place for the time $\Delta\tau$ is $1 - \kappa\Delta\tau$. Therefore, the probability that the particle desorbs from the sorption place during the time interval from τ to $\tau + \Delta\tau$ is

$$(1 - \kappa\,\Delta\tau)^{\tau/\Delta\tau}\kappa\,\Delta\tau. \tag{45}$$

We denote by T_m the random event that the protocolloid particle desorbs from the sorption place during the time interval (τ_m, τ_{m+1}) where $\tau_m = m \times \Delta\tau$. The probability that the sorption place at the approaching of the contaminant plume is occupied by the protocolloid particle, which desorbs for the time τ_m, and the particle of the pseudocolloid of the type n sorbs on that place by the time moment t' is denoted by P_n^+. All events T_m are antithetical. Therefore,

$$P_n^+ = \sum_{m \geq 0} P(T_m)P(n, t'|T_m), \tag{46}$$

where $P(T_m)$ is the probability of T_m and $P(n,t'|T_m)$ is the probability that the pseudocolloid particle of the type n sorbs by the time moment t' under condition of T_m.

It can be written in accepted notation

$$P(n,t'|T_m) = P(n,t'-\tau_m|0,0). \tag{47}$$

Since $(1-\kappa\,\Delta\tau)^{\tau_m/\Delta\tau} \to \mathrm{e}^{-\kappa\tau_m}$ at $\Delta\tau \to 0$, we obtain from (46) in the limit at $\Delta\tau \to 0$

$$P_n^+ = \int_0^{t'} \kappa \mathrm{e}^{-\kappa\tau} P(n,t'-\tau|0,0)\mathrm{d}\tau. \tag{48}$$

The probability that the pseudocolloid particle of the type n precipitates at the sorption place by the time moment t' under condition that the sorption place is empty by the moment of the plume approaching is $P(n,t'|0,0)$. We denote by $P(n,t')$ the probability that the sorption place is occupied by the pseudocolloid particle of the type n at the time moment t'. Then

$$P(n,t') = p_0 P(n,t'|0,0) + (1-p_0)P_n^+ = \frac{k_n}{k_\Sigma + \kappa}\left(1 - \mathrm{e}^{-\kappa t'}\right). \tag{49}$$

The probability that the potential sorption place is occupied by a pseudocolloid particle from the ith group can be construed as the ratio of the number of the colloid particles sorbed on the rock w_i to the total number of the potential sorption places on the rock $w_{\lim}$

$$\frac{w_i}{w_{\lim}} = \frac{k_i}{k_\Sigma + \kappa}\left(1 - e^{-\kappa t'}\right). \tag{50}$$

It follows from (37) that the ratio of w_i to $w_{\lim}$ at sorption on initially free sorption places takes the form

$$\frac{w_i}{w_{\lim}} = \frac{k_i}{k_\Sigma + \kappa}. \tag{51}$$

The known models consider sorption of the pseudocolloid on the host rock as sorption on the initially free surface of the channels of the groundwater flow. In fact sorption of the pseudocolloid is substitution of the previously sorbed protocolloid. From the practical point of view, it is worthwhile to estimate the difference between the forecasts obtained with use of these two models of the pseudocolloid sorption. With that end in view, we consider transport of the pseudocolloid by the groundwater in one-dimensional approximation. It is assumed that dispersion is minor, and the transport of the pseudocolloid is caused predominantly by convection. This modeling approximation is valid for aquifers formed by homogeneous slightly cemented sedimentary rocks (Bochever and Oradovskaya 1972). For the sake of simplicity we assume that the colloid is homogeneous, i.e., it consists of equal particles. This is not the case in general; however, this approximation simplifies the analysis of the role that the colloid sorption mechanism plays in the simulation of radioactive pollution propagation.

Transport of pseudocolloid in this case is governed by the following equation

$$\rho\varphi\frac{\partial C}{\partial t} + \rho_r(1-\varphi)\frac{\partial C_r}{\partial t} + \rho v\frac{\partial C}{\partial z} = 0, \tag{52}$$

where ρ is density of the groundwater, ρ_r is density of the rock, φ is rock porosity, v is Darcy velocity of the groundwater, t is time, and z is the spatial coordinate.

Since the concentrations of the pseudocolloid in the groundwater and on the host rock are equal to zero at $t \leq 0$ and $z > 0$, initial conditions for (52) are written as

$$t = 0, \quad z > 0, \quad C = C_r = 0. \tag{53}$$

We assume that the contaminant plume front comes at $t = 0$ to the left boundary of the considered domain with the spatial coordinate $z = 0$. Then the boundary condition for (52) takes the form

$$z = 0, \quad t \geq 0, \quad C = C_0, \tag{54}$$

where C_0 is certain specified value of the pseudocolloid concentration in the polluted groundwater that comes into the considered domain of rocks (this domain corresponds to spatial coordinates $z \geq 0$).

Let us consider first the model where the sorption of the pseudocolloid is considered as substitution of the previously sorbed protocolloid. As the colloid is homogeneous, expression (50) is written as

$$\frac{C_r}{C_{\lim}} = \frac{k}{k+\kappa}\{1 - \exp[-\kappa t']\}, \tag{55}$$

where C_r is concentration (fraction by weight) of the pseudocolloid on the host rock and $C_{\lim}$ is sorption capacity of the host rock relative to the pseudocolloid.

Because v is constant at one-dimensional flow, the contaminant front comes to the rock at some value of spatial coordinate z in the time z/V, where $V = v/\varphi$ is flow velocity. Therefore,

$$t' = t - z/V. \tag{56}$$

This implies

$$\frac{C_r}{C_{\lim}} = \frac{k}{k+\kappa}\{1 - \exp[-\kappa(t - z/V)]\}. \tag{57}$$

Since the colloid is homogeneous, the parameter k in (57) is without the index of the group of the pseudocolloid particles.

If the sorption of the pseudocolloid takes place on the surface of the flow channels which is initially free of the protocolloid, the ratio of the pseudocolloid concentration in the groundwater and on the host rock satisfies (51) which is written in the considered case as

$$\frac{C_{\mathrm{r}}}{C_{\mathrm{lim}}} = \frac{k}{k+\kappa}. \tag{58}$$

Since $n \sim C$, the rate of the pseudocolloid precipitation can be represented as $k = AC$ where A is constant.

Let us introduce the dimensionless variables

$$Z = \frac{z}{L}, \quad \vartheta = \frac{tV}{L}, \quad c = \frac{\rho\varphi C}{\rho_r(1-\varphi)C_{\mathrm{lim}}}, \quad c_{\mathrm{r}} = \frac{C_{\mathrm{r}}}{C_{\mathrm{lim}}}. \tag{59}$$

Then (52) in the case (57) (i.e., if the sorption of the pseudocolloid is considered as substitution of the protocolloid) is written as

$$\frac{\partial c}{\partial \vartheta}\left\{1 + \frac{K_{\mathrm{A}}}{(1+K_{\mathrm{A}}c)^2}\left[1 - \mathrm{e}^{-Da(\vartheta - Z)}\right]\right\} + \frac{\partial c}{\partial Z} = -\frac{DaK_{\mathrm{A}}}{1+K_{\mathrm{A}}c}\mathrm{e}^{-Da(\vartheta - Z)}. \tag{60}$$

where

$$K_{\mathrm{A}} = \frac{AC_{\mathrm{lim}}\rho_{\mathrm{r}}(1-\varphi)}{\kappa\rho\varphi}, \quad Da = \frac{\kappa L}{V}. \tag{61}$$

The dimensionless number K_{A} characterizes sorption properties of the rock relative to the pseudocolloid. If the concentrations of the pseudocolloid in the groundwater are low (what, as a rule, takes place in fact), the value $K_{\mathrm{A+1}}$ is approximately equal to the retardation coefficient of the colloid particles as the surface of the flow channels is free of the protocolloid. The dimensionless number Da is an analog of the Damköhler number and characterizes the ratio of the rates of two processes: desorption of the colloid particles from the surface of the flow channels and movement of the groundwater through the considered domain of the rock; ϑ is dimensionless time.

If sorption of the pseudocolloid is considered as precipitation on the free surface of the flow channels (i.e., in the case of (58)), (52) is written as

$$\frac{\partial c}{\partial \vartheta}\left[1 + \frac{K_{\mathrm{A}}}{(1+K_{\mathrm{A}}c)^2}\right] + \frac{\partial c}{\partial Z} = -\frac{K_{\mathrm{L}}K_{\mathrm{A}}}{1+K_{\mathrm{A}}c}. \tag{62}$$

The initial condition (53) in the dimensionless form is written as

$$\vartheta = 0, \quad Z > 0, \quad c = c_r = 0. \tag{63}$$

The boundary condition (54) in the dimensionless form is written as

$$Z = 0, \quad \vartheta \geq 0, \quad c = c_0, \tag{64}$$

where

$$c_0 = \frac{\rho \varphi C_0}{\rho_r (1 - \varphi) C_{\text{lim}}}. \tag{65}$$

Equations (60) and (62) at the initial condition (63) and the boundary condition (64) are integrated numerically by the semi-implicit method with the upwind finite difference approximation of the convective terms (Roache 1976).

Let us analyze the results of the numerical simulation at K_A=3, i.e., at relatively high sorption properties of the rock relative to the particles of the pseudocolloid (and protocolloid). We consider first the case when the Damköhler number is low ($Da = 0.1$), i.e., the velocity of the groundwater flow is much higher than the rate of the colloid desorption from the rock. The distributions of the dimensionless concentration of the pseudocolloid in the groundwater throughout the whole considered domain of the rock (i.e., at $0 \leq z \leq L$, which can be written in the dimensionless form as $0 \leq Z \leq 1$) at different values of time, are shown in Figs. 10–12 for both the models of pseudocolloid precipitation on the host rock.

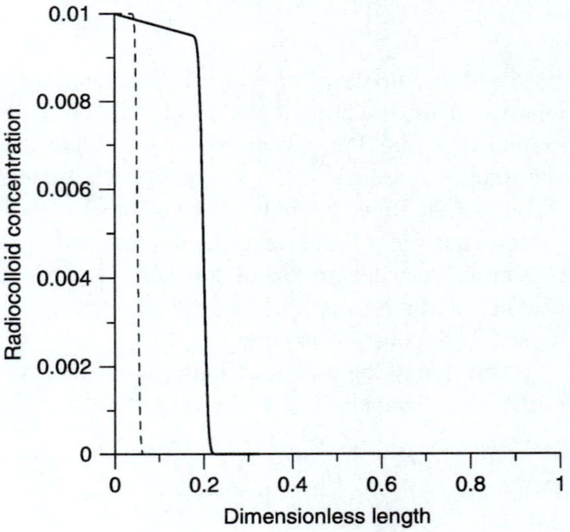

Fig. 10 Distribution of the dimensionless concentration of the pseudocolloid in the groundwater along the rock massif length at $\vartheta = 0.2$; $K_A = 3$, Da $= 0.1$, $c_0 = 0.01$. The solid line denotes solution of (60) (i.e. corresponds to the case when sorption of the pseudocolloid is considered as substitution of the protocolloid); the dashed line denotes solution of (62) (i.e. corresponds to the case when presence of the previously sorbed protocolloid is not taken into account, and the host rock is assumed free of the previously sorbed colloid)

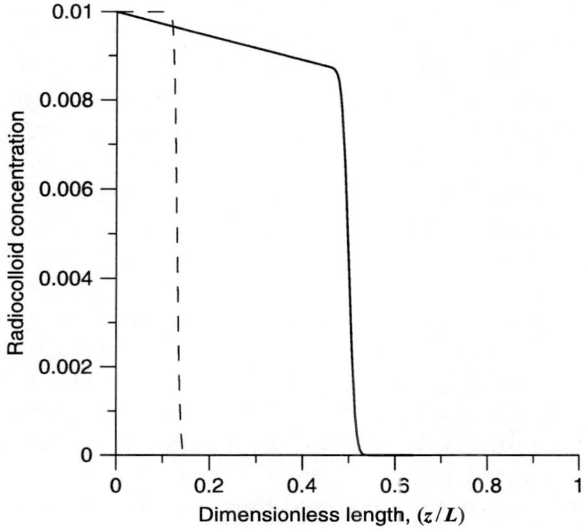

Fig. 11 Distribution of the dimensionless concentration of the pseudocolloid in the groundwater along the rock massif length at $\vartheta = 0.5$; $K_A = 3$, $\mathrm{Da} = 0.1$, $c_0 = 0.01$. The *solid line* denotes solution of (60); the dashed line denotes solution of (62)

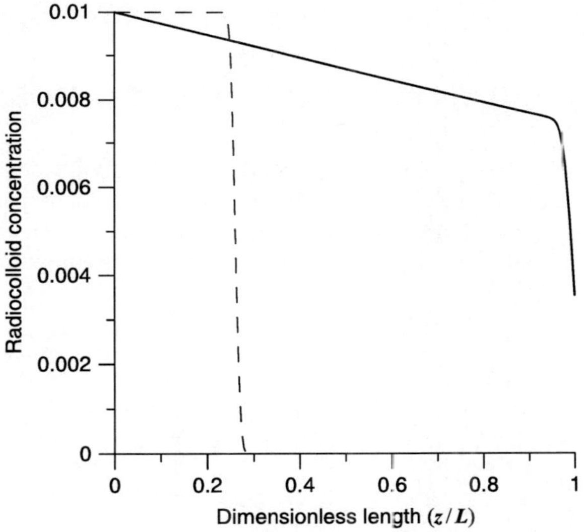

Fig. 12 Distribution of the dimensionless concentration of the pseudocolloid in the groundwater along the rock massif length at $\vartheta = 1$; $K_A = 3$, $\mathrm{Da} = 0.1$, $c_0 = 0.01$. The *solid line* denotes solution of (60); the *dashed line* denotes solution of (62)

One can see that the front of the contaminant in the colloid form moves much faster in the case when sorption of the pseudocolloid is considered with regard for the nonradioactive colloid sorbed previously on the host rock than in the case when the presence of the previously sorbed nonradioactive colloid on the host rock is not taken into account in the model of the pseudocolloid sorption. The contaminant front reaches practically the right boundary of the rock domain (Fig. 11) at $\vartheta = 1$ in the first case, whereas the travel of the contaminant front calculated with use of the traditional model (where the rock is assumed to be initially free of the sorbed nonradioactive colloid particles) is approximately a quarter of this distance for the same time. Therefore, calculations on the basis of the traditional model at Damköhler number of the order of 0.1 lead to substantial underestimation of the colloid-facilitated transport of radionuclides by groundwater.

Let us now consider the case when the Damköhler number is high ($K_L = 10$), i.e., the groundwater moves with low velocity relative to the rate of desorption of the protocolloid from the rock. Distributions of the dimensionless concentration of the pseudocolloid in the groundwater throughout the whole considered domain of the rock at different time values are shown in Figs. 13 and 14 for both the models of pseudocolloid sorption on the host rock.

One can see that calculation of the pseudocolloid distribution at Damköhler number of the order of 10 gives near results. The only substantial difference between the pseudocolloid distributions calculated with use of two models is that the distributions obtained on the basis of the model where the pseudocolloid precipitation is considered as substitution of the previously sorbed protocolloid are fuzzier. Therefore, the presence of the previously sorbed protocolloid at high Damköhler number exerts approximately the same influence on pseudocolloid transport by the groundwater as the hydraulic dispersion.

5 Effect of Elevated Velocity of Colloid Migration

The ecological hazard caused by pseudocolloids depends on their mobility. Particles are carried by groundwater, but their velocity can be less than the velocity of the groundwater due to their sorption on the immovable solid phase. If sorption of the colloid or the solute transported by the groundwater on the host rocks is weak or

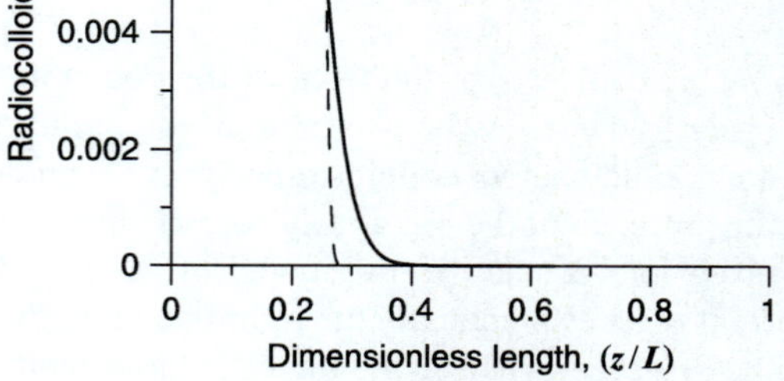

Fig. 13 Distribution of the dimensionless concentration of the pseudocolloid in the groundwater along the rock massif length at $\vartheta = 1$; $K_A = 3$, Da $= 10$, $c_0 = 0.01$. The *solid line* denotes solution of (60); the *dashed line* denotes solution of (62)

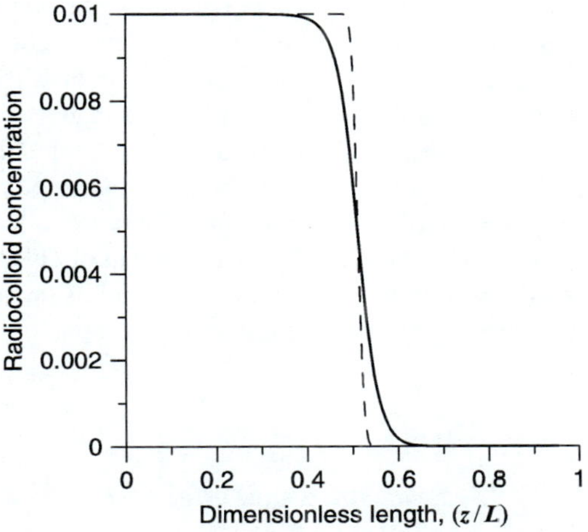

Fig. 14 Distribution of the dimensionless concentration of the pseudocolloid in the groundwater along the rock massif length at $\vartheta = 2$. $K_A = 3$, Da $= 10$, $c_0 = 0.01$. The *solid line* denotes solution of (60); the *dashed line* denotes solution of (62)

absent, one can expect that the velocity of their migration is close to the interstitial velocity of the groundwater. Such solutes (tracers) are in common use in field tests for estimation of interstitial velocity of groundwater (tracer tests). It seems to be reasonable that migration velocity of any substance transported by groundwater should be less than the groundwater velocity. Therefore, migration velocity of any substance should be less than tracer velocity. However, on the basis of the results of their laboratory and field tests, many authors have come to the conclusion that the velocity of colloid migration in their experiments was higher than tracer velocity (Champ and Schroeter 1988; Harvey et al. 1989; Harvey and Garabedian 1991; Kim et al. 1994; Shonnard et al. 1994; Ryan and Elimelech 1996; Artinger et al. 1998; Knapp et al. 2000), moreover, sometimes this excess in velocity was multiple (Champ and Schroeter 1988; Knapp et al. 2000).

At first glance, particle migration at a velocity higher than that of the groundwater is doubtful because the only mechanism of the particles' migration is their transport by the groundwater. However, most authors explain this effect by giving the following reason, which seems to be plausible. Groundwater flows in geological medium through connected pore and fracture voids that form a set of flow channels. Permeability of these channels depends on their apertures, curvatures, interconnections with other channels, and so on. Since distribution of these properties in geological medium is highly heterogeneous, flow velocities in different channels can differ markedly. One can assume that hydrodynamic effects caused by heterogeneity of the flow velocity field in the system of flow channels (for example, Magnus effect) or influence of electrostatic repulsion between particles and walls of the channels may result in movement of the particles to such domains of the channel cross-sections and such channels where the local velocity of the flow is maximum. Velocity of the particles in this case will be close to the maximum velocity of the groundwater in the most permeable channel. The interstitial velocity of the groundwater in geological medium is defined by averaging of the velocities in all flow channels in an elementary representative volume. Hence, the ratio of the particle velocity to the interstitial velocity of water can be equal to the ratio of the maximum velocity of water in the most permeable flow channels in the representative volume to the average velocity in all flow channels in this volume. Therefore, the more heterogeneous are flow channels in geological medium, the more the ratio of the colloid particle velocity to the interstitial velocity of the groundwater. This explanation is used, for example, in (Harvey et al. 1989; Puls and Powell 1992; Knapp et al. 2000). The model is plausible but not obvious. The objective of this study is a more rigorous analysis of the effect and its quantitative estimation for mathematical modeling of radioactive contaminant transport by groundwater.

5.1 Hydrodynamic Model of the Transverse Drift of the Particles

Let us consider nature of forces responsible for drift of colloidal particles to such domain of percolation channel where velocity of groundwater flow is maximum. One can assume that these forces are caused by electrostatic repulsion between the surfaces of the colloid particles and the channel walls (Puls and Powell 1992). This hypothesis seems to be implausible for the following reason. As it is mentioned in (Spielman 1977; Ryan et al. 1999), the surfaces of the percolation channels in geological medium are highly heterogeneous (which implies heterogeneity of ζ potential sign). Therefore, both repulsion and attraction can exert influence on the particles during their movement along the percolation channel. In addition to that, the electrostatic forces can exert influence on the particle movement only in immediate proximity to the channel walls where the velocity of the groundwater is weak. A significant excess of the particle velocity relative to the mean velocity of the groundwater in the channel can take place only at a drift of the particle to the core of the channel cross-section where the flow velocity is close to its maximum in this cross-section, i.e., electrostatic forces should exert influence on the whole cross-section what is unlikely.

Some authors explain the drift of the particles in the cross-section of the percolation channel by an influence of purely hydrodynamic effects. Parameters of the groundwater flow in geological medium correspond, as a rule, to a laminar flow regime. Let us consider a channel of a cylindrical shape, as a model approximation. The flow velocity v_f in this case is a function of distance from the channel axis r. The function $v_f(r)$ in the cylindrical channel satisfies the Poiseuille's law

$$v_f(r) = v_{max}\left[1 - (r/r_w)^2\right], \tag{66}$$

where v_{max} is the maximum value of the velocity of the flow (which is reached at the channel axis) and r_w is the radius of the channel.

The maximum velocity v_{max} can be calculated from the formula

$$v_{max} = \frac{r_w^2}{4\mu}\frac{\partial P}{\partial z}, \quad (67)$$

where μ is the dynamic viscosity of the groundwater, P is the pressure, and z is the longitudinal coordinate axis.

The gradient of the flow velocity is normal to the axis in this case. Let us consider a spherical particle of the radius r_p ($r_p << r_w$) in the channel. The velocity of the flow at the particle surface on the side facing the channel wall is lower than that at the opposite side (Fig. 15). As a result, the particle will rotate. The rotation axis is normal to the longitudinal section of the channel passing through the particle and the channel axis (i.e., the rotation axis is normal to the plane of Fig. 15). It follows from a general reasoning (Saffman 1965) that the angular velocity of this rotation ω can be expressed as

$$\omega = \frac{1}{2}\left|\frac{\partial v_f}{\partial r}\right|. \quad (68)$$

Expression (68) is in good agreement with the experimental data from (Aoki et al. 1979).

Let us assume that the longitudinal component of the particle center velocity differs from the flow velocity by δv. We consider first the case $\delta v < 0$. As a result of the particle rotation, the liquid (velocity of which relative to the particle center is $-\delta v$) slows down on the particle side facing to the channel wall. Deceleration of the flow at this part of the particle surface causes an excess pressure under action of which the particle moves toward the channel axis. The same effect in the contrary case (at $\delta v > 0$) causes drift of the particle toward the wall (i.e., in the opposite

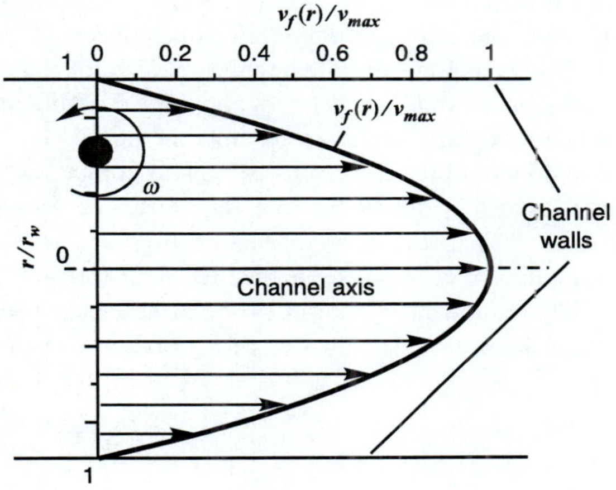

Fig. 15 Rotation of a solid particle due to heterogeneity of local velocities of a liquid phase in a percolation channel

direction). This drift of rotating particle normal to velocity of gas or liquid flowing round the particle is called Magnus effect (Loitsiansky 1973).

For modeling of the particle movement with the Magnus effect taken into account, we introduce cylindrical coordinates z, r where z axis is directed along the channel axis. We assume that movement of the particles is symmetric about the channel axis; therefore, it is sufficient to consider their movement only in coordinates z, r. Trajectories of the particles can be represented in this case in the form

$$z = z_{\mathrm{c}}(t), \quad r = r_{\mathrm{c}}(t), \tag{69}$$

where t is the time.

Only the axial component of the liquid flow velocity is not equal to zero in an ideal case of a smooth cylindrical channel. The values of this component can be determined from expression (66). Then the components of the particle velocity relative to the ambient liquid δv_z and δv_r can be represented as

$$\delta v_z = z_{\mathrm{c}}'(t) - v_{\mathrm{f}}(r_{\mathrm{c}}), \quad \delta v_r = r'(t)_{\mathrm{c}}. \tag{70}$$

According to (Aoki et al. 1979), the radial component F_r of the force acting on the particle due to Magnus effect (Fig. 16) at the particle rotation and nonzero δv_z can be determined by the following expression

$$F_r = \pi r_{\mathrm{p}}^3 \rho_{\mathrm{f}} \omega \delta v_z, \tag{71}$$

where r_{p} is the particle radius and ρ_{f} is the liquid density.

A different expression for F_r was obtained theoretically by Saffman (1965) under some additional assumptions. However, Saffman (1965) did not carry out a quantitative comparison of the derived formula with the available data. Good agreement between the trajectories of particles calculated with the use of expression (71) and the corresponding trajectories observed in the experiments in (Aoki et al. 1979) gives preference to expression (71) for the determination of F_{r}.

Magnus effect is caused by the force $\vec{F}$ acting on particle due to its rotation and nonzero velocity relative to ambient liquid. Let us call $\vec{F}$ a drift force. Radial component of the drift force is F_r. This component is caused by different velocities of the particle and the liquid in the z direction. However, as it follows from (70), the radial drift of the particle under the action of F_r leads to a difference in radial velocities of the particle and the liquid. Therefore, the drift force can have a nonzero axial component F_z caused by the same mechanism as F_r (Fig. 16). Therefore, the expression for determination of F_z should be analogous to (71), i.e.,

$$F_z = -\pi r_{\mathrm{p}}^3 \rho_{\mathrm{f}} \omega \, \delta v_{\mathrm{r}}. \tag{72}$$

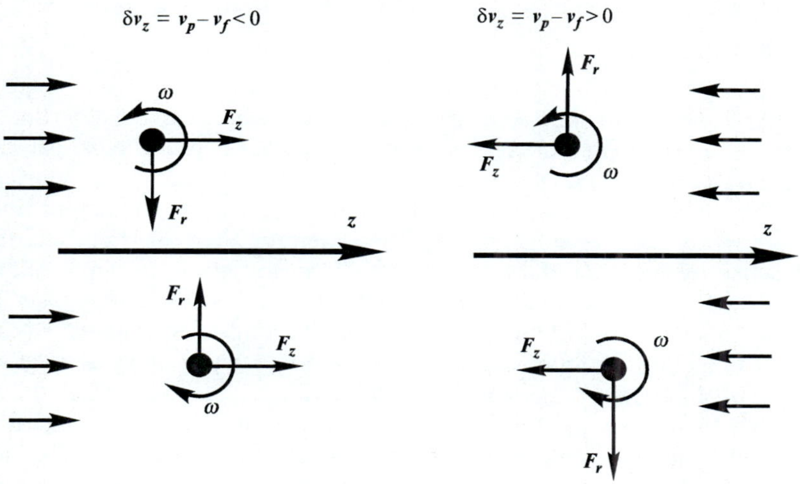

Fig. 16 Drift forces caused by rotation of a particle

It is worth mentioning that the axial component of the drift force was not taken into account in the model used in (Aoki et al. 1979) for the calculation of particle movement.

It follows from (71) that F_r is nonzero only in the case when velocities of the particle and the flow are different. This can be caused by an influence of the gravity force if the densities of the particle and the liquid are different. However, this influence can lead to a difference of axial components of the velocities of the particles and the liquid only in the case when the channel axis is close to a vertical. It follows also from (71) that the drift toward the channel axis (where the flow velocity is maximum) can take place only at negative δv_z. The density of pseudo-colloids is usually higher than the groundwater density. Therefore, δv_z can be negative only at an ascending flow in near-vertical channels. Hence, the drift of the particles to domains of the maximum velocity in percolation channels, resulting in the movement of the particles at velocities exceeding the interstitial velocity of the groundwater, can take place only at ascending flow in near-vertical channels.

Hereafter, we will consider only the case when the direction of contaminant transport by groundwater is predominantly near horizontal. The influence of the gravity force on axial components of the particle velocity in this case can be neglected. The influence of the gravity force on the transverse motion of the particles in the percolation channels can lead in this case only to precipitation of the particles and has no concern with their redistribution in the channel cross-section toward domains of maximum velocity of the groundwater flow. We assume that the influence of the gravity force on the transverse component of the particle velocity is weak and does not lead to a noticeable precipitation of the particles.

5.2 *Influence of the Drift Force on Movement of Relatively Large Particles*

It should be kept in mind in studies of colloid-facilitated transport of a contaminant by groundwater that the influence of the drift forces on colloidal particles' migration can depend on chaotic Brownian motion of the particles. Velocity of Brownian motion and extent of its influence on the particle migration depend on the particle size. The larger the particle at certain values of temperature and particle density, the less is the velocity of its Brownian motion. The term colloid is usually used to refer to particles in a liquid phase with sizes ranging from 1 to 450 nm (Vilks and Degueldre 1991). These limits are defined conditionally to a certain extent and can be specified otherwise. For example, in accordance with (Mercier et al. 2001), particles are referred to as colloid if their size is from 1 nm to 1 μm, and Mironenko and Rumynin (1998) specified the lower and upper limits of the colloid particles' size range as 1 nm and 10 μm, respectively. Therefore, both small and relatively large particles can participate in the colloid-facilitated transport of radionuclides.

Let us first consider the problem concerning a role of the drift forces in the case when the particles are relatively large, so the influence of Brownian motion on their migration can be neglected. Let us consider the simplest case of migration of a relatively large spherical particle in a horizontal cylindrical channel. Movement equations of the particle in cylindrical coordinates take the form:

$$\begin{aligned}\frac{4}{3}\pi r_{\mathrm{p}}^3\rho_{\mathrm{p}}\frac{d^2 r_{\mathrm{c}}}{dt^2} &= \pi r_{\mathrm{p}}^3\rho_{\mathrm{f}}\omega\delta v_z - 6\pi\mu r_{\mathrm{p}}\frac{dr_{\mathrm{c}}}{dt}{}' \\ \frac{4}{3}\pi r_{\mathrm{p}}^3\rho_{\mathrm{p}}\frac{d^2 z_{\mathrm{c}}}{dt^2} &= -\pi r_{\mathrm{p}}^3\rho_{\mathrm{f}}\omega\delta v_{\mathrm{r}} - 6\pi\mu r_{\mathrm{p}}\frac{dz_{\mathrm{c}}}{dt}\end{aligned}, \tag{73}$$

where ρ_{p} is the particle density, μ is dynamic viscosity of the liquid phase, and t is the time.

Variables δv_z and δv_r in (73) can be determined from (70). It follows from (66) and (68) that

$$\omega = \frac{v_{\max} r_{\mathrm{c}}}{r_{\mathrm{w}}^2}. \tag{74}$$

Let us consider initial conditions in the form

$$r_{\mathrm{c}}(0) = r_0, \quad r_{\mathrm{c}}'(0) = 0, \quad z_{\mathrm{c}}(0) = z_0, \quad z_{\mathrm{c}}'(0) = v_{\mathrm{f}}(r_0). \tag{75}$$

Solutions of Cauchy problem (73) and (75) take the form

$$r_{\mathrm{c}}(t) = r_0, \quad z_{\mathrm{c}}(t) = z_0 + v_{\mathrm{f}}(r_{\mathrm{c}})t. \tag{76}$$

The solutions (76) correspond to rectilinear trajectories of the particles parallel to the channel axis. Therefore, immediate effect of the drift forces does not lead to a redistribution of the particles in the channel cross-section and to an increase of the average velocity of the particles migration along the channel. However, one should keep in mind that these solutions were obtained at conditions (75) where the initial velocity of the particle is assumed to be equal to the velocity of the liquid phase.

Let us assume that at least one of two variables δv_z, δv_r is not equal to zero at the initial time moment. If the system (73) is stable, the solution at sufficiently small δv_z and δv_r remains arbitrary close to (76). Therefore, if the system is stable, and initial values of δv_z, δv_r are small, the influence of the drift forces does not lead to a redistribution of the particles in a cross-section of the channel.

Let us assume that the system (73) is unstable for some solutions (76). This implies that arbitrary small perturbations of the initial values of δv_z and δv_r can lead to considerable deviations in the trajectories of the particles from the straight lines corresponding to the solutions (76). As the right-hand sides of the equations set (73) do not depend explicitly on z, the stability domain (as well as the instability domain) should be cylindrical with a simply or multiply connected base and an axis which coincides with the channel axis. If a particle is in the instability region, it will move in a trajectory with significant variation in the transverse coordinate r_c until the particle crosses the boundary of the instability domain. After that, the movement of the particle in the transverse direction becomes stable and its trajectory asymptotically approaches a straight line parallel to the channel axis. Stability analysis of the linearized system (73) for solutions (76) showed that the instability domain was determined by the following expression

$$r > r_s \tag{77}$$

where

$$r_s = \frac{r_w^2}{r_p}\sqrt{\frac{18}{\sqrt{5}Re}}, \qquad Re = \frac{v_{max} r_w \rho_f}{\mu}. \tag{78}$$

This implies under condition

$$r_s < r_w \tag{79}$$

that the instability domain is a cylinder with the annular base $r_s < r < r_w$, the axis of the cylinder coincides with the channel axis (Fig. 17).

The condition (79) can be satisfied at relatively high Reynolds' number Re. In this case, the particles from the peripheral domain of the flow (i.e., adjacent to the channel wall) move due to instability to the central domain where the flow velocities are higher than in the peripheral domain. As a result, the average velocity of the particles' migration increases.

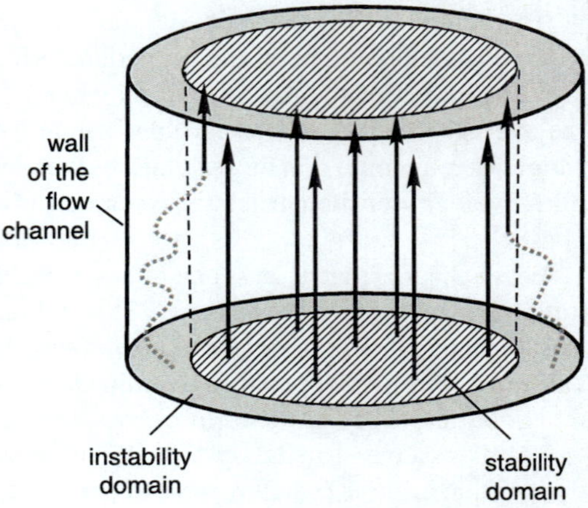

Fig. 17 Stability of the trajectories of the non-Brownian particles. The stability domain in the transverse cross-section of the channel is shown with hatching. The annular instability domain in the transverse cross-section of the shown in *gray color*

As the main mechanism of radioactive contaminant migration in the underground water is transport of the radionuclides by the groundwater, one of the main requirements placed upon selection of the site for development of the repository of the RW is the relatively low velocity of the groundwater flow. Quantitative estimations of *Re* show that the inequality (79) is most likely not satisfied under conditions suitable for underground disposal of the RW. Therefore, transport of radionuclides by relatively large particles under conditions typical of repository site does not lead to a significant increase of radionuclides migration (in comparison with groundwater flow velocity).

5.3 *Influence of the Drift Forces on Migration of Brownian Particles*

Let us now consider the influence of the drift forces on the migration of relatively small particles for which the effect of Brownian motion can be significant. The probabilistic approach is most natural in this case for theoretical modeling of the particles' migration. Let us assume that concentration of the particles is low, and processes of their movement can be considered to be independent. State of the particle is characterized in this case by its coordinates and components of its velocity. Therefore, the state space of the particle is six-dimensional in a general case. As the movement of the particle system is symmetric about the channel axis, the consideration can be confined to four-dimensional state space where coordinates of the state are z, r, v_z, and v_r. Let us designate probability density for some state of the particle at the instant t as $p(t, z, r, v_z, v_r)$. According to (Gardiner 1983), probability density in the case of one-dimensional Brownian motion satisfies Kramers' equation. Much as it was done in (Gardiner 1983) for the one-dimensional case, one can easily obtain the Kramers' equation for three-dimensional Brownian motion as well. This equation in the considered case of the cylindrically symmetric migration of the particles in the percolation channel can be written in the cylindrical coordinates as:

$$\frac{\partial p}{\partial t} = -v_r \frac{\partial p}{\partial r} - (\delta v_z + v_f) \frac{\partial p}{\partial z} + \frac{6\pi r_p}{m} \left[\frac{\partial}{\partial v_r}(v_r p) + \frac{\partial}{\partial \delta v_z}(\delta v_z p) \right] + \frac{3 v_{\max} r}{4 r_w^2} \left[\frac{\partial}{\partial v_r}(\delta v_z p) - \frac{\partial}{\partial \delta v_z}(v_r p) \right] + \frac{6\pi r_p k_B T}{m^2} \left(\frac{\partial^2 p}{\partial v_r^2} + \frac{\partial^2 p}{\partial \delta v_z^2} \right), \quad (80)$$

where m is the particle mass, k_B is the Boltzmann constant, and T is the temperature.

Let us assume that percolation channel is sufficiently long. This assumption is quite reasonable because length of the particle path in the underground medium is much more than transverse dimensions of the percolation channels. In this case, one can expect that p becomes practically independent of z at sufficiently large z. In exactly the same way, one can expect that probability density becomes stationary at sufficiently large t. It is easy to check that solution of (80) in the steady-state approximation independent of z is the Boltzmann' distribution, i.e., the same distribution is in the absence of the drift forces. Therefore, in the case of significant influence of Brownian motion on the particles' migration, the effect of the drift forces does not lead to a redistribution of the particles in the channel cross-section and, as a result, to an increase in the particles' migration velocity as compared with the velocity of the groundwater.

5.4 *Analysis of Available Experimental Data and Theoretical Models*

The results of the presented theoretical analysis of the influence of drift forces on migration of the particles in the percolation channels are obtained under the assumption that the influence of the gravity force on the longitudinal component of the particle velocity is weak. As the density of the particles can differ from the density of groundwater, these results relate to migration in the percolation channels predominantly in the near-horizontal direction.

The analysis of the particles' migration in the case of insignificant influence of Brownian motion shows that the influence of the drift forces can lead to instability of the particles' movement. As a result, the particles can be redistributed in the cross-section of the channel. The particles from the periphery of the cross-section adjacent to the channel walls move to the central part where velocity of the flow is higher than in the near-wall part. As a result, the average velocity of the particle migration can increase. However, quantitative estimation shows that this effect can take place only at relatively high velocity of the liquid phase, which is not characteristic of groundwater flow.

Chaotic and relatively sharp oscillations are usually typical of unstable movement of the particles. Similar oscillations are typical of Brownian motion as well. As the instability of the particles' movement can lead to some increase in the velocity of their migration in the percolation channels, one can expect that the same effect can take place due to Brownian motion as well. The analysis shows that the instability of the particles' movement calls for relatively high velocity of the liquid phase whereas this condition is not necessary in the case of the Brownian motion. Therefore, an increase in the velocity of the particles' migration at a substantial role of Brownian motion due to the influence of the drift forces seems to be probable (which is important for the assessment of an ecological hazard caused by the transport of radioactive contaminant by groundwater). An attempt was made in (McTigue et al. 1986) to take into account Brownian motion in the analysis of the influence of drift forces on the particles' movement in the case of Poiseuille flow of a liquid. On the basis of their analysis, the authors of (McTigue et al. 1986) came to the conclusion that particles shift to the central part of the channel if their velocities are less than the local velocity of the liquid; otherwise they shift to the peripheral part of the channel. As a matter of fact, this conclusion is in a good agreement with present notion about the influence of the drift forces. However, it is difficult to use results from (McTigue et al. 1986) for practical quantitative estimations. The movement of the particles was analyzed in (McTigue et al. 1986) in an approximation of a continuous system, i.e., the disperse phase was considered as a continuum. As a result, the set of governing equations was not completely defined. For that reason, the authors introduced a relationship between the liquid pressure and a "pressure of the disperse phase" to close the set of the governing equations. Validity of the form of this additional relationship is not evident; in addition to that it includes unknown parameters determination of which represents a practically unsolvable problem.

The approach used in this chapter to determine the influence of the drift forces on the migration of the particles in percolation channels taking into account Brownian motion seems to be more substantiated. The effect was considered on the basis of the Kramers equation for the function of the distribution of particles throughout their velocity components and the channel volume. The analysis of this equation with additional terms caused by the presence of the drift forces shows that their effect in the case of substantial influence of Brownian motion does not lead to an increase in the particles' migration velocity in comparison with the average velocity of the liquid in the percolation channel.

Therefore, the theoretical analysis of transport of the colloid particles by the liquid shows that average velocity of colloid migration in near-horizontal direction does not exceed average velocity of liquid phase. This conclusion should be compared with available data of laboratory and field experiments on transport of colloid particles by liquid phase in porous medium. The velocity of particle migration was determined in the following way in these studies. In laboratory experiments, water (or an aqueous solution) enters a column filled by crushed rock through the inlet cross-section. At some instant of time t_0, the concentration (fraction by weight) of colloid in the solution increases stepwise from zero to some value C_0 and remains constant for a time period Δt; then the concentration decreases stepwise to zero and remains zero up to the end of the experiment. Concentration of the colloid at the outlet cross-section $C_b(t)$ measured during the experiment increases from zero, reaches some maximum value C_{max} at the instant of time t_{max} and then decreases again to zero. The experimental dependence $C_b(t)$ (the breakthrough curve) is compared with theoretical solutions of a one-dimensional transient problem of particles' transport obtained at different values of particle migration velocity v_p and hydraulic dispersion D, in approximation of the advection–dispersion model described, e.g., in (deMarsily 1986). True values of v_p and D are determined from the best fit of the theoretical solution and the experimental breakthrough curve. The velocity of the particles traveling through the layer of the thickness Z can also be estimated approximately from the following simple expression

$$v_p = \frac{Z}{t_{max} - (t_0 + \Delta t/2)}. \tag{81}$$

The velocity of the neutral tracer v_t is determined in the same manner from analogous experiment with the tracer. As it is assumed that precipitation of the tracer and its interaction with the immovable solid phase are absent, one can expect that v_t is approximately equal to the average interstitial velocity of the liquid phase in the layer. Authors of the experimental studies usually consider the possibility that the particle migration velocity exceeds the average interstitial velocity of the groundwater, reasoning from the ratio v_p/v_t.

The field experiments are carried out similarly. An injection well is used in the field test as an inlet into considered domain of rocks, with a pumping well usually serving as its outlet.

It is worth mentioning that determination of v_p/v_t in different laboratory and field experiments gave conflicting results. Insignificant (of the order of 10^{-1}) or near-zero values of the parameter $\gamma = v_p/v_t - 1$ were obtained in the laboratory experiments (Eichholz et al. 1982; Puls and Powell 1992; Kim et al. 1994; Shonnard et al. 1994; Artinger et al. 1998; Zhuang et al. 2003; Schäfer et al. 2003) and in the field tests (Harvey et al. 1989; McCarthy et al. 1998b). These results are in full agreement with the results of the theoretical analysis presented earlier. The insufficient apparent lag of the tracer can be explained by uncertainty in the measurement of v_t. The fact is that some part of the tracer is absorbed during its migration in rocks due to molecular diffusion in relatively small pores in which diffusion of larger particles is impossible. The diffusion sink of the tracer into the rock matrix was not taken into account in the determination of v_t in the experimental studies. For this reason, v_t can be underestimated and γ can be slightly overestimated.

At the same time, values of γ of the order 10^0–10^1 were obtained in the laboratory experiment in (Knapp et al. 2000) and in the field test (Champ and Schroeter 1988). These results bear witness to the fact that particle migration velocity can far exceed the average interstitial velocity of the liquid phase that carries these particles.

Let us consider techniques used in these experiments. Knapp et al. (2000) studied transport of monodisperse colloid particles from polystyrene latex by water flowing through a tuff column with a longitudinal fracture. Most values of γ were obtained in the experiments where the fracture aperture was not uniform. One should keep in mind in assessment of the results of the experiment that the particles were manufactured from the material density of which was higher than water density, the column was vertical, and water with colloid entered the column from below. Therefore, local velocity of the particles was less than the local velocity of water due to the influence of gravity force. As it is shown in experiments described in (Aoki et al. 1979), the influence of the drift forces leads in this case to redistribution of the particles to a domain of the fracture where the local velocity of the flow was a maximum. A noticeable effect of this kind was observed in the experiments described in (Aoki et al. 1979) where differences between the densities of the particles and the liquid were less than 1%. The ratio of these densities in (Knapp et al. 2000) was 1.055. Therefore, in spite of the statement in (Knapp et al. 2000) that these densities are practically equal, their difference was sufficient for the redistribution of the particles to the domain of the flow where the local velocity was at its maximum (Fig. 18). The range of local velocities of the flow in fractures with heterogeneous aperture can be rather wide, and the maximum velocity of the flow can be much higher than the average velocity. As a result of the redistribution of the particles in fracture cross-section, their migration velocities must be close to the maximum velocity of the flow, whereas the tracer migration velocity must be equal to the average velocity of the flow. Therefore, relatively high values of γ could be actually obtained in (Knapp et al. 2000). However, one should bear in

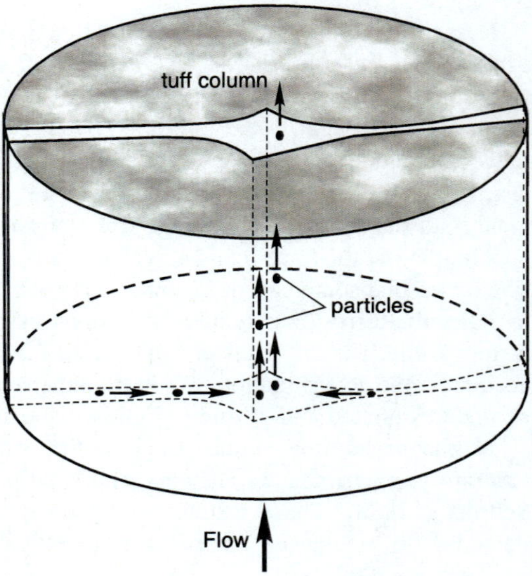

Fig. 18 Migration of the particles through the fracture with heterogeneous aperture in experiment from (Knapp et al. 2000)

mind that this effect could take place only due to vertical orientation of the percolation channel (the longitudinal fracture in the vertical column) and difference between the densities of the particles and the liquid. If direction of radioactive contaminant transport by groundwater is predominantly near-horizontal (i.e., through near-horizontal percolation channels), practical importance of the effect observed in (Knapp et al. 2000) is insignificant.

Results of field studies using a neutral tracer and colloid particles represented by artificial beads and bacteria *E. coli* (Fig. 19) are given in (Champ and Schroeter 1988). Reasoning from the estimation of migration velocities of the neutral tracer and the colloid particles by formula (81), one can come to the conclusion that the particles' migration velocity far exceeds the average interstitial velocity of the groundwater. Since the flow velocity in this case is oriented predominantly in near-horizontal direction, influence of the gravity force can be neglected. A characteristic feature of the obtained results was that values of C_{max}/C_0 in the tests with the colloid particles were by an order of magnitude less than the values of C_{max}/C_0 in the tests with the neutral tracer. Taking into account this fact one can show that the values of γ of the order of 10^1 can be imaginary and are caused by not quite correct interpretation of the experimental data. With that end in view let us consider the flow of groundwater with the tracer or the colloid particles through a rock column oriented along the direction from the injection well to the pumping well. The column is confined in length by two cross-sections. They are normal to the direction of the column, and their positions correspond approximately to the positions of the injection and pumping wells. We assume that the transport of the indicator (represented by the tracer or the colloid particles) in the field tests can be modeled as a one-dimensional process in the considered rock column.

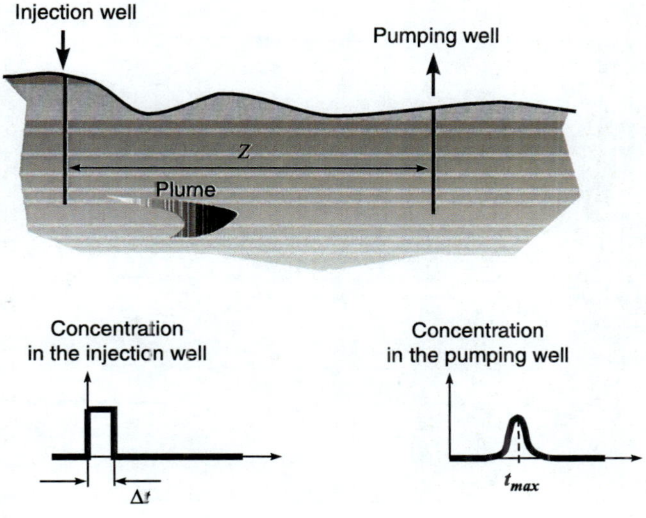

Fig. 19 Field test for migration of colloid and neutral tracer

For that let us consider a system of percolation channels represented by parallel round capillaries (Fig. 20) radii of which satisfy logarithmically normal distribution, i.e.,

$$n = \frac{N}{\sigma\sqrt{2\pi}} \exp\left(\frac{l^2}{2\sigma^2}\right), \quad l = \ln\frac{r_w}{r_0}, \tag{82}$$

where N is the number of the capillaries that pass through a unit area in a cross-section normal to the capillaries, n is the distribution density of logarithm of the capillary radius, r_0 is the radius of a capillary corresponding to expectation of $\ln r_w$, and σ is the standard deviation of the logarithmically normal distribution.

Let us assume that concentration of an indicator in the groundwater at the inlet cross-section of the rock column varies with time in the following way:

$$C_{in}(t) = \begin{cases} 0, & t<0; \\ C_0 = const, & 0 \le t<\Delta t; \\ 0, & t \ge \Delta t. \end{cases} \tag{83}$$

This expression can be represented in the more simple form as:

$$C_{in}(t) = C_0[\theta(t) - \theta(t - \Delta t)] \tag{84}$$

where $\theta(t)$ is the unit step function which is equal to zero at $t < 0$ and unity at $t \ge 0$.

Since the capillaries are round, the flow velocity distribution in them satisfies the Poiseuille formula (66). The average velocity of water flow through a capillary of radius r_w can be determined in this case from the formula

$$v_{av} = v_0\left(\frac{r_w}{r_0}\right)^2 = v_0 e^{2l}, \quad v_0 = -\frac{r_0^2}{8\mu}\frac{\partial p}{\partial z}. \tag{85}$$

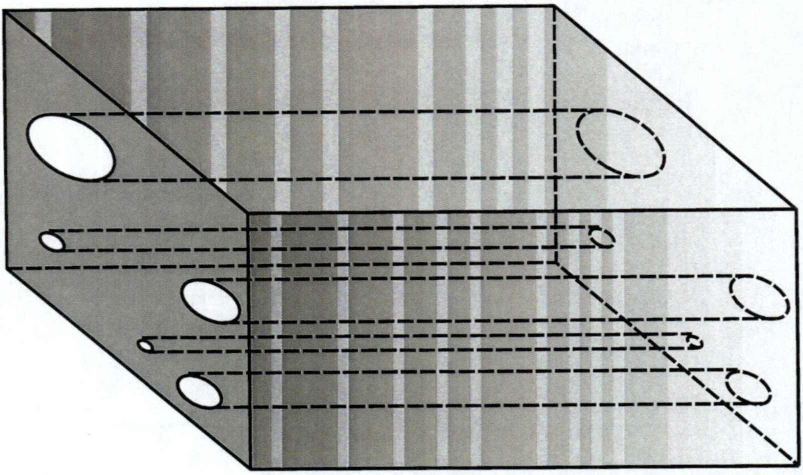

Fig. 20 A simplified model of the system of flow channels

Let us designate G the volume of the water flowing through a cross-section of the capillary per unit time. Then it follows from the previous expression that

$$G = \pi r_0^2 v_0 \left(\frac{r_w}{r_0}\right)^4 = \pi r_0^2 v_0 e^{4l}. \tag{86}$$

If the indicator is a neutral tracer, then its concentration at the outlet cross-section at the instant t is equal to the amount of the indicator transported by water through the channels that satisfy the following condition

$$v_{av}(l) \geq v_{av}(L) = Z/t, \tag{87}$$

where Z is the length of the rock column.

Therefore, one can obtain

$$C(t) = C_0 \int_{L(t)}^{\infty} G(l)n(l)\mathrm{d}l \bigg/ \int_0^{\infty} G(l)n(l)\,\mathrm{d}l. \tag{88}$$

Substitution of expressions (82), (84), and (86) into (88) gives

$$\frac{C(\tau)}{C_0} = \frac{1}{2}\left[\theta(\tau)\mathrm{erfc}\left(\frac{-0.5\ln\tau - 4\sigma^2}{\sigma\sqrt{2}}\right) - \theta(\tau - \Delta\tau)\mathrm{erfc}\left(\frac{-0.5\ln(\tau - \Delta\tau) - 4\sigma^2}{\sigma\sqrt{2}}\right)\right], \tag{89}$$

where τ is dimensionless time,

$$\tau = v_0 t/Z, \quad \Delta\tau = v_0 \Delta t/Z. \tag{90}$$

Let us consider the case when the indicator is the colloid particles. Then the particles can be transported to the outlet cross-section only through the capillaries radii of which are larger than the particle radius, i.e., $r_w > r_p$. One can obtain in this case:

$$C(t) = C_0 \int_{\Lambda(t)}^{\infty} G(l)n(l)\,\mathrm{d}l \bigg/ \int_0^{\infty} G(l)n(l)\,\mathrm{d}l, \tag{91}$$

where

$$\Lambda(t) = \max\{L(t), l_p\}, \quad l_p = \ln\frac{r_p}{r_0}. \tag{92}$$

Substitution of expressions (82), (84), and (86) into (91) gives

$$\frac{C(\tau)}{C_0} = \frac{1}{2}\left[\theta(\tau)\mathrm{erfc}\left(\frac{A - 4\sigma^2}{\sigma\sqrt{2}}\right) - \theta(\tau - \Delta\tau)\mathrm{erfc}\left(\frac{B - 4\sigma^2}{\sigma\sqrt{2}}\right)\right], \tag{93}$$

where

$$A = \max\{-0.5 \ln \tau, l_p\}, \quad B = \max\{-0.5 \ln(\tau - \Delta\tau), l_p\}. \tag{94}$$

The breakthrough curves for concentrations of the neutral tracer and the particles determined from expressions (89) and (93), respectively, are shown in Fig. 21. It is significant that at the beginning of the process only the particles carried by water through the largest capillaries reach the outlet cross-section of the rock column. These particles move with the same velocity as the tracer transported by the water through these capillaries. As a result, the concentrations of the tracer and the particles practically coincide at the beginning of the process. Later on, the particles that entered the large capillaries during the time interval Δt passed through the column, while the rest of the particles were mechanically retained in smaller capillaries. As a result, the concentration of the particles at the outlet cross-section diminishes up to zero. Since the tracer continues to move to the outlet cross-section through the small capillaries, its concentration at the outlet cross-section increases for some time and can reach the maximum when the particles' concentration is already close to zero. Let us represent the breakthrough curves $C(t)/C_0$ for the tracer and the particles on different scales (Fig. 22), as it was done in (Champ and Schroeter 1988). The shape of the obtained plot is practically identical to one of the plots presented in (Champ and Schroeter 1988) on the basis of which the authors made the conclusion that velocity of the particles is much higher than the velocity of the tracer. Indeed, if the velocities of the tracer and the particles are estimated from expression (81), then the wrong conclusion can be made that the velocity of the particles exceeds the tracer velocity. However, the presented analysis shows that lower value of tmax implies only that the colloid particles are mechanically retained in most of the percolation channels.

As evidenced by the foregoing analysis, velocity of the colloid particles cannot exceed the interstitial velocity of the groundwater at near-horizontal direction of the groundwater flow. Therefore, colloid-facilitated transport of radionuclides by groundwater cannot cause elevated velocity of radionuclides migration if the

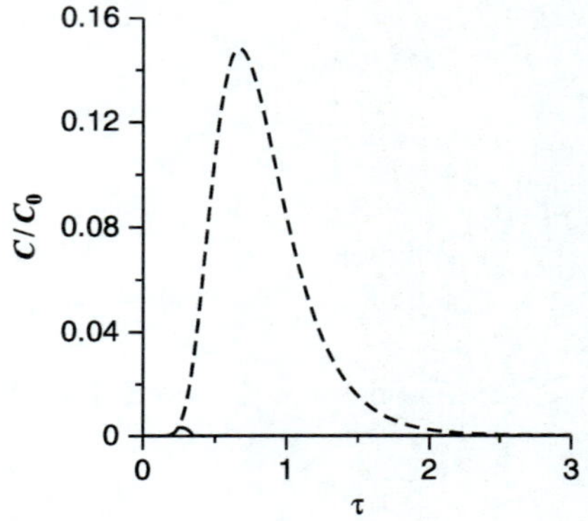

Fig. 21 Break-through curves of a colloid and a neutral tracer

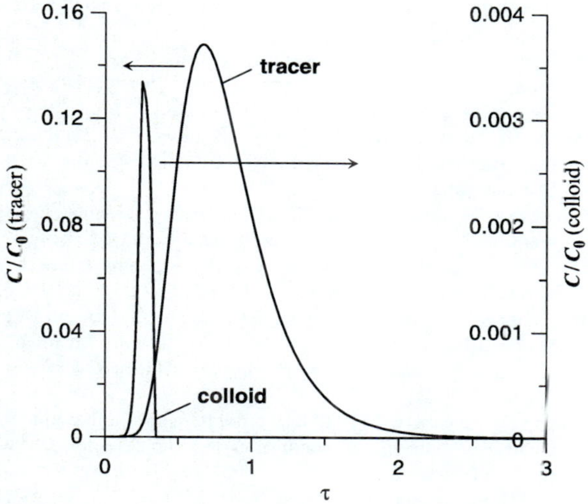

Fig. 22 Break-through curves of a colloid and a neutral tracer at different scales of the ordinate axes

direction of the groundwater flow is predominantly near-horizontal. Let us consider a case when vertical component of the velocity should be taken into account. As a rule, density of colloidal particles of natural origin is higher than groundwater density. The redistribution of the particles in cross-section of percolation channels to the domain, where local velocity of the flow is a maximum, is possible only if the local velocity of the flow is higher than the local velocity of the particles. Since the density of the particles is higher than the groundwater density, these conditions can arise from the influence of the gravity force at the ascending flow. Therefore, one can expect that the migration velocity of the colloid particles is higher than the interstitial velocity of the groundwater in the ascending branch of the groundwater flow in a discharge zone on a territory with rugged topography (Malkovsky et al. 1997) and in an ascending branch of a thermal convection cell developed as a result of heat generation in a loaded part of an underground repository of high-level RW (Malkovsky et al. 1995).

Acknowledgments The author acknowledges support from International Science & Technology Center under Award Number 3290 and from Russian Foundation of Basic Research under Award Number 09-05-00347.

References

Airey PL (1986) Radionuclide migration around uranium ore bodies in the Alligator Rivers region of the northern territory of Australia – analogue of radioactive waste repositories – a review. Chem Geol, 55:255–268

Amme M, Aldave de las Heras L, Betti M, Lang H, Stöckl M (2004) Effects of colloidal and dissolved silica on the dissolution of UO_2 nuclear fuel in groundwater leaching tests. J Radioanalyt Nucl Chem, 261:327–336

Aoki H, Kurosaki Y, Anzai H (1979) Study on the tubular pinch effect in a pipe flow. Bull. Jpn Soc Mech Eng, 22:206–212

Artinger R, Kienzler B, Schüssler W, Kim JI (1998) Effects of humic substances on the ^{241}Am migration in a sandy aquifer: column experiments with Gorleben ground-water/sediment systems. J Contamin Hydrol, 35:261–275

Bates JK, Bradley JP, Teetsov A, Bradley CR, Buchholtz ten Brink M (1992) Colloid formation during waste form reaction: Implication for nuclear waste disposal. Science, 256:649–651

Bochever FM, Oradovskaya AE (1972) Hydrogeological substantiation of protection of ground-water and water intakes from pollutions. Nedra, Moscow

Buck EC, Bates JK (1999) Microanalysis of colloids and suspended particles from nuclear waste glass alteration. Appl Geochem, 14:635–659

Buesseler KO, Bauer JE, Chen RF, Eglinton TI, Gustafsson O, Landing W, Mopper K, Moran SB, Santschi PH, VernonClark R, Wells ML (1996) An intercomparison of cross-flow filtration techniques used for sampling marine colloids: Overview and organic carbon results. Mar Chem, 55:1–31

Champ DR, Schroeter J (1988) Bacterial transport in fractured rock – a field-scale tracer test at the Chalk River nuclear laboratories. Water Sci Technol, 20:81–87

Davis JA (1982) Adsorption of natural dissolved organic matter at the oxide/water interface. Geochim Cosmochim Acta, 46:2381–2393

Davis JA, Gloor R (1981) Adsorption of dissolved organics in lake water by aluminum oxide. Effect of molecular weight. Environ Sci Technol, 15:1223–1229

deMarsily G (1986) Quantitative hydrogeology. Academic, Orlando

Ebert WL, Bates JK (1993) A comparison of glass reaction at high and low glass surface/solution volume. Nucl Tech, 104:372–384

Eichholz GG, Wahlig BG, Powell GF, Craft TF (1982) Subsurface migration of radioactive waste materials by particulate transport. Nucl Tech, 58:511–520

Elimelech M, Nagai M, Ko CH, Ryan JN (2000) Relative insignificance of mineral grain zeta potential to colloid transport in geochemically heterogeneous porous media. Environ Sci Technol, 34:2143–2148

Fairhurst AJ, Warwick P, Richardson S (1995) The influence of humic acid on the absorption of europium onto inorganic colloids as a function of pH. Coll Surf A, 99:187–199

Fane AG, Fell CJD, Waters AG (1981) The relationship between membrane surface pore characteristics and flux for ultrafiltration membranes. J Membr Sci, 9:245–262

Finn PA, Buck EC, Gong M, Hoh JC, Emery JW, Hafenrichter LD, Bates JK (1994) Colloidal products and actinide species in leachate from spent nuclear fuel. Radiochim Acta, 66/67: 189–195

Fortner JA, Mertz CJ, Wolf SF, Jemian PR (2003) Natural groundwater colloids from the USGS J-13 well in Nye county, NV: a study using SAXS and TEM. In: Finch RJ, Bullen DB (eds) Scientific basis for nuclear waste management XXVIV. Mater Res Soc Proc, 757:483–488

Frolov YuG (1988) Colloid chemistry course. Surface effects and dispersed system. Khimiya, Moscow

Gardiner CW (1983) Handbook of stochastic methods for physics, chemistry, and the natural sciences. Springer, Berlin

Glagolenko YuV, Dzekun EG, Drozhko EG (1996) Strategy of radioactive waste management at production association "Mayak". Voprosy radiatsionnoy bezopasnosti, 2:3–10

Gregg SJ, Sing KSW (1967) Adsorption, surface area, and porosity. Academic Press. London

Hara K, Takeda S, Masuda S (1996) Research and development program of geological disposal of high-level radioactive waste in Japan. In: Proc. Int. conf. on deep geological disposal of radioactive waste, September 16-19, 1996. Winnipeg, Canada, 1-3 – 1-21

Harrington RV, Apps JA (1982) Solidification of high-level waste. Undergr Space, 6:259–263

Harvey RW, Garabedian SP (1991) Use of colloid filtration theory in modeling movement of bacteria through contaminated sandy aquifer. J Contamin Hydrol, 25:178–185

Harvey RW, George LH, Smith R, LeBlanc DR (1989) Transport of microspheres and indigenous bacteria through a sandy aquifer: results of natural- and forced-gradient tracer experiments. Environ Sci Technol, 23:51–56

Inagaki Y, Sakata H, Idemitsu K, Arima T, Banda T, Maeda T, Matsumoto S, Tamura Y, Kikkawa S (1998) Effects of eatwr redox conditions and presence of magnetite on leaching of Pu and Np from HLW glass. In: McKinley IG, McCombie C (eds) Scientific basis for nuclear waste management XXI. Mater Res Soc Proc 506:177–184

Kersting AB, Efurd DW, Finnegan DL, Rokop DJ, Smith DK, Thompson JL (1999) Migration of plutonium in ground water at the Nevada Test Site. Nature, 397:56–59

Keswick BH, Gerba CP (1980) Viruses in groundwater. Environ Sci Technol 14:1290–1297

Kim JI, Delakowitz B, Zeh P, Klotz D, Lazik D (1994) A column experiment for the study of colloidal radionuclide migration in Gorleben aquifer system. Radiochim Acta, 66/67:165–171

Klyachko VA, Apel'tsyn IE (1971) Purification of natural water. Stroyizdat, Moscow

Knapp RB, Chiarappa ML, Durham WB (2000) An experimental exploration of the transport and capture of abiotic colloids in a single fracture. Water Resour Res, 36:3139–3149

Krauskopf KB (1988) Geology of high-level nuclear waste disposal. Annu Rev Earth Planet Sci, 16:173–200

Kretzschmar R, Schäfer Th (2005) Metal retention and transport on colloidal particles in the environment. Elements 1:205–210

Laverov NP (2006) Fuel-energy resources: state and conservation. In: National reference book "Fuel and energy complex of Russia". Center of strategic programs, Moscow

Laverov NP, Kantsel AV, Lisitsyn AK (1991) Main problems of radiogeoecology concerning disposal of radioactive waste. Atomnaya Energiya, 71:523–534

Laverov NP, Omelyanenko BI, Velichkin VI (1994) Geoecological aspects of radioactive waste disposal. Geoecologiya, 6:3–20

Laverov NP, Velichkin VI, Omel'yanenko BI, Yudintsev SV (2003) Geochemistry of actinides during the long-term storage and disposal of spent nuclear fuel. Geol Ore Depos, 45:1–18

Ledin A, Karlsson S, Dücker A, Allard B (1994) The adsorption of Europium to colloidal iron oxyhydroxides and quartz – the impact of pH and an aquatic fulvic acid. Radiochim Acta, 66/67:213–230

Loitsiansky LG (1973) Mechanics of fluids and gases. Nauka, Moscow.

Malkovsky VI, Pek AA, Omelyanenko BI (1995) Influence of the inetrwell distance on the thermoconvective transport of radionuclides by groundwater from a two-well high-level nuclear waste repository. In: Slate S, Feizollahi F, Creer J (eds.) Cross-cutting issues and management of high-level waste and spent fuel, 5th International conference on radioactive waste management and environmental remediation, Berlin, September 3–7, 1995. ASME, New York

Malkovsky VI, Pek AA, Velichkin VI (1997) Transport of radionuclides from HLW repository by regional flow of groundwater. Voprosy radiatsionnoy bezopasnosti, 4:9–15

Malkovsky VI, Dikov YuP, Kalmykov SN, Buleev MI (2009) Structure of colloid particles in groundwaters on the territory of the Mayak Production Association and its impact on the colloid transport of radionuclides in subsoil environments. Geochem Int, 47:1100–1106

McCarthy J, Czerwinski KR, Sanford WE, Jardine PM, Marsh JD (1998a) Mobilization of transuranic radionuclides from disposal trenches by natural organic matter. J Contamin Hydrol, 30:49–77

McCarthy JF, Sanford WE, Stafford PL (1998b) Lanthanide field tracers demonstrate enhanced transport of transuranic radionuclides by natural organic matter. Environ Sci Technol, 32:3901–3906

McTigue DF, Givler RC, Nunziato JW (1986) Rheological effects of nonuniform particle distribution in dilute suspensions. J Rheol, 30:1053–1076

Mercier F, Moulin V, Barre N, Casanova F, Toulhoat P (2001) Study of the repartition of metallic trace elements in humic acids colloids: potentialities of nuclear microprobe and complementary technique. Anal Chim Acta, 427:101–110

Mironenko VA, Rumynin VG (1998) Problems of geoecology. Vol. 1. Moscow Mining University, Moscow
Missana T, Turrero MJ, Adell A (2000) Surface cherge and electrophoretic properties of colloids obtained from homoionic and natural bentonite. In: Smith RW, Shoesmith DW (eds). Scientific basis for nuclear waste management XXIII. Mater Res Soc Proc, 608:255–260
Morel FMM, Gschwend PM (1987) The role of colloids in the partitioning of solutes in natural waters. In: Stumm W (ed) Aquatic surface chemistry. Wiley, New York
Nyhan JW, Drennon BJ, Abeele WV, Wheeler ML, Purtymun WD, Trujillo G, Herrera WJ, Booth JW (1985) Distribution of plutonium and americium beneath a 33-yr-old liquid waste disposal site. J Environ Qual, 14:501–509
O'Melia CR (1987) Particle-particle interaction. In: Stumm W (ed) Aquatic surface chemistry. Wiley, New York
Olofsson U, Allard B, Torstenfelt B, Andersson K (1982) Properties and mobilities of actinide colloids in geologic systems. In: Lutze W (ed) Scientific basis for nuclear waste management V. Elsevier, New York
Orlandini KA, Penrose WR, Harvey BR, Lovett MB, Findlay MW (1990) Colloidal behavior of actinides in an oligotrophic lake. Environ Sci Tech, 24:706–712
Penrose WR, Polzer WL, Essington EH, Nelson DM, Orlandini KA (1990) Mobility of plutonium and americium through a shallow aquifer in a semiarid region. Environ Sci Technol, 24:228–234
Pham MK, Garnier JM (1998) Distribution of trace elements associated with dissolved compounds ($<$0.45 μm – 1nm) in freshwater using coupled (frontal cascade) ultrafiltration and chromatographic separations. Environ Sci Technol, 32:440–449
Puls RW, Powell RM (1992) Transport of inorganic colloids through neutral aquifer material: implications for contaminant transport. Environ Sci Technol, 26:614–621
Ringwood AE (1980) Safe disposal of high-level radioactive wastes. Fortschritt in Mineralogie, 58:149–168
Roache PJ (1976) Computational fluid dynamics. Hermosa, Albuquerque
Ryan JN, Elimelech M (1996) Colloid mobilization and transport in groundwater. Coll Surf A, 107:1–56
Ryan JN, Elimelech M, Ard RA, Harvey RW, Johnson PR (1999) Bacteriophage PRD1 and silica colloid transport and recovery in an iron oxide-coated sand aquifer. Environ Sci Technol, 33:63–73
Saffman PG (1965) The lift on a small sphere in a slow shear flow. J Fluid Mech, 22:385–400
Salbu B, Bjørnstaad HE (1990) Analytical techniques for determining radionuclides associated with colloids in waters. J Radioanal Nucl Chem, 138:337–346
Schäfer Th, Artinger R, Dardenne K, Bauer A, Schuessler W, Kim JI (2003) Colloid-borne americium migration in Gorleben groundwater: significance of iron secondary phase transformation. Environ Sci Tech, 37:1528–1534
Shonnard DR, Taylor RT, Hanna ML, Boro CO, Duba AG (1994) Injection-attachment of Methylosinus trichosporium OB3b in a two-dimensional miniature sand-filled aquifer simulator. Water Resour Res, 30:25–35
Short SA, Lowson RT (1988) $^{234}U/^{238}U$ and $^{230}Th/^{234}U$ activity ratios in the colloidal phases of aquifers in lateritic weathered zones. Geochim Cosmochim Acta, 52:2555–2563
Smith PA, Degueldre C (1993) Colloid-facilitated transport of radionuclides through fractured media. J Contamin Hydrol, 13:143–166
Smith DK, Finnegan DL, Bowen SM (2003) An inventory of long-lived radionuclides residual from underground nuclear testing at the Nevada test site, 1951–1992. J Environ Rad, 67:35–51
Spielman LA (1977) Particle capture from low-speed laminar flows. In: Van Dyke M, Wehausen JV, Lumley JL (eds) Annual review of fluid mechanics. vol. 9. Annual Review, Palo Alto
Tait JC, Hayward PJ, Devgun JC (1989) Technologies for contaminant immobilization, and disposal of radioactive wastes. Can J Civ Eng, 16:444–458

Tipping E (1981) The adsorption of aquatic humic substances by iron oxides. Geochim Cosmochim Acta, 45:191–199
Vilks P, Degueldre C (1991) Sorption behaviour of ^{85}Sr, ^{131}I and ^{137}Cs on colloids and suspended particles from the Grimsel test site, Switzerland. Appl Geochem, 6:553–563
Vilks P, Miller HG, Doern DC (1991) Natural colloids and suspended particles in the Whiteshell Research Area, Manitoba, Canada, and their potential effect on radiocolloid formation. Appl Geochem, 6:565–574
Voloshuk VM (1984) Kinetic theory of coagulation. Gidrometeoizdat. Moscow
Zachara JM, Smith SC, Liu Ch, McKinley JP, Serne RJ, Gassman PL (2002) Sorption of Cs^+ to micaceous subsurface sediments from the Hanford site, USA. Geochim Cosmochim Acta, 66:199–211
Zhuang J, Flury M, Jin Y (2003) Colloid-facilitated Cs transport through water-saturated Hanford sediment and Ottawa sand. Environ Sci Technol, 37:4905–4911

Section 3
Interfacial Phenomena and Formation of Actinide Nanoparticles

Nanoscale Chemistry of Uranyl Selenates

Sergey V. Krivovichev, Vladislav V. Gurzhiy, Ivan G. Tananaev, and Boris F. Myasoedov

Abstract Due to their high solubility and ability to form crystals of good quality, uranyl selenates can be considered as a kind of model systems for studies of at least some of the nanoscale processes occurring in actinide-bearing natural and technological systems. Uranyl selenates display an exceptional structural diversity that can be studied using topological and computational methods such as graph theory and cellular automata. This allows to suggest models of nanoscale self-assembly that occurs during crystallization of uranyl selenates, which is most likely based upon successive condensation of cyclic tetramers. The most interesting feature is the ability of uranyl selenates to form nanotubular structures of at least two types. In organically templated systems, the formation of nanoscale composites and molecular control upon structural architecture is governed by the delicate balance of hydrophilic/hydrophobic interactions of organic molecules that results in the formation of nanoscale supramolecular aggregates and the interactions of these aggregates with inorganic complexes.

S.V. Krivovichev (✉)
Department of Crystallography, Faculty of Geology, St. Petersburg State University, University Emb.7/9, 199034 St. Petersburg, Russia
and
Nanomaterials Research Center, Kola Science Center, Russian Academy of Sciences, Fersmana 14, 184209 Apatity, Russia
e-mail: skrivovi@mail.ru

V.V. Gurzhiy
Department of Crystallography, Faculty of Geology, St. Petersburg State University, University Emb.7/9, 199034 St. Petersburg, Russia

I.G. Tananaev and B.F. Myasoedov
A.N. Frumkin Institute of Physical Chemistry and Electrochemistry, Russian Academy of Sciences, Leninsky pr. 31, Moscow 119991, Russia

S.N. Kalmykov and M.A. Denecke (eds.), *Actinide Nanoparticle Research*,
DOI 10.1007/978-3-642-11432-8_9,

1 Introduction

The structure, properties, and mechanisms of formation of actinide nanoscale particles are relatively new fields of research compared, for instance, to transition metal oxides, silica, aluminophosphates, etc. Along with actinide peroxide nanospheres [1–4] and plutonium colloids [5], uranyl selenate nanotubules [6–9] are one of the few known purely inorganic nanoscale actinide-based structures (several metal-organic tubular and cluster structures) are also known that have been recently reported [10–12]. In general, among actinyl oxysalts, uranyl selenates provide a unique opportunity to study self-assembly of complex structures, due to a number of reasons. First of all, uranyl selenates are easily soluble in water under ambient conditions. Second, they crystallize (and recrystallize) nicely, forming crystals of good quality, which makes their structural investigation possible. This is especially important for hybrid organic–inorganic systems, where the formation of solids usually requires hydrothermal conditions. Finally, uranyl selenates form amazingly diverse range of topologies, from finite clusters to chains, layers, nanotubules, and frameworks. All topological variations observed can coherently be described within one classification scheme that allows formulation of several important concepts such as topological and geometrical isomerism. Generally speaking, structural chemistry of uranyl selenates is similar to that of other actinyl oxysalts with tetrahedral oxyanions (molybdates, sulfates, chromates [13], phosphates and arsenates [14], silicates, etc.). For this reason, uranyl selenates can be considered as a kind of model systems for studies of at least some of the processes occurring in actinide-bearing natural and technological systems.

In this chapter, we provide an overview of nanoscale structural chemistry of uranyl selenates and related compounds on the basis of our recent results obtained in this field [15–41]. First, we discuss basic features of molecular polymerization in uranyl selenates and outline general approaches to the classification and characterization of structural topologies. Then, we describe some examples of in situ studies of structural transformations in uranyl selenate aqueous solutions and hypothesize on the molecular and nanoscale mechanisms of their crystallization. In order to model these mechanisms, we implement some theoretical techniques known as cellular automata and suggest that they could provide a modeling environment for description of growth of complex uranyl-based structures at the molecular and nano scales. In the final section, we discuss topology, geometry, and packing principles of uranyl selenate nanotubules.

2 Topology of Structural Polymerization: Basic Graph, Its Variations and Isomerism

In this chapter, we shall consider structures based upon linear uranyl complexes $[O{=}U^{6+}{=}O]^{2+}$ with valence requirements of terminal O atoms more or less satisfied (Fig. 1). As a result, linkage of uranyl ions to other structure elements proceeds

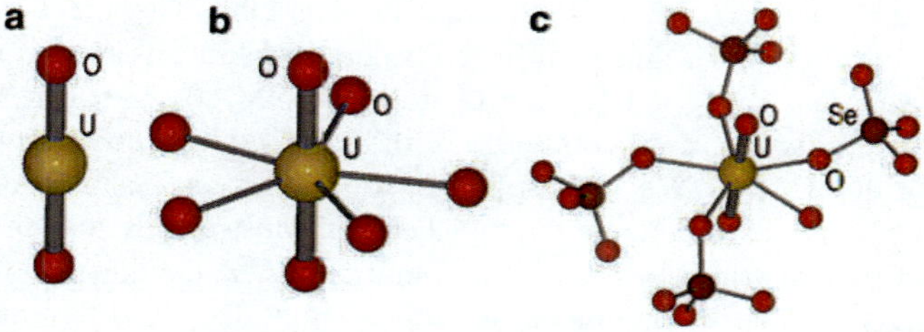

Fig. 1 Linear actinyl ion $[O{=}U^{6+}{=}O]^{2+}$ (**a**) and its coordination by equatorial anions (**b**) and tetrahedral oxyanions (**c**)

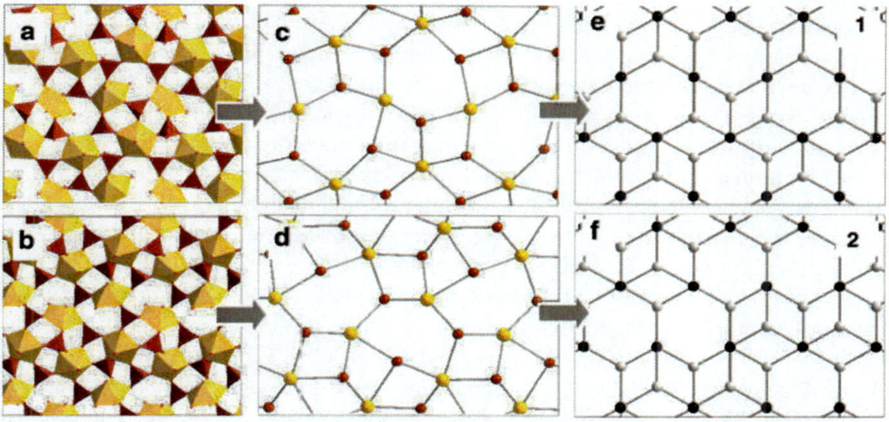

Fig. 2 Uranyl selenate layers $[(UO_2)_2(SeO_4)_3(H_2O)]^{2-}$ with different topologies of linkage of U (*light*) and Se (*dark*) coordination polyhedra (**a**, **b**): presentation as graphs (**c**, **d**) and as idealized graphs (**e**, **f**)

along bonding to uranyl ions in their equatorial planes. In uranyl selenates, uranyl ions are equatorially coordinated by O atoms of tetrahedral selenate complexes or OH groups and H_2O molecules. The most frequently observed coordination is fivefold and leads to the formation of a flattened pentagonal bipyramid, UO_7, centered by a uranium ion (Fig. 1b). Usually, coordination of uranyl ion by a SeO_4 tetrahedron is monodentate (UO_7 and SeO_4 groups share one O atom only), though bidentate coordination is quite common in uranyl sulfates [42–46]. Polymerization of actinyl ions via tetrahedral oxyanions results in the formation of a whole family of extended complexes with different compositions and topologies (Fig. 1c). The general formula of such complexes may be written as $[(UO_2)_p(SeO_4)_q(H_2O)_r]$.

Two examples of 2D layers with the same composition $[(UO_2)_2(SeO_4)_3(H_2O)]^{2-}$ are shown in Fig. 2a, b. The topology of the complexes is very similar, but one cannot be transformed into another by continuous topological transformation, i.e., without breaking of chemical bonds. The easiest way to identify topological differences between two layers is to draw their graphs. The graph can be constructed by elimination of anions from consideration and linking those U and Se atoms that share at least one common O atom (Fig. 2c, d). As it can be seen, the graphs shown in Fig. 2c, d are composed of the same relative numbers of 4- and 6-membered rings (MRs). We identify these topologies as **1** and **2**, respectively. Description of complexes in terms of graphs demonstrates clearly that, in topology **1**, 6-MRs form edge-sharing chains, whereas, in topology **2**, they form corner-sharing chains. Therefore, these topologies should be regarded as different, and their corresponding uranyl selenate complexes should be considered as *topological* isomers.

We note that graphs shown in Fig. 2c, d can be topologically transformed into graphs shown in Fig. 2e, f, respectively. In turn, the latter two graphs are derivatives of the ideal graph **3** shown in Fig. 3. This graph consists of 6-connected black and 4-connected white vertices and was identified as basic graph, as it is a parent graph to several dozens of topologies observed in the structures of inorganic oxysalts [47]. Figure 3 shows some examples of the topological derivatives observed in the structures of uranyl compounds. These examples show that topological isomerism is a common phenomenon for uranyl complexes occurring in the structures of not only selenates but also sulfates, chromates, and molybdates. In the following section, we discuss some illustrative examples of isomeric units that occur in sulfates and chromates as well as in selenates.

Topological structure of a heteropolyhedral complex observed in uranyl oxysalts cannot be completely characterized by its graph. Figure 4 shows three 1-D chains with identical formula, $[UO_2(TO_4)(H_2O)_2]$ (T = Se, S, Cr), and identical graphs. Within the chains, the $UO_5(H_2O)_2$ pentagonal bipyramids share three equatorial O atoms with TO_4 tetrahedra, whereas TO_4 tetrahedra have three vertices each shared with actinyl pentagonal bipyramids and one apical ligand that points either up (**u**) or down (**d**) relative to the plane of the chain. In different structures, the orientations of

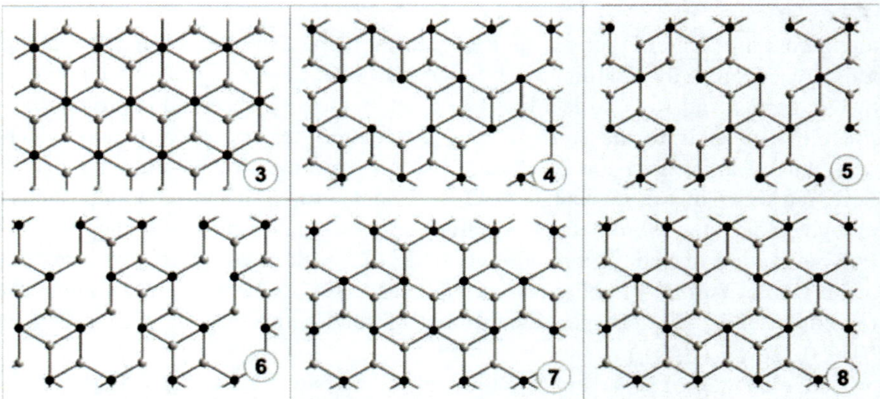

Fig. 3 Basic graph (**3**) and its derivative graphs that appear as underlying topologies for actinyl oxysalts with tetrahedral oxyanions TO_4 (T = S, Cr, Se, and Mo)

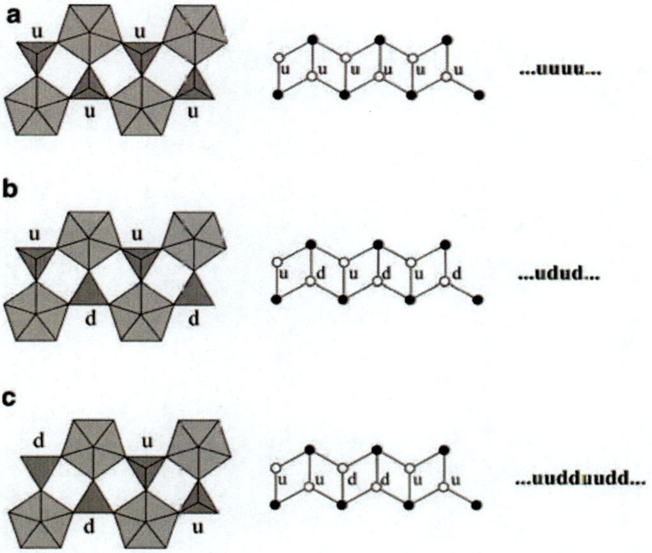

Fig. 4 To the description of geometrical isomerism of the $[UO_2(TO_4)(H_2O)_2]$ chains

apical ligands along the chain extension may be different. For example, in the structures of $[UO_2(CrO_4)(H_2O)_2]$ and $[UO_2(CrO_4)(H_2O)_2](H_2O)$ [48], the orientations alternate in the order "...up-down-up-down..." or ...**udud**... [or $(\mathbf{ud})_\infty$] (Fig. 4b). In contrast, the structure of $[UO_2(CrO_4)(H_2O)_2]_4(H_2O)_9$ [49] contains two types of chains: one with the sequence ...**ududud**... [$=(\mathbf{ud})_\infty$] and one with the sequence ...**uuuu**...$=(\mathbf{u})_\infty$, with the nonshared corners of chromate tetrahedra on one side of the chain only (Fig. 4a). The structure of β-$[UO_2(SO_4)(H_2O)_2](H_2O)_3$ [50] contains chains with the sequence ...**uudduudd**... or $(\mathbf{uudd})_\infty$. We note that, despite different orientations of tetrahedra, the three chains shown in Fig. 4 correspond to the same graph. The three chains should be considered as geometrical isomers, i.e., chemical structures that have the same global topology (e.g., graph) but are different due to some specific properties of linkage (e.g., *cis-trans*-isomerism). It is likely that, in actinyl oxysalts, the occurrence of a certain type of isomer is impacted by the specificity of interactions of polyhedral units with each other and with other structural elements. In the case of uranyl chromate hydrates mentioned above, it seems that occurrence of isomers is related to the number of water molecules present in the structure, i.e., by features of the hydrogen bonding system.

In the case of 2-D layers, arrangement of orientations of tetrahedral oxyanions relative to the layer plane can be quite complex. Figure 5 shows three 2-D graphs that represent topology **7** (Fig. 3) of linkage of uranyl pentagonal bipyramids and chromate tetrahedra in structures of uranyl selenates and chromates with $[(UO_2)_3(TO_4)_5]^{4-}$ (T = Se, Cr) anionic layers [22, 32, 51].

Symbols **u** and **d** written near the white vertices indicate orientations of chromate tetrahedra up or down. Transferring lattice of these symbols onto 2-D a

b

c

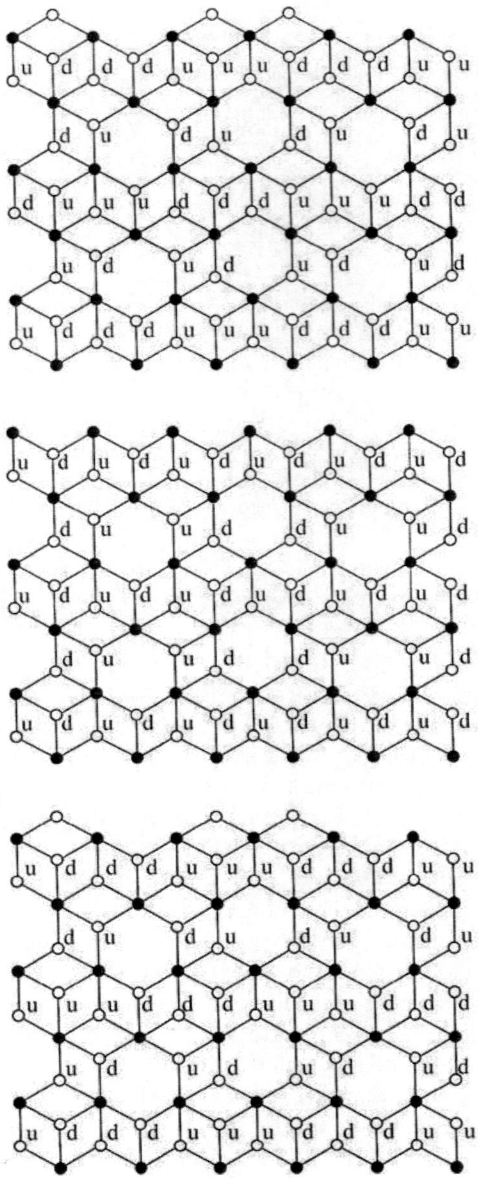

d

u	d	d	d	u	u	u	d	d	d	u	u
□	d	u	□	d	u	□	d	u	□	d	u
d	u	u	u	d	d	d	u	u	u	d	d
□	u	d	□	u	d	□	u	d	□	u	d
u	d	d	d	u	u	u	d	d	d	u	u
□	d	u	□	d	u	□	d	u	□	d	u
d	u	u	u	d	d	d	u	u	u	d	d
□	u	d	□	u	d	□	u	d	□	u	d

e

u	d	u	d	u	d	u	d	u	d	u	d
□	d	u	□	u	d	□	d	u	□	u	d
u	d	u	d	u	d	u	d	u	d	u	d
□	d	u	□	u	d	□	d	u	□	u	d
u	d	u	d	u	d	u	d	u	d	u	d
□	d	u	□	u	d	□	d	u	□	u	d
u	d	u	d	u	d	u	d	u	d	u	d
□	d	u	□	u	d	□	d	u	□	u	d

f

u	d	d	d	u	u	u	d	d	d	u	u
□	d	u	□	d	u	□	d	u	□	d	u
u	u	u	d	d	d	u	u	u	d	d	d
□	u	d	□	u	d	□	u	d	□	u	d
u	d	d	d	u	u	u	d	d	d	u	u
□	d	u	□	d	u	□	d	u	□	d	u
u	u	u	d	d	d	u	u	u	d	d	d
□	u	d	□	u	d	□	u	d	□	u	d

Fig. 5 *Black*-and-*white* graphs with the **u**, **d**, and □ symbols of the $[(UO_2)_3(CrO_4)_5]$ sheets in the structures of $Mg_2[(UO_2)_3(CrO_4)_5](H_2O)_{17}$ (**a**), $Ca_2[(UO_2)_3(CrO_4)_5](H_2O)_{19}$ (**b**), and $K_4[(UO_2)_3(CrO_4)_5](H_2O)_8$ (**c**); their corresponding **u**, **d,** and □ symbolic tables (**d**, **e**, and **f**, respectively). The matrices of orientation of tetrahedra are indicated in (**d**), (**e**), and (**f**) by *bold* lines

plane and using symbol "□" to denote vacancy (absence of white vertex in the specific place of the graph), one may obtain a 2-D square array of symbols that can be characterized by its "unit cell" confined in orthogonal boxes in Fig. 5d–f. This "unit cell" or matrix of tetrahedra orientation is a unique descriptor of a particular geometrical isomer.

It should be noted that the system of orientation of tetrahedra in different geometrical isomers may be chiral that can be clearly identified by analysis of their matrix descriptors. For instance, topology **1** observed in uranyl selenates and sulfates has two geometrical isomers observed so far, with one of them being chiral (the corresponding compounds crystallize in the chiral space group $P2_1$) [37].

3 Crystallization Mechanisms: Prenucleation Building Units

3.1 General Considerations

There are many models of crystallization mechanisms in aqueous solutions of inorganic oxysalts, in particular, of zeolites. Most of them are still a matter of continuous debates [52]. The most important problem is the existence of prenucleation building units with structures similar to that observed in the resulting solids. In metal oxysalts, a considerable research was concentrated on metal phosphates, due to their importance as microporous materials and catalysts [53–57]. The existing models are based upon the crystallization sequence of phases with special attention to metastable transitional phases [54, 55, 58, 59]. If the structures of the latter are known, a certain model of inheritance of structural architectures may be constructed. If the same fundamental building block (FBB) shows up in all phases that appear over the crystallization sequence in the system, this points out that this block is stable in solution. For metal phosphates, the following structural units are distinguished: (a) primary building units (PBUs) are coordination polyhedra of cations; (b) secondary building units (SBUs) that are built by condensation of PBUs; (c) prenucleation building units (PNBUs) that exist in aqueous solutions before crystallization of solids. Note that PNBUs may have a structure different from that of SBUs, because reconstruction may occur when the unit is attached to the crystal surface. In metal phosphate systems, the most common SBU is a tetramer formed by two metal (Al, Ga, Zn, Ti, Fe) polyhedra and two phosphate anions. From structural point of view, these tetramers represent either 4-membered rings (cyclic tetramers) or two fused 3-membered rings (bicyclic tetramers). Existence of such tetramers in phosphate-containing metal aqueous solutions is confirmed by nuclear magnetic resonance (NMR) studies [53].

As to our knowledge, no extensive studies have been done on crystallization mechanisms in actinide-containing aqueous solutions that involve larger than mononuclear actinide complexes. In particular, actinide sulfate and carbonates were a subject of comprehensive investigations that compare structures of complexes in solid state compounds and in aqueous solutions [60–65]. Below, we outline possible models of crystallization of uranyl selenates based upon consideration of structures of metastable intermediate phases that appear in uranyl selenate solutions with variable organic and inorganic cations.

3.2 *Hydrated Mg Uranyl Selenates*

Crystals of two modifications of $Mg_2[(UO_2)_3(SeO_4)_5](H_2O)_{16}$ and $Mg[(UO_2)(SeO_4)_2(H_2O)](H_2O)_4$ were synthesized from aqueous solution prepared by reactions of MgO and uranyl nitrate in selenous acid [15]. The solution was left to evaporate in a fume hood under ambient conditions. Green lamellar crystals of β-$Mg_2[(UO_2)_3(SeO_4)_5](H_2O)_{16}$ started to form after 24 h. After another 24 h, these crystals dissolved completely and almost all crystallization space was filled by isometric greenish-yellow crystals of α-$Mg_2[(UO_2)_3(SeO_4)_5](H_2O)_{16}$, together with small amounts of $Mg[(UO_2)(SeO_4)_2(H_2O)](H_2O)_4$. Thus, β-$Mg_2[(UO_2)_3(SeO_4)_5](H_2O)_{16}$ represents a metastable intermediate phase, which was also manifested in the imperfect quality of its crystals in air. Crystals of all three Mg uranyl selenate hydrates were extracted from solution and their structures were studied. Structures of two modifications of $Mg_2[(UO_2)_3(SeO_4)_5](H_2O)_{16}$ are shown in Fig. 6a, b. Both contain $[(UO_2)_3(SeO_4)_5]^{4-}$ layers of UO_7 pentagonal bipyramids sharing O corners with SeO_4 tetrahedra (Fig. 6c, d). Topologies of the layers are shown as graphs in Fig. 6e, f. It can be seen that they are built by condensation of 4- and 6-MRs but still are topologically different, which means that the two graphs correspond to the two different topological isomers. In contrast to two $Mg_2[(UO_2)_3(SeO_4)_5](H_2O)_{16}$ modifications, the structure of $Mg[(UO_2)(SeO_4)_2(H_2O)](H_2O)_4$ is based upon chains with topology shown in Fig. 7a. It is worthy to note that all three topologies contain the same building unit: the cyclic tetramer formed by cyclic condensation of U and Se coordination polyhedra (highlighted in Fig. 3).

3.3 *1,10-Diaminodecane–UO_2SeO_4–H_2O System*

Crystals of $[C_{10}H_{26}N_2][(UO_2)(SeO_4)_2(H_2O)](H_2SeO_4)_{0.85}(H_2O)_2$ and $[C_{10}H_{26}N_2]$-$[(UO_2)(SeO_4)_2](H_2SeO_4)_{0.50}(H_2O)$ [31] were obtained from aqueous uranyl selenate solutions in the presence of 1,10-diaminododecane. The crystallization sequence of phases was as following. First, the structure with 1D uranyl selenate chains formed (Fig. 7a); then, crystals of this phase dissolved and the phase based upon 2D layers crystallized (Fig. 7b). Similar effect of 1D → 2D transformation in aqueous solutions was frequently observed in the course of hydrothermal crystallization of metal phosphates [66–75]. In particular, Oliver et al. [76] proposed the model of crystallization of aluminophosphate that involves formation of 1D chains with the topology shown in Fig. 7a as a prenucleation building block or precursor structure for a wide range of aluminophosphate complexes with extended

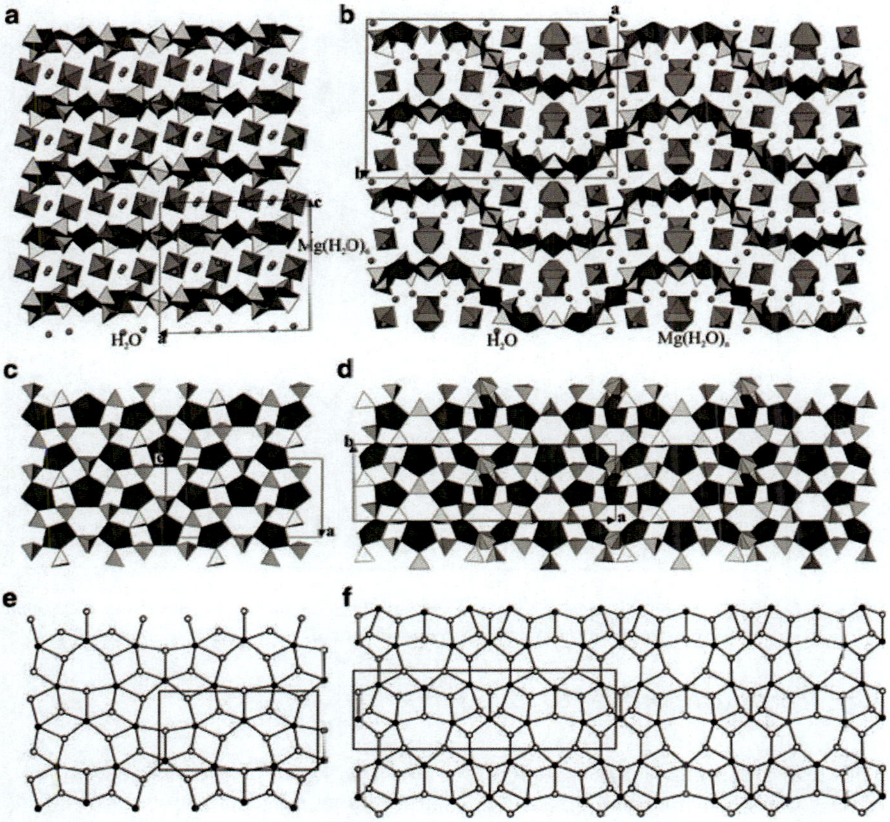

Fig. 6 Projections of the structures of α- (**a**) and β-$Mg_2[(UO_2)_3(SeO_4)_5](H_2O)_{16}$ (**b**) along the *b* or *c* axis, respectively; the $[(UO_2)_3(SeO_4)_5]^{4-}$ sheets in the crystal structures of α- (**c**) and β-$Mg_2[(UO_2)_3(SeO_4)_5](H_2O)_{16}$ (**d**) and their nodal representations (**e** and **f**, respectively). Legend: $[UO_7]^{8-}$ bipyramids = *black circles*; $[SeO_4]^{2-}$ tetrahedra = *white circles*

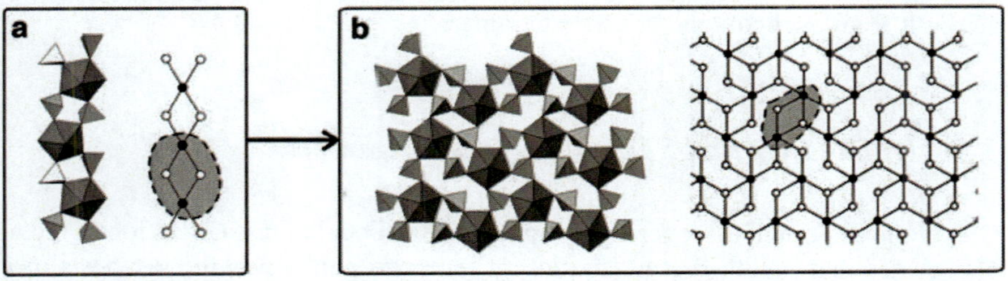

Fig. 7 Evolution of structural topologies in the 1,10-decanediamine-templated system: initial 1D topology (**a**) transforms into 2D topology (**b**)

dimensionality. Later, this model received additional indirect experimental evidence [74]. It is noteworthy that both 1D and 2D structures may be considered to be composed of cyclic tetramers.

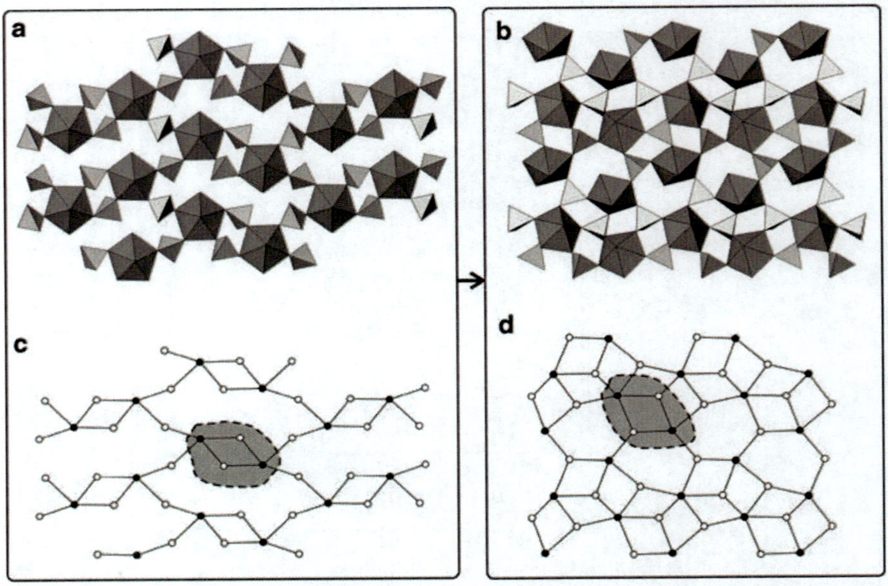

Fig. 8 Evolution of structural topologies in the *tert*-butylamine-templated system: initial 2D open topology (**a** and **b**) transforms into 2D dense topology (**c** and **d**)

3.4 *Tert-butylamine–UO_2SeO_4–H_2O System*

The first metastable phase that forms in this system is $[C_4H_{12}N]_2$–$[(UO_2)(SeO_4)_2(H_2O)]$, which has the structure based upon 2D layers shown in Fig. 9a. Several crystals of this phase were extracted from solution and dissolved in distilled water, after which the resulting solution was left to evaporate. After several hours, crystals of the phase with composition $[C_4H_{12}N]_2$–$[(UO_2)_2(SeO_4)_3(H_2O)]$ formed. Its structure (Fig. 8b) also contains 2D layers, but with topology different from that of the precursor phase. Although the two topologies are obviously different, both are built from cyclic uranyl selenate tetramers.

3.5 *Isolated Uranyl Selenate Cyclic Tetramers*

One of the important discoveries in support of the model of crystallization of metal phosphates that involved condensation of tetrameric units was the synthesis of a crystalline phase containing isolated (i.e., not condensed) tetramers [77]. In the course of our studies of uranyl selenate crystallization, we were able to prepare a compound with the formula $[C_3H_{12}N_2][(UO_2)(SeO_4)_2(H_2O)_2](H_2O)$. Its structure contains a $[(UO_2)_2(SeO_4)_4(H_2O)_4]^{4-}$ branched cyclic tetramer ([1 + 4 + 1] unit) shown in Fig. 9 [16].

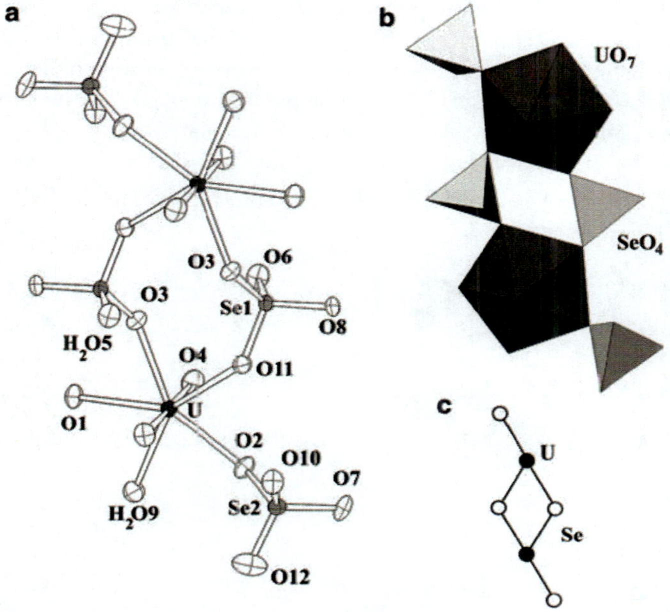

Fig. 9 Branched cyclic tetramer $[(UO_2)_2(SeO_4)_4(H_2O)_4]^{4-}$ complex in $[C_3H_{12}N_2][(UO_2)$-$(SeO_4)_2(H_2O)_2](H_2O)$ (**a** and **b**) and its graph (**c**)

In fact, most of the uranyl selenate structures, from 1D chains to 3D frameworks, may be considered as formed by successive condensation of cyclic tetramers or [1 + 4 + 1] units in the course of Aufbau process [78]. This fact may serve as an indirect evidence for the existence of these units as PNBUs in aqueous solutions of uranyl selenates. No doubt that this conclusion needs further confirmation by independent spectroscopic methods.

4 Crystallization Mechanisms: Cellular Automata Modeling

4.1 Methylamine-Templated Uranyl Selenates: Ordered and Disordered Structures

Gurzhiy et al. [79] investigated phase formation in the system $UO_2(NO_3)_2$–H_2SeO_4–methylamine–H_2O and found at least eight different crystalline uranyl selenates with U:Se = 1:2, 2:3, 3:5 and 5:8, some of them metastable or unstable in air. Of particular interest is the family of structures with general formula $(CH_3NH_3)[(UO_2)(SeO_4)_2(H_2O)](H_2O)_n$, where $n = 0, 0.5, 1$. The structures are based upon complex units with the composition $[(UO_2)(SeO_4)_2(H_2O)]^{2-}$, but with different topology of linkage of U and Se polyhedra for different values of n. For $n = 1$, the structure contains 1-D chains (Fig. 10a); for $n = 0$, the structure is based upon 2-D layers (Fig. 10b). For $n = 0.5$, a disordered structure was observed that represents a superposition of the 1-D and 2-D structures (Fig. 11). Transition between two overlapping configurations can be achieved by translation of the [1 + 4 + 1] unit (see above) by the 1/2**a**. It is important to note that all three structures shown in Figs. 10 and 11 can be obtained by successive polymerization of the [1 + 4 + 1]

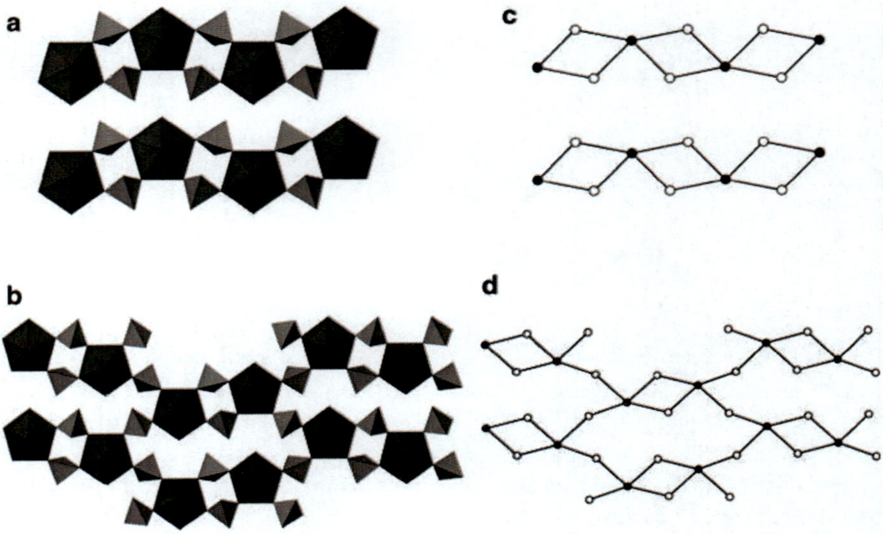

Fig. 10 Uranyl selenate layers in the structures of $(CH_3NH_3)[(UO_2)(SeO_4)_2(H_2O)](H_2O)_n$, where $n = 0$ (**a**) and 1 (**b**) and their graphs (**c** and **d**, respectively)

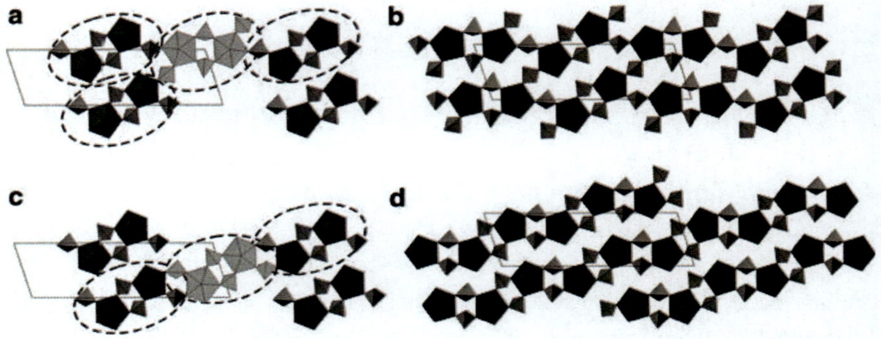

Fig. 11 Uranyl selenate layer in the structure of $(CH_3NH_3)[(UO_2)(SeO_4)_2(H_2O)](H_2O)_{0.5}$ can be represented as a superposition of two different structures related to each other by translation of the [1 + 4 + 1] complex (highlighted) by 1/2 translation (**a** and **c**), which results in the formation of either 2-D (**b**) or 1-D (**d**) ordered structures

cluster, which supports our suggestion concerning its existence as a prenucleation building block in aqueous solutions. Thus, in the $(CH_3NH_3)[(UO_2)(SeO_4)_2(H_2O)](H_2O)_n$ family, the appearance of disordered structure with $n = 0.5$ can be represented as an oscillation of growth process between the structure types with $n = 0$ and 1. This process can be modeled by the construction of an abstract dynamical system known as cellular automata.

4.2 Cellular Automata: Basic Concepts

Cellular automata (CA) have been introduced for simulation of self-reproductive biological systems [80] and have attracted considerable attention as a possible environment for modeling of a broad range of physical objects and processes [81–83], in particular, of periodic growth of complex chemical structures [84–87].

From the formal point of view, CA is defined as a collection of five basic components: CA: $= <Z, S, N, f, B>$,

where Z is a lattice (discrete working space of the CA consisting of cells; the simplest example is a 2-D plane filled by square cells);

$S = \{0, 1, 2, ...\}$ is a finite number of values that the cells may take (usually, these values are associated with colors, e.g., $S = \{0, 1\}$ characterizes a binary (2-color) CA;

$N = \{-k_1, -k_1 + 1, ..., -1, 0, 1, ..., k_2 - 1, k_2\}$ is a neighborhood of CA action; for 1-D CA, the value of the cell x_0 at the time $t = 1$ is determined by the values of k_1 and k_2 cells on the left and right sides at the time $t = 0$ (in the simplest case, the neighborhood is symmetrical $k_1 = k_2 = 1$ and has a radius of $k_0 = 1$, i.e., it consists of three cells ($k = 3$), x_{-1}, x_0, x_1; the value of the cell x_0 at the time $t = 1$ is determined by the values x_{-1}, x_0, x_1 at the time $t = 0$);

f is a local transition function that works for a certain neighborhood (usually written as a set of rules of the form $010 \rightarrow 1$);

B is a boundary conditions.

The example of a CA is shown in Fig. 12. According to the Wolfram classification, the CA has a number of 90. Its local transition function can be written as $f = \{111 \rightarrow 0, 110 \rightarrow 1, 101 \rightarrow 0, 100 \rightarrow 1, 011 \rightarrow 1, 010 \rightarrow 0, 001 \rightarrow 1, 000 \rightarrow 0\}$. Using one black cell as an initial condition, this CA results in the formation of a branching pattern known as the Sierpinski triangle. It is obvious that, for modeling of periodic structures (that may work as diffraction lattices), one needs a specific class of CA.

4.3 CA Model of Uranyl Selenate Structures and Their Growth

In order to construct a CA that reproduces the formation of uranyl selenate structures with U:Se = 1:2, we topologically (i.e., without breaking of bonds)

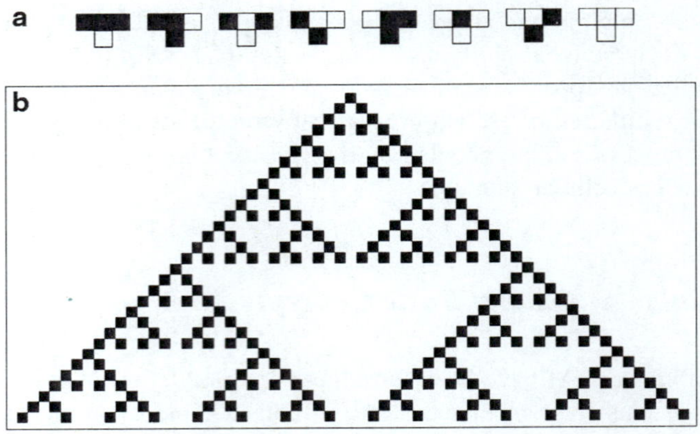

Fig. 12 1-D cellular automaton 90: its transition rules (a) and the result of its work by the initial conditions given as a single black cell (b)

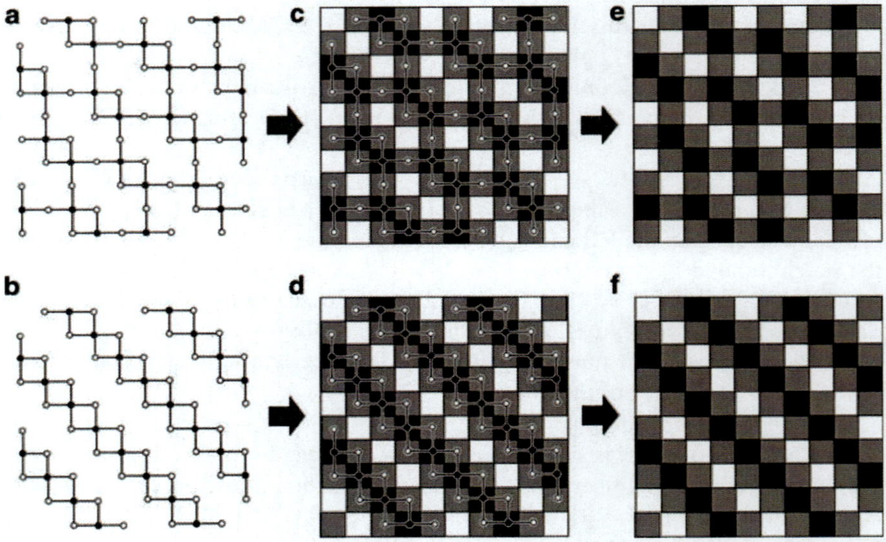

Fig. 13 2-D (a) and 1-D (b) graphs isomorphous to the graphs shown in Fig. 10c, d, respectively, and their transformation to tricolor cellular structures (c–f)

transform the graphs shown in Fig. 10c, d into graphs shown in Fig. 13a, b, respectively. In turn, these graphs may be replaced by a 2-D cellular structure, where black and white vertices are replaced by black and grey cells, respectively (white cells are reserved for "empty" space regions) (Fig. 13c, d). The cells that share common edges correspond to the vertices linked by an edge. As a result, we have two tricolored cellular structures as shown in Fig. 13e, f. Analysis of these

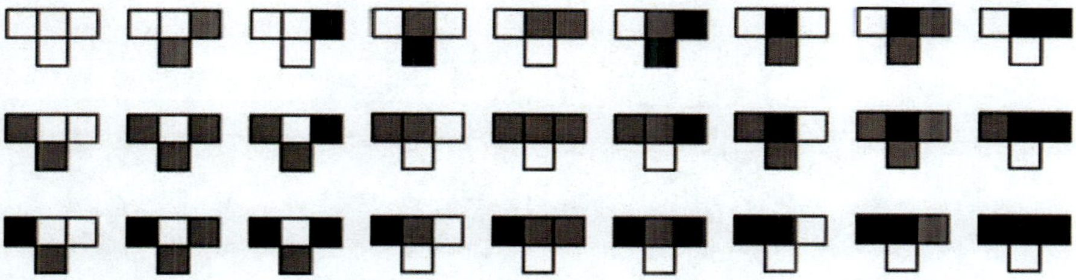

Fig. 14 Visual representation of the local transition functions that can be used to generate patterns shown in Fig. 13e, f

structures indicate that they can be successively constructed using trinary CA with the transition rules shown in Fig. 14. By assignment of values of 0, 1, and 2 to white, grey, and black cells, respectively, the set of transition rules can be written down as $f = \{000 \rightarrow 0,\ 001 \rightarrow 1,\ 002 \rightarrow 0,\ 010 \rightarrow 2,\ 011 \rightarrow 0,\ 012 \rightarrow 2,\ 020 \rightarrow 1,\ 021 \rightarrow 1,\ 022 \rightarrow 0,\ 100 \rightarrow 1,\ 101 \rightarrow 1,\ 102 \rightarrow 1,\ 110 \rightarrow 0,\ 111 \rightarrow 0,\ 112 \rightarrow 0,\ 120 \rightarrow 1,\ 121 \rightarrow 1,\ 122 \rightarrow 0,\ 200 \rightarrow 1,\ 201 \rightarrow 1,\ 202 \rightarrow 1,\ 210 \rightarrow 0, 211 \rightarrow 1, 212 \rightarrow 0, 220 \rightarrow 0, 221 \rightarrow 0, 222 \rightarrow 0\}$.

Most of these rules are excessive in our case; however, they are used for the reasons of generality. This CA can be used to obtain cellular structures shown in Fig. 13e, f. This means that *the same* CA generates *different* structural topologies. The resulting topology is therefore determined by the initial conditions, i.e., by the structure of the first row. In the language of chemistry, this could suggest that molecular-level growth mechanism of uranyl selenate layers is the same, and the topology of the structure is controlled by the structure of the nucleus spontaneously formed in solution.

In order to investigate the dependence of structure topology from the structure of initial conditions, a computer experiment was performed by means of the Mathematica 6.0 program package. The structure of the input row was given as an infinite periodic sequence of numbers 0, 1, and 2. Some results of the modeling are shown in Fig. 15. The chain topology (Fig. 15a) can be produced using the sequence {0121}, whereas the layer topology (Fig. 15b) is the result of input of the sequence {01012121}. The cellular structure shown in Fig. 15c and resulting from input {12} deserves special attention. It corresponds to the structural unit, which has the composition $[(AnO_2)(TO_4)_2(H_2O)]$ and is observed in a large number of actinyl oxysalts, including uranyl selenates [24, 34]. It is of interest that this unit can also be produced using the CA described above.

The results of computer modeling may shed some light into process of self-assembly in uranyl selenate systems. In the case of ordered nuclei, either chain or layer topology is formed, whereas, in the case of disordered nuclei or growth faults (in computer language, errors of CA function), a disordered topology is formed that represents a superposition of chain and layer topologies.

In general, rather simple examples of CA modeling of self-assembly processes in actinyl-based systems indicate that the theory of CA and finite automata may

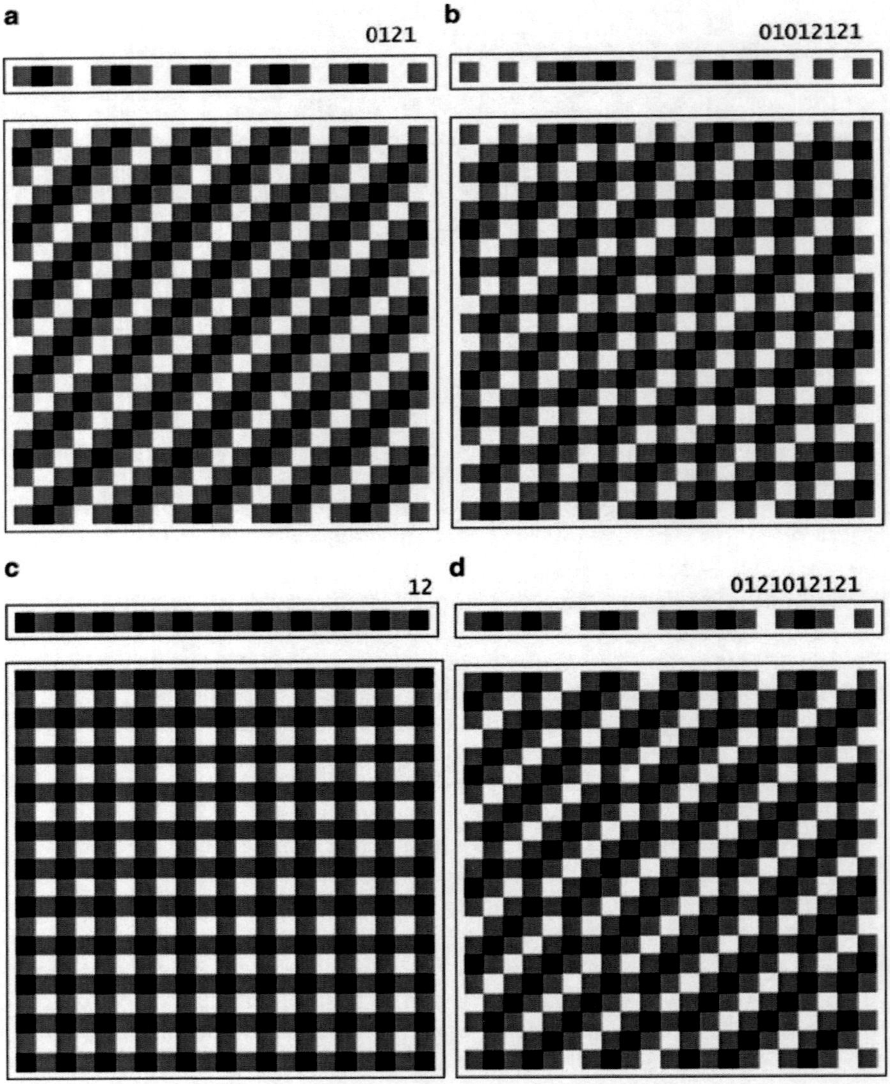

Fig. 15 Examples of patterns generated by the automaton with local transition function indicated in Fig. 14 from initial conditions given as a periodic sequence of cells (periods are shown as numbers in upper right parts of patterns). See text for details

provide a coherent framework for description of dynamics of chemical systems, provided that an appropriate abstract model is constructed. Moreover, it seems feasible that, in the deep sense, growth of periodic structures is similar to the growth of cellular structures during the development of a CA and, therefore, crystal growth can be viewed as a computation.

The parallels between chemical structures and such formal objects as CA can also be extended further. For instance, the CA theory contains the concept of a "Garden-of-Eden" configuration, which is a configuration that can never appear as a result of the specific CA work. In the case of 1-D automaton, this is the row of cells, which has no predecessors. There could be an infinite number of "Garden-of-Eden" configurations for a particular CA, and there are specific elementary configurations that any "Garden-of-Eden" contains. These configurations are called orphans. By inspection of the CA rules shown in Fig. 14, it is easy to conclude that, for the given transition functions, sequences {11} and {22} are orphans, i.e., they may not form as the result of the CA work under any circumstances. Translating these sequences into the language of chemistry, this could imply that, in our system, no uranyl dimers or diselenate groups Se_2O_7 may form. Indeed, the formation of uranyl dimers is generally possible (with bridging hydroxyl of fluoride ions), but not under the acidic conditions employed in our particular experiment. The formation of diselenate groups is also rather impossible, taking into account their instability relative to the monoselenate groups.

Application of CAs allows the prediction of possible topologies that may form in a specific system. It may also provide a computational basis for studying complexity and dynamics of topologically and chemically similar structures forming under similar physicochemical conditions.

5 Uranyl Selenate Nanotubules: Topology, Geometry, and Packing Principles

As it was noted in the Introduction, uranyl selenate nanotubules are ones of the few known purely inorganic nanoscale actinide structures. The tubules represent one-dimensional porous structures formed by linking uranyl pentagonal bipyramids to SeO_4 tetrahedra through common O atoms. There are two types of nanotubules reported so far. "Small" tubules (Fig. 16) have been found in two structures [6, 9]. Their composition is characterized by the U:Se ratio equal to 3:5, so the chemical formula is $[(UO_2)_3(SeO_4)_5]^{4-}$, and the structure is stabilized by the presence of positively charged species, namely K^+ cations. Local topology of the tubules is isomorphous to the topology **8** shown in Fig. 3. This topology displays a tendency toward the formation of geometrical isomers with different systems of orientations of tetrahedra relative to the plane of the sheets. For 2-D topologies, the ratio **u:d** of "up" and "down" orientations is invariably 1:1. However, the situation is different for uranyl selenate nanotubules, where the **u:d** ratio is equal to 4:1 (implying "up" orientation as pointing out of the tubules and "down" orientation as pointing inside the tubule). Thus, while there is no difference between orientations of tetrahedra for 2-D structures (two half-spaces separated by the sheet are chemically indistinguishable), for cylindrical (tubular) topologies, the space is separated into exterior and interior, which are both

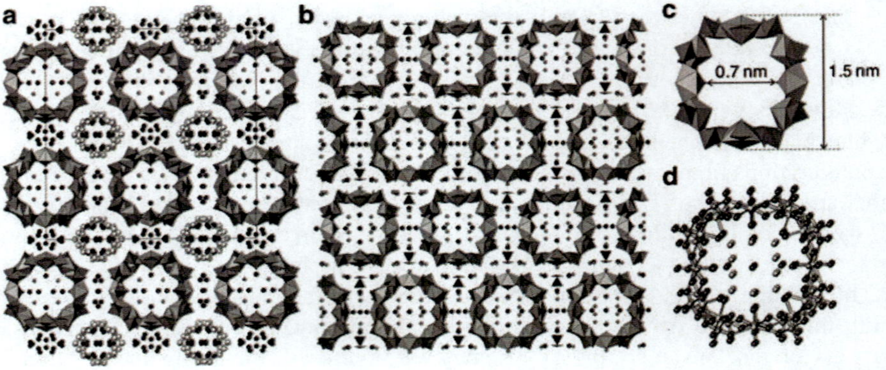

Fig. 16 Crystal structures of $K_5[(UO_2)_3(SeO_4)_5](NO_3)(H_2O)_{3.5}$ (**a**) and $(H_3O)_2K[(H_3O)$@-(18-crown-6)]$[(UO_2)_3(SeO_4)_5](H_2O)_4$ (**b**) containing "small" uranyl selenate nanotubules with the U:Se ratio of 3:5 (**c** and **d**). Uranyl and selenium coordination polyhedra are shown as *yellow* and *light-brown*, respectively

geometrically and chemically different. The geometry of the 3:5 uranyl selenate nanotubules is almost identical in the two structures reported so far.

The parameter along the tubule is equal to 11.293 Å in $K_5[(UO_2)_3(SeO_4)_5](NO_3)(H_2O)_{3.5}$ and 11.240 Å in $(H_3O)_2K[(H_3O)$@(18-crown-6)]$[(UO_2)_3(SeO_4)_5](H_2O)_4$. Figure 17 shows a diagram that demonstrates geometrical similarity of the tubules in the two structures in terms of the U–O–Se valence angles. The dotted line corresponds to the ideal case, when all corresponding angles are equal, whereas the deviation of points from this line shows the difference between the corresponding angles in the two structures. It is clear from the diagram that this difference does not exceed 5–6°, which means that the tubules have very similar geometrical parameters in different structures. Remarkably, local rod symmetry group of the tubules is the same (*mm*2) and consists of two vertical mutually perpendicular mirror planes intersecting along the twofold symmetry axis.

The observed arrangements of 3:5 uranyl selenate nanotubules are schematically illustrated in Fig. 18. It can be rationalized in terms of a sum of axial rotations and vertical shifts of tubules relative to each other. The tubules may be shifted relative to each other by ±1/2 translations along the tubule extension. Thus, there are two different vertical positions of tubules, indicated as 0 and 1/2 in Fig. 18. The tubules also allow axial positions rotated relative to each other by 90°; these are distinguished in Fig. 18 by white and black triangles. The schemes of arrangement of tubules in the two structures are shown in Fig. 18c, d. In the structure of $K_5[(UO_2)_3(SeO_4)_5](NO_3)(H_2O)_{3.5}$, the tubules are packed in a hexagonal fashion (tubules are parallel, and their cross-sections imitate hexagonal packing of circles), whereas in $(H_3O)_2K[(H_3O)$@(18-crown-6)]$[(UO_2)_3(SeO_4)_5](H_2O)_4$, the tubules are arranged in a tetragonal packing, which is the result of the presence of single columns of stacked $[(H_3O)$@(18-crown-6)$]^+$ complexes.

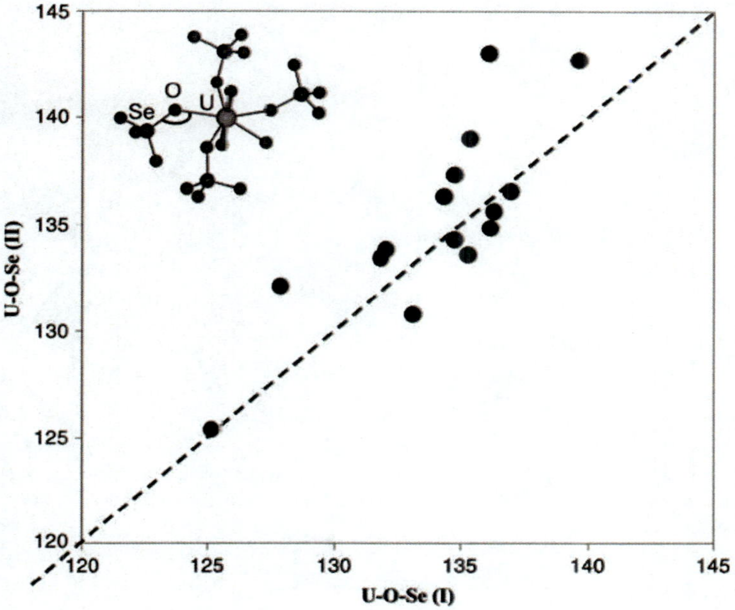

Fig. 17 Diagram showing difference of the U–O–Se valence angle values in uranyl selenate nanotubules in the structures of $K_5[(UO_2)_3(SeO_4)_5](NO_3)(H_2O)_{3.5}$ (I) and $(H_3O)_2K[(H_3O)@(18\text{-crown-6})][(UO_2)_3(SeO_4)_5](H_2O)_4$ (II)

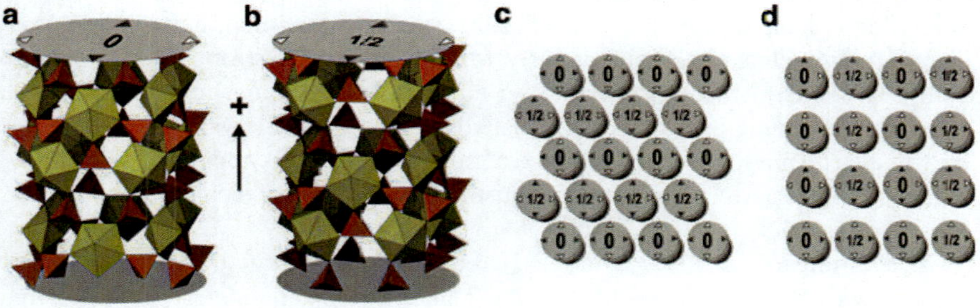

Fig. 18 Two orientations of uranyl selenate nanotubules rotated relative to each other by 90° and shifted by 1/2-translation along the axis of a tubule (**a** and **b**) and schemes of arrangements of nanotubules in the structures of $K_5[(UO_2)_3(SeO_4)_5](NO_3)(H_2O)_{3.5}$ (**c**) and $(H_3O)_2K[(H_3O)@(18\text{-crown-6})][(UO_2)_3(SeO_4)_5](H_2O)_4$ (**d**)

The two structures with 3:5 tubules possess K^+ cations located inside the tubules and showing remarkably similar local coordination to the tube walls (Fig. 19). In both structures, K^+ cations are coordinated by six anions of the uranyl selenate wall: two O atoms of the uranyl ions, two nonbridging O atoms of selenate

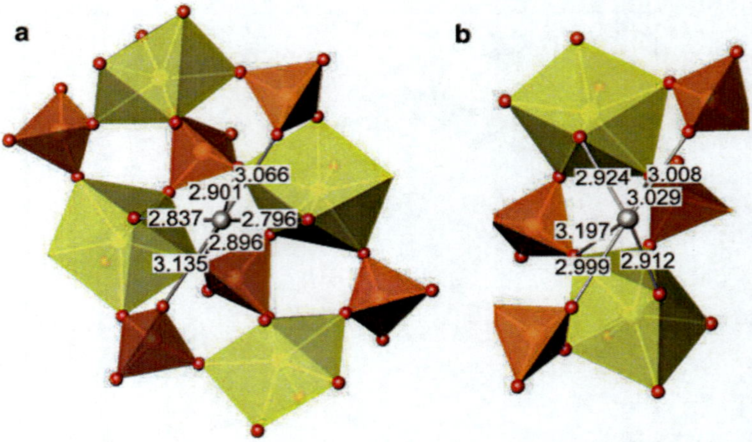

Fig. 19 Coordination of K^+ cations relative to the walls of uranyl selenate nanotubules in the structures of $K_5[(UO_2)_3(SeO_4)_5](NO_3)(H_2O)_{3.5}$ (**a**) and $(H_3O)_2K[(H_3O)@(18\text{-crown-}6)][(UO_2)_3(SeO_4)_5](H_2O)_4$ (**b**)

tetrahedra (namely, those oriented *inside* the tubule), and two O atoms bridging between uranyl ions and selenate tetrahedra. As a whole, K^+ ion coordinates the [1 + 4 + 1] complex that was proposed as a basic unit present in uranyl selenate aqueous solution. It can be hypothesized that hydrated $\{K[(UO_2)_2(SeO_4)_4]\}$ units played a role of precrystallization building units from which the tubules assembled during the crystallization process. From the structural viewpoint, it is beyond any doubt that it is the K^+ ion that is responsible for the formation of curved topologies in 3:5 uranyl selenate nanotubules. By forming relatively strong K^+–O bonds to uranyl O atoms, it forces inclination of adjacent linear uranyl ions towards each other, thus inducing curvature of the usually planar uranyl oxysalt units. The fact that tubules do not form in the presence of other alkali metal cations (Na^+, Rb^+, Cs^+) points out to the perfect geometrical adjustments provided by the size of the K^+ ions (Figs. 20 and 21).

In addition to "small" 3:5 tubules, uranyl selenates provide another type of tubules with the ratio U:Se of 14:17 [7] and external diameter of ca. 2.5 nm (Fig. 10). This tubule does not have isomorphous topology among 2-D uranyl selenate sheets known to date. The mechanism of formation of this type of tubules is not clear; however, some possible directions can be outlined.

The tubules are most probably templated by aggregates of protonated butylamine molecules driven by delicate balancing between hydrophilic–hydrophobic interactions. Though the organic substructure in the original 14:17 nanotubular structure was not located, some similarities may be inferred by the analysis of the structure of $(H_3O)_2[C_{12}H_{30}N_2]_3[(UO_2)_4(SeO_4)_8](H_2O)_5$ [23] shown in Fig. 11. In this structure, protonated 1,12-diaminododecane molecules form cylindrical

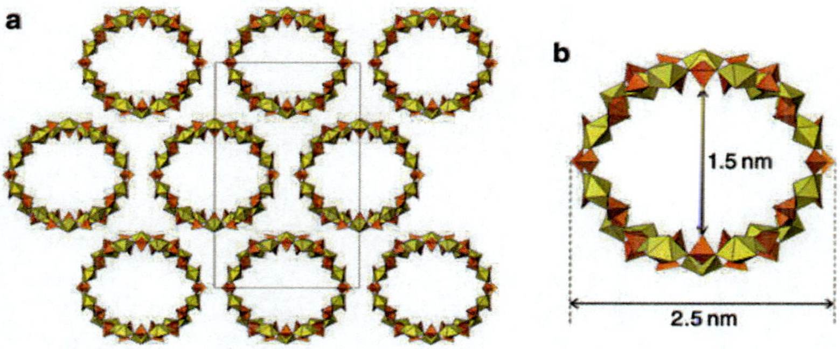

Fig. 20 Arrangement of the 14:17 uranyl selenate nanotubules in the structure of $(C_4H_{12}N)_{14}[(UO_2)_{10}(SeO_4)_{17}(H_2O)]$ (**a**) and cross-section of the nanotubule (**b**)

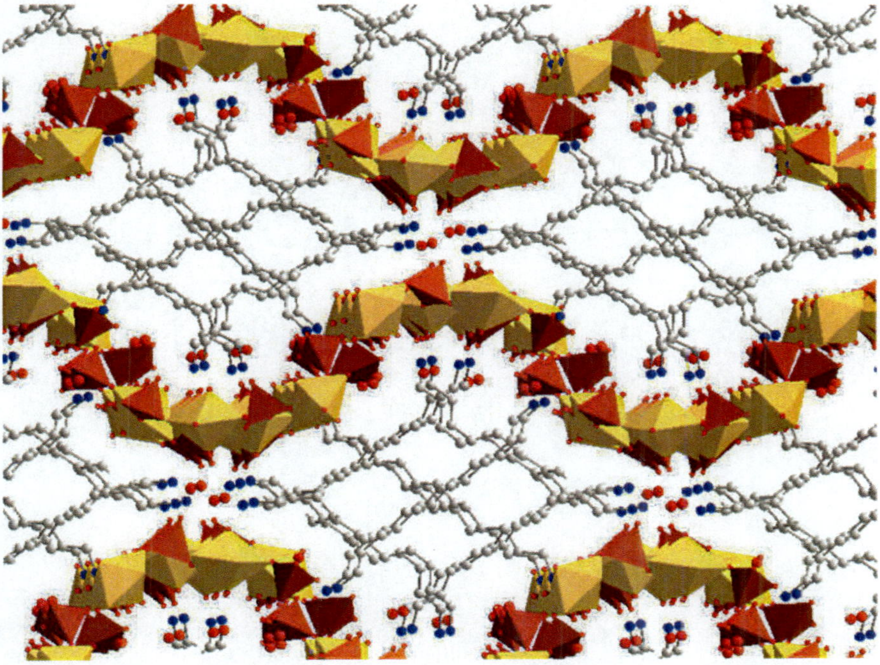

Fig. 21 Structure of $(H_3O)_2[C_{12}H_{30}N_2]_3[(UO_2)_4(SeO_4)_8](H_2O)_5$ with modulated uranyl selenate sheets and micellar aggregates of protonates 1,12-diaminododecane molecules (C, N, and H_2O are shown as *grey*, *blue*, and *red*, respectively; H atoms are omitted for clarity)

micelles with amine head groups on the surface that are surrounded by uranyl selenate sheets. The micelle dimensions are approximately elliptical with a cross-section of 2.2 × 2.5 nm. In order to adapt to the micelles, inorganic sheets

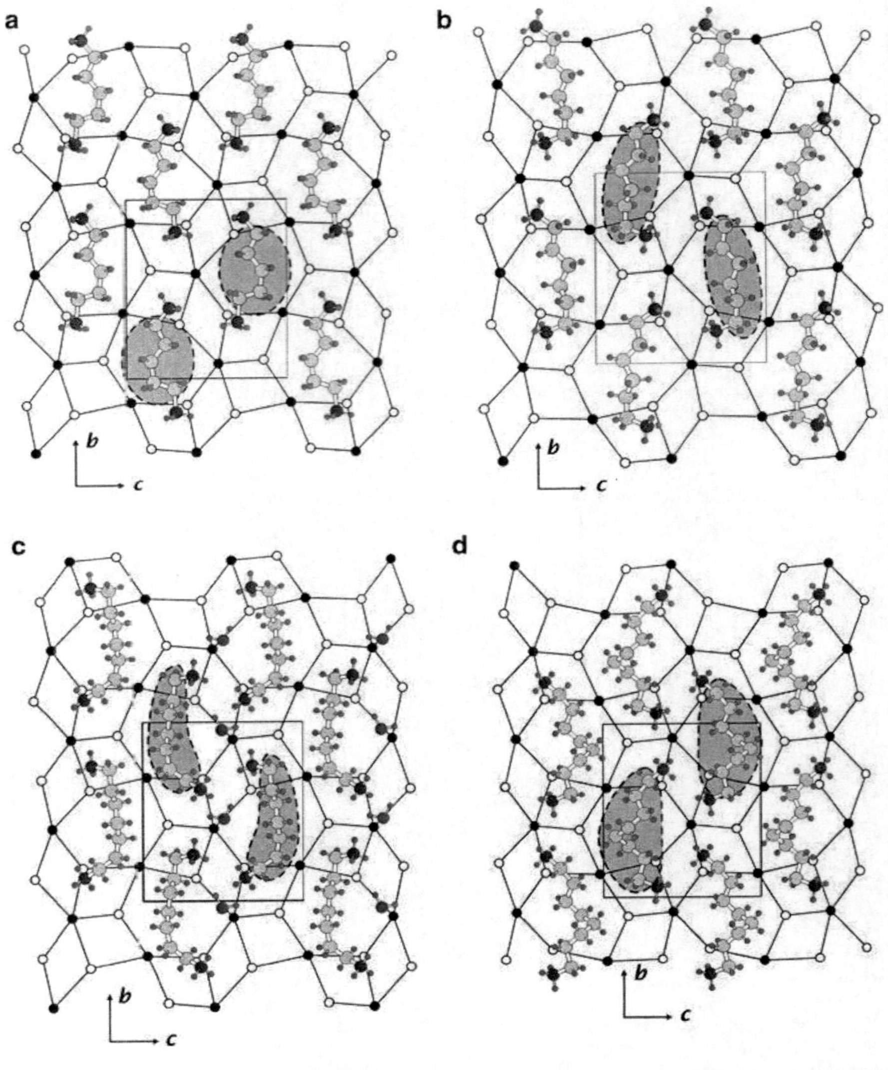

Fig. 22 Location of the protonated amine molecules in the structures of $[C_5H_{16}N_2]$ $[(UO_2)_2(SeO_4)_3(H_2O)]$, $[(UO_2)_2(SeO_4)_3(H_2O)](H_2O)$, $[C_7H_{20}N_2][(UO_2)_2(SeO_4)_3$–$(H_2O)](H_2O)$, and $[C_8H_{22}N_2][(UO_2)_2(SeO_4)_3(H_2O)]$ relative to the *black*-and-*white* graph of the inorganic layer (**a**, **b**, **c**, **d**, respectively)

possess a strong curvature with modulations of the adjacent sheets being in anti-phase. Taking into account that 1,12-diaminododecane is about twice as long as the butylamine chain, it seems likely that the 14:17 nanotubule owes its formation to micellar aggregates of protonated butylamine molecules.

6 Organic–Inorganic Composite Structures: Molecular Control of Structural Architectures

There are many known organically templated uranyl sulfates, molybdates, chromates, and selenates. However, systematic studies of the influence of template structure upon topology and geometry of inorganic substructure have been (at least partially) done for uranyl selenates only, due to the relatively simple synthesis conditions and their ability to form crystals of enough good quality for structural characterization.

In particular, interactions of uranyl selenates with chain diamines, $NH_2(CH_2)_nNH_2$, have been the subject of several studies [23, 26, 31, 39, 40]. It was found that, for $n < 8$, 2-D (or pseudo-2-D) topologies are favored with chain molecules oriented parallel to the inorganic layers. Starting from $n = 9$, the length of the aliphatic chain becomes a structure-driving force in the self-assembly of organic molecules that form either 1-D micelles [23, 39] or 2-D blocks [31, 40] (in both cases, hydrophobic fragments are hidden inside the micelle or block, and amine head groups are located on the surface). It is important to note that surface charge density of uranyl oxysalt layers is usually lower than that of the other oxysalts (e.g., transition metal phosphates, etc.), due to the large surface area occupied by equatorial base planes of uranyl bipyramids. The surface charge density of diamine blocks is two times higher than that of uranyl oxysalt layers (Fig. 22).

This discrepancy results in the inclusion of acid-water complexes into the organic blocks, which modifies its charge density in order to achieve structure stability [31]. In turn, 1-D micelles with cross-like orientation of long diamine chains maintain their charge density without inclusion of additional components.

The influence of shape and composition of organic templates upon the topology of uranyl selenate layers with U:Se ratio of 2:3 has been studied in [37]. As it was mentioned above, there are two different topologies based upon $[(UO_2)_2(SeO_4)_3(H_2O)]^{2-}$ layers shown in Fig. 2. Experimental studies of the formation of these topologies in the presence of different amines revealed that aliphatic components of amine molecules tend to associate with 6-MRs of the inorganic layers. As a result, molecules with longer and more spacious aliphatic components favor the formation of layers with topology **2**, whereas those with shorter aliphatic components prefer layers with the topology **1**. Several particular examples of mutual orientation of chain diamine molecules with the topology of inorganic layers is shown in Fig. 12.

Acknowledgments This work was supported by the internal budget grant of St. Petersburg State University (# 3.37.84.2011) and the Programme of Presidium of the Russian Academy of Sciences.

References

1. Burns PC, Kubatko KA, Sigmon G, Fryer BJ, Gagnon JE, Antonio MR, Soderholm L (2005) Actinyl peroxide nanospheres. Angew Chem Int Ed 44:2135–2139
2. Forbes TZ, McAlpin JG, Murphy R, Burns PC (2008) Metal-oxygen isopolyhedra assembled into fullerene topologies. Angew Chem Int Ed 47:2824–2827
3. Sigmon GE, Ling J, Unruh DK, Moore-Shay L, Ward M, Weaver B, Burns PC (2009a) Uranyl-peroxide interactions favor nano-cluster self-assembly. J Am Chem Soc 131:16648–16649
4. Sigmon GE, Unruh DK, Ling J, Weaver B, Ward M, Pressprich L, Simonetti A, Burns PC (2009b) Symmetry vs. minimal pentagonal adjacencies in uranium-based polyoxometalate fullerene topologies. Angew Chem Int Ed 48:2737–2740
5. Soderholm L, Almond PM, Skanthakumar S, Wilson RE, Burns PC (2008) The structure of a 38-plutonium oxide nanocluster: $[Pu_{38}O_{56}Cl_{54}(H_2O)_8]^{14-}$. Angew Chem Int Ed 47:298–302
6. Krivovichev SV, Kahlenberg V, Kaindl R, Mersdorf E, Tananaev IG, Myasoedov BF (2005) Nanoscale tubules in uranyl selenates. Angew Chem Int Ed 44:1134–1136
7. Krivovichev SV, Kahlenberg V, Tananaev IG, Kaindl R, Mersdorf E, Myasoedov BF (2005) Highly porous uranyl selenate nanotubules. J Am Chem Soc 127:1072–1073
8. Krivovichev SV, Tananaev IG, Kahlenberg V, Kaindl R, Myasoedov BF (2005) Synthesis, structure, and properties of inorganic nanotubes based on uranyl selenates. Radiochemistry 47:525–536
9. Alekseev EV, Krivovichev SV, Depmeier W (2008) A crown ether as template for microporous and nanostructured uranium compounds. Angew Chem Int Ed 47:549–551
10. Nocton G, Burdet F, Pécaut J, Mazzanti M (2007) Self-assembly of polyoxo clusters and extended frameworks by controlled hydrolysis of low-valent uranium. Angew Chem Intl Ed 46:7574–7578
11. Thuéry P (2008) Two uranyl-organic frameworks with formic acid. A novel example of a uranyl-based nanotubular assemblage. Inorg Chem Commun 11:616–620
12. Thuéry P (2009) A nanosized uranyl camphorate cage and its use as a building unit in a metal – organic framework. Cryst Growth Des 9:4592–4594
13. Krivovichev SV, Burns PC (2007) Actinide compounds containing hexavalent cations of the VI group elements (S, Se, Mo, Cr, W). In: Krivovichev SV, Burns PC, Tananaev IG (eds) Structural Chemistry of Inorganic Actinide Compounds. Elsevier, Amsterdam, pp. 95–182.
14. Locock AJ (2007) Crystal chemistry of actinide phosphates and arsenates. In: Krivovichev SV, Burns PC, Tananaev IG (eds) Structural Chemistry of Inorganic Actinide Compounds. Elsevier, Amsterdam, pp. 217–278.
15. Krivovichev SV, Kahlenberg V (2004) Synthesis and crystal structures of α- and β-$Mg_2[(UO_2)_3(SeO_4)_5](H_2O)_{16}$. Z Anorg Allg Chem 630:2736–2742
16. Krivovichev SV, Kahlenberg V (2005) Low-dimensional structural units in amine-templated uranyl oxoselenates(VI): Synthesis and crystal structures of $[C_3H_{12}N_2][(UO_2)(SeO_4)_2(H_2O)_2]$-$(H_2O)$. Z Anorg Allg Chem 631:2352–2357
17. Krivovichev SV, Kahlenberg V (2005) Crystal structure of $(H_3O)_6[(UO_2)_5(SeO_4)_8(H_2O)_5]$-$(H_2O)_5$. Radiochemistry 47:456–459
18. Krivovichev SV, Kahlenberg V (2005) Crystal structure of $(H_3O)_2[(UO_2)_2(SeO_4)_3(H_2O)_2]$-$(H_2O)_{3.5}$. Radiochemistry 47:452–455
19. Krivovichev SV, Kahlenberg V (2005) Synthesis and crystal structures of $M_2[(UO_2)_3(SeO_4)_5](H_2O)_{16}$ (M = Co, Zn). J Alloys Compd 395:41–47
20. Krivovichev SV, Kahlenberg V (2005) Preparation and crystal structures of $M[(UO_2)(SeO_4)_2(H_2O)](H_2O)_4$ (M = Mg, Zn). Z Naturforsch 60b:538–542
21. Krivovichev SV, Kahlenberg V (2005) Structural diversity of sheets in rubidium uranyl oxoselenates: Synthesis and crystal structures of $Rb_2[(UO_2)(SeO_4)_2(H_2O)](H_2O)$, $Rb_2[(UO_2)_2(SeO_4)_3(H_2O)_2](H_2O)$ and $Rb_4[(UO_2)_3(SeO_4)_5(H_2O)]$. Z Anorg Allg Chem 631:739–744

22. Krivovichev SV, Kahlenberg V (2005) Synthesis and crystal structure of $Zn_2[(UO_2)_3(SeO_4)_5](H_2O)_{17}$. J Alloys Compd 389:55–60
23. Krivovichev SV, Kahlenberg V, Avdontseva EY, Mersdorf E, Kaindl R (2005) Self-assembly of protonated 1,12-dodecanediamine molecules and strongly undulated uranyl selenate sheets in the structure of amine-templated uranyl selenate: $(H_3O)_2[C_{12}H_{30}N_2]_3[(UO_2)_4(SeO_4)_8](H_2O)_5$. Eur J Inorg Chem 2005:1653–1656
24. Krivovichev SV, Kahlenberg V, Tananaev IG, Myasoedov BF (2005) Amine-templated uranyl selenates with layered structures. I. Structural diversity of sheets with a U:Se ratio of 1:2. Z Anorg Allg Chem 631:2358–2364
25. Krivovichev SV, Tananaev IG, Kahlenberg V, Myasoedov BF (2005) Synthesis and crystal structure of the first uranyl selenite(IV)-selenate(VI) $[C_5H_{14}N][(UO_2)(SeO_4)(SeO_2OH)]$. Dokl Phys Chem 403:124–127
26. Krivovichev SV, Gurzhiy VV, Tananaev IG, Myasoedov BF (2006) Topology of inorganic complexes as a function of amine molecular structure in layered uranyl selenates. Dokl Phys Chem 409:228–232
27. Krivovichev SV, Tananaev IG, Myasoedov BF (2006) Geometric isomerism of layered complexes of uranyl selenates: Synthesis and structure of $(H_3O)[C_5H_{14}N]_2$-$[(UO_2)_3(SeO_4)_4(HSeO_4)(H_2O)]$ and $(H_3O)[C_5H_{14}N]_2[(UO_2)_3(SeO_4)_4(HSeO_4)$. Radiochemistry 48 552–560
28. Krivovichev SV, Tananaev IG, Myasoedov BF (2006) Nanostructures in uranium oxocompounds. Mater Res Soc Symp Proc 893:325–335
29. Krivovichev SV, Tananaev IG, Kahlenberg V, Myasoedov BF (2006) Synthesis and crystal structure of a new uranyl selenite(IV)-selenate(VI), $[C_5H_{14}N]_4[(UO_2)_3(SeO_4)_4(HSeO_3)(H_2O)]$-$(H_2SeO_3)(HSeO_4)$. Radiochemistry 48:217–222
30. Krivovichev SV, Burns PC, Tananaev IG, Myasoedov BF (2007) Nanostructured actinide compounds. J Alloys Compd 444–445:457–463
31. Krivovichev SV, Tananaev IG, Myasoedov BF (2007) Charge-density matching in organic–inorganic uranyl compounds. C R Chim 10:897–904
32. Gurzhiy VV, Krivovichev SV (2008) Synthesis and crystal structure of new uranyl selenate $Ni_2(UO_2)_3(SeO_4)_5(H_2O)_{16}$. Vestnik Sankt Peterb Univ Ser Geol 3:33–40
33. Krivovichev SV (2008) Crystal chemistry of selenates with mineral-like structures. VI. Hydrogen bonds in the crystal structure of $[(H_5O_2)(H_3O)(H_2O)][(UO_2)(SeO_4)_2]$. Geol Ore Dep 50:795–800
34. Krivovichev SV (2008) Crystal chemistry of selenates with mineral-like structures: V. Crystal structures of $(H_3O)_2[(UO_2)(SeO_4)_2(H_2O)](H_2O)_2$ and $(H_3O)_2[(UO_2)(SeO_4)_2(H_2O)](H_2O)$, new compounds with rhomboclase and goldichite topology. Geol Ore Dep 50:789–794
35. Krivovichev SV, Gurzhiy VV, Tananaev IG, Myasoedov BF (2009) Uranyl selenates with organic templates: Principles of structure and characteristics of self-organization. Russ J Gen Chem 79:2723–2730
36. Krivovichev SV (2009) Crystal chemistry of selenates with mineral-like structures: VII. The structure of $(H_3O)[(UO_2)(SeO_4)(SeO_2OH)]$ and some structural features of selenite-selenates. Geol Ore Dep 51:663–667
37. Krivovichev SV, Gurzhiy VV, Tananaev IG, Myasoedov BF (2009) Amine-templated uranyl selenates with chiral $[(UO_2)_2(SeO_4)_3(H_2O)]^{2-}$ layers: topology, isomerism, structural relationships. Z Kristallogr 224:316–324
38. Gurzhiy VV, Bessonov AA, Krivovichev SV, Tananaev IG, Armbruster T, Myasoedov BF (2009) Crystal chemistry of selenates with mineral-like structures: VIII. Butlerite chains in the structure of $K(UO_2)(SeO_4)(OH)(H_2O)$. Geol Ore Dep 51:833–837
39. Gurzhiy VV, Krivovichev SV, Burns PC, Tananaev IG, Myasoedov BF (2010) Supramolecular templates for the synthesis of new nanostructured uranyl compounds: Crystal structure of $[NH_3(CH_2)_9NH_3][(UO_2)(SeO_4)(SeO_2OH)](NO_3)$. Radiochemistry 52:1–6
40. Krivovichev SV, Gurzhiy VV, Burns PC, Tananaev IG, Myasoedov BF (2010) Partially ordered organic-inorganic nanocomposites in the system UO_2SeO_4–H_2O–$NH_3(CH_2)_9NH_3$. Radiochemistry 52:7–11

41. Krivovichev SV (2010) Actinyl compounds with hexavalent elements (S, Cr, Se, Mo): structural diversity, nanoscale chemistry and cellular automata modeling. Eur J Inorg Chem.
42. Mikhailov YN, Kokh LA, Kuznetsov VG, Grevtseva TG, Sokol SK, Ellert GV (1977) Synthesis and crystal structure of potassium trisulfatouranylate $K_4(UO_2(SO_4)_3)$. Koord Khim 3:508–513
43. Hayden LA, Burns PC (2002) A novel uranyl sulfate cluster in the structure of $Na_6(UO_2)(SO_4)_4(H_2O)_2$. J Solid State Chem 163:313–318
44. Norquist AJ, Doran MB, O'Hare D (2005) The role of amine sulfates in hydrothermal uranium chemistry. Inorg Chem 44:3837–3843
45. Krivovichev SV (2008) Crystal structure of $KNa_3[(UO_2)_5O_6(SO_4)]$. Radiochemistry 50:450–454
46. Hennig C, Ikeda A, Schmeide K, Brendler V, Moll H, Tsushima S, Scheinost AC, Skanthakumar S, Wilson R, Soderholm L, Servaes S, Görrler-Walrand C, Van Deun R (2008) The relationship of monodentate and bidentate coordinated uranium(VI) sulfate in aqueous solution. Radiochim Acta 96:607–611
47. Krivovichev SV (2008) Structural Crystallography of Inorganic Oxysalts, Oxford University Press, Oxford.
48. Krivovichev SV, Burns PC (2003) Geometrical isomerism in uranyl chromates I. Crystal structures of $(UO_2)(CrO_4)(H_2O)_2$, $[(UO_2)(CrO_4)(H_2O)_2](H_2O)$ and $[(UO_2)(CrO_4)(H_2O)_2]_4(H_2O)_9$. Z Kristallogr 218:568–574
49. Serezhkin VN, Trunov VK (1981) Crystal structure of uranyl chromate 5.5 hydrate $(UO_2CrO_4 \cdot 5.5H_2O)$. Kristallografiya 26:301–304
50. Brandenburg NP, Loopstra BO (1973) Uranyl sulphate hydrate, $UO_2SO_4(H_2O)_{3.5}$. Cryst Struct Commun 2:243–246
51. Krivovichev SV, Burns PC (2003) Geometrical isomerism in uranyl chromates II. Crystal structures of $Mg_2[(UO_2)_3(CrO_4)_5](H_2O)_{17}$ and $Ca_2[(UO_2)_3(CrO_4)_5](H_2O)_{19}$. Z Kristallogr 218:683–690
52. Cundy CS, Cox PA (2005) The hydrothermal synthesis of zeolites: Precursors, intermediates and reaction mechanism. Microporous Mesoporous Mater 82:1–78
53. Taulelle F, Pruski M, Amoureux JP, Lang D, Bailly A, Huguenard C, Haouas M, Gerardin C, Loiseau T, Férey G (1999) Isomerization of the prenucleation building unit during crystallization of AlPO4-CJ2: An MQMAS, CP-MQMAS, and HETCOR NMR study. J Am Chem Soc 121:12148–12153
54. Serre C, Lorentz C, Taulelle F, Férey G (2003) Hydrothermal synthesis of nanoporous metalofluorophosphates. 2. In situ and ex situ ^{19}F and ^{31}P NMR of nano- and mesostructured titanium phosphates crystallogenesis. Chem Mater 15:2328–2337
55. Serre S, Taulelle F, Férey G (2003) Rational design of porous titanophosphates. Chem Commun 2003:2755–2765
56. Vistad O, Akporiaye DE, Taulelle F, Lillerud KP (2003) In situ NMR of SAPO-34 crystallization. Chem Mater 15:1639–1649
57. Loiseau T, Beitone L, Millange F, Taulelle F, O'Hare D, Férey G (2004) Observation and reactivity of the chainlike species $([Al(PO_4)_2]^{3-})_n$ during the X-ray diffraction investigation of the hydrothermal synthesis of the super-sodalite sodium aluminophosphate MIL-74 $(Na_2Al_7(PO_4)_{12} \cdot 4trenH_3 \cdot Na(H_2O)_{16})$. J Phys Chem B108:20020–20029
58. Férey G (1995) Oxyfluorinated microporous compounds ULM-n: Chemical parameters, structures and a proposed mechanism for their molecular tectonics. J Fluor Chem 72:187-193
59. Férey G (1998) The new microporous compounds and their design. C R Acad Sci Ser IIc 1:1–13
60. Moll H, Reich T, Hennig C, Rossberg A, Szabo Z, Grenthe I (2000) Solution coordination chemistry of uranium in the binary UO_2^{2+}-SO_4^{2-} and the ternary UO_2^{2+}–SO_4^{2-}–OH– system. Radiochim Acta 88:559–566
61. Neuefeind J, Skanthakumar S, Soderholm L (2004) Structure of the UO_2^{2+}–SO_4^{2-} ion pair in aqueous solution. Inorg Chem 43:2422–2426

62. Hennig C, Schmeide K, Brendler V, Moll H, Tsushima S, Scheinost AC (2007) EXAFS investigation of U(VI), U(IV), and Th(IV) sulfato complexes in aqueous solution. Inorg Chem 46:5882–5892
63. Ikeda A, Hennig C, Tsushima S, Takao K, Ikeda Y, Scheinost AC, Bernhard G (2007) Comparative study of uranyl(VI) and -(V) carbonato complexes in an aqueous solution. Inorg Chem 46:4212–4219
64. Hennig C, Kraus W, Emmerling F, Ikeda A, Scheinost AC (2008) Coordination of a uranium (IV) sulfate monomer in an aqueous solution and in the solid state. Inorg Chem 47:1634–1638
65. Hennig C, Ikeda-Ohno A, Emmerling F, Kraus W, Bernhard G (2010) Comparative investigation of the solution species $[U(CO_3)_5]^{6-}$ and the crystal structure of $Na_6[U(CO_3)_5]\cdot 12H_2O$. Dalton Trans 39:3744–3750
66. Francis RJ, Price SJ, O'Brien S, Fogg AM, O'Hare D, Loiseau T, Férey G (1997) Formation of an intermediate phase during the hydrothermal synthesis of ULM-5 studied using time-resolved *in situ* X-ray powder diffraction. Chem Commun 1997:521–522
67. Walton RI, Millange F, Le Bail A, Loiseau T, Serre C, O'Hare D, Férey G (2000) The room-temperature crystallisation of a one-dimensional gallium fluorophosphate, $Ga(HPO_4)_2F\cdot H_3N(CH_2)_3NH_3\cdot 2H_2O$, a precursor to three-dimensional microporous gallium fluorophosphates. Chem Commun 2000:203–204
68. Walton RI, Norquist AJ, Neeraj S, Natarajan S, Rao CNR, O'Hare D (2001) Direct in situ observation of increasing structural dimensionality during the hydrothermal formation of open-framework zinc phosphates. Chem Commun 2001:1990–1991
69. Ayi AA, Choudhury A, Natarajan S, Neeraj S, Rao CNR (2001) Transformations of low-dimensional zinc phosphates to complex open-framework structures. Part 1: Zero-dimensional to one-, two- and three-dimensional structures. J Mater Chem 2001:1181–1191
70. Choudhury A, Neeraj S, Natarajan S, Rao CNR (2001) Transformations of the low-dimensional zinc phosphates to complex open-framework structures. Part 2: One-dimensional ladder to two- and three-dimensional structures. J Mater Chem 2001:1537–1546
71. Choudhury A, Neeraj S, Natarajan S, Rao CNR (2002) Transformations of two-dimensional layered zinc phosphates to three-dimensional and one-dimensional structures. J Mater Chem 2002:1044–1052
72. Millange F, Walton RI, Guillou N, Loiseau T, O'Hare D, Férey G (2002) Two chain gallium fluorodiphosphates: Synthesis, structure solution, and their transient presence during the hydrothermal crystallisation of a microporous gallium fluorophosphate. Chem Commun 2002:826–827
73. Millange F, Walton RI, Guillou N, Loiseau T, O'Hare D, Férey G (2002) Synthesis and structure of low-dimensional gallium fluorodiphosphates seen during the crystallization of the three-dimensional microporous gallium fluorophosphate ULM-3. Chem Mater 14:4448–4459
74. Wang K, Yu J, Song Y, Xu R (2003) Assembly of one-dimensional $AlP_2O_8^{3-}$ chains into three-dimensional $MAlP_2O_8\cdot C_2N_2H_9$ frameworks through transition metal cations (M = Ni^{2+}, Co^{2+} and Fe^{2+}). Dalton Trans 2003:99–103
75. Norquist AJ, O'Hare D (2004) Kinetic and mechanistic investigations of hydrothermal transformations in zinc phosphates. J Am Chem Soc 126:6673–6679
76. Oliver S, Kuperman A, Ozin GA (1998) A new model for aluminophosphate formation: Transformation of a linear chain aluminophosphate to chain, layer, and framework structures. Angew Chem Int Ed 37:46–62
77. Neeraj S, Natarajan S, Rao CNR (2000) Isolation of a zinc phosphate primary building unit $[C_6N_2H_{18}]^{2+}[Zn(HPO_4)(H_2PO_4)_2]^{2-}$ and its transformation to open-framework phosphate $[C_6N_2H_{18}]^{2+}[Zn_3(H_2O)_4(HPO_4)_4]^{2-}$. J Solid State Chem 150:417–422
78. Rao CNR, Natarajan S, Choudhury A, Neeraj S, Ayi AA (2001) Aufbau principle of complex open-framework structures of metal phosphates with different dimensionalities. Acc Chem Res 34:80–87
79. Gurzhiy VV, Kovrugin VD, Krivovichev SV (2010) Submitted.

80. von Neumann J (1951) A general and logical theory of automata. In: Jeffress LA (ed) Celebral Mechanisms in Behaviour: The Hixon Symposium. Wiley, New York, pp. 1–32
81. Toffoli T, Margolus N (1987) Cellular Automata Machines: A New Environment for Modeling. MIT Press, Boston
82. Ilachinski A (2001) Cellular Automata: A Discrete Universe. World Scientific, Singapore
83. Wolfram S (2002) A New Kind of Science. Wolfram Media, Inc., Urbana
84. Mackay A (1976) Crystal symmetry. Phys Bull 1976:495
85. Krivovichev S (2004) Crystal structures and cellular automata. Acta Crystallogr A60: 257–262
86. Shevchenko VY, Krivovichev SV (2008) Where are genes in paulingite? Mathematical principles of formation of inorganic materials on the atomic level. Struct Chem 19:571–577
87. Shevchenko VY, Krivovichev SV, Mackay AL (2010) Cellular automata and local order in the structural chemistry of the lovozerite group minerals. Glass Phys Chem 36:1–9

Actinide Thin Films as Surface Models

A. Seibert, S. Stumpf, T. Gouder, D. Schild, and M.A. Denecke

Abstract Reducing the dimensionality of a material can effectively influence its properties. Compared to the knowledge on actinide bulk materials concerning their structure, behavior, and reactivity, much less is known regarding the materials with reduced dimensions (thin films, colloids, nanoparticles, clusters, isolated atoms in a matrix). The surface science laboratory installations at the Institute for Transuranium Elements (ITU) provide a combination of a reactive sputter deposition set-up for thin film production with photoemission spectroscopy instrumentation. This allows for characterization of the surface elemental composition of thin film materials, to draw conclusions regarding the chemical environment of the constitutive elements and to gather information on the electronic structure of the materials. X-ray photoelectron spectroscopy (XPS), probing the sample up to tens of monolayers (ML) deep, and UV photoelectron spectroscopy (UPS), with a probing depth of one to four MLs, are used. In this chapter, we focus on the investigation of surface reactions, and some examples are given to show how the electronic structure of 5*f* materials is influenced by reducing the materials' dimensions. Additional characterization methods are applied to show that the films can also serve as structural models: characterization of the film surfaces by SEM–EDX (scanning electron microscopy combined with energy dispersive X-ray spectroscopy) and AFM (atomic force microscopy) are carried out. In the case of bulk characterization, X-ray diffraction (XRD) is used to give insight in crystallinity. The designed thin film models are used in gas adsorption (O_2, O_{atom}, H_2, H_{atom}) and electrochemical experiments probing their redox behavior. Specific examples are presented to illustrate this approach to mechanistic understanding.

A. Seibert (✉), S. Stumpf, and T. Gouder
Institute for Transuranium Elements, European Commission, Joint Research Centre, P.O. Box 2340, 76125 Karlsruhe, Germany
e-mail: Alice.SEIBERT@ec.europa.eu

D. Schild and M.A. Denecke
Institute for Nuclear Waste Disposal, Karlsruhe Institute of Technology (KIT), P.O. Box 3640, 76021 Karlsruhe, Germany

S.N. Kalmykov and M.A. Denecke (eds.), *Actinide Nanoparticle Research*,
DOI 10.1007/978-3-642-11432-8_10,

1 Introduction and Basic Principles

1.1 Why Surfaces?

Surfaces and interfaces constitute a particularly important part of condensed matter systems. Chemical processes between different systems typically operate at their interfaces. The reactions involve adsorption, diffusion, dissociation, and desorption of atoms and molecules, which in turn depend on the surface properties. These may strongly differ from the bulk properties because of the exposed nature of atoms (terraces, steps, kinks, defects) and the break in symmetry and chemical bonding. In this sense, surface is considered the uppermost atomic layer of a solid. Often, surface reactions involve only a small active part of the surface, typically defects or adatoms, so they are difficult to predict on the basis of the average surface properties probed by most spectroscopies.

In the context of our research, the long-term prediction of corrosion properties of spent nuclear fuel, surface processes play a central role. The fuel itself is very complex, with up to 25 chemical elements, intermixed or occurring in different phases. Describing surface reactions in such systems is difficult, and one tries instead to break them down into simple model reactions, which are easier to investigate. Thus, there is a need for simplified model surfaces exposed to well-defined experimental conditions (molecular or atomic gases, radiation, temperature, aqueous medium, etc). This chapter describes recent advances in the study of actinide-based thin film models. As nanoparticles, the topic of this book, inherently have large surface to bulk ratios, our thin films may find application as surface models to advance our understanding and modeling of nanoparticle structure and reactivity.

The rapid development of surface-sensitive analytical techniques over the last few years has been the prerequisite for a better atomistic and mechanistic insight into processes occurring at interfaces [1]. Such molecular-level surface studies led to new understanding and concepts, for example, of the reactivity of surfaces or how catalysts work on a molecular scale [2]. It is common to divide surface chemistry into different areas involving solid–gas, solid–liquid, and solid–solid interfaces. Although many of the physical–chemical phenomena to be studied are the same at different interfaces, the experimental techniques are quite different. In our discussion of actinide surfaces, we focus on the solid–gas interface as many of the techniques providing molecular-scale information are applicable only under ultrahigh vacuum (UHV) or low pressure conditions. Clearly, the two other areas also play a role in spent nuclear fuel behavior and are briefly mentioned. Solid–liquid interactions are probed, e.g., by electrochemistry experiments, and are a key issue in the corrosion behavior of spent nuclear fuel during (accidental) contact with groundwater. Solid–solid interactions are important in the structuring of the fuel during reactor operation, when fission products are generated and incorporated in the fuel matrix as metallic inclusions, mixed or separated oxides, etc. These structures are also found at the surface, where they may influence the corrosion behavior by catalytic action, galvanic coupling, and specific dissolution behavior. Specific examples are discussed below.

UHV vacuum conditions corresponding to 10^{-8} mbar are required to obtain and maintain clean surfaces (for time spans in the range of an hour). Unintentionally adsorbed layers may induce different chemical properties at the surface by blocking sites or changing the oxidation states of the surface atoms and therewith the nature of bonding of subsequently incoming reactants.

To extricate the various properties of a given surface/bulk system, several complementary techniques (see list of abbreviations for definition of acronyms) can be used in parallel, e.g.,

- Methods sensitive to atomic geometry at surfaces (LEED, RHEED, TEM, atomic scattering or diffraction, SEXAFS, STM, AFM)
- Methods sensitive to electron distributions at surfaces (work function measurements such as Kelvin probe)
- Methods sensitive to chemical composition at surfaces (AES, XPS/UPS, TDS, SIMS, XSW)
- Methods sensitive to the vibrational structure of surfaces (IR, Raman spectroscopy, HREELS)
- Methods sensitive to bulk structural and compositional properties: XRD, SEM–EDX, XAS

In addition to this concept of complementary techniques, we also follow the concept of model experiments. This means that either single or combinations of reactants are used (for example, in adsorption experiments) together with specially designed surface models. These surface models are well characterized and can be produced in the same UHV system that also hosts most of the surface science analytical techniques used in our investigations. The sophisticated surface science instrumentation together with the possibility to use it for investigations on the entire actinide series including the highly radiotoxic elements (glove box installation) at ITU is unique.

1.2 Why Actinide Thin Films?

Actinides have an especially interesting electronic structure, with respect to their unfilled $5f$ electron shell. The $5f$ electrons are located in small orbitals, which only weakly overlap with neighboring atoms. In metallic actinide systems, where the actinide atoms share their $6d7s5f$ valence electrons in covalent or metallic bonds, the $5f$ states form narrow bands of highly correlated electrons. They are responsible for the peculiar physical properties of such actinide compounds, for example, the coexistence of magnetic and superconducting states.

In oxidized compounds, which are especially important in the nuclear field, early actinides can occur in a number of different oxidation states, again due to the $5f$ electrons, which are easily transferred to neighboring atoms in ionic chemical bonding. Chemical properties such as actinide solubility in water depend on the oxidation state.

Actinide thin films are accessible surface models to study electronic structure, oxidation state, and their effect on reactivity and structure. The thin films can be produced with varying thickness, size, and composition. Their thickness can range from submonolayers up to several micrometers, thus allowing investigations of surfaces with minimal amount of underlying bulk, as well as bulk-like systems. Reaction areas of the order of a square centimeter can be generated on samples weighing only a few micrograms. In addition, modern thin film synthesis methods allow for the preparation of systems, which are not stable as bulk phases. In particular, finely dispersed phases, off-stoichiometric compounds, surface structures, and layered compounds can be prepared by codeposition and sequential deposition.

1.3 The Instrumentation

In the following chapter, we will briefly describe the surface science instrumentation used by ITU. Figure 1 schematically presents the general approach in our investigations.

The synthesis of the thin film surfaces takes place in an UHV preparation chamber (base pressure $<10^{-8}$ mbar), which is directly connected to a second UHV chamber for surface characterization by XPS and UPS. The freshly prepared film surfaces can be measured without delay or contact with ambient laboratory atmosphere. Inside the same UHV system, gas adsorption experiments can be performed with direct measurement of the reacted surfaces. Characterization methods other than XPS/UPS are carried out outside the UHV system under ambient

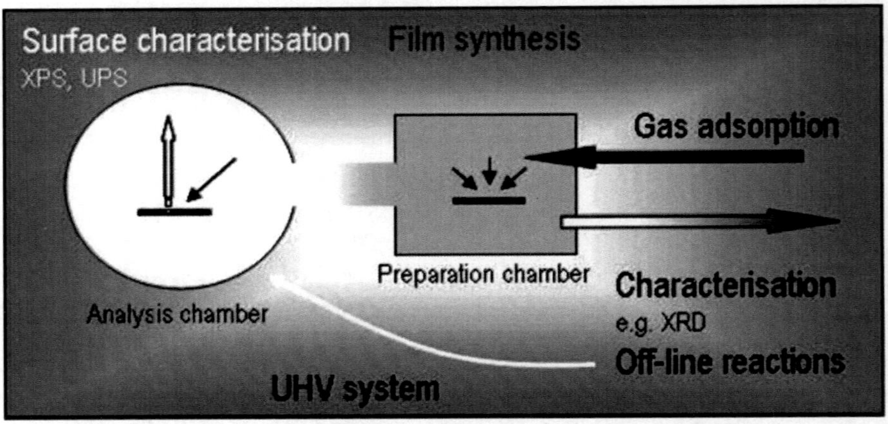

Fig. 1 Experimental approach to our surface science investigations

conditions (XRD, AFM, Raman spectroscopy) or in a different vacuum system (SEM–EDX). Also, methods such as electrochemical investigations may be combined with the surface science techniques in an off-line approach by connecting a high-pressure reaction cell with the UHV system for the surface-sensitive investigations before and after the reaction at high pressure. Investigations with electrochemical methods are carried out under ambient conditions. For examination of the surface after these off-line reactions, the samples are introduced into the UHV system again for XPS/UPS analysis or are investigated with other techniques.

1.3.1 Characterization Methods

XPS/UPS are the main characterization methods for the thin film surface analysis. Additional methods are applied if they provide ancillary information on specific characteristics of the thin films or in understanding processes taking place at their surfaces. Examples for this approach are presented later; here only a short description of the methods and the equipment used is given.

X-Ray Photoelectron Spectroscopy and Ultraviolet Photoelectron Spectroscopy

Photoelectron spectroscopy is a (semi-)quantitative electron spectroscopic technique that allows determination of the elemental composition and chemical state of a sample surface. Information on the electronic structure of the elements can also be obtained. X-ray photoelectron spectroscopy (XPS) spectra are recorded by irradiating a material with X-ray photons while measuring the kinetic energy (KE) and intensity of (mainly core level) electrons that escape from the top 1 to 10 nm of the material as function of the KE. UPS refers to the measurement of the photoelectron spectra emitted upon excitation with ultraviolet photons and characterize molecular energy levels in the valence band region. The obtained KE can be easily transferred into binding energy (BE) values. As no two elements have the same set of electronic energy levels, XPS enables elemental analysis. Any changes in the chemical environment of an atom are reflected in a change of the respective BE and therewith provide chemical information [3, 4].

Mg K_α (1,253.6 eV) and Al K_α (1,486.6 eV) excitation radiation is used for the XPS experiments. UPS spectra are taken with He I (21.22 eV) and He II (40.81 eV) UV light produced by a high intensity windowless helium discharge lamp. Taking advantage of the different mean free paths (λ) for He I and He II compared to Mg K_α and Al K_α excited photoelectrons, bulk and surface properties can be accessed [5]. The techniques allow investigation of the uppermost surface layer (He II excitation) and also provide information on "bulk properties" from regions down to some tens of monolayers (Al K_α/Mg K_α excitation). In addition, the energy-dependent photoionization cross-sections for electrons [6] from different orbitals (*p*, *d* or *f*) can be utilized for identification of the orbital character of the valence band region by varying the excitation energy and following the change of intensity for the different peaks.

Electron spectroscopy methods require UHV conditions for electron energy analysis. The background pressure in the analysis chamber is better than 2×10^{-9} mbar. Spectra are recorded in this work with a hemispherical analyzer from Omicron (EA 125 U5) or Leyboldt (EA-10) or Specs (Phoibos 150 CCD). The composition of the thin films is determined by analyzing the intensity (area) ratios of the corresponding photoemission peaks, for example, An-4*f*, Pd-3*d*, and O-1*s*.

Atomic Force Microscopy

Atomic force microscopy (AFM) is a high-resolution type of scanning probe microscopy, which allows imaging, measuring, and manipulating matter at nanoscale or even atomic dimensions. In this technique, different forces (mainly short-range interatomic repulsive forces, but other forces are also encountered such as Coulomb or van der Waals forces, dipole–dipole interactions, etc., which can be repulsive or attractive) between a sharp tip and the sample surface are measured. The tip is raster-scanned over the surface by piezoelectric drive that allows tiny but accurate and precise movements down to the nanometer to subnanometer range. The imaging is accomplished by measuring the deflection of a bendable cantilever onto which the tip is fixed [3].

Three modes of measurement are usually applied: contact mode, tapping mode, and noncontact mode. For the present investigations, only contact mode measurements are applied to the sample surface. In this mode, the tip is in perpetual contact with the surface during the measurement and is gently driven over the sample in *X–Y* direction. The bending of the cantilever in *Z*-direction is counteracted (to maintain a constant cantilever deflection) by an electronic loop on the piezoelectric element, and the data are processed internally to generate the image.

For the AFM measurements, films of about 1 μm thickness are deposited on polycrystalline gold disks (10 mm diameter, 0.25 mm thickness, 99.99% purity) that can be further used as electrodes in electrochemical AFM (ECAFM) measurements. A commercial AFM (Topometrix, TMX 2000, Explorer) is used for the topographic characterization of the thin films. For combined electrochemical–topographic measurements, the microscope is equipped with an electrochemical cell designed by Römer et al. [7]. In this cell, the coated gold disk is embedded in an epoxy resin using an electrically contacted holder of stainless steel (working electrode). Additionally, flexible silicon tubes are embedded in the system for the liquid exchange and as a salt bridge. The counter electrode is a platinum-coated titanium ring, which forms the wall of the liquid cell. A Pt reference electrode is contacted with the electrolyte via one of the embedded tubes. The potential is adjusted by a bipotentiostat (type MP81, Bank Elektronik, Clausthal-Zellerfeld, Germany). The microscopic pictures are evaluated using WSxM 4.0 Develop 9.3 software (Nanotec Electronica S.L.) [8].

X-Ray Diffraction

X-ray diffraction (XRD) as a method for the determination of the arrangement of atoms in a crystal can be also be applied in the crystallographic characterization of thin film samples. The technique requires the deposition of a film of ~1 μm thickness. XRD experiments are performed on a θ/2θ goniometer using Cu Kα X-rays. Lattice parameters (**a, b, c**) are calculated from measured angles of reflection θ according to Bragg's law, with the simplification for cubic lattices (**a** = **b** = **c**; 90°): $a = (h^2 + k^2 + l^2)^{1/2} \times \lambda \times (2\sin\theta)^{-1}$.

Scanning Electron Microscopy Combined with Energy-Dispersive X-Ray Spectroscopy

Scanning electron microscopy (SEM) is a type of electron microscopy that images the sample surface by scanning it with a high-energy beam of electrons in a raster scan pattern. The electrons interact with the atoms of the sample surface, producing signals that contain information about the sample surface topography, composition, and other properties such as electrical conductivity. SEM measurements are performed in an environmental scanning electron microscope (Philips ESEM XL 30 FEG) equipped with a backscatter electron detector for imaging. Vacuum conditions are also required for this electron microscopy technique. The prepared films are electrically conducting, so they can be analyzed without a conducting surface coating, which may impair backscatter diffraction bands. The generation of fluorescence X-rays, as a secondary process following ionization of an atom by the electron beam, is used for spatially resolved identification of the elemental composition of the sample (EDX) [3].

1.3.2 Preparation of Thin Films

Actinide thin films are prepared by sputter (co-)deposition from one to three elemental targets with argon or argon/reactive gas mixtures as sputter gas. The sputter codeposition relies on an Ar plasma to knock material from elemental targets (U, Am, Np, Pd...). Hereby, Ar atoms are ionized by electrons (50–100 eV) emitted from a hot W cathode and accelerated towards the target in an electric field. Since the process is not one of evaporation, sputtering is one of the most flexible deposition techniques and especially useful for compounds or mixtures; a sputter-source can be designed to contain a number of elemental targets. Furthermore, the deposition can be performed in the presence of additional "reaction" gases such as nitrogen or oxygen. For example, for the preparation of U–Pd mixed thin films, a block of elemental U (4×6 mm^2) is fixed in the center of the sputter source and a Pd wire loop is placed 1 mm above it (Fig. 2). Both targets are connected to a DC power supply, allowing applying a voltage up to 1,000 V. As can be seen in Fig. 2a, the W-filament serving as electron emitter is fixed above the

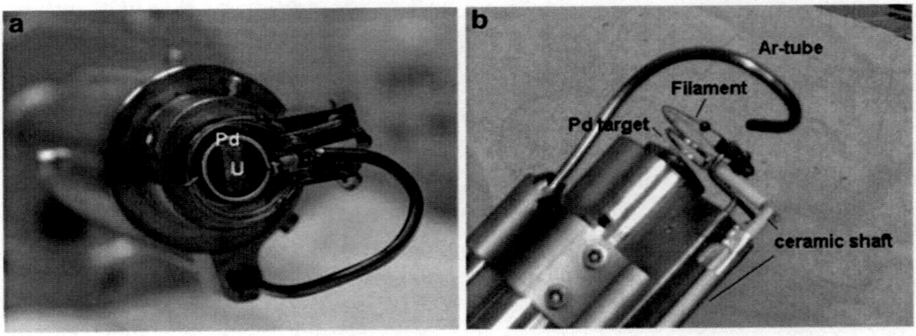

Fig. 2 Set-up of the U–Pd sputter source for codeposition. (**a**) U metal block in the center with a Pd wire loop fixed 1 mm above, *top view*. (**b**) assembly seen from side with view of filament, gas tube, and electrical connections

targets. All electrical connections are isolated by hollow ceramic shafts to avoid direct contact to mass and exposure of the wires to the plasma. The Ar-gas is introduced by a capillary directly into the plasma chamber to achieve high local pressures, while keeping the background pressure and the load of the pumping system as low as possible.

Deposition rates are about 0.5 monolayers (MLs) per second. The thickness of the films is checked by weight difference before and after sputter deposition. The target sputter rate is determined by the applied target voltage. For the single U target and otherwise similar conditions, the target current is determined to correlate linearly with the deposited mass. For multiple target set-ups, the ratio of the respective target currents corresponds (for otherwise similar conditions) to a specific elemental composition of the deposited film. The composition of the thin films is controlled by the respective target voltages, the partial pressure of reactive gas, and the deposition temperature.

The sputter source is mounted in the preparation chamber of a photoemission spectroscope (Fig. 3). This preparation chamber is separated from the analysis chamber and photoemission spectroscopy set-up by a valve. This set-up allows the in-situ preparation of samples without contamination of the analysis chamber and direct transfer to the spectrometer for surface analysis.

1.3.3 In-Situ Characterization of the Thin Films

The elemental composition of thin films sputtered in the presence of reaction gases such as oxygen correlates with the gas partial pressure (for otherwise similar conditions). As an example, the XPS spectra of U (U-4*f*) sputtered at varying oxygen pressures are given in Fig. 4. The U-4*f* spectrum is characterized by two main peaks with an energy separation of about 11 eV; these are attributed to the spin–orbit split $4f_{5/2}$ and $4f_{7/2}$ lines. The energy positions, as well as the shape of

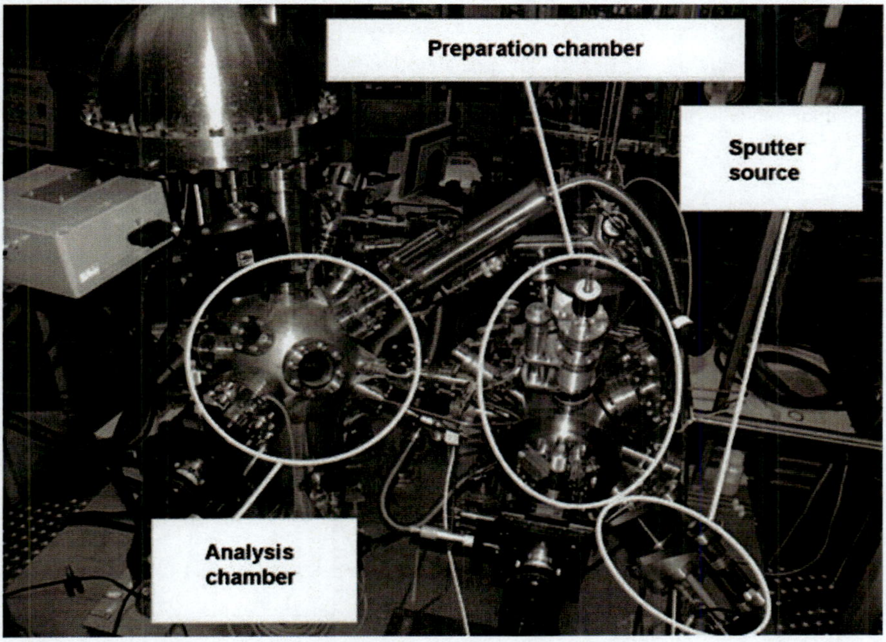

Fig. 3 Photoemission spectroscopy with supporting tools

satellite peaks at ~7 eV higher BE, allow the qualitative speciation of the uranium oxide [10–15]. Evaluation of the U-4*f*/O-1*s* area ratios allows a quantitative speciation. According to the literature, the U-4$f_{7/2}$ BE in metallic uranium is 377.4 eV [16]. The films deposited at an O_2 partial pressure of 5×10^{-9} mbar (Fig. 4, lowest curve) can therefore be attributed to the pure metal (measured BE: 377.4 eV). With increasing O_2 partial pressure, the intensity of these metal lines decreases and new features appear on their high BE sides, at 380.9 eV and 391.8 eV, but still showing the energy separation typical for the U-4*f* doublet (10.9 eV). The new energy positions correspond to UO_2, which is deposited simultaneously with elemental uranium. At an O_2 partial pressure of 1×10^{-7} mbar, the lines at 377.4 eV and 388.3 eV disappear showing that all of the deposited uranium is oxidized. The energy separation between satellite and main line, 6.7 eV, is characteristic for UO_2. The satellites are due to an internal excitation of the U atoms by the outgoing photoelectrons, which appear at higher BE. This excitation consists of the transfer of an O-2*p* valence electron into an empty U-5*f* level [17]. The transfer energy depends on the position of the U-5*f* level, which itself changes with the oxidation state. In stoichiometric UO_2, the satellite is observed at 6.7 eV high BE, while in U_4O_9, it appears at 8.2 eV [15]. The shape and position of the satellites thus give direct information about the U oxidation state [14]. In UO_{2+x}, the additional interstitial oxygen atoms change the local oxidation state of the surrounding U atoms so that even in $UO_{2.06}$, the satellite

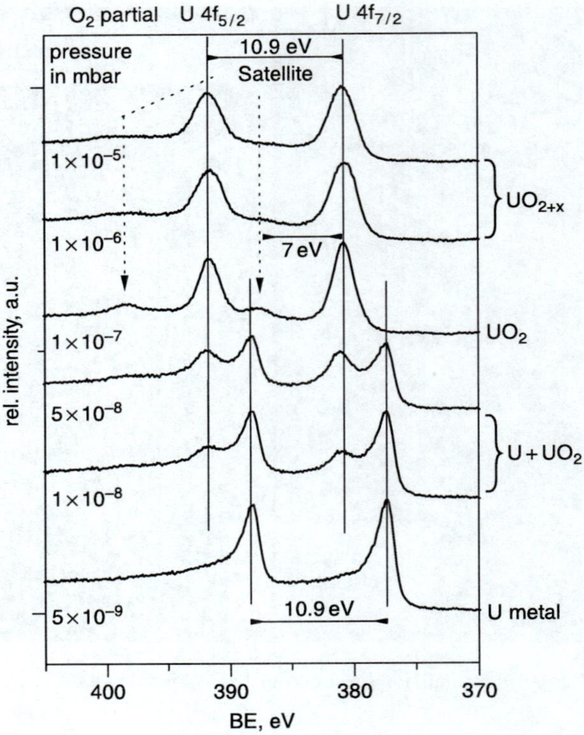

Fig. 4 U-4*f* photoemission spectra of thin films deposited at different O_2 partial pressures, adapted from [9]. Excitation source is Mg Kα (1,253.6 eV)

peak broadens significantly compared to that in $UO_{2.0}$ [14, 15]. Formation of this oxide is confirmed by the U-4*f*/O-1*s* intensity (area) ratios. With further increase of the oxygen partial pressure, the broadened satellites thus clearly indicate the presence of a higher stoichiometric oxide UO_{2+x} [14], which is quantified as having a composition near $UO_{2.33}$.

Another important sputter parameter is the temperature of the substrate onto which the film is deposited. For substrates at RT, diffusion of film components is strongly constricted. As a result, the structure and composition of the film does not necessarily correspond to that of the thermodynamically stable system but rather to a snapshot of the cooled plasma composition. This is overcome when the deposition is performed on substrates at higher temperature, where diffusion of film components after their deposition is possible. For pure $UO_{2(+x)}$ films on gold substrates, heating (up to 320°C) is applied during the deposition process. For temperatures above 200°C, this leads to stoichiometric UO_2 and better crystallized films are produced (see below).

The effect of elevated substrate temperature, on the other hand, can have some drawbacks. An example is shown in the following for the deposition of a U–O–Pd film on a silicon substrate heated (in this case after deposition) to different

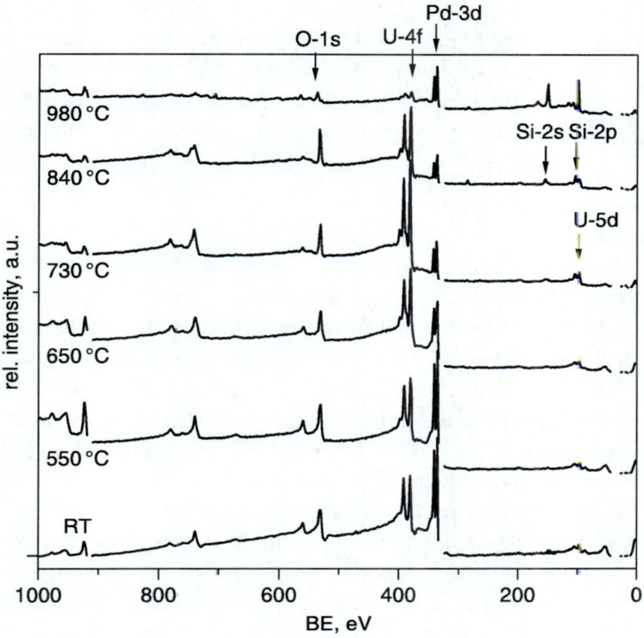

Fig. 5 Photoemission survey spectra of a UO_2 thin film doped with 70% Pd heated to different temperatures, adapted from [9]. MgKα (1,253.6 eV) is used for excitation

temperatures. Figure 5 shows the XPS survey spectra of a UO_{2+x} film doped with 70% Pd and heated from RT to successively higher temperatures, up to 980°C. The continuous increase of the U/Pd ratio with increasing temperature is striking and points to the diffusion of thin film components even at temperatures of 550°C. The evolution of the U-4*f*, O-1*s*, and Pd-3*d* intensity (area) ratios indicates that U segregates to the surface, replacing Pd [9]. The oxygen concentration at the surface stays unchanged. During diffusion, U transforms initial UO_{2+x} into UO_2, which is confirmed by the shape of the corresponding U-4*f* detailed spectra. The initial Pd-3*d* spectrum at RT shows that metallic Pd (BE 336.0 eV) is the main species. With increasing temperature, this peak intensity decreases and its shape changes. At 650°C, a shoulder appears in the spectrum at 336.9 eV, which completely replaces the main line with further temperature increase.

The overall intensity decrease of the Pd-3*d* lines indicates that the surface is depleted in Pd up to a temperature of 840°C, probably due to Pd diffusion out of the UO_2–Pd film into the underlying Si-substrate. The Pd-3*d* line at 1.7 eV higher BE can be either attributed to Pd alloying with U [18, 19], Pd oxidation, or Pd alloying with Si. In fact, XRD measurements have confirmed that the Pd depletion of the surface is associated with Pd–Si compound formation [9]. For even higher temperatures, the U concentration at the surface decreases significantly, while a strong Si signal appears. This is possibly due to diffusion into the substrate or the onset of evaporation into the vacuum.

The results of such systematic studies of the effects of substrate temperature on thin film composition and structure show that moderate heating between 150°C and 200°C in several cycles leads to better results than heating once to very high temperatures. The moderate conditions can be used to produce thin films for modeling UO_2 matrices containing noble metal particles (see below).

1.3.4 Reactions

After preparation and characterization, the actinide thin films are investigated with regard to their reactivity by gas adsorption experiments and electrochemical methods. The preparation chamber provides additional gas inlets for molecular gases such as O_2, H_2, water, and others. Also, the exposure of the surfaces to atomic oxygen and hydrogen (via an atom source) is possible. With this set-up, gas adsorption experiments in the low and medium Langmuir range (< 1 L up to 10^4 L) can be accomplished; one Langmuir, 1 L, is the dosage needed for ML coverage if all gas atoms or molecules stick to the surface, 10^{-6} Torr s.

An electrochemistry set-up (an electrochemical quartz crystal microbalance, EQCM) is installed for U and inactive samples. This set-up consists of a quartz crystal microbalance arranged as working electrode for electrochemical experiments, e.g., corrosion potential measurements or cyclovoltammetry (CV). This allows for securing in-situ mass information during the electrochemical experiment and thereby helps to classify single reaction steps observed in a cyclovoltammogram according in, e.g., oxidation/reduction without significant mass change of the electrode, cathodic reaction with mass increase of the electrode, and others.

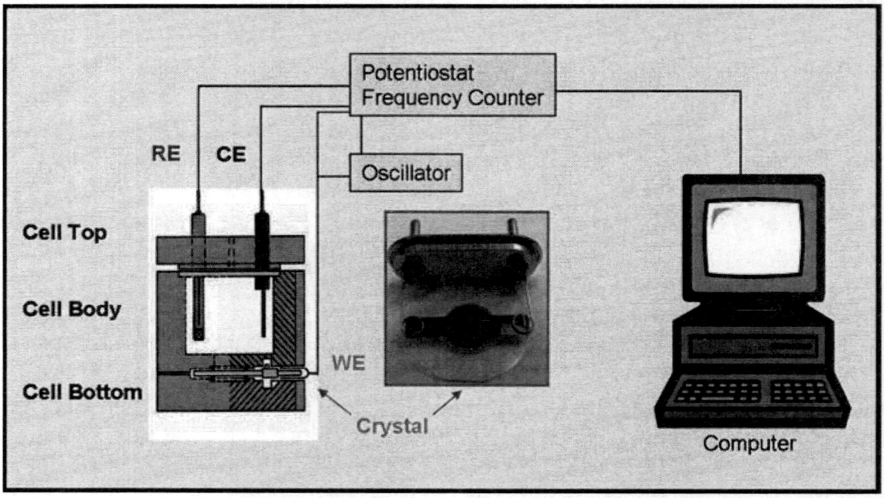

Fig. 6 Sketch of the EQCM set-up for electrochemical experiments. *RE* reference electrode; *CE* counter electrode; *WE* working electrode

Figure 6 shows a general sketch of the set-up and a detailed picture of the electrodes used in the electrochemical experiments described. The working electrode consists of the thin film deposited on one of the two gold electrodes on a common microbalance quartz crystal (an AT-cut crystal with 8 MHz resonance frequency is generally used). This crystal with the thin film electrode is fixed with O-rings between cell bottom and cell body building the electrolyte container (~5 ml volume). Contact between working electrode and electrolyte solution is provided by a 1 cm hole in the cell body tightened by the O-rings. The mass change on the electrode is calculated from the measured frequency signal according to the Sauerbrey equation [20]: $\Delta m = -C_f \times \Delta f$. The calibration factor, C_f, was determined for the actual set-up in a set of independent experiments [21].

2 Investigating the Electronic Structure of Actinides

Thin films of the actinide metals are used to study the evolution of chemical bonding, when the actinides are confined to the top-surface or when they are strongly diluted in a metallic matrix. Some results are described in the following. Also the interaction of the early actinides with reactive gases (e.g., oxygen, hydrogen, nitrogen) is a matter of interest both from a fundamental and applied point of view. Their electronic structure allows a set of bonding configurations with only minimal changes in energy that lead to a complex (physico-)chemical behavior. The more applied field of research is described in the second part of investigations of the solid–gas interface below.

2.1 *The 2D Confinement*

In case of a surface phase, decreasing layer thickness leads to suppression of 5*f* bonding simply due to the decreasing number of neighboring atoms. This is especially true, where the actinide phase is deposited on a substrate with which it only weakly interacts. When the layer thickness is below one ML, actinide atoms ideally form isolated adatoms. In reality, aggregation of surface atoms in islands or clusters will, at least partially, occur when the atoms are allowed to diffuse on the surface at elevated temperature.

Figure 7 shows the evolution of the valence band of Pu atoms deposited on Mg at RT [22]. Pu and Mg are not miscible, nor do they form any intermetallic compound; their interaction can be considered to be weak. A thick overlayer of five MLs has a valence band similar to that for bulk α-Pu metal. It forms a peak with a maximum directly at the Fermi-level, E_F, which continuously decreases at higher BE up to around 4 eV BE. This is the common peak-shape for delocalized, i.e., bonding 5*f* electrons found in U, Np, and α-Pu. The occupied part of the band observed in photoemission at E_F forms the boundary to the unoccupied part at higher energy

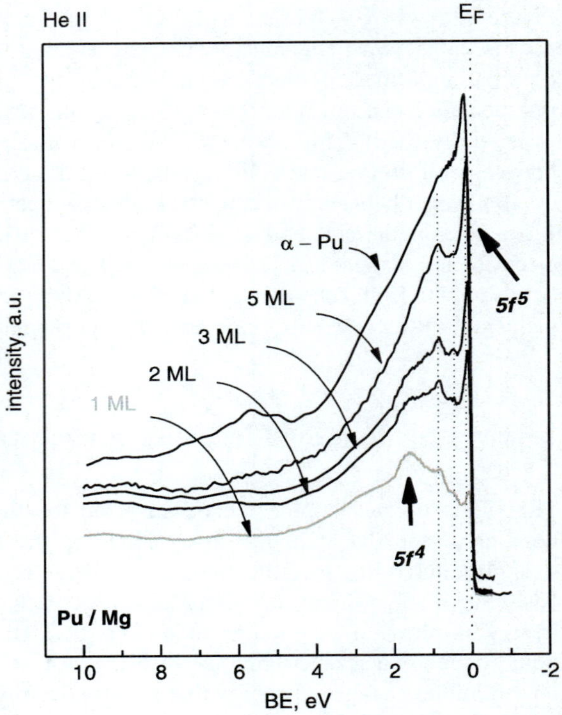

Fig. 7 UPS spectra of metallic Pu thin films of different thickness deposited on magnesium [22]. He II (40.8 eV) radiation is used for excitation

(lower BE). With decreasing layer thickness, a three peak multiplet structure emerges ($5f^5$ with peak maxima at around 0.1 eV, 0.5 eV and 0.8 eV) replacing part of the band emission. The multiplet is still located at E_F. The same multiplet has been observed in very different Pu compounds (PuSe, PuTe, PuSi, etc) pointing to the atomic rather than band character of the feature [23, 24]. At about one ML, the peak at E_F decreases and is replaced by another emission at 1.6 eV BE, which is now attributed to a nonresolved $5f^4$ multiplet structure well removed from E_F. This spectral evolution has been attributed in the past to gradual localization of the Pu-5*f* states with decreasing overlayer thickness [22].

2.2 *The Dilution Scenario*

5*f* localization can also be induced by diluting the actinide in a weakly interacting matrix. Thick films of binary system are produced by sputter codeposition from the elemental targets. Even for nonmiscible elements, e.g., U and Ag, well-dispersed films without agglomeration can be prepared by sputtering at room temperature (RT). This was shown by the gradually evolving valence band spectra (e.g., the

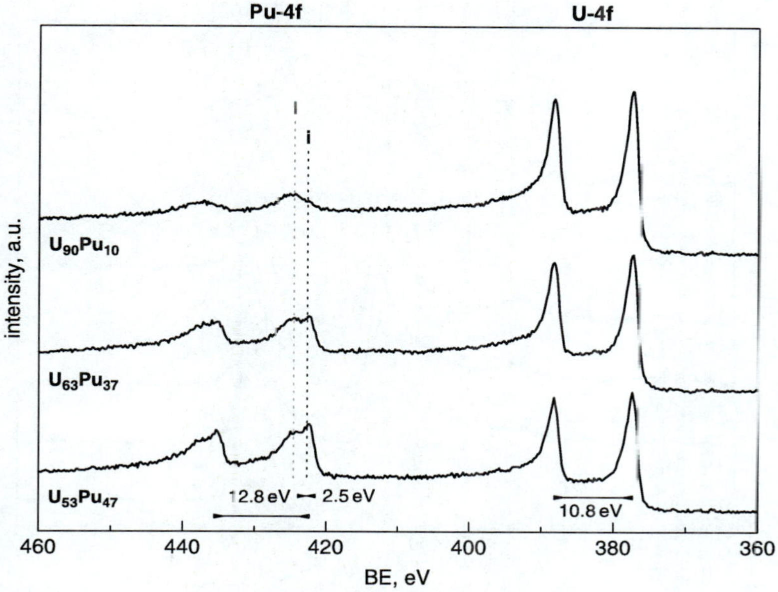

Fig. 8 Core level photoemission spectra of U–Pu alloy thin films of different composition produced at RT. Excitation radiation is Al K_{α} (1,486.6 eV)

Ag-4*d* band), which continuously narrow upon dilution of the element (Ag) [25]. In the case of phase separation, broader and composition-independent valence band emissions would have been observed. The fast cooling of impinging atoms at RT, together with the high deposition rate, favor immobilization of the atoms and fine dispersion, even in systems that are thermodynamically unstable.

Figure 8 shows the U-4*f* and Pu-4*f* core level lines for three U–Pu alloys. Both An-4*f* peaks are split in the characteristic spin–orbit doublet (ΔBE ~10.8 eV and 12.8 eV for U and Pu, respectively). The Pu-4*f* shows a further splitting (ΔBE ~2.5 eV) in a low and high BE component. This is attributed to Pu final states with itinerant (i; bonding) and localized (l; nonbonding) Pu-5*f* valence electrons [26]. The itinerant 5*f*-electrons screen the photoelectron-hole (left after the photoemission process) better than the localized ones. Therefore, the photoemission peaks appear at lower BE when the 5*f*-electrons are itinerant. The spectra can be described as sort of intermediate valence final state (i.e., after the photoemission event), having both an itinerant and a localized component. With decreasing Pu concentration in the alloy, the itinerant component is replaced by the localized one. This is a clear indication for Pu-5*f* localization upon dilution in the U metal. From Fig. 8, it is also clear that the U-4*f* level does not change. The U-5*f* electrons are well delocalized so that even dilution in Pu cannot force them into localization.

When Pu is diluted with Am (Fig. 9), the transition from itinerant to localized 5*f* states occurs much earlier, already at $Pu_{84}Am_{16}$ (in contrast to $U_{90}Pu_{10}$). The Am-5*f*

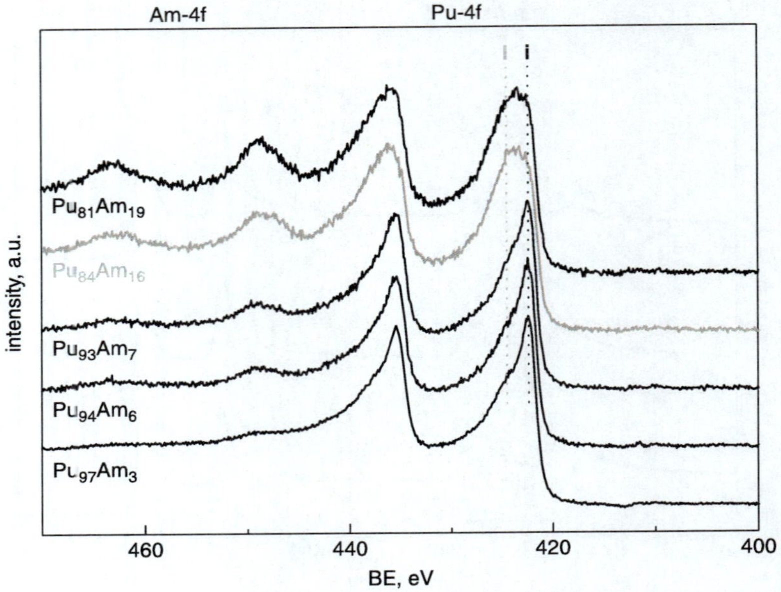

Fig. 9 Core level photoemission spectra of Am–Pu alloy thin films of different composition produced at RT. Excitation radiation is Al K_{α} (1,486.6 eV)

states are always localized, even in the pure element (except at high pressure). The explanation for the contrasting behaviors in the U versus the Am matrix is related to the difference in 5*f* electronic structure between U and Am. Uranium has well itinerant 5*f* states, which can bond with the Pu-5*f* states and keep them itinerant even when Pu is highly diluted. Am has localized 5*f* states, which do not hybridize with the Pu-5*f* states [27]. Therefore, even at modest dilution (~16% Am), the Pu-5*f* states become localized.

These examples demonstrate some interesting physics of simple actinide systems. The effects observed in these simple systems form the basis for investigations of more complex compounds (oxides, nitrides, doped fuels), which are in the focus of our corrosion research. As will be shown below, these more complex compounds are produced with the same sputter source set-up by simply sputtering in the presence of a reactive gas (oxygen, nitrogen). The phenomena of atomic dispersion, homogeneity, electronic structure, and surface composition are exactly the same as for the physics of the simple systems.

2.3 *Investigations at the Solid–Gas Interface*

As the actinide dioxides, AnO_2, are commonly used as reactor fuel (An = Th, U, Pu) and as actinides are formed in nuclear reactors during fuel irradiation

(An = Np, Pu, Am, etc.), or, in case of Pu, remain from the reduction of nuclear weapon stockpiles (mainly as Pu_{metal}), their behavior under final waste repository or storage conditions is of major interest for performance assessment. An important scenario in particular is the dissolution of spent nuclear fuel once in contact with intruded groundwater in a nuclear waste repository. This will lead to dissolution and leaching processes. For the fuel matrix, UO_2, and the higher actinides Np and Pu, solubility increases in the higher oxidation states [U(VI), Pu(VI, V), and Np(V) show higher solubility than their lower valent counterparts U(IV), Pu(IV), and Np (IV)]. Higher solubility can lead to an enhanced risk for the release into the biosphere. To assess and extrapolate these risks, reliable thermodynamic data bases and a thorough knowledge of acting reaction mechanisms is indispensable.

Our investigations aim at elucidating basic reactions and mechanisms at the actinide surfaces in a systematic way: on the one hand, looking at the different processes throughout the actinide series, and on the other, looking at single effects influencing these processes by increasing the complexity of the systems in a systematic, step-by-step manner.

2.3.1 Oxidation Processes Throughout the Actinide Series

Starting with a simple system, we investigate the oxidation behavior of the early actinide elements in experiments with molecular oxygen, which requires chemisorption and dissociation prior to reaction, and with atomic oxygen, a most reactive oxidizing species. These studies aim at determining the highest possible oxidation state that can be reached in these binary oxides. Photoemission spectroscopy is an ideal tool to perform chemical analysis of the surface/near surface (bulk) region of the thin films. It gives insight into the redox processes of these surfaces, as well as providing direct evidence on the electronic structure of the surface, and allows identification of reaction intermediates and stable reaction products. The experimental lay-out is similar for all sets of experiments; the scheme is depicted in Fig. 10. A metallic actinide surface is initially exposed to doses of molecular oxygen up to saturation at RT, whereby an oxidic phase forms (step I). Molecular O_2 reacts readily with metallic actinide surfaces to form the sesquioxides or dioxides. In step II, the oxidic surface is exposed to atomic oxygen at RT, leading to the highest possible surface oxidation state. Step III involves a heating step, leading either to oxygen diffusion into the bulk of the film or to decomposition of the oxide releasing oxygen into the vacuum. Both possibilities are accompanied by a partial reduction of the surface layer. To distinguish between these two possibilities, step IV combines heating and oxidation with atomic oxygen either leading to complete oxidation of the entire film (increased diffusion + formation of stable bulk oxide in highest oxidation state) or to saturation of the entire film at a lower oxidation state (increased diffusion + instability of the higher oxide as bulk oxide). The intermediate phases and end products for the different actinides are identified straightforwardly by UPS [29].

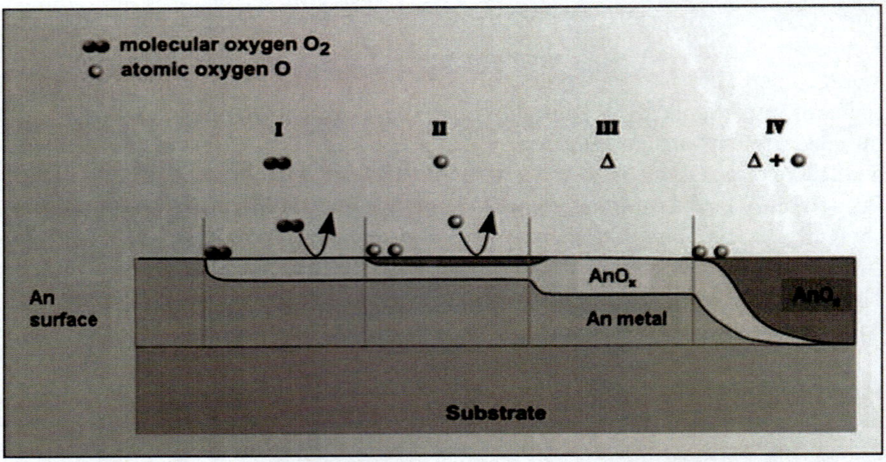

Fig. 10 Schematic layout of the gas adsorption experiments adapted from [28]

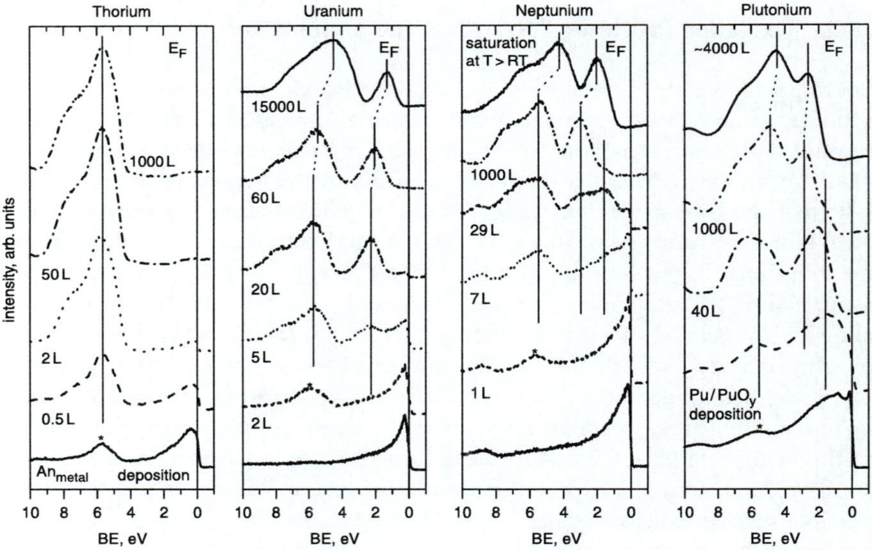

Fig. 11 UPS spectra for different O_2 dosages on An_{metal} surfaces [29]. Deposited metal films of Th and Pu show a slight oxygen contamination (~6 eV BE, *asterisk* mark explained in the text). He II (40.8 eV) radiation is used for excitation

Figure 11 shows the UPS-He II spectra recorded on Th, U, Np, and Pu thin films exposed to different O_2 dosages (corresponding to varying Langmuir). The initial actinide films (Fig. 11, bottom spectra) are typical metal spectra with the peak at the E_F (marked by the vertical line) and a featureless tailing towards higher BE.

The peak at E_F (0 eV BE) is attributed to the conduction band formed by itinerant, i.e., bonding 5*f* states (U, Np, and Pu). In Th, which has no *f*-states, the maximum at E_F is due to its 6*d*7*s* conduction band. After an O_2 addition of several Langmuir for all metal films, a peak grows at ~5 to 6 eV BE (asterisk mark in Fig. 11). In the low dosage regime, this peak is symmetrical, typical for isolated, chemisorbed oxygen atoms forming local bonds, and not the broad valence band indicative of bulk oxides [30, 31]. These bulk phases develop for all actinides at higher O_2 dosage, as shown by the asymmetrical O-2*p* shape, with a maximum intensity at the low BE side (Fig. 11, uppermost spectra). For Th, the intensity around E_F decreases with increasing O_2 dosage, according to the transfer of the $(6d7s)^3$ valence electrons into the Th–O bond of ThO_2.

For the remaining An elements in Fig. 11, successive O_2 adsorption leads to the appearance of one or two peaks between the O-2*p* band and the Fermi-level for U and for Np and Pu, respectively. These peaks are attributed to localized An-5*f* states because they are suppressed in HeI excited spectra (not shown here), which is typical for *f*-like cross-section suppression [26]. For U, only one localized *f*-peak is observed, while in Np and Pu, two peaks coexist. The *f*-peak at lower BE is attributed to the localized 5*f*-level of An_2O_3, while the high BE *f*-peak is ascribed to the 5*f* level in AnO_2.

For Np, the sesquioxide appears as a transient near-surface species (up to a layer thickness of about nine ML) and seems to be stable only in the presence of metal [32, 33]. This behavior is in accordance with known trends observed for the lighter actinide elements, where the behavior for Np lies between Pu, which is the first actinide in the series exhibiting a stable bulk sesquioxide, and U, which does not exhibit the sesquioxide at all. No further oxidation above AnO_2 (An = Th, U, Np, Pu) or An_2O_3 is achieved with molecular O_2 up to a dosage range of ~1,000 L.

Information on the thermal stability of the high oxidation state species is obtained by exposing the surface to atomic $O^\bullet$ at elevated temperatures. Under these conditions, oxygen diffuses into the deeper layers, thereby saturating the entire film (about 30 atomic layers) at the high oxidation state. This is expected and experimentally observed for uranium [28]. Figure 12 shows that the UPS-He II spectra of U, Np, and Pu dioxide surfaces further reacts with atomic oxygen at RT and at higher temperature (200°C to 300°C). The intensity of the 5*f* peak (~2 eV BE) can be directly associated with the $5f^n$ count and can thus be used to determine the oxidation state of the actinide ($5f^2$ for U(IV), $5f^3$ for Np(IV), and $5f^4$ for Pu(IV)). With atomic oxygen, $O^\bullet$, uranium spontaneously reacts with UO_3, as seen by the complete disappearance of the U-5*f* line according to the different electronic configuration of the two oxides (UO_2: $5f^2$, UO_3: $5f^0$). Also, the O-2*p* peak broadens and becomes more symmetrical.

NpO_2 also oxidizes to a higher oxide (even higher than Np_2O_5, judging from the Np-$5f^n$ intensity). This oxide is also found to be a stable compound under similar conditions as for U. PuO_2 also reacts with atomic oxygen, as seen by the suppression of the $5f^4$ emission at RT. However, upon $O^\bullet$ exposure at 300°C, the 5*f* line intensity increases again, indicating decomposition of the higher oxide. For both Np

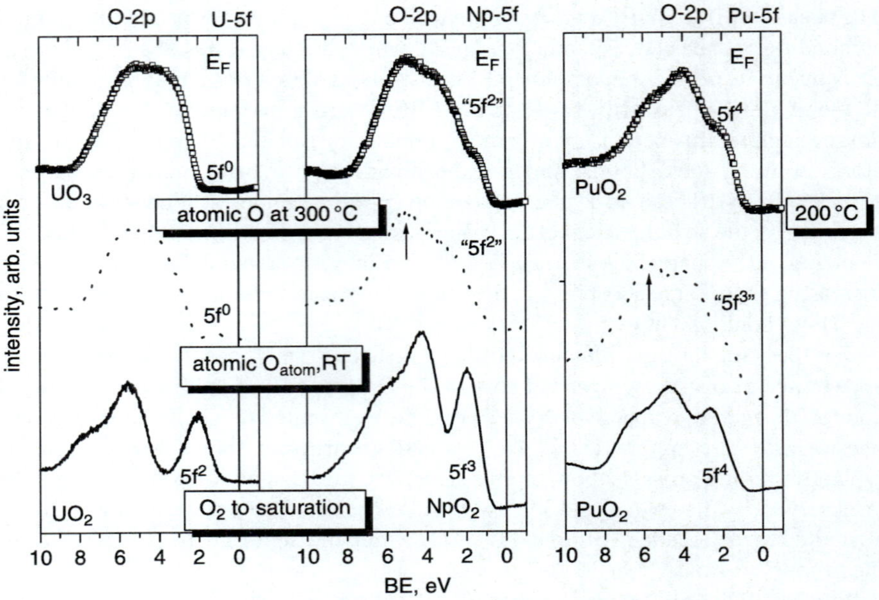

Fig. 12 UPS spectra for reaction of AnO_2 (An = U, Np, Pu) surfaces exposed to O• [29]. He II (40.8 eV) radiation is used for excitation. *Arrow marks* are explained in the text

and Pu, greater surface oxide formation at RT is also reflected in the change in shape of the O-2*p* peak, which acquires additional intensity at ~5 to 6 eV BE (marked with arrows in Fig. 12). This feature is more pronounced in He II than in He I. We explain this by the higher surface sensitivity of He II and attribute this peak to supplementary (chemisorbed) oxygen forming a local bond at the top-surface of the oxides (Np_2O_5, PuO_2) [28, 33]. The results of this experimental series show that plutonium does not form any stable bulk oxide compound higher than PuO_2 under these conditions. Neptunium forms a bulk (or at least near-surface) Np_2O_5 phase, while U forms stable higher-valent bulk oxides ($UO_{2.67}$, UO_3). The higher Pu oxide "PuO_{2+x}" (even at low temperature) is exclusively limited to the top surface and is better described as a local bond between a Pu surface atom and the chemisorbed oxygen.

In conclusion, this comparative study shows the gradual increase of lower oxide (An_2O_3) stability in the actinide series from uranium to plutonium (also to americium, data not shown), whereas a decreasing tendency of higher oxide formation (AnO_x, $x > 2$) is also observed. The same explanation applies for both these phenomena: increasing localization of 5*f* states upon increasing Z. The ability of 5*f* states to contribute to bonding decreases with the increasing localization so that An_2O_3, involving An-$(6d7s)^3$ electrons, is stabilized. For neptunium and plutonium, the nonbonding 5*f*-states at the surface atoms of the higher oxides (Np_2O_5 and PuO_2) readily participate in bonding. This behavior is not observed for thorium dioxide and uranium trioxide (with formally no occupied 5*f* states) and is also strongly suppressed for americium dioxide [29].

Similar investigations are undertaken in the reductive regime with hydrogen species (molecular and atomic) on different thin film surfaces (UO_2, UO_2/Pd, PuO_2) [34, 35] with the goal of identifying mechanisms responsible for the suppression of spent nuclear fuel matrix dissolution observed in different laboratory experiments [36–40] and redox phenomena observed on PuO_2 surfaces in contact with water (see chapter 2.3.2 The System PuO_2 + Water) [34].

2.3.2 The System PuO_2 + Water

Only few direct surface investigations of the PuO_2/water interaction are reported [41, 42]. Earlier reviews by Waber [43] and Colmenares [14, 44], mainly dealing with the oxidation of Pu metal with different oxidants, also suggest that previous surface science investigations are rare. This is on one hand astonishing, as direct investigations of surface reactions are possible with surface science techniques. On the other hand, especially for experiments with water that could bridge the gap from UHV conditions to aqueous systems, specific instrumentation such as high pressure cells are desirable but not yet available in many laboratories.

Our study [34] aiming at directly following the water adsorption on PuO_2 surfaces and identifying the surface reaction products with XPS and UPS is done by exposing freshly deposited surfaces with water dosages below 1,000 L. To achieve high water concentrations under the experimentally required UHV conditions, experiments with ice layers deposited on the film samples (cooled down to around −200°C with liquid N_2) are carried out. The main advantage of the in-situ production of the films by sputter deposition is that it avoids contamination of the surfaces with carbon and other contaminants from the laboratory atmosphere that may cause artifacts in investigations of the surface region [42].

Although for systems containing water the formation of mixed-valent Pu(IV/V or VI) binary oxide is observed [41, 45–48], in our own strictly dry experiments concerning oxidation behavior of actinides [28, 29], no further (bulk) oxidation of PuO_2 was detected even when exposing these surfaces to very reactive atomic oxygen. This may be considered substantiation of reports indicating water to be directly involved in the oxidation reaction [45]. In fact, the underlying reaction route to a hyperstoichiometric PuO_{2+x} is still not clear: water or oxygen are proposed as oxidizing agents or water is suggested to act as either as a catalyst or simply as an agent to produce the thermodynamically less stable amorphous hydrous oxide from a crystalline phase [45–48].

Figure 13 shows the Pu-4*f* and O-1*s*, and valence band data of a PuO_2 film exposed in-situ to a water vapor dosage of 188 L. For the freshly deposited film (Fig. 13a), the Pu-$4f_{7/2}$ and Pu-$4f_{5/2}$ peaks lie at 425.8 eV and 438.5 eV, each with a satellite peak at ~7 eV higher BE, typical in all early actinide dioxides (from ThO_2 to AmO_2) [26, 49]. The O-1*s* line is observed at 529.6 eV BE, which also is typical for these oxides. The additional component in the O-1*s* spectrum at ~531 eV is

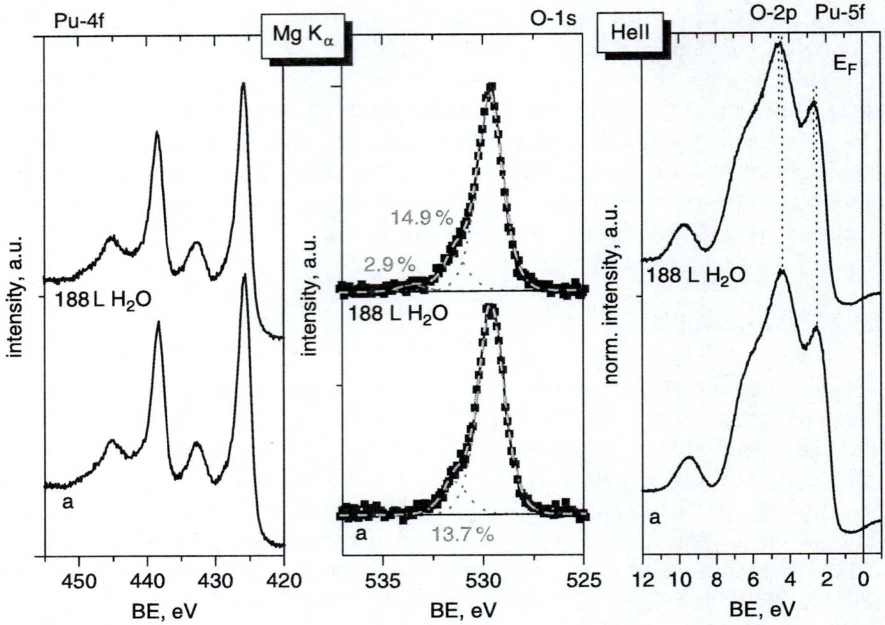

Fig. 13 Core level and valence band photoemission spectra of a PuO_2 thin film after production at RT and exposed to 188 L H_2O dosage [34]. (**a**) freshly prepared film. Excitation radiation is Mg K_α (1,253.6 eV) or He II (40.81 eV). He II spectra are normalized to the same height (slight intensity decrease to ~70% for the water contacted surface is therefore not reflected). Curve fitting is applied for the O-1*s* peaks: *dotted lines* give the peak shapes used for the fitting, *light gray, dashed line* gives the fitted envelope. Percentages are given for the nonoxidic O-1*s* contributions (% of the total area from fit)

usually attributed to hydroxyl species [31, 42, 50]. Its relative intensity of 13.7% points to the presence of one ML of OH. Both the Pu-4*f* and the O-1*s* lines stay virtually unchanged after a 188 L dosage of water. Only the hydroxyl component at high BE increases to 14.9% in intensity and a small intensity increase (2.9%) at 533.5 eV BE is seen, corresponding to molecular water [51]. Obviously, the freshly deposited film surfaces are saturated with water/hydroxyl, which is presumably due to the fact that we are working with high water dosages during this study and water displaced from the chamber walls by oxygen can coadsorb on the surface during sputter deposition of the films. The valence band region (Fig. 13, right graph) probed with UPS corroborates this. The valence band of the freshly prepared film shows three distinct features. The peak at ~2.5 eV is assigned to localized Pu-5*f* states, the broad feature between 4 eV and 7 eV is the O-2*p*/Pu-6*d* bonding band and a smaller intensity ~9.6 eV is attributed to surface hydroxyl species [31]. Again, exposure to 188 L water does not change the spectra significantly (except for a small BE shift and intensity decrease, not visible in Fig. 13 due to normalization).

When PuO_2 surfaces with sub-ML dosages of water/ice at −200°C (liquid N_2 cooled sample) are warmed up and the water desorbs, the PuO_2 surface reappears

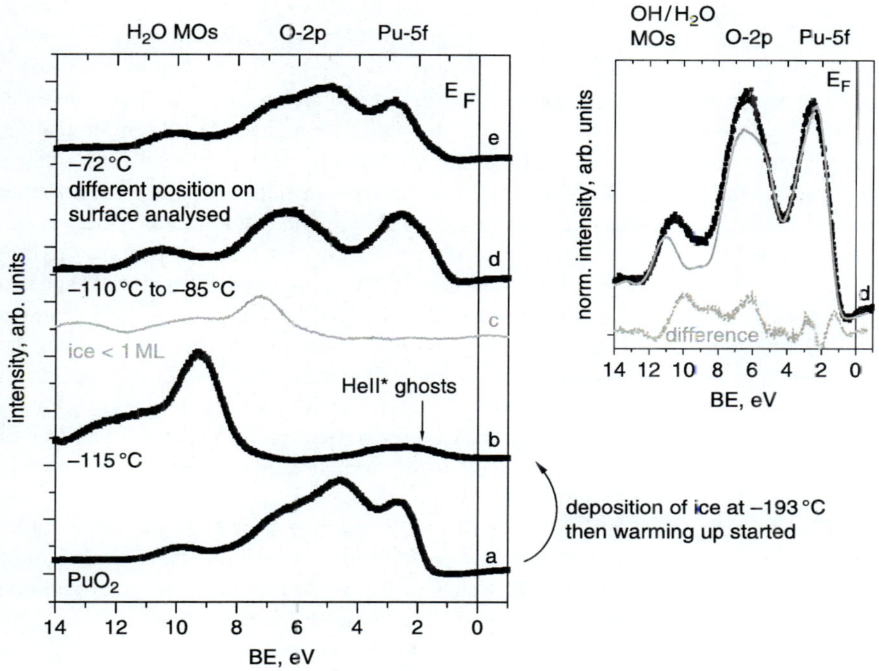

Fig. 14 Valence band photoemission spectra of a PuO_2 thin film after production at RT and exposed to high H_2O dosages at −193°C. PuO_2 film before (**a**) and after adsorption of 172 L ice (**b**) and after desorption in the presence and absence of UV light (**d**, **e**, respectively). Difference spectrum of a PuO_2 thin film exposed to a sub-ML ice, with the PuO_2 part removed (**c**). *Arrow mark* indicates intensity due to photoelectrons from He II* excitation. Excitation radiation is He II (40.81 eV) with He II* admixture. The *inset* shows spectrum "d" in comparison with a Pu_2O_3 reference spectrum (*thin gray line*) and the corresponding difference spectrum (*dotted gray line*)

unaltered. In contrast, when the PuO_2 surface is subjected to thicker layers of ice (higher dosages of water), a strong influence on the surface is observed following water desorption. Figure 14 depicts the valence band spectra measured throughout an adsorption/desorption experiment of ~172 L water vapor on a cold PuO_2 surface. The freshly deposited Pu thin film (Fig. 14a) shows the typical features of PuO_2 described above. Addition of ~172 L water vapor at a sample temperature of around −193°C leads to a strong intensity decrease for all spectral features (not shown). Upon warming to −115°C (Fig. 14b), at which a pressure maximum in the vacuum chamber is observed, all photoelectron peaks of the H_2O molecular orbitals (MOs) shift to lower BE. Their position approaches the region (e.g., from >12 eV to 9.4 eV for the He II excited $1b_1$ line of the H_2O MO) corresponding to those for sub-ML films (Fig. 14c; here, a difference spectrum with the PuO_2 part subtracted is shown). The small peak at ~2.5 eV in the spectrum Fig. 14b (marked with an arrow and "He II* ghosts") is assigned to the $1b_1$ level of H_2O excited by He II* light. In addition to other wavelengths, the UV source produces both He II (40.81 eV) and He II* (48.4 eV) light so that the associated photoemission spectra superimpose. The He II* excited lines are found at ~7.6 eV lower BE than the He II excited lines.

After most water has desorbed at temperatures above −110°C (Fig. 14d), the underlying Pu oxide layer has undergone a significant change. The Pu-5*f* intensity is strongly increased compared to the O-2*p* line. This peak also broadens, showing a new signal at lower BE (1.8 eV). The O-2*p* band changes its shape from the double peak appearance to a much more symmetric shape with a maximum at higher BE compared to the original PuO_2 surface. The spectrum becomes more similar to that of Pu_2O_3 than PuO_2 [29, 32]. The identification of these uppermost surface species as mainly Pu_2O_3 with some admixture of PuO_2 and additional H_2O/OH features (e.g., the peak at ~10 eV) is clearly visible upon comparison of the spectrum of the Pu oxide film after water desorption in Fig. 14d with a Pu_2O_3 reference spectrum (see Fig. 14 inset). The residual remaining following subtraction of the Pu_2O_3 reference (inset; light gray, dotted line) shows some similarities to the features of the sub-ML ice (Fig. 14c), with an additional pronounced peak at ~10 eV (probably indicating OH^-).

This is unexpected behavior for the PuO_2 surface based on thermodynamic considerations. Further investigation is performed to determine the mechanism of this thermodynamically unfavorable reduction. We find that the observed reduction here is not caused by the simple absence of an environment oxidative enough to maintain the Pu(IV) state at the surface (UHV conditions). On comparing the spectra of a surface warmed up under UV light and one warmed up in the absence of light (Fig. 14d, e), we observe that in the absence of light, virtually no reduction takes place and the spectrum appears more like that of PuO_2, with only a small fraction of reduced species present, visible in the spectral feature at 1.8 eV. We therefore conclude that a necessary requirement for the reduction process to take place is the irradiation with the UV-light during the warm-up process.

Reiterating what we know about the reduction from these observations, (1) UV light plays a crucial role, (2) the ice layer thickness plays a role (no reduction for sub-ML ice vs. reduction for thick ice layers), and (3) the thawing time (and therewith the irradiation time) is similar for thin and thick layers, so it cannot likely be the determining factor for the difference between thin and thick ice layers. Considering these facts, a photochemically induced reaction in the 3D ice layer producing reductive species may be assumed to prevail. Primary photolysis products of ice can be considered: excited molecules, radical species such as $H^•$, $HO^•$, $O^•$, and ionic species H_2O^+, OH^+, H^+, and e^-, as well as molecular products such as H_2, O_2, and reaction products such as H_2O_2 [52]. In this scenario, the diffusion of the reducing species must be fast enough to reach the interface between ice and PuO_2 surface before it is lost or consumed via release to the vacuum upon thawing or before other reactions occur. What species is involved and the reaction mechanism responsible for this reduction is currently under investigation.

Conceptually, it is in any case remarkable that the radiation-induced process does not lead to surface oxidation, as commonly observed with water subjected to radiolysis in contact with UO_2 surfaces (spent nuclear fuel, α-doped UO_2 or external γ-irradiation). Instead, surface reduction is induced. Although we aimed at determining the impact of water on the oxidation process up to PuO_{2+x}, our results raise more questions than they answer. From the results of the different methods of producing PuO_2 and PuO_2/OH, we conclude that initial addition of water in the low Langmuir range does not lead to a significant oxidation. Instead, dissociative adsorption giving rise to a completely hydrolyzed surface is observed. If such a hydrolyzed surface is oxidized more easily (by O_2 or atomic O) than the "dry" surface remains to be seen.

3 Investigations with Combination to Off-Line Methods

Several complementary methods are used to further characterize the produced thin film systems or to investigate specific reaction mechanisms. Especially, the uranium-based films are produced as models for spent nuclear fuel. Therefore, it is indispensable to show their applicability for this purpose by demonstrating their comparability in structure and reactivity to real systems.

3.1 The Systems UO_2 and UO_2/Pd: Comparative Characterization

Model thin layers for pure UO_2 fuel matrix and for UO_2 + Pd, to simulate noble metal ε-particles in spent nuclear fuel, are produced, their structural and morphological properties characterized by means of XPS/UPS, XRD, SEM-EDX, TEM, and AFM and results compared to the literature data for real (spent) nuclear fuel. The pure UO_2 system shows a strong dependency with respect to its composition and morphology on the deposition temperature, with other deposition parameters such as O_2 pressure, argon pressure, and deposition rate held constant. With increasing deposition temperature from RT to 300°C, the composition changes from hyperstoichiometric to stoichiometric. Congruently, their XRD diffraction peaks become more pronounced (with a calculated lattice parameter, **a**, similar to that for bulk powder samples) and the size of the deposited crystallites increases. Figure 15 summarizes the XRD, AFM, and XPS results for thin films of $UO_{2(+x)}$ deposited on gold at RT and at 300°C.

XRD results show a cubic fluorite lattice of UO_2 for all thin films. At high deposition rate and/or high temperature, a significant preference of the (111) orientation is observed. Quantitative evaluation shows that the reflexes for the UO_{2+x} RT samples are shifted significantly to higher 2θ values compared to literature values [14, 53] and to a reference powder sample of UO_2 measured in a similar geometry. The determination of **a** from the observed diffraction peaks results in values close to 5.404 Å. This is much lower than expected for $UO_{2.0}$

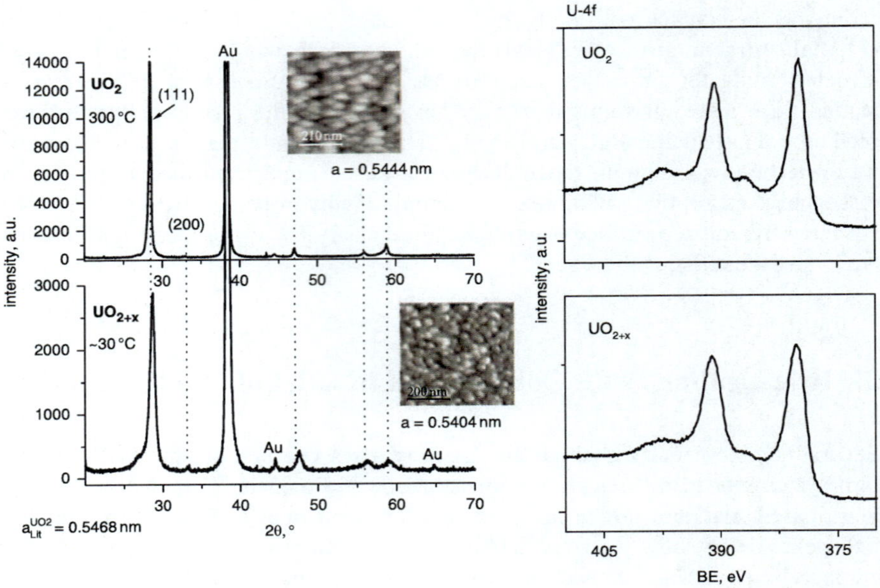

Fig. 15 XRD, AFM, and XPS results for UO_2 thin film depositions on Au substrates at different substrate temperature. *Lower graphs* show results for RT deposition with high deposition rate, *upper graphs* results for 300°C depositions with high deposition rate. AFM pictures have an edge length of 1 μm. Excitation source for XPS was Al K_α (1,486.6 eV)

(a = 5.468 Å) and also for higher oxidized samples (UO_{2+x}) [14], where values down to 5.44 Å ($UO_{2.2}$) or 5.41 Å ($UO_{2.34}$ [54]) are reported. The RT samples show high full width at half maximum (FWHM) values between 0.4° and 0.8°; therefore, a small crystallite size and a high concentration of defects are expected. Indeed, AFM investigation shows crystallites with sizes below 100 nm for high and low deposition rates at RT. For deposition at 300°C (upper, left plot in Fig. 15), the intensity of the XRD reflections is drastically increased compared to the low temperature samples. The determination of **a** in this case yields ~5.445 Å. This is somewhat lower than expected for bulk UO_2 (5.468 Å), but significantly higher than the values obtained for the low temperature samples. The 300°C samples show FWHMs between 0.2° and 0.54°, suggesting a higher degree of crystallization in the film. This is corroborated by AFM, which reveals an increased crystallite size (up to 150 nm) compared to the low temperature samples .

XPS data show that hyperstoichiometric UO_{2+x} is produced in the RT depositions, while the 300°C depositions usually result in thin films of UO_2 near stoichiometry (calculated O/U intensity (area) ratios vary between 1.93 and 2.04). This is clearly seen in the U-4*f* spectra (Fig. 15, right), where the characteristic "7 eV satellites" for UO_2 are distinct and both peak and satellite FWHM are decreased compared to the hyperstoichiometric oxide. The production of a hyperstoichiometric oxide, in case of deposition at lower temperature, seems reasonable since the

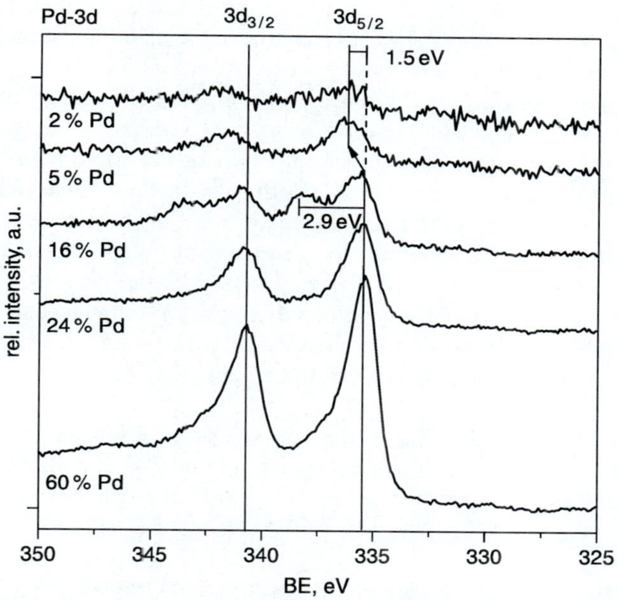

Fig. 16 Pd-3*d* spectra from UO_2/Pd thin films with varying Pd concentration (adapted from [9]). Mg K_α (1,253.6 eV) is used for excitation

sticking probability of O_2 (on the developing U–UO_2 surface) can be assumed to be higher at lower temperature.

Similar results are obtained for the UO_2 + Pd system. The codeposition of U and Pd in the presence of oxygen results in the formation of $UO_{2(+x)}$, whereas the speciation of Pd depends on its concentration [35]. With varying Pd concentration, a change of the Pd speciation is observed by XPS (Fig. 16). The BE of the Pd-3$d_{5/2}$ line at Pd concentrations ~60% and above is 335.4 eV and exhibits additional intensity at 346.7 eV. Both features can be attributed to metallic Pd [16]. With decreasing palladium concentration, the Pd signal intensity decreases, the Pd-3*d* spectrum slightly broadens and a second peak doublet appears (with the characteristic spacing of Pd-3*d*), which is shifted towards higher BE ($3d_{5/2}$ = 338.3 eV), indicating the formation of a new Pd species (assumed to be a Pd–O species) in addition to Pd metal.

Oxidation of Pd is reported by several authors, and comparable BE shifts are obtained for the oxidic species. The Pd $3d_{5/2}$ peak chemical shift for the PdO_y species are quoted between 1.3 and 1.9 eV [18, 55] for PdO and between 2.1 and 2.9 eV for PdO_2 [16]. Even the formation of PdO_3 has been assumed by some authors (with a BE shift ~3.5 eV) [19, 56]. The observed peak at 338.3 eV in Fig. 16 is tentatively attributed to a PdO_2 species. With even lower Pd concentration (down to 2%), this oxidic signal as well as the peaks attributed to the metallic species vanish in favor of a Pd-3*d* spectrum that is shifted to higher BE by 1.5 eV (compared to the pure Pd spectrum). This indicates that under suitable deposition conditions, the formation of a PdO-like phase could also be possible. The oxidation of Pd is not observed when solely sputtering the noble metal in the presence of O_2. Obviously, codeposition with U enforces such oxidation process.

XRD measurements on the Pd-doped thin films deposited at RT show that the codeposition of U and Pd in presence of oxygen results in the formation of a crystalline UO_2 matrix (typical diffraction peaks for the cubic fluorite lattice are found). With Pd concentrations (from 2% to ~20%), the UO_2 diffraction peaks shift slightly to higher 2θ values (i.e., decrease in **a**). With increasing Pd concentration, the intensity of the UO_2 reflexes, decreases or even disappears completely (from Pd concentrations ~30% and above). Under the same deposition conditions, no reflections for metallic Pd are observed below a Pd dopant concentration of 70%. Obviously with increasing Pd doping, the UO_2 matrix becomes distorted, culminating in a complete matrix amorphization for Pd concentrations above 30% [35].

TEM and SEM images of thin films produced at RT with up to ~16% Pd TEM reveal homogeneous films consisting of small crystallites (<100 nm; Fig. 17). SEM–EDX measurements (Fig. 18) reveal both elements to be distributed

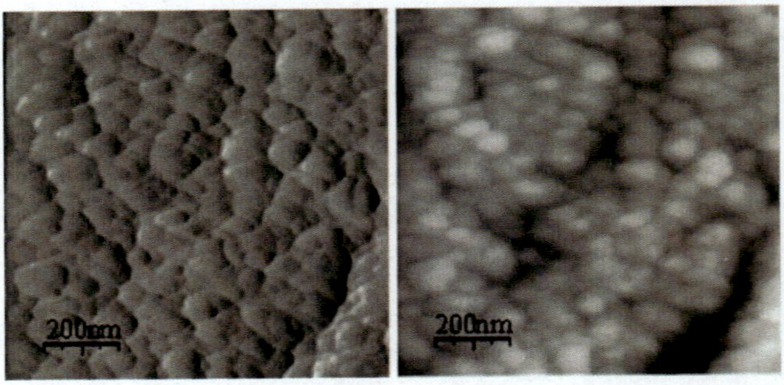

Fig. 17 AFM pictures of a UO_2/Pd (16%) thin film. AFM pictures have an edge length of 1 μm. *Left plot*: deflection data, *right plot*: height data

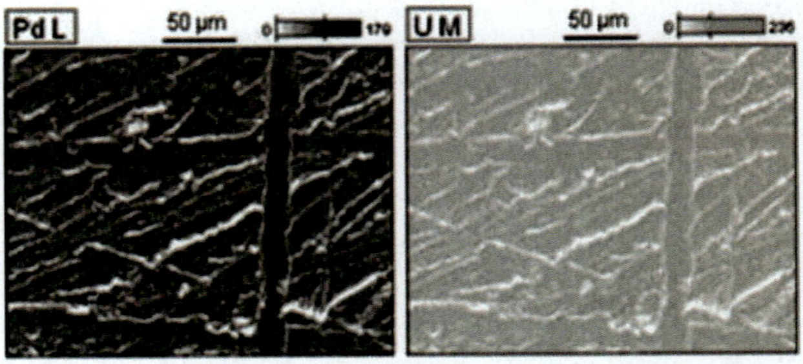

Fig. 18 Image of an EDX mapping of a UO_2/Pd (40%) film deposited at RT. Pd (*left picture*) and U (*right picture*) are mapped over a 250 × 200 μm^2 area [35]

homogeneously over the entire area. Local enrichment or depletion of one element is not observed. Codeposition of U, Pd, and O at RT results in the formation of a homogeneous $UO_{2(+x)}$/Pd/PdO_y mixture, and no indication of a separated metallic Pd phase (at low Pd concentration) is observed, which could be used to simulate the noble metal ε-particles in spent nuclear fuel [35].

In order to obtain Pd nanoparticles embedded in UO_2 as model ε-particles, a specific heating procedure is applied: the films are repeatedly heated for a few minutes to moderate temperatures (150–200°C) after deposition. The XPS spectra of a UO_{2+x}/Pd (16%) film annealed in several cycles for maximum 2 min at temperatures between 150°C and 200°C are shown in Fig. 19. Changes in the Pd-3*d* and U-4*f* spectral features indicate diffusion of film components already at such moderate temperatures. The initial Pd 3*d* spectrum (from RT deposition) exhibits two features at 336 and 337.8 eV for the Pd-$3d_{5/2}$ line, which are attributed to Pd metal and Pd oxide, respectively. With an increase in the number of heating cycles, the PdO_y features at higher BE decrease and eventually disappear; the peak shape narrows congruently. This is interpreted as evidence that the films deposited at RT exhibit defect structures with atoms in high energy positions showing decreased bonding ability. Photoelectrons from such atoms usually leave behind less screened holes that reduce their KE, therewith leading to photoemission peaks at higher BE. After the final cycling in the series, the $3d_{5/2}$ line is shifted to lower BE by 0.3 eV, indicating greater screening of the photoelectron-hole, which suggests that the Pd cluster size has increased. Uranium is deposited as hyperstoichiometric UO_{2+x}, which transforms into stoichiometric UO_2 during heating (see description of associated spectral features above). The XRD pattern of the heated UO_2/Pd films (in the range 2% to ~30% Pd) show diffraction peaks comparable to UO_2 powder reference data. The heating steps lead to a structural reorganization/recrystallization process and development of a less distorted UO_2 lattice and a metallic Pd phase. This metallic phase (observed as single phase with XPS) is not seen in the XRD, possibly due to low concentration or poor crystallinity (small cluster size).

AFM and SEM images of the films corroborate the presence of noble metal particles, as model ε-particles, in a thin film UO_2 matrix. Exemplary images of a heated UO_2/Pd (16%) film are shown in Fig. 20. Compared to the AFM images for the RT samples (Fig. 17), the heated sample (Fig. 20) consists of larger crystallites of about 350 nm in size. In addition, nearly spherical particles with a diameter of up to 3 μm on top of the surface are observed. The SEM image of one such particle is shown in Fig. 20, right. This particle is an agglomeration of smaller 300–800 nm spheres. EDX analysis confirms that these agglomerates consist of Pd. The EDX raster image of a film heated at 200°C is shown in Fig. 21. In contrast to the EDX image of the nonannealed film (Fig. 18), a local enrichment in Pd is observed, while the uranium distribution remains homogeneous. This supports that sputter codeposition of U + Pd + O with subsequent heating to moderate temperature results in segregation processes and formation of metallic Pd particle embedded in UO_2 thin film. This provides preparation of a model system to simulate ε-particles in the UO_2 (fuel) matrix for further study.

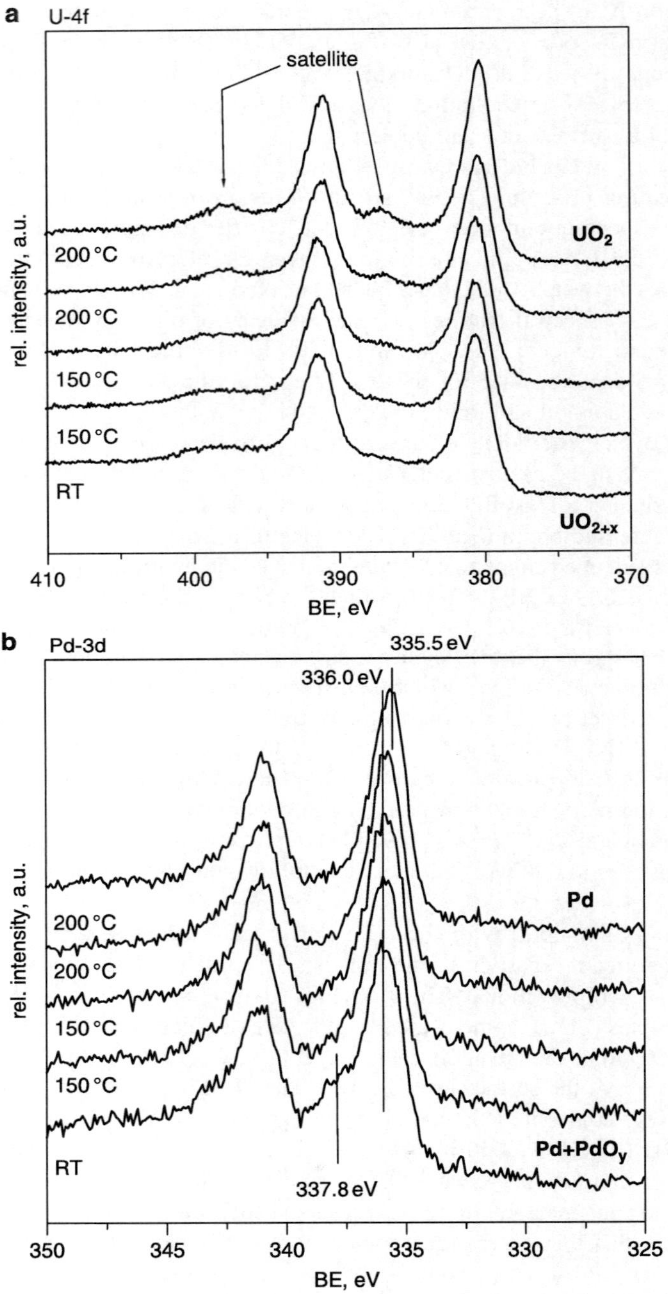

Fig. 19 U-4*f* (**a**) and Pd-3*d* (**b**) detail spectra of a thin film of UO_2/Pd (17%) throughout several heating cycles (1–2 min each), adapted from [35]. Mg K_{α} (1,253.6 eV) is used for excitation. *Arrow mark* in (**a**) indicates characteristic "7 eV satellites" of stoichiometric UO_2

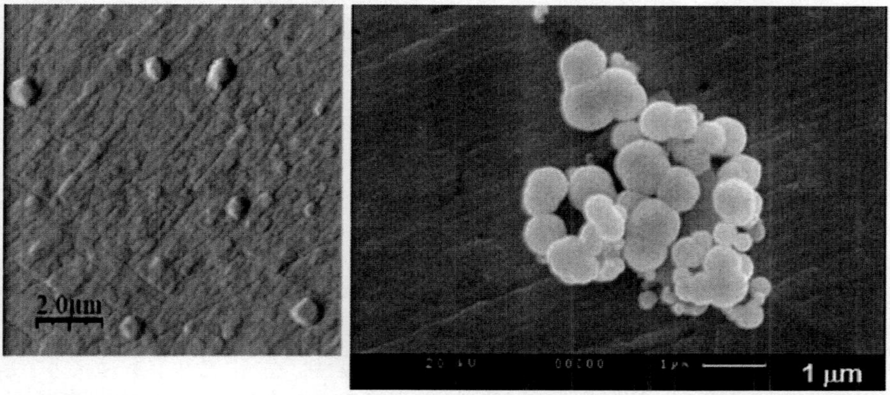

Fig. 20 AFM (*left*) and SEM (*right*) pictures of a UO_2/Pd (16%) thin film. AFM pictures have an edge length of 10 μm (deflection data shown), adapted from [35]

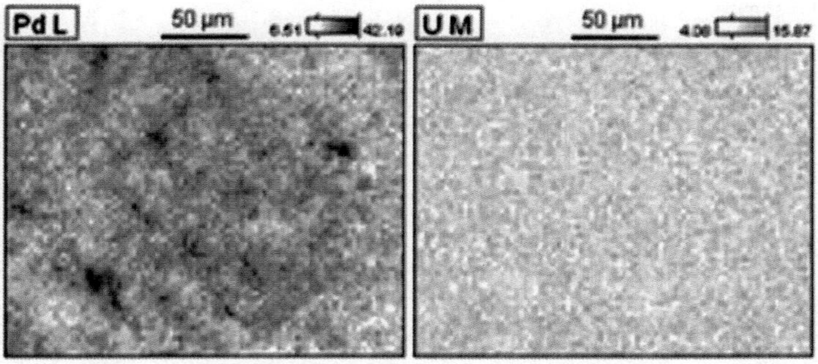

Fig. 21 Image of an EDX mapping of a UO_2/Pd (16%) film deposited at RT and heated afterwards to ~ 200°C [35]. Pd (*left picture*) and U (*right picture*) are mapped over a 250 × 200 μm^2 area

3.2 Investigations at the Solid–Solution Interface: UO_2 and UO_2/Pd

The thin films described above are used in investigations concerning their reactivity in aqueous solution. Preferentially, electrochemical methods using the EQCM are applied to determine redox processes at the exposed surfaces. As a simple approach, we begin with the comparative investigation of pure and Pd-doped UO_2 thin films in aerated 0.01 M NaCl at pH 5–6 (as a simplistic ground water simulate) with voltammetric methods.

3.2.1 Electrochemistry + XPS + Off-Line AFM

After a corrosion potential measurement ($E_{corr} = 0.37$ V) and a set of sequentially applied constant potentials up to 0.508 V (not shown here), a linear sweep

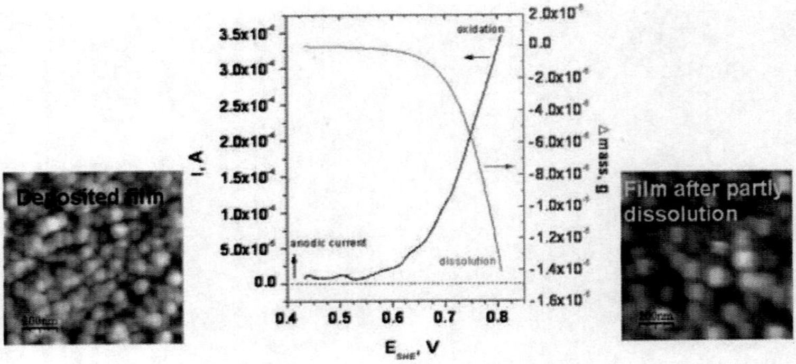

Fig. 22 Linear sweep voltammogram (*current data in black*) and mass data (*in gray*) for a UO_{2+x} thin film electrode in 0.01 M NaCl at pH 6. The UO_{2+x} electrode was preconditioned by corrosion potential measurement and a forced corrosion (sequentially applied potentials) up to 0.508 V. Scan velocity is 10 mV s^{-1}, potentials are referenced versus standard hydrogen electrode, SHE. AFM pictures of the film surface after production (*left*) and after the LSV experiment (*right*) are additionally shown

voltammetry (LSV) to $E = +0.808$ V is applied to a pure UO_2 film electrode, which is characterized with AFM and XPS prior to and following LSV. The current and mass responses of the preoxidized electrode in the LSV are shown in Fig. 22. Clearly, an oxidative dissolution process is observed; the mass of the electrode decreases and an anodic current is measured. From the mass and current data, charge over mass ratios, Q/m, are determined and compared to theoretical values calculated for specific reactions possibly taking place at the UO_2 surface. For the described LSV experiment, a Q/m ratio of -442 C g^{-1} (without off-set) is calculated. This corresponds well to the theoretical Q/m ratio of -466 C g^{-1} for the dissolution reaction of a preoxidized uranium oxide surface:

$$UO_{2.33} + 0.33H^{+} \leftrightarrow (UO_2{}^{2+})_{sol} + 0.33\ OH^{-} + 1.33 \cdot e.$$

We therefore conclude that the observed reaction is the direct dissolution of a higher oxide with the approximate composition of $UO_{2.33}$ or slightly higher. The surface left over from the partially dissolved film is measured with AFM and compared to the AFM image of the pristine film directly after production. Both images are shown in Fig. 22 along with the LSV voltammogram. Only subtle differences in film morphology are observed before and after the dissolution reaction; crystallite size remains about the same, pit depth appears to have increased only slightly, and the film remains homogeneous, with a decreased crystallite structure height. No observations indicative for preferential dissolution are made. Instead dissolution occurs homogeneously over the entire surface. No significant precipitation of any secondary phase is indicated in the morphological data or in the current and mass LSV data.

The UO_{2+x}/Pd ($x = 0.12$) film is also investigated by XPS directly after production and after the initial corrosion potential measurement. A O/U ratio of

2.5 is determined following initial corrosion. In addition, a U-4*f* signal with a significant satellite structure at about 8.9 eV higher BE as the main peak is observed. Determination of the oxidic O-1*s* corresponding to the U–O solid phase is difficult due to overlap of water/hydroxyl species with the oxidic O-1*s* signal, thereby introducing uncertainties in calculated O_{oxidic}/U ratios from peak fitting. However, we may still conclude that the starting point for the LSV experiment can be assumed to be of $UO_{2.33}$ composition or higher for the reaction equation above.

This example shows that electrochemical methods are well suited to investigate the corrosion behavior of the thin films. Our thin film approach, for which the EQCM set-up can be applied, specifically helps in identification of reaction mechanisms. This has been shown in investigation of the well-known system of pure UO_2 [21]. Other complementary methods such as AFM (as shown above) can help to support findings from electrochemical investigations.

In further studies, CV is used to determine specific differences between the pure UO_2 system and the Pd-doped UO_2 in the same electrolyte system, as described above (aerated 0.01 M NaCl solutions, pH 5–6). The CV results for the pure UO_2 film electrodes (Fig. 23, left) are comparable with previous results for bulk UO_2 electrodes [57, 58].

The distinct features identifiable in the cyclovoltagramms are (A) the reversible surface oxidation, (C) a superimposed bulk oxidation and dissolution, (D) a strong dissolution reaction in the region above 0.5 V, and (E) reprecipitation of material under reductive conditions (potential region 0.25 to −0.5 V). Region F at very reductive potentials is possibly reduction of the matrix back to near stoichiometric UO_2, as it is accompanied not by a distinct mass change, but only a slight mass decrease.

Initial results for the Pd-doped UO_2 systems show significant differences from this behavior (Fig. 23, right). The total mass changes are significantly lower for the doped system (at least for the early scans). Also, the distinct features in the current signal show some differences compared to the pure system: features C and D are strongly suppressed for early scans (solid, black curve), but they increase with successive scans (dashed, black curve). Feature E, observed with suppressed current, is found to increase with successive scans as well. An additional feature (e^1) is observed in the region ~0.55 V with a slight mass increase, but a much more significant reduction current than for the pure system. This feature increases with the Pd content of the system. In principle, one can say that with increasing Pd doping (2–50%), a stronger deviation from UO_2 behavior is observed, which does not resemble the pure Pd system (light gray curve). The redox properties of Pd are discussed in the literature [59–63], where the CV features are attributed to the kinetically controlled adsorption of solvent components (H_2O, O_2, H_2) onto the noble metal surface. Hereby, adsorption is described as being dissociative, and the adsorbed species (H_{ads}, O_{ads}, OH_{ads}) are reported to undergo further surface reactions including oxidation, reduction and recombination reactions. Our results indicate an inhibitory effect of the Pd particles on the dissolution of UO_2. With increasing number of scans, however, the electrode response increasingly

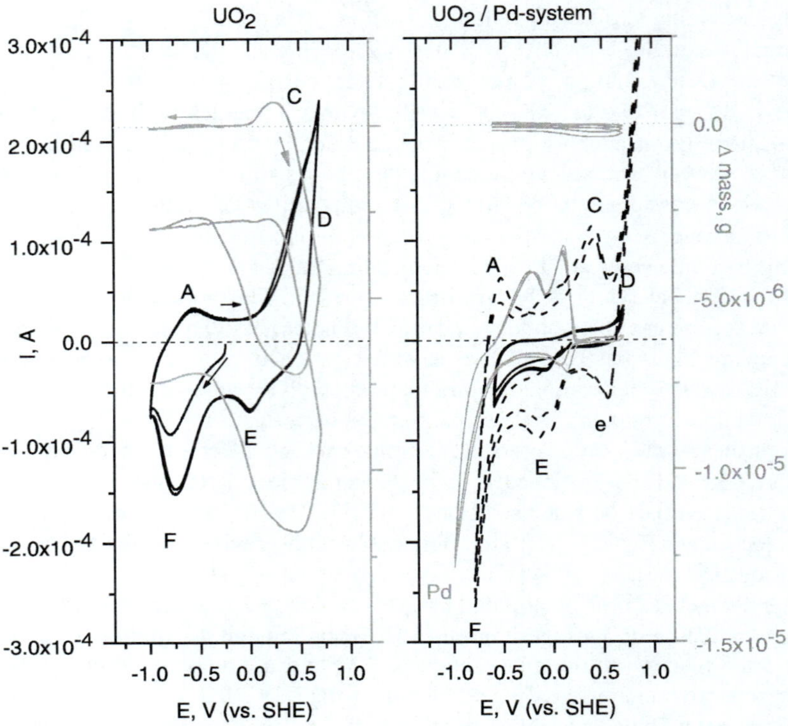

Fig. 23 CV results for pure UO_2 (*left graph*) and UO_2/Pd (16%) (*right graph*) thin film electrodes. *Black*: current response (*left axis*), *gray*: mass signal for early scans (*right axis*). For comparison, CV results for a pure Pd thin film electrode are additionally given (*light gray* current signal). Scan velocity is 10 mV s^{-1}, potentials are referenced versus standard hydrogen electrode, SHE. See text for details of marked regions

resembles the pure UO_2 case, except for the distinct feature seen in region e′. This effect indicates passivation of Pd particle dissolution inhibition. Shoesmith et al. [58, 64] propose the formation of secondary phases as being responsible for blocking of donor–acceptor sites at the UO_2 surface. In fact, the precipitation of such phases is directly revealed by the mass increase detected on the electrodes in region E. Also, AFM measurements in a combined AFM/electrochemistry set-up (ECAFM) [7] show this very distinctly.

From our CV results, we can deduce a general impact of Pd on the overall redox behavior of UO_2 as follows: In an initial reaction phase, prior to the buildup of secondary phases on the UO_2 surface, Pd seems to inhibit dissolution of the UO_2 surface. Also, specific redox reactions are observed that are not seen for the pure system. These are possibly related to the dissociative adsorption of solvent components onto the noble metal surface and their further reaction.

3.2.2 In-Situ Electrochemistry + AFM

An interesting method to visualize the processes on a nanoscopic-scale is the combination of AFM with electrochemical methods, which was recently used for bulk UO_2 electrodes [7] and is now applied to our thin film electrodes. With this electrochemical AFM (ECAFM) method, local dissolution processes can be directly observed under the electrochemical conditions of interest. First ECAFM experiments with the UO_2/Pd thin film electrodes in aerated 0.01 M NaCl, pH ~6 showed good agreement with the results from CV experiments with the EQCM.

AFM images obtained in an electrochemical experiment are displayed in Fig. 24. At a potential of −1 V, the electrode shows a low structured surface, which is supposed to be due to a reprecipitated hydrolyzed film at this low potential. Application of a 0.3 V potential on the electrode results in partial dissolution of the surface, which becomes more pronounced with a further potential increase up to 0.5 V. The AFM image in this case clearly reveals that some particulate structures do not dissolve during the oxidative scan. These two features (to the very left part of the images) remain fixed on the surface and are likely the Pd ε-particle simulate (characterized by SEM–EDX measurements; not shown here).

The application of a negative potential induces a massive precipitation of secondary phases with no distinct crystallite structures, which also cover the metallic Pd particles. Such precipitates are composed of hexavalent uranium as

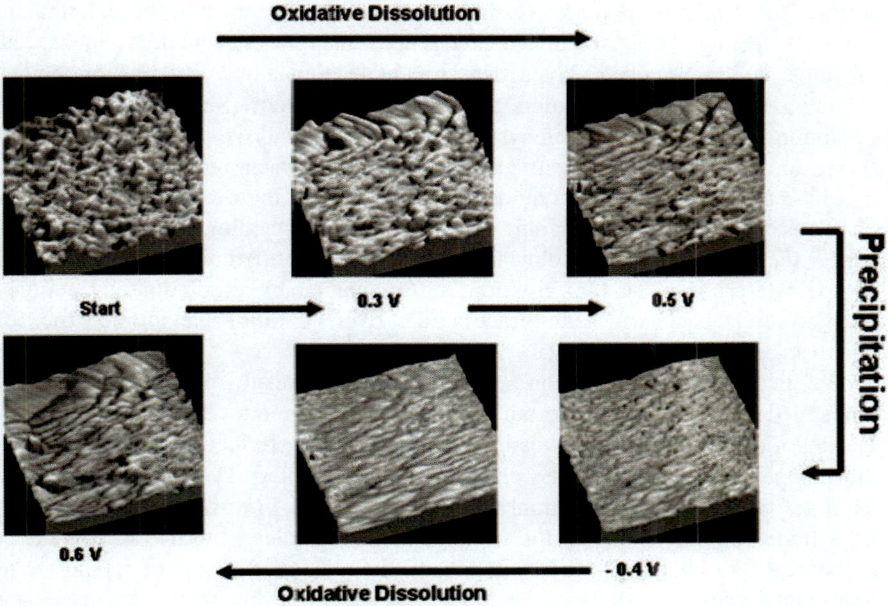

Fig. 24 AFM images of a UO_2/Pd (~16%) thin film monitored for a redox scan from −1 V up to 0.6 V (measured versus a Pt reference electrode). The edge length of the AFM scans is 5 μm

shown by XPS analysis on similar samples (not shown). A subsequent oxidative scan results in the dissolution of the poorly crystalline precipitate, leaving a structured uranium oxide surface and reappearing Pd-particles. It becomes clear that the precipitation of these secondary phases not only leads to Pd coverage but also masks the inhibitory effect of Pd particles on surface dissolution. Moreover, this precipitate is the main surface component in contact with the solution involved in oxidation/dissolution, as well as reduction/precipitation processes. This means that this secondary phase, not the matrix UO_2, ultimately determine the electrochemical behavior of the surface at this stage. This is in accord with similar observations in the CV experiments described above.

4 Conclusions and Outlook

This chapter provides the reader with an introductory overview of the utility and versatility of thin layer (co-)deposition of actinide-containing systems. This is, to our knowledge, the first such overview published. A number of examples are presented, which demonstrate that by varying the elemental targets and the experimental parameters (including pressure, temperature, target voltages, etc.) thin films with widely varying characteristics such as, e.g., thickness, crystallinity, composition can be produced. These thin films can be used to investigate reactivity, structure, and bonding of surfaces and to characterize surface/interfacial reactions. In the context of nanoparticles, surfaces and interfaces are often the determinate feature effecting their fate. In a wider sense, surfaces (as junctions, interfaces, coatings, coverings, etc.) play a determinant role in a whole range of themes spanning economical standpoints to environmental ones. Surface reactions are predominant in a plethora of systems including catalysis, solution/dissolution (corrosion), redox reactions, to name just a few. Systematic investigations of specified thin layer systems as models for surfaces and their reactions allows their characterization, thereby providing mechanistic understanding, and their quantification, thereby ultimately yielding access to kinetic and thermodynamic parameters for their description. This information, in turn, can enable prediction of reactions/reaction products at surfaces/interfaces under given conditions, an ultimate tool for chemists and material scientists.

We include in our discussion a short summary of results on ε-particle models embedded in UO_2. This is important for studying noble metals in spent nuclear fuel. Related films as models for investigating other fission products or other fuel types are planned. This study can be considered an exemplary system, which can be extended to nearly an infinite number of model systems. For example, investigation of actinide thin films with specific embedded nanoparticles, or thin films containing embedded actinide nanoparticles, or nanoparticles situated on top of crystalline or amorphous substrates or embedded between matrix layers (the possibilities are virtually endless) can be envisaged. The greatest challenge herein lies in the thin film synthesis and characterization.

Acknowledgments S. Stumpf would like to thank the European Commission for a grant in the framework of the program "Mobility for young researchers." The research leading to part of these results received funding from the European Atomic Energy Community Seventh Framework Program (FP7/2007-2013) under grant agreement No 212287, Collaborative Project Recosy.

References

1. Somorjai GA, Park JY (2009) Concepts, instruments, and model systems that enabled the rapid evolution of surface science. Surf Sci 603:1293–1300
2. Somorjai GA, Contreras AM, Montano M, Rioux RM (2006) Clusters, surfaces, and catalysis. PNAS 103:10577–10583
3. Bubert H, Riviere JC (2002) Photoelectron spectroscopy. In: Bubert H, Jenett H (eds.) Surface and thin film analysis, 1st edn. Wiley-VCH, Weinheim
4. Briggs D, Seah MP (1990) Practical surface analysis. Wiley, Chichester
5. Seah MP, Dench WA (1979) Quantitative electron spectroscopy of surfaces: a standard data base for electron inelastic mean free paths in solids. Surf Interf Anal 1:2–11
6. Yeh JJ, Lindau I (1985) Atomic subshell photoionisation cross-sections and asymmetry parameters: $1 \leq Z \leq 103$. Atomic Data Nucl Data Tables 32:1–155
7. Römer J, Plaschke M, Beuchle G, Kim JI (2003) In situ investigation of U(IV)-oxide surface dissolution and remineralization by electrochemical AFM. J Nucl Mater 322:80–86
8. Horcas I, Fernandez R, Gomez-Rodriguez JM, Colchero J, Gomez-Herrero J, Baro AM (2007) WSXM: a software for scanning probe microscopy and a tool for nanotechnology. Rev Sci Instrum 78:013705
9. Stumpf S, Seibert A, Gouder T, Huber F, Wiss T, Römer J (2009) Development of fuel-model interfaces: investigations by XPS, TEM, SEM and AFM. J Nucl Mater 385:208–211.
10. Teterin YuA, Kulakov VM, Baev AS, Nevzorov NB, Melnikov IV, Streltsov VA, Mashirov LG, Suglobov DN, Zelenkov AG (1981) A study of synthetic and natural uranium oxides by X-ray photoelectron spectroscopy. Phys Chem Min 7:151–158
11. van den Berghe S, Laval JP, Gaudreau B, Terryn H, Verwerft M (2000) XPS investigationson cesium uranates: mixed valency behaviourof uranium. J Nucl Mater 277:28–36
12. Bera S, Sali S, Sampath SV, Narasimhan V, Venugopal V (1998) Oxidation state of uranium: an XPS study of alkali and alkaline earth uranates. J Nucl Mater 255:26–33
13. Santos BG, Nesbitt HW, Noel JJ, Shoesmith DW (2004) X-ray photoelectron spectroscopy study of anodically oxidized SIMFUEL surfaces. Electrochim Acta 49:1863–1873
14. Colmenares CA (1984) Oxidation mechanisms and catalytic properties of the actinides. Prog Solid State Chem 15:257–364
15. Allen GC, Tucker PM, Lewis RA (1984) X-ray photoelectron spectroscopy study of the initial oxidation of uranium metal in oxygen-water-vapour mixtures. J Chem Soc Faraday Trans 80:991–1000
16. Wagner CD, Naumkin AV, Kraut-Vass A, Allison JW, Powell CJ, Rumble JR Jr (2003) NIST X-ray photoelectron spectroscopy database NIST. Standard Reference Database 20, Version 3.5. http://srdata.nist.gov/xps
17. Yamazaki T, Kotani A (1991) Systematic analysis of 4f core photoemission spectra in actinide oxides. J Phys Soc J 60:49–52
18. Tura JM, Regull P, Victori L, de Castellar DM (1988) XPS and IR (ATR) analysis of Pd oxide films obtained by electrochemical methods. Surf Interf Anal 11:447–449
19. Genesca J, Victori L (1981) Titel. Rev Coat Corros 4:325–348
20. Sauerbrey GZ (1959) Verwendung von Schwingquarzen zur Wägung dünner Schichten und zur Mikrowägung. Zeitsch Phys 155:206–222

21. Seibert A, Wegen DH, Gouder T, Huber F, Römer J, Wiss T, Glatz J-P (2010) The use of the electrochemical quartz crystal microbalance (EQCM) in corrosion studies of UO_2 thin film models. RA 98:647–654
22. Gouder T, Havela L, Wastin F, Rebizant J (2001) Evidence for the 5 f localization in thin Pu layers. Europhys Lett 55:705–711
23. Gouder T, Wastin F, Rebizant J, Havela L (2000) 5-f electron localization in PuSe and PuSb. Phys Rev Lett 84:3378–3381
24. Gouder T, Eloirdi R, Rebizant J, Boulet P, Huber F (2005) Multiplet structure in Pu-based compounds: a photoemission case study of PuSix ($0.5<x<2$) films. Phys Rev B 71:165101
25. Eloirdi R, Gouder T, Korzhavyi P, Wastin F, Rebizant J (2004) Dilution effect on the U-5f states: U in an Ag matrix. J Alloys Compd 386:70–74
26. Naegele JR, Ghijsen J (1985) Localisation and hybridisation of 5f States in the metallic and ionic bond as investigated by photoelectron spectroscopy. In: Manes L (ed.) Actinides – chemistry and physical properties, structure and bonding 59/60. Springer, Berlin
27. Gouder T, Havela L, Rebizant J (2005) Photoemission study of δ-plutonium with Am doping. Physica B 359–361:1090–1092
28. Gouder T, Seibert A, Havela L, Rebizant J (2007) Search for higher oxides of Pu: a photoemission study. Surf Sci 601:L77–L80
29. Seibert A, Gouder T, Huber F (2009) Formation and stability of actinide oxides: a valence band photoemission study. Radiochim Acta 97:247–250
30. Zangwill A (1988) Physics at surfaces. University Press, Cambridge
31. Wandelt K (1982) Photoemission studies of adsorbed oxygen and oxide layers. Surf Sci Rep 2:1–21
32. Naegele JR, Cox LE, Ward JW (1987) Photoelectron spectroscopy (UPS/XPS) study of Np_2O_3 formation on the surface of neptunium metal. Inorg Chim Acta 139:327
33. Seibert A, Gouder T, Huber F (2009) Reaction of Neptunium with molecular and atomic oxygen: formation and stability of surface oxides. J Nucl Mater 389:470–478
34. Seibert A, Gouder T, Huber F (2010) Interaction of PuO_2 thin films with water. Radiochim Acta 98:647–654. doi: 10.1524/ract.2010.1765
35. Stumpf S, Seibert A, Gouder T, Huber F, Wiss T, Römer J, Denecke MA (2010) Development of fuel-model interfaces: characterisation of Pd containing UO_2 thin films. J Nucl Mater 397:19–26
36. Carbol P, Fors P, Gouder T, Spahiu K (2009) Hydrogen suppresses UO_2 corrosion. Geochim Cosmochim Acta 73:4366–4375
37. Nilsson S, Jonsson M (2008a) On the catalytic effects of UO_2(s) and Pd(s) on the reaction between H_2O_2 and H_2 in aqueous solution. J Nucl Mater 372:160–163
38. Nilsson S, Jonsson M (2008b) On the catalytic effect of Pd(s) on the reduction of UO_2^{2+} with H_2 in aqueous solution. J Nucl Mater 374:290–292
39. Trummer M, Nilsson S, Jonsson M (2008) On the effects of fission product noble metal inclusions on the kinetics of radiation induced dissolution of spent nuclear fuel. J Nucl Mater 378:55–59
40. Trummer M, Roth O, Jonsson M (2009) H_2 inhibition of radiation induced dissolution of spent nuclear fuel. J Nucl Mater 383:226–230
41. Stakebake JL, Larson DT, Haschke JM (1993) Characterisation of the plutonium-water reaction II: formation of a binary oxide containing Pu(VI). J Alloys Compd 202:251–263
42. Farr JD, Schulze RK, Neu MP (2004) Surface chemistry of Pu oxides. J Nucl Mater 328:124–136
43. Waber JT (1967) Corrosion and oxidation. In: Wick OJ (ed.) Plutonium handbook–a guide to technology, Vol. 1. Gordon and Breach, New York
44. Colmenares CA (1975) The oxidation of Thorium, Uranium and Plutonium. Prog Solid State Chem 9:139–239
45. Haschke JM, Allen TH, Morales LA (2000) Reaction of plutonium dioxide with water: formation and properties of PuO_{2+x}. Science 287:285–287

46. Haschke JM, Allen TH (2002) Equilibrium and thermodynamic properties of the PuO2+x solid solution. J Alloys Compd 336:124–131
47. Neck V, Altmaier M, Fanghänel Th (2007) Thermodynamic data for hydrous and anhydrous PuO_{2+x}(s). J Alloys Comp 444–445:464–469
48. Neck V, Altmaier M, Seibert A, Yun JI, Marquardt CM, Fanghänel Th (2007) Solubility and redox reactions of Pu(IV) hydrous oxide: evidence for the formation of PuO_{2+x}(s, hyd). Radiochim Acta 95:193–207
49. Veal BW, Lam DJ, Diamond H, Hoekstra HR (1977) X-ray photoelectron spectroscopy study of oxides of the transuranium elements Np, Pu, Am, Cm, Bk, and Cf. Phys Rev B 15:2929–2942
50. Dupin J-C, Gonbeau D, Vinatier P, Levasseur A (2000) Systematic XPS studies of metal oxides, hydroxides and peroxides. Phys Chem Chem Phys 2:1319–1324
51. Henderson MA (2002) The interaction of water with solid surfaces: fundamental aspects revisited. Surf Sci Rep 46:1–308
52. Spinks JWT, Woods RJ (1990) An introduction to radiation chemistry. Wiley, New York
53. Cooper MJ (1982) The analysis of powder diffraction data. Acta Crystallogr B 38:264–269
54. Gronvold FH (1948) Crystal structure of uranium oxide. (U_3O_8) Nature 162:69–70
55. Kim KS, Gossman AF, Winograd N (1974) X-ray photoelectron spectroscopic studies of palladium oxides and the palladium-oxygen electrode. Anal Chem 46:197–200
56. Bolzan AE, Chialvo AC, Arvia AJ (1984) Fast faradaic processes observed during the potentiodynamic polarization of polycrystalline palladium in acid electrolyte. J Electroanal Chem 179:71–82
57. Miserque F, Gouder T, Wegen DH, Bottomley PDW (2001) Use of UO_2 films for electrochemical studies. J Nucl Mater 298:280–290
58. Shoesmith DW, Sunder S, Hocking WH (1994) Electrochemistry of UO_2 nuclear fuel. In: Lipkowski J, Ross PN (eds.) Electrochemistry of novel materials. VCH, New York
59. Kibler LA, El-Aziz AM, Kolb DM (2003) Electrochemical behaviour of pseudomorphic overlayers: Pd on Au(111). J Mol Catal A Chem 199:57–63
60. Martin MH, Lasia A (2008) Study of the hydrogen absorption in Pd in alkaline solution. Electrochim Acta 53:6317–6322
61. Duncan H, Lasia A (2007) Mechanism of hydrogen adsorption/absorption at thin Pd layers on Au(111). Electrochim Acta 52:6195–6205
62. Naohara H, Ye S, Uosaki K (2000) Electrocatalytic reactivity for oxygen reduction at epitaxilly grown Pd thin layers of various thickness on Au(111) and Au(100). Electrochim Acta 45:3305–3309
63. Cao Y, Chen Z-X (2006) Theoretical studies on the adsorption and decomposition of H_2O on Pd(111) surface. Surf Sci 600:4572–4583
64. Shoesmith DW (2000) Fuel corrosion processes under waste disposal conditions. J Nucl Mater 282:1–31

Sorption and Speciation of Uranium on Silica Colloids

Dmitry N. Batuk, Andrei A. Shiryaev, Stepan N. Kalmykov, Olga N. Batuk, Anna Yu. Romanchuk, Eugeni A. Shirshin, and Yan V. Zubavichus

Abstract Sorption behavior of uranium onto silica colloids is studied to derive surface complexation data. Several spectroscopic and structural methods, i.e., Time-resolved Laser-induced Fluorescence (TRLIF), X-ray Absorption Fine Structure (XAFS), and Small-angle X-ray Scattering (SAXS) are employed for characterization of colloid suspensions and surface-sorbed species. TRLIF and EXAFS allowed identification of two principal uranium species on the silica surface – bidentate coordinated uranyl complexes of the following stoichiometries $(\equiv SiO)_2UO_2$ and $(\equiv SiO)_2UO_2OH^-$. No polynuclear species or surface precipitates are formed as shown by EXAFS and SAXS examination of suspensions at various silica/uranium ratios. However, when uranium coverage of the silica surface exceeds 7%, the colloid stability as determined by electrophoretic mobility increases. Potentially, this could affect colloid-facilitated transport.

1 Introduction and Theoretical Background

Sorption of radionuclides onto rocks, soils, bottom sediments, colloids, etc. is the main chemical reaction that governs their migration behavior. It is commonly characterized by the distribution coefficient, K_d. However, the use of this parameter

D.N. Batuk, S.N. Kalmykov (✉), O.N. Batuk, and A. Yu. Romanchuk
Chemistry department, Lomonosov Moscow State University, Leninskie Gory, 119991 Moscow, Russia
e-mail: stepan@radio.chem.msu.ru

A.A. Shiryaev
Frumkin Institute of Physical chemistry and electrochemistry, RAS, Leninsky av., 31, 119991 Moscow, Russia

E.A. Shirshin
Physics department, Lomonosov Moscow State University, Leninskie Gory, 119991 Moscow, Russia

Y.V. Zubavichus
Kurchatov Institute, Academician Kurchatov sq., 1, 123182 Moscow, Russia

S.N. Kalmykov and M.A. Denecke (eds.), *Actinide Nanoparticle Research*,
DOI 10.1007/978-3-642-11432-8_11, © Springer-Verlag Berlin Heidelberg 2011

to describe sorption of trace concentration of radionuclides could result in erroneous interpretations. At very low surface coverage, where the electrostatic repulsion between sorbing cations in the electric double layer could be neglected, the sorption does not depend on the sorbent to sorbate ratios in the broad range of this ration. At the same time, variations of this ratio lead to changes of the K_d.

Another approach to describe sorption is by using surface complexation modeling (SCM) based on the application of mass action laws to the system under consideration. Many studies rely on this method, which is also reviewed in several reports [1–6]. However, no sufficiently accurate database of equilibrium constants for sorption reactions on various mineral phases that could be used by performance assessment engineers exists. By combining equilibrium constants of reactions that radionuclides undergo in suspension, the K_d for any given system can theoretically be calculated. The simple chemical system involving cation hydrolysis in solution and its sorption onto a mineral surface with formation of a monodentate species is presented below with corresponding expressions for their equilibrium constants and the resulting K_d:

Surface reactions

Acid–base equilibrium:

$\equiv XOH + H^+ \leftrightarrow \equiv XOH_2^+$ $$K_{a1} = \frac{[\equiv XOH_2^+]}{[\equiv XOH][H^+]}\exp(-F\Psi/RT)$$

$\equiv XOH \leftrightarrow \equiv XO^- + H^+$ $$K_{a2} = \frac{[\equiv XO^-][H^+]}{[\equiv XOH]}\exp(-F\Psi/RT)$$

Formation of monodentate complexes on the surface:

$Met^{3+} + \equiv XOH \leftrightarrow \equiv XOMet^{2+} + H^+$ $$\beta_1 = \frac{[\equiv XOMet^{2+}][H^+]}{[Met^{3+}][\equiv XOH]}\exp(-F\Psi/RT)$$

$MetOH^{2+} + \equiv XOH \leftrightarrow \equiv XOMetOH^+ + H^+$ $$\beta_2 = \frac{[\equiv XOMetOH^+][H^+]}{[MetOH^{2+}][\equiv XOH]}\exp(-F\Psi/RT)$$

$Met(OH)_2{}^+ + \equiv XOH \leftrightarrow \equiv XOMet(OH)_2 + H^+$ $$\beta_3 = \frac{[\equiv XOMet(OH)_2][H^+]}{[Met(OH)_2^+][\equiv XOH]}\exp(-F\Psi/RT)$$

where $\equiv XOH$, $\equiv XOH_2{}^+$, $\equiv XO^-$ denote neutral, protonated, and deprotonated surface sites

Reactions in solution

Cation hydrolysis:

$Met^{3+} + OH^- \leftrightarrow MetOH^{2+}$ $$\beta_4 = \frac{[MetOH^{2+}]}{[Met^{3+}][OH^-]}$$

$Met^{3+} + 2OH^- \leftrightarrow Met(OH)_2{}^+$ $$\beta_5 = \frac{[Met(OH)_2^+]}{[Met^{3+}][OH^-]^2}$$

$Met^{3+} + 3OH^- \leftrightarrow Met(OH)_3$ $$\beta_6 = \frac{[Met(OH)_3]}{[Met^{3+}][OH^-]^3}$$

Expression for K_d:

$$K_d = \frac{[\equiv X-OH]\{\beta_1 + \beta_2\beta_4[OH^-] + \beta_3\beta_5[OH^-]^2\}}{[H^+]\{1 + \beta_4[OH^-] + \beta_5[OH^-]^2 + \beta_6[OH^-]^3\}}\exp(F\Psi/RT)$$

where K_d is the distribution coefficient, describing the metal retention to the surface under particular conditions.

Regardless of its complexity, any chemical system in which several sorption and solution complexation reactions occur could be described by this approach. However, the absence of reliable surface complexation data limits its application in performance assessment.

This paper describes application of the surface complexation approach to describe sorption of uranium onto silica colloids. This study is performed for the needs of performance assessment at the liquid nuclear waste injection site at Siberian Chemical Combine, SCC. The wastes at SCC are injected at two sites [7, 8]. Low-level radioactive wastes are deposited at the first site in the geological horizons II and III located at the depths of 349–386 m and 270–320 m below sea level, respectively. Intermediate-level radioactive wastes are deposited at the second site in the horizon II located at the depths of 314–341 m. The reservoir horizons are composed of quartz (40–70%), clay minerals belonging to the illite, montmorillonite and kaolinite groups (10–40%), feldspars (5–20%); micas, chlorites, ferric oxides, and hydroxides; carbonates and sulfides are also present. Weakly cemented permeable sand–clay rocks are separated by clay strata that prevent vertical migration of waste components. The distance from the center of the site to its boundaries is 1.6–2.0 km. The types of liquid nuclear wastes that are injected are alkaline low-level wastes with pH 8.0–10.5 [8], neutral low-level liquid nuclear wastes with a total salt concentration below 1 g/L, and acidic intermediate-level wastes, with pH 2–3. Fission products such as ^{90}Sr, ^{134}Cs, ^{137}Cs, ^{95}Zr, ^{95}Nb, ^{103}Ru, ^{106}Ru, ^{141}Ce, ^{144}Ce, as well as trace amounts of uranium and $^{239}Pu(IV)$, $^{241}Am(III)$, and $^{237}Np(V)$ are present [8]. Upon the interaction of the wastes with quartz sands, amorphous silica is formed, which could be responsible for the radionuclide sorption.

For uranium sorption, different spectroscopic speciation methods such as TRLIF [9–11], FT-IR spectroscopy [12, 13], EXAFS [11, 13, 14], and XPS [15] could be applied because of the high solubility of U(VI) and the possibility of performing sorption experiments at relatively high total metal concentration. This would be impossible for Pu(IV) or other low soluble radionuclides. In the current paper, we apply TRLIF as a spectroscopic tool to study U(VI) speciation on the silica surface, SAXS to assess possible surface precipitations of U(VI) hydroxide at high metal concentration, and EXAFS to determine the local atomic environment for uranium on the surface.

Silicon-bearing minerals comprise the major part of Earth's crust, and SiO_2 accounts for nearly 60% of it; this justifies the interest in sorption of radionuclides onto silica. However, the available information on the molecular-level speciation of radionuclides on silica is limited. Both spectral studies of uranium on silica and conventional sorption techniques have been used by several research groups to provide data for the modeling. However, certain contradictions can be found and ambiguity remains concerning both the structure of surface complexes and their stability constants [5, 6, 9, 11, 13, 14].

The goal of our work is to study the interaction of U(VI) with silica nanospheres by a combination of conventional batch sorption method with various spectroscopic methods (TRLIF, EXAFS, SAXS).

2 Experimental

2.1 Colloidal Amorphous Silica Spheres Preparation and Characterization

The colloidal amorphous silica spheres (CASS) are synthesized by the Stöber reaction [16] by hydrolysis of tetraethoxysilane, TEOS (Merch, 98% grade), in the presence of ammonia (analytical grade) in ethanol. This reaction enables synthesis of spherical colloidal silica particles with a narrow size distribution, depending on the temperature, reagent concentrations, and ratio. The procedure is as follows: 3 ml of TEOS is added to the reaction vessel with 62 ml of stirred ethanol, 30 ml of deionized water (18 MΩ/cm), and 2.5 ml of ammonia. The suspension is continuously stirred during 2 h at 25°C; the synthesized CASS are then purified by dialysis using 10 kDa membrane (Servapor) against deionized water, which is renewed numerous times to reach a dilution of about 10^{10}.

The size and surface morphology of the synthesized CASS are characterized by transmission electron microscopy (TEM) using LEO 912 AB OMEGA microscope (at 120 kV acceleration voltages); some images are shown in Fig. 1. The TEM study confirms uniform spherical shape of the particles. According to selected area electron diffraction (SAED), no crystalline phases are present in the samples. During the dialysis, partial dissolution of CASS occurs, resulting in a roughening

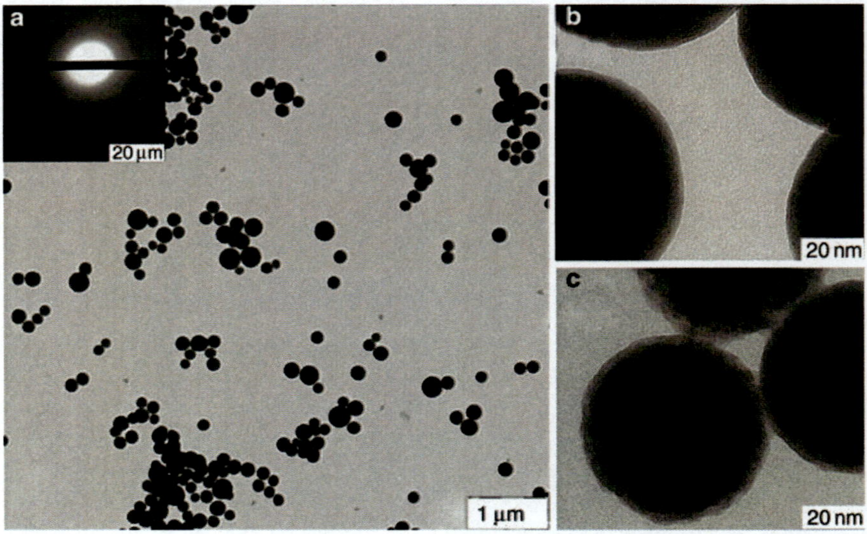

Fig. 1 The images of the synthesized CASS (**a**) TEM image and SAED pattern showing the absence of crystalline phases, (**b**) TEM image prior to the dialysis, (**c**) TEM image after the dialysis

Table 1 The main characteristic of the synthesized CASS

Parameter	Value
Average particle size, nm	80 ± 18
Free surface area, m^2/g	95 ± 8
pHiep	4.36

of initially smooth surface. This is evidenced by TEM images taken with a higher magnification (Fig. 1b, c).

The main parameters of CASS are given in Table 1. The particle size distribution is calculated using TEM images, the free surface area is determined by BET technique (Quantachrome NOVA 4200e), and the value of the pH of isoelectric point (pHiep) is measured by zeta potential titration (Malvern ZETASIZER nano-ZS).

2.2 Sorption Experiments

^{233}U with purity controlled by alpha spectrometry (Si PIPS detectors) and standardized by alpha liquid scintillation spectrometry (Tri-Carb 2700TR, Packard) is used in all sorption experiments.

The sorption of U(VI) onto the CASS surface is studied in a batch mode at 25 ± 1°C in plastic vials. All the experiments are handled in N_2-atmosphere glove box to avoid CO_2 contamination, and all the solutions are additionally purged with nitrogen. The pH values are measured using SevenEasy pH S20-K (Mettler Toledo) ionomer equipped with the combination electrode InLab Expert Pro (Mettler Toledo) and adjusted by diluted NaOH and $HClO_4$ solutions.

A number of test experiments have shown that the steady state conditions are reached faster than 30 min of equilibration; therefore, we set the duration of all further experiments to 30 min. Various sets of sorption experiments are performed at different uranium/silica ratios and various pH values (pH sorption edges). The phase separation after sorption is performed by centrifugation at 25,000 g using Allegra 64R (Beckman Coulter) centrifuge for 10 min. After centrifugation, an aliquot of the solution is taken for the liquid scintillation counting of ^{233}U that remains unsorbed. The percentage of sorbed uranium (Γ) is calculated as the difference between the initial (I_0) and the measured (I) activities of ^{233}U:

$$Q = \frac{I_0 - I}{I} \times 100.$$

2.3 Small Angle X-Ray Scattering

In order to perform SAXS measurements on the aged CASS suspensions with uranium, the suspensions are concentrated by high speed centrifugation at 20,000 rpm. About 10–15 μL of concentrated suspension is placed into a 1 mm glass capillary (wall thickness 10 μm). The measurements are performed using a dedicated AMUR-K diffractometer equipped with linear position sensitive detector OD3. Monochromatic Cu Kα_1 radiation and Kratki collimation scheme are employed. The obtained SAXS data are desmeared using standard routines. The size distribution of the scatterers is calculated in the hard sphere approximation using GNOM [17] and Polymix [18] software.

2.4 Uranium Speciation on CASS Surface by TRLIF

TRLIF investigation is conducted on the suspensions placed to quartz cell. The samples are excited with the fourth harmonic (266 nm) of a Nd:YAG pulsed laser (pulse length 10 ns, repetition rate 1 kHz, power 1 mW). Luminescence is registered using an optical multichannel analyzer (a CCD camera attached to a polychromator) with 0.2 nm spectral resolution. Kinetic curves are obtained with the use of gating system (gate width 1 μs, gate step 100 ns) and then processed using multiexponential model:

$$I(t) = \sum_i A_i \times e^{\frac{-t}{\tau_i}},$$

where A_i and τ_i are the amplitude and excited state lifetime of the ith component, correspondingly.

Up to 100 spectra are recorded for each sample, which allows derivation of the kinetic curves showing the intensity of luminescence versus time.

2.5 Uranium Speciation on Silica Surface by EXAFS

To reach the sensitivity of XAFS measurements, 2 L of SiO_2 suspension is equilibrated with U(VI) using the same reagent ratio as in the sorption experiment and for the TRLIF measurements. The suspensions are equilibrated with U(VI) at pH 4.7, 5.2, 5.6, and 6.5 in order to reach steady state conditions. After equilibration, the suspensions are filtered through 50 nm nuclear pore filters (JINR, Dubna). A wet precipitate from the filter is carefully transferred to the sample holder for XAFS measurements. The sample holder is made from a 5 mm thick aluminum plate with the 2 × 4 mm opening in the center. The samples are sealed by Kapton tape.

The XAFS measurements are performed at the U LIII edge at "Structural Material Science" beamline at the "Kurchatov Center of Synchrotron Radiation and Nanotechnologies" (Moscow, Russia). The Siberia-2 storage ring operates at 2.5 GeV with the current of 50–80 mA. The Si(111) channel-cut crystal is applied to monochromatize X-rays. The spectra are recorded in the transmission mode at room temperature. The ionization energy (E_0), the origin for calculation of the chi(k)-function, is held at the maximum of the most intense absorption feature, the white line (WL) in the individual spectra. In order to extract metric parameters [neighboring atomic distances (*R*), mean square radial displacements or EXAFS Debye- Waller factors, coordination numbers (*N*)] from the recorded data, model backscattering amplitude and phase shift functions for single and double scattering paths are calculated using the theoretical modeling code FEFF8 and optimized as implemented in the IFEFFIT program package [19]. Before fitting data to the EXAFS equation, k^3-weighted EXAFS spectra are Fourier transformed using $\Delta k = 0.1$ Å^{-1} and a hanning window. Fits are performed in *k*-space for all the spectra. The amplitude reduction factor (S02) is held at 0.85. This value reproduces the theoretical coordination number of uranyl oxygen atoms for uranyl solution and the coordination number (*N*) for the U-Oax shell constant at 2 for fits to all sample spectra.

3 Results and Discussion

3.1 pH Sorption Edges

Most of the sorption experiments are performed at a total uranium concentration of 1×10^{-5} and 1×10^{-4} M, which allow us to apply various spectroscopic techniques such as EXAFS, TRLIF, and SAXS for speciation investigations.

The obtained pH sorption edges are S-shaped, typical for cation sorption on oxide/hydroxides (Fig. 2). For a silica concentration of 35 m^2/L and total U(VI) concentration of 10^{-5} M, the pH_{50} (the pH value of 50% sorption) is 4.6. At very low

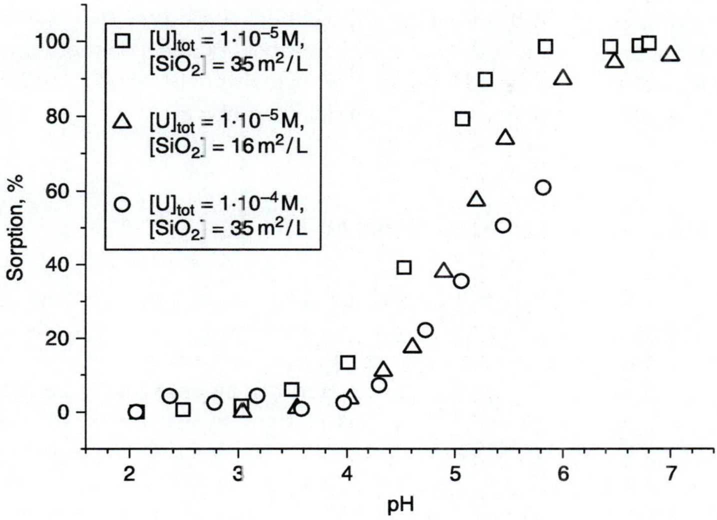

Fig. 2 The pH sorption edges of U(VI) onto silica at various U_{tot}/silica ratio

surface coverage, where the electrostatic repulsion between the sorbing ions can be neglected, the pH_{50} does not depend on the surface to sorbate ratio. In this study, the sorption experiments are done at relatively high surface coverage, reaching 7%. This explains the pH_{50} increase upon the increase of surface to sorbate ratio. The pH sorption edge is shifted to the higher pH values with increasing the U/silica ratio.

Various speciation methods are applied in this work to determine the speciation of sorbed uranium, possible formation of polynuclear species, and the way this would affect the properties of silica colloids, i.e., their elecrophoretic mobility. For the silica suspensions with sorbed uranium, the ζ-potential dependence on pH is studied. At pH <6 at which uranium sorption is not quantitative, the decrease of the ζ-potential with pH follows the same trend as for pure silica colloids (Fig. 3). In contrast, at pH >6, when the maximum uranium sorption is observed, the ζ-potential sharply increases and reaches positive values at pH 6.5. This trend of the ζ-potential dependence with pH is controlled by modification of the silica surface properties due to uranium sorption. This important observation indicates an increase of silica colloid stability at neutral pH and therefore an increase in the potential role of colloidal facilitated transport of elements such as uranium.

Additional information about the U(VI) surface layer on silica colloids is provided by SAXS. This technique provides information about nanoscale heterogeneities of electron density and is complementary to TEM since it probes the whole sample volume, thus providing averaged information. The structural information is extracted from the angular dependence of the scattered radiation. Figure 4 demonstrates the normalized size distribution of scatterers in the initial sample and after its equilibration with U(VI) solution at various pH. For all studied samples, the size distribution of the scattering centers has two maxima. The maximum around 4 nm corresponds to internal heterogeneities of silica colloids, consisting of primary nanoparticles of 8–10 nm in size. The second maximum corresponds to the whole colloidal particles. The distribution of the scatterers depends neither on the pH nor on the amount of sorbed uranium. This indicates that uranium does not form surface precipitates and is distributed homogenously on the surfaces of silica colloids even at pH >6, at which the increase of ζ-potential is observed.

3.2 *Sorbed Uranium Speciation by TRLIF*

The fluorescence spectra are recorded for suspensions with different silica to uranium concentrations and the degree of uranium sorption. The fluorescence intensity strongly correlates with the sorption, as presented in Fig. 5. However, the features of the spectra (peak positions and resolution) remain the same. This behavior is somewhat different from observations by Gabriel et al. [9]. These authors found that the fluorescence spectra of amorphous silica suspensions containing 1 μM total uranyl depend on pH. However, the total uranium concentration in the cited work is one-to-two orders of magnitude lower than in our case, and their spectra have much lower statistics.

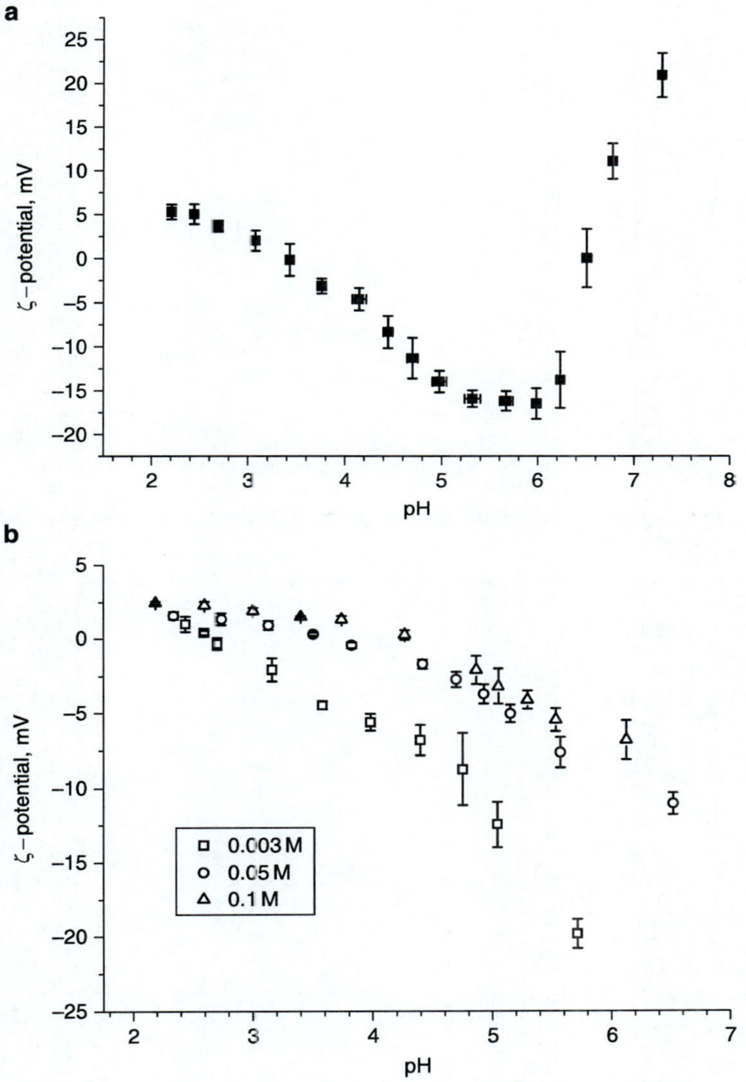

Fig. 3 Measured pH dependencies of ζ-potential for silica colloids (**a**) in presence of U(VI) at total concentration 10^{-5} M, (**b**) in absence of U(VI), at various $NaClO_4$ concentrations in suspension

The recorded fluorescence spectra have the features typical for uranyl complexes (Fig. 6). Upon fluorescence decay, the spectra do not change – the maxima positions and peak resolution remain the same (Fig. 6); only the intensity changes. Two components with lifetimes of 46 ± 9 μs and 200 ± 24 μs are extracted from the decay curve, corresponding to two different uranyl species in the silica surface. The

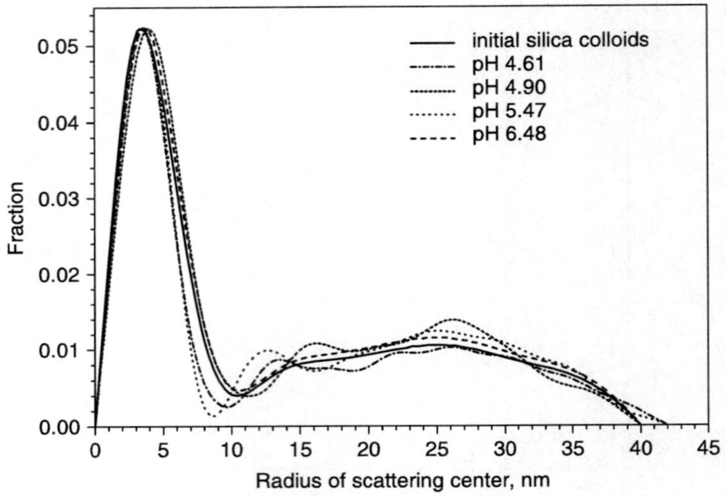

Fig. 4 The normalized size distribution of the scattering objects as determined by small angle X-ray scattering

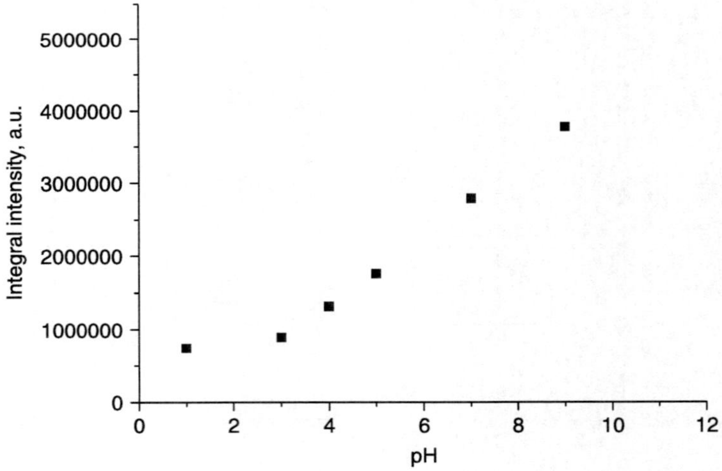

Fig. 5 The correlation of fluorescence intensity with pH of suspension

lifetimes of the two uranium complexes on the surfaces quoted by Gabriel et al. [9] are 170 ± 25 μs and 360 ± 50 μs and are longer than those found here. This discrepancy maybe explained by difficulties in extraction of the lifetimes from multiexponential decays. For example, for the conditions of our experiment (1×10^{-5} M total uranium and 0.087 g/L of silica), the pure UO_2^{2+} species is the dominant complex in the solutions at pH up to 5.5, at which the sorption is ~60%. The intensity of the pure uranyl ion fluorescence is low, the decay life time is only

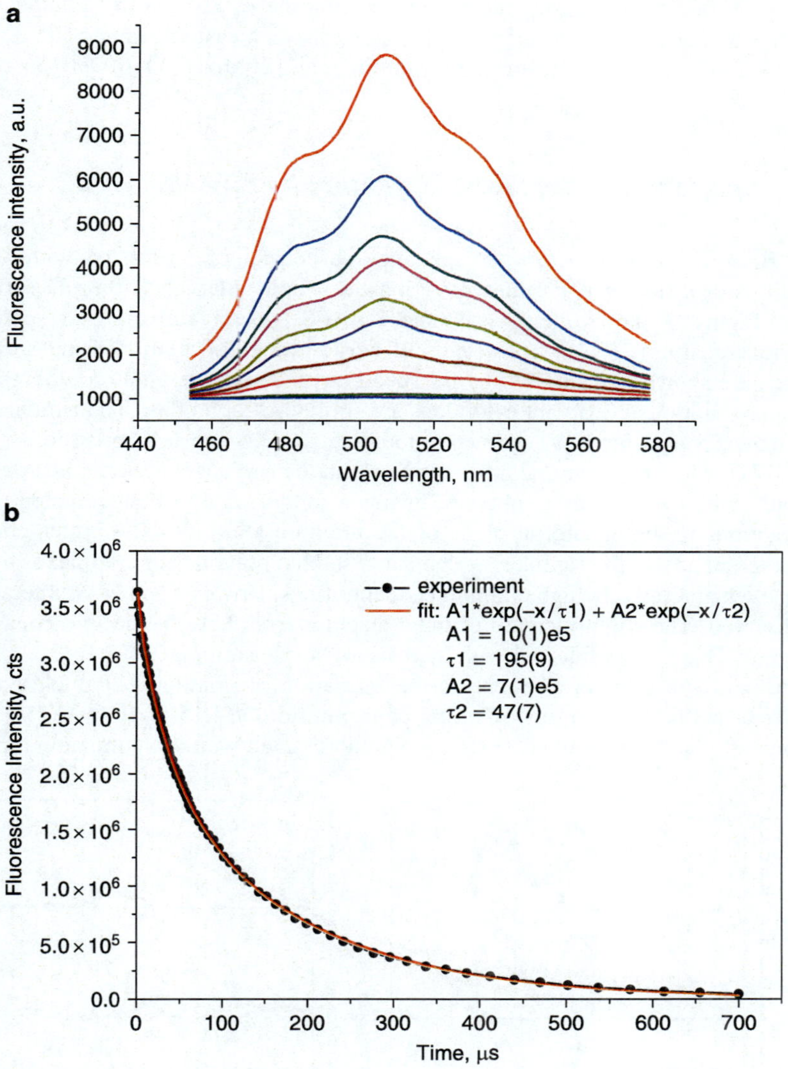

Fig. 6 The decay of fluorescence spectrum of U(VI) surface species. (**a**) Spectra evolution, (**b**) typical biexponential decay curve

2.5 ± 0.1 μs and is negligible compared with much more intense fluorescence from sorbed uranyl-silica species with much longer lifetimes.

We test the same suspension at a total U concentration of 1×10^{-4} M. Under these conditions, the fraction of the $(UO_2)_3(OH)_5{}^+$ complex is significantly larger. This complex has a much higher fluorescence yield and lifetime in comparison with pure uranyl. For the silica colloids equilibrated with 1×10^{-4} M uranium, we observe a fluorescence component with the lifetime of 25 ± 3 μs, together with 170 μs component from the uranyl-silica species. The measured value of 25 ± 3 μs is close to the 23 ± 3 μs reported by Moulin et al. [20] for $(UO_2)_3(OH)_5^+$.

3.3 Speciation of the Sorbed Uranium by EXAFS

The EXAFS study is performed on silica suspensions equilibrated with U(VI) solutions at different pH values covering the whole interval of the pH sorption edge. Figure 7 shows the k^3-weighted EXAFS spectra and the corresponding Fourier transforms (FTs). The results of curve-fitting and extracted metric parameters are shown in Table 2. The FT spectra of the silica samples equilibrated in uranium solution at different pH values are similar to each other. The similarity of LIII EXAFS data for U(VI) sorbed onto silica at pH 3 and 5 is reported by Dent et al. [21], while no structural information about the surface complexes is provided. In our case, the best fit is obtained using a model of two nonequivalent U–O interactions in the equatorial plane of the uranium atom. No U–U interaction is observed in any of the samples, which suggests that polynuclear complexes do not form even at relatively high uranium concentrations. However, no U–Si interaction is observed. The slight increase of bond distance from U to the second equatorial oxygens, Oeq2, is established with increase of suspension pH.

Earlier, Sylwester et al. reported the formation of uranium polynuclear complexes at pH 6.46 and a total uranium concentration of 4.15×10^{-5} M [14]. The authors find an average of ~1 silicon atom coordinated with uranium atom through

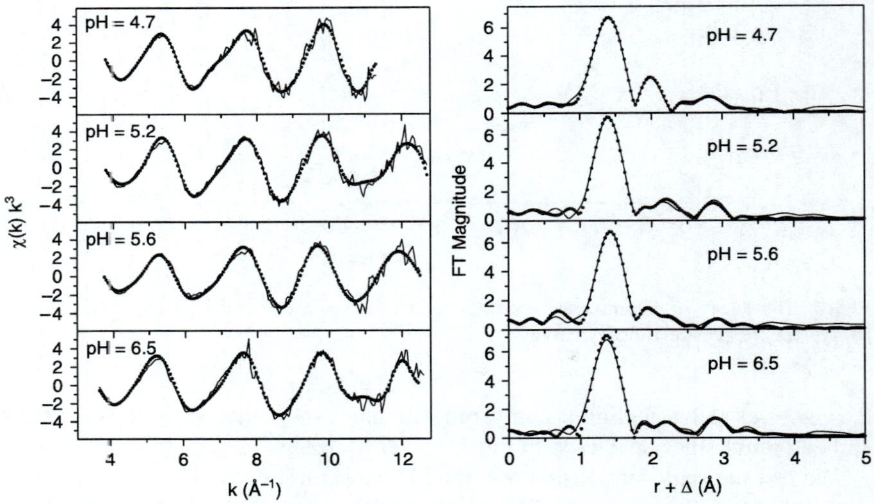

Fig. 7 (**a**) k^3-weighted EXAFS spectra and (**b**) corresponding Fourier transforms (FTs) of indicated samples and fit curves

Table 2 The curve-fitting results

Sample	Coord. shell	N	R, Å	σ^2, Å^2	ΔE_0, eV
4.7	U-Oax	2[a]	1.80	0.0008	−10.9
	U-Oeq1	2.2(6)	2.23	0.0090	
	U-Oeq2	3.2(6)	2.42	0.0076	
5.2	U-Oax	2[a]	1.79	0.0010	−11.0
	U-Oeq1	2.1(4)	2.24	0.0055	
	U-Oeq2	2.7(6)	2.47	0.0074	
5.6	U-Oax	2[a]	1.80	0.0012	−9.4
	U-Oeq1	1.7(4)	2.24	0.0054	
	U-Oeq2	3.1(9)	2.48	0.0120	
6.5	U-Oax	2[a]	1.79	0.0011	−10.3
	U-Oeq1	1.7(5)	2.24	0.0071	
	U-Oeq2	2.54(7)	2.50	0.0074	

[a]Held constant during the fit

two oxygen atoms at 3.1 Å and approximately two uranium atoms at 4.0 Å. This is the only reported case where polynuclear uranium species are detected in experiments on uranium sorption on silica. Sylwester et al. demonstrate that at pH 3.14 the equatorial oxygen atoms remain split into two distinct shells, but no near-neighbor silicon or uranium atoms are detected, indicating that surface precipitation may occur only at high pH. Both FT EXAFS spectra show a split equatorial shell with distances of 2.27 and 2.47 Å. A similar split of equatorial oxygen shell is reported in other papers [11, 22, 23]. According to Reich et al. [22], the best fit of the experimental data for uranium(VI) sorbed onto silica gel at pH 3.5–4 is obtained using two oxygen atoms in the equatorial uranyl plane separated by approximately 0.24 Å. Similar splitting of the oxygen shell is also observed for EXAFS analysis of schoepite and uranium(VI) precipitates at pH ≥7 Å on silica [23]. Walter et al. [11] also observed a split equatorial oxygen shell in the study of the uranium(VI) uptake by the aluminum-silicate albite before and after leaching. The equatorial oxygen shell of uranium(VI) sorbed on the leached sample is split into two 2.23 and 2.44 Å distances. This indicates the formation of an inner-sphere surface complex and allowed the authors to conclude that uranium(VI) sorption on alumosilicate minerals at pH about 6 is dominated by the formation of mononuclear bidentate inner-sphere surface complexes.

According to Riech et al. [14], the FT spectra recorded at U LIII edge for uranium(VI) sorbed on silica samples do not show any significant intensity above 3 Å. This means that even at high uranyl loading (59 mgU/g), no U–U interatomic distances are detected, which would indicate formation of polynuclear complexes or surface precipitation. Not more than one silicon atom ($N = 0.5 \pm 0.4$) is located at 2.72 ± 0.02 Å from uranium. The U–Si distance is much smaller than described for uranium(VI)-silica bearing minerals [24]; this is explained in the frame of a model in which one silica tetrahedron is linked via two of its atoms to the uranyl group (edge-sharing coordination). Harfouche et al. [25] investigated uranium(VI) uptake by calcium silicate hydrates and also observed a split equatorial oxygen shell (R1 = 2.23–2.77 Å, R2 = 2.36–2.4 Å) and neighboring silicon atoms at both short (3.07–3.11 Å) and long (3.71–3.77 Å) distances.

Based on the crystallographic data described for silica mineral phases [24], we consider a probability of the following types of U complexes being formed on the silica surface: mononuclear bidentate uranium(VI) surface complex, $=Si(O)_2=UO_2$ (Fig. 8a); mononuclear monodentate uranium(VI) surface complex, $\equiv SiO{-}UO_2$ (Fig. 8b); binuclear bidentate uranium(VI) surface complex, $=(SiO)_2=UO_2$ (Fig. 8c). The significant difference in bond distances U-Oeq1 and U-Oeq2 and coordination numbers extracted from EXAFS curve-fit analysis deny the formation of mononuclear monodentate uranium(VI) surface complex presented in Fig. 8c.

The splitting of U-Oeq bond distances and the corresponding coordination numbers enable us to conclude that bidentate complex of U(VI) is formed on the surface. In fact, if a mononuclear bidentate uranium(VI) surface complex were formed, its formation would result in a significant distortion of the Oeq2-U-Oeq2 angle from the 72° typical for uranyl ion in water solution to at least 90°; this would necessarily lead to a significant decrease of coordination numbers in equatorial plane. Reich et al. [14] and Walter et al. [11] established U–Si interaction with distances 2.72 Å and 3.09 Å, correspondently, while no splitting in equatorial oxygen shell was observed. Sylwester et al. [22] explain their experimental data for U(VI) sorbed on silica at pH 6.46 by including an interaction U–Si at a distance of about 3.09 Å; in their case, calculated coordination numbers are 3.49 for U-Oeq1 and 1.68 for U-Oeq2. As mentioned above, EXAFS curve-fitting in case of our sample enabled us to conclude the formation of binuclear bidentate uranium(VI) surface complex.

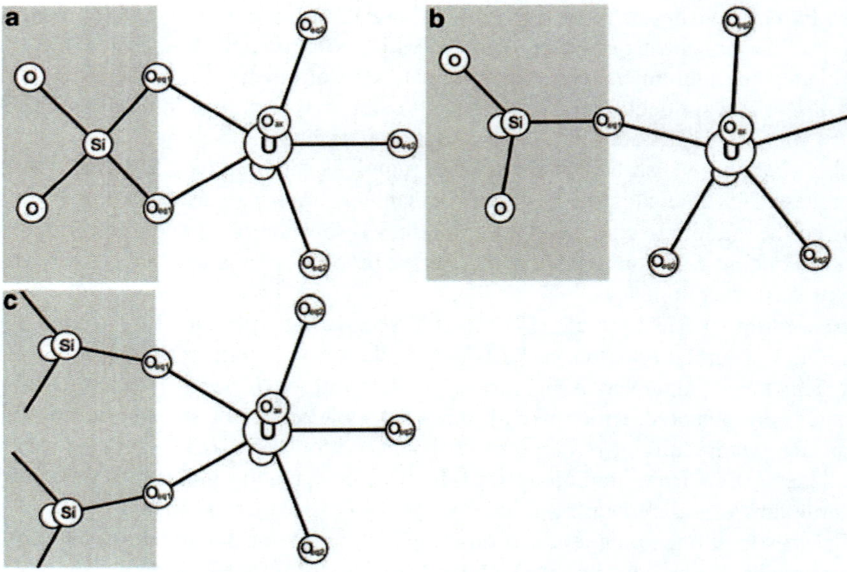

Fig. 8 (**a**) Mononuclear bidentate uranium(VI) surface complex, $=Si(O)_2=UO_2$; (**b**) mononuclear monodentate uranium(VI) surface complex, $\equiv SiO{-}UO_2$; (**c**) binuclear bidentate uranium(VI) surface complex, $\equiv(SiO)_2=UO_2$

The compilation of our results extracted from TRLIF and EXAFS analysis and crystallographic knowledge of silica mineral phase led to the following conclusion. Two structurally similar uranium species with different fluorescent lifetimes are present on the silica surfaces. Both types of complexes are bidentate coordinated uranyl species with two silicon atoms that differ in H_2O and OH group around uranyl moiety, i.e., one water molecule is substituted by an OH group in one of the complexes.

The observed increase in fluorescence life time at higher pH, which was also reported for uranyl upon hydrolysis in solution by Moulin et al. [20], supports our conclusion regarding the hydroxylated surface species.

3.4 Surface Complexation Modeling

SCM is applied to derive the stability constants of the proposed uranium species on the surface. The general nonlinear least-squares optimization program FITEQL version 4.0 [26] is applied in order to calculate equilibrium constants based on the experimental sorption data. We use diffuse layer model (DLM) to describe uranium behavior in the double layer of CASS. The thermodynamic data of uranium species in the solutions is taken from NEA database [27]. The concentration of sorption sites on the silica surface is defined as 4.9 nm^{-2}, the value from studies of more than 100 silica samples each synthesized via different routes and having different specific surface areas [28].

Based on the spectroscopic data, the following simplified reactions are considered on the surface:

$$2\equiv \mathrm{SiOH} + \mathrm{UO_2^{2+}} \rightleftarrows (\equiv \mathrm{SiO})_2\mathrm{UO_2} + 2\mathrm{H^+} \qquad K_1$$

$$2\equiv \mathrm{SiOH} + \mathrm{UO_2^{2+}} + \mathrm{H_2O} \rightleftarrows (\equiv \mathrm{SiO})_2\mathrm{UO_2OH^-} + 3\mathrm{H^+} \qquad K_2.$$

The best curve fitting results are obtained using the following equilibrium constants: $\log K_1 = -2.22 \pm 0.03$ and $\log K_2 = -5.85 \pm 0.05$. These calculated values appear reliable, as they provide an adequate description of experimental data obtained under sufficiently different conditions (various uranium concentration and uranium to silica surface ratio). The results of modeling together with experimental data are shown in Fig. 9.

4 Conclusions

This paper demonstrates the utility of combining different spectroscopic methods, providing molecular level speciation with SCM using an example of U(VI) sorption onto silica nanoparticles under varying conditions to provide reliable equilibrium

Fig. 9 The results of SCM (**a**) C(U(VI)) = 1×10^{-5} M, C(SiO_2) = 35 m^2/L, (**b**) C(U(VI)) = 1×10^{-5} M, C(SiO_2) = 16 m^2/L, (**c**) C(U(VI)) = 1×10^{-4} M, C(SiO_2) = 35 m^2/L

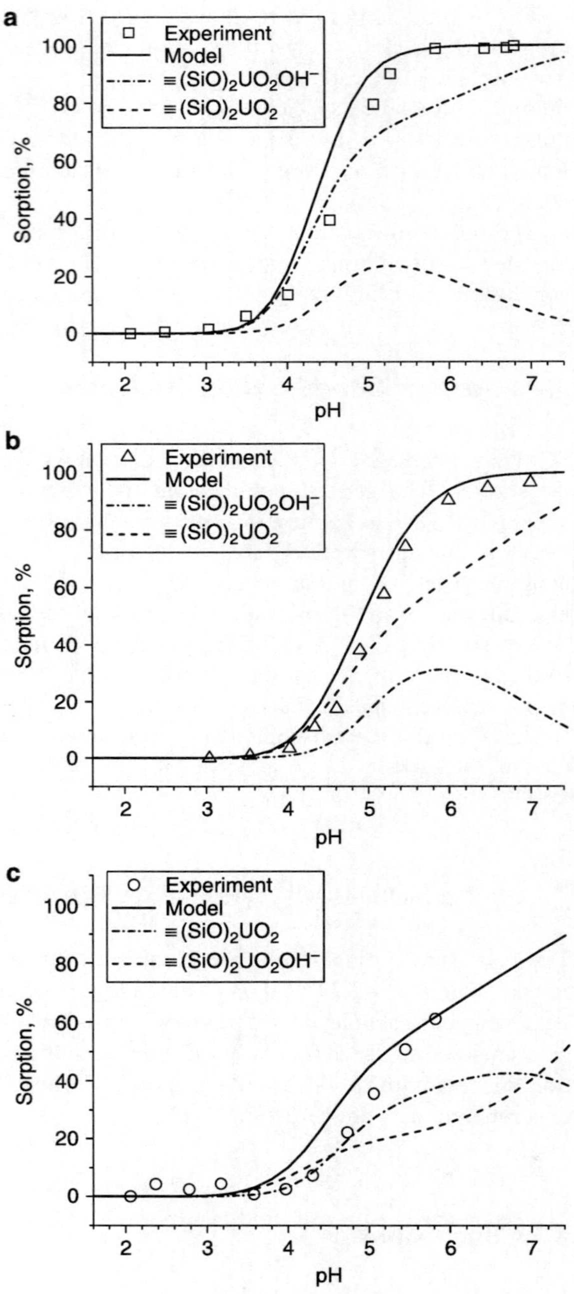

constants. The solubility of U(VI) is sufficiently high to allow application of several in situ spectroscopic methods for surface speciation and then to employ results from the experimental data in SCM. For other actinides with lower solubility, e.g., Pu(IV), the choice of direct speciation methods is limited by concentration constraints, and speciation must often be determined by the best fit of experimental sorption data to different surface complexation models.

Acknowledgments We acknowledge financial support as a Joint Research Group between the German Helmholtz Society and the Russian Basic Research Foundation (HRJRG-011 and 07-03-92280-SIG_a) and from the President of Russian Federation Grant (to SNK, # MD- 7150.2010.3).

References

1. Davis JA, Kent DB (1990) Surface complexation modeling in aqueous geochemistry in mineral-water interface geochemistry: reviews in mineralogy. Miner Soc Am 23:177–260
2. Bradbury MH, Baeyens B (2006) A quasi-mechanistic non-electrostatic modeling approach to metal sorption on clay minerals, In: J. Lutzenkirchen (ed) Surface complexation modeling, Academic Press, Amsterdam
3. Um W, Serne RJ (2005) Sorption and transport behavior of radionuclides in the proposed low-level radioactive waste disposal facility at the Hanford site, Washington. Radiochim Acta 93:57–63
4. Huber F, Lützenkirchen J (2009) Uranyl retention on quartz new experimental data and blind prediction using an existing surface complexation model. Aquatic Geochem 15:443–456
5. Guo Z, Su HY, Wu W (2009) Sorption and desorption of uranium(VI) on silica: experimental and modeling studies. Radiochim Acta 97:133–140
6. Pikryl JD, Jain A, Turner DR, Pabalan RT (2001) UraniumVI sorption behavior on silicate mineral mixtures. J Contam Hydrol 47:241–248
7. Rybal'chenko AI, Pimenov MK, Kostin PP, Balukova VD, Nosukhin AV, Mikerin EI, Egorov NN, Kaimin EP, Kosareva IM, Kurochkin VM (1998) Deep injection disposal of liquid radioactive waste in Russia. Battelle Press, Richland
8. Zubkov AA, Ryabov AS, Sukhorukov VA, Danilov VV, Rybal'chenko AI (2005) Results of long-term deep liquid radioactive waste injection site operation at the Siberian Chemical Combine. Dev Water Sci 52:487–491
9. Gabriel U, Charlet L, Schläpfer CW, Vial JC, Brachmann A, Geipel G (2001) Uranyl surface speciation on silica particles studied by time-resolved laser-induced fluorescence spectroscopy. J Colloid Interf Sci 239:358–368
10. Kowal-Fouchard A, Drot R, Simoni E, Ehrhardt JJ (2004) Use of spectroscopic techniques for uranium(VI)/montmorillonite interaction modeling. Environ Sci Technol 38:1399–1407
11. Walter M, Arnold T, Geipel G, Scheinost A, Bernhard G (2005) An EXAFS and TRLFS investigation on uranium(VI) sorption to pristine and leached albite surfaces. J Colloid Interf Sci 282:293–305
12. Lefèvre G, Kneppers J, Fédoroff M (2008) Sorption of uranyl ions on titanium oxide studied by ATR-IR spectroscopy. J Colloid Interf Sci 327:15–20
13. Lefèvre G, Noinville S, Fédoroff M (2006) Study of uranyl sorption onto hematite by in situ attenuated total reflection–infrared spectroscopy. J Colloid Interf Sci 296:608–613
14. Sylwester ER, Hudson EA, Allen PG (2000) The structure of uranium (VI) sorption complexes on silica, alumina, and montmorillonite. Geochim Cosmochim Acta 64:2431–2438
15. Drot K, Roques J, Simoni E (2007) Molecular approach of the uranyl/mineral interfacial phenomena. Compt Rendus Chim 10:1078–1091

16. Stöber W, Fink A, Bohn E (1968) Controlled growth of monodisperse silica spheres in the micron size range. J Colloid Interf Sci 26:62–69
17. Svergun DI (1992) Determination of the regularization parameter in indirect-transform methods using perceptual criteria. J Appl Crystallogr 25:495–503
18. Svergun DI, Konarev PV, Volkov VV, Koch MHJ, Sager WFC, Smeets J, Blokhuis EM (2000) A small angle x-ray scattering study of the droplet–cylinder transition in oil-rich sodium bis(2-ethylhexyl) sulfosuccinate microemulsions. J Chem Phys 113:1651–1665
19. Ravel B, Newville M (2005) ATHENA, ARTEMIS, HEPHAESTUS: data analysis for X-ray absorption spectroscopy using IFEFFIT. J Synchrotron Radiat 12:537–541
20. Moulin Ch, Laszak I, Moulin V, Tondre Ch (1998) Time-resolved laser-induced fluorescence as a unique tool for low-level uranium speciation. Appl Spectrosc 52:528–535
21. Dent AJ, Ramsay JDF, Swanton SW (1992) An EXAFS study of uranyl ion in solution and sorbed onto silica and montmorillonite clay colloids. J Colloid Interf Sci 150:45–60
22. Riech T, Moll H, Arnold T, Denecke MA, Hennig C, Geipel G, Bernhard G, Nitsche H, Allen PG, Bucher JJ, Edelstein NM, Shuh DK (1998) An EXAFS study of uranium(VI) sorption onto silica gel and ferrihydrite. J Electron Spectrosc 96:237–243
23. Allen PG, Shuh DK, Bucher JJ, Edelstein NM, Palmer CEA, Silva RJ, Nguyen SN, Marquez LN, Hudson EA (1996) Determinations of uranium structures by EXAFS: schoepite and other U(VI) oxide precipitates. Radiochim Acta 75:47–53
24. Soderholm L, Skanthakumar S, Gorman-Lewis D, Jensen MP, Nagy KL (2008) Characterizing solution and solid-phase amorphous uranyl silicates. Geochim Cosmochim Acta 72:140–150
25. Harfouche M, Wieland E, Dähna R, Fujita T, Tits J, Kunz D, Tsukamoto M (2006) EXAFS study of U(VI) uptake by calcium silicate hydrates. J Colloid Interf Sci 303:195–204
26. Herbelin AL, Westall JC (1999) FITEQL 4.0: a Computer Program for Determination of Chemical Equilibrium Constants from Experimental Data; Report 99–01. Department of Chemistry, Oregon State University, Corvallis
27. Guillaumont R, Fanghaenel Th, Fuger J, Grenthe I, Neck V, Palmer DA, Rand MH (2003) Chemical thermodynamics, In: Update on the chemical thermodynamics of uranium, neptunium, plutonium, americium and technetium, Elsevier, North-Holland, Amsterdam
28. Zhuravlev LT (1987) Concentration of hydroxyl groups on the surface of amorphous silica. Langmuir 3:316–318

Section 4
Environmental Behavior of Actinide Colloids and Nanoparticles

Radioactive Particles Released into the Environment from Nuclear Events

Brit Salbu and Ole Christian Lind

Abstract A major fraction of refractory radionuclides, such as uranium (U) and plutonium (Pu) that are released during severe nuclear events including nuclear weapon tests, reactor explosions or fires, and use of depleted uranium ammunition, is present as particles, often ranging from submicrons to fragments. Reprocessing facilities and civil reactors are also sources of radioactive particles and colloids via their authorized or accidental releases. Furthermore, radioactive particles are identified in sediments in the close vicinity of radioactive waste dumped at sea. Thus, whenever refractory radionuclides are released following severe nuclear events, radioactive particles should also be expected. Substantial amounts of refractory fission products, activation products and transuranics can be carried by radioactive particles that are heterogeneously distributed in the environment. In case of radioactive particle contamination, representative sampling is difficult, and complete digestion of inert particles is a challenge. Thus, the estimated inventories can be underestimated. When radioactive particles enter the environment, weathering processes occur and, subsequently, associated radionuclides are mobilized. Therefore, particle contaminated soils or sediments can act as diffuse sources for future ecosystem transfer. In cases where significant radioactive particle releases occur, environmental impact assessment should therefore include detailed studies dedicated to provide information on particle characteristics such as size distributions, crystallographic structures, oxidation states, these being variables influencing weathering, mobility, and biological uptake. The present paper provides a summary on sources reported to have contributed to radioactive particle contamination in the environment and the characteristics of observed particles.

B. Salbu (✉) and O.C. Lind
Isotope Laboratory, Department of Plant and Environmental Sciences, Norwegian University of Life Sciences, PO Box 5003, 1432 Aas, Norway
e-mail: brit.salbu@umb.no

S.N. Kalmykov and M.A. Denecke (eds.), *Actinide Nanoparticle Research*,
DOI 10.1007/978-3-642-11432-8_12,

1 Introduction

Radionuclides released from a source may be present in different physico-chemical forms, such as low molecular mass (LMM) species, colloids and pseudocolloids, particles as well as fragments. A major fraction of refractory radionuclides such as uranium (U) and plutonium (Pu) released during historical nuclear events has been associated with particles (Salbu 2000a). Radioactive particles are defined as localised aggregates of radioactive atoms that give rise to an inhomogeneous distribution of radionuclides significantly different from that of the matrix background (IAEA CRP 2001). Radioactive particles in the aquatic environment are defined as entities having diameters larger than 0.45 μm, i.e. will settle due to gravity, while particles larger than 2 mm are referred to as fragments (Salbu 2009). In air, radioactive particles ranging from submicrons in aerosols to fragments are classified according to the aerodynamic diameters, where particles less than 10 μm are considered respiratory. The presence of radioactive species ranging from particles to fragments can easily be identified by imaging techniques such as digital autoradiography (Salbu 2000b), reflecting their inhomogeneous distributions in water, sediments, and soils (Fig. 1).

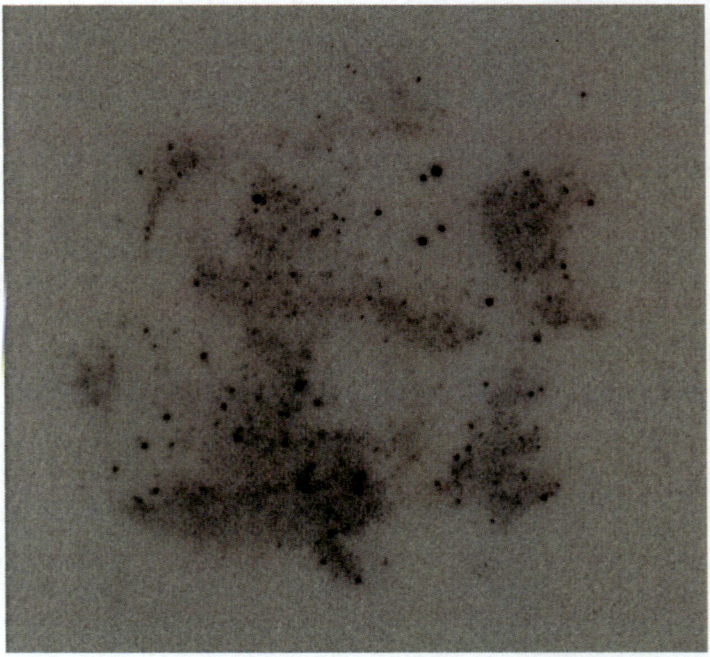

Fig. 1 Digital phosphor imaging (autoradiography) of sediments from Reservoir 11 (Station 1, 0–10 cm) at Mayak PA

After the Chernobyl accident, radioactive particles were identified close to the source (Kuriny et al. 1993) as well as more than 2,000 km from the damaged reactor (Devell et al. 1986). During that time the "hot" particle phenomenon was considered to be a peculiarity of the Chernobyl accident. However, radioactive particles were identified already during the 1950s–1960s in areas where nuclear weapons were tested, for instance within the Nevada Test Site (Crocker et al. 1966). As the technical capabilities to characterize particles were limited, literature data from the 1960s were mainly limited to large particles or fragments. Thus, the scientific field related to the characterization of radioactive particles was re-established when the impact of the Chernobyl particles was realized. Since then, new advanced techniques have been developed. Using sensitive analytical techniques, radioactive particles containing refractory radionuclides have been identified at a series of sites affected by radioactive contamination from different nuclear or radioactive sources, and not at all limited to the Chernobyl accident. Therefore, these sites should act as field laboratories, where knowledge on the characteristics and ecosystem behaviour of radioactive particles from previous events can be utilized to provide information that can be implemented in impact assessments for existing contaminated areas as well as for future events.

Radioactive particles containing refractory radionuclides such as U and Pu are identified within nuclear weapon test sites (e.g., Nevada, US; Maralinga, Australia; Mururoa, French Polynesia; Semipalatinsk, Kazakhstan), in fallout from nuclear reactor explosions (Chernobyl, Ukraine) or fires (e.g., Windscale, UK), and in areas contaminated from accidents with reactor-powered satellites (e.g., Cosmos 954, Canada) or with aircrafts carrying nuclear weapons (e.g., Thule, Greenland; Palomares, Spain). Particles have also been shown to be present in authorised discharges and accidental releases from reprocessing plants (e.g., Sellafield, UK; Krasnoyarsk and Mayak, Russia) and in leakages from radioactive wastes dumped in the Kara Sea. Therefore, releases of radioactive particles have occurred more frequently than usually anticipated.

The radionuclide composition of released particles will depend on the source, while particle characteristics such as particle size distribution, crystallographic structures and oxidation states will also depend on the release scenarios, in particular on the temperature, pressure and the presence of air/oxygen (Salbu et al. 2001). Thus, the activity concentration, isotope ratios and burn-up will reflect the emitting source, while release conditions (high temperature and pressure, presence of air) will influence the activity concentrations of volatiles as well as particle characteristics (i.e., size distributions, crystallographic structures, oxidation states) of importance for their ecosystem behaviour.

2 Methods for Characterization of Radioactive Particles

A number of analytical tools have been applied in radioactive particle research. For radionuclides in waters, filtration (<0.45 μm) and ultrafiltration (nm to μm range membranes) combined with radioanalytical techniques may be applied to distinguish ionic species, colloidal and particulate material (Salbu 2000b). To separate particles from air and provide estimates on the size distribution, cascade impactors with membranes having different cut-off levels (μm to mm) are frequently used in combination with radioanalytical methods. In contaminated areas, portable detectors can be utilized to identify hot spots. In the laboratory, radioactive heterogeneities may be identified using imaging techniques; repeated sample mixing (Bunzl 1998; Bunzl and Tschiersch 2001) or splitting of samples (Bunzl 1997) combined with gamma spectrometry. For contaminated soils and sediments, biological material and filters from air or water fractionation procedures, imaging techniques such as digital phospor imaging (digital autoradiography) have proved especially useful for locating radioactive particles (Fig. 1). Hot spots reflect particles and further separations of subsamples using radiation detectors and microscopes are needed prior to the use of surface (and to some extent subsurface) sensitive EM-techniques. Using conventional (SEM) or environmental scanning electron microscopes (ESEM) in Backscattered Electron Imaging (BEI) mode, bright areas reflect the presence of high atomic number elements, while the distribution of elements in the upper particle surface layer is attained using X-ray mapping. Energy dispersive X-ray analysis (EDX) measurements provide information on the elemental composition. The structure of particles is characterised from images acquired in Secondary Electron Imaging mode. SEM and ESEM also serve as a screening technique prior to further detailed studies using other nanometer or micrometer sized beam techniques. To obtain information on the distribution of elements and species below the surface of particles, microanalytical techniques with greater penetration depths such as proton (or particle) induced X-ray emission (μ-PIXE) and secondary ionization mass spectrometry (SIMS) as well as synchrotron radiation (SR) based micro X-ray techniques are needed. Using high flux monochromatic X-ray microbeams, synchrotron radiation (SR) based X-ray microscopic techniques have proved most useful for characterisation of individual particles. Microscopic X-ray fluorescence analysis (μ-XRF) provides information on 2 dimensional or 3 dimensional (μ-tomography) elemental distribution within individual particles, while micro-X-ray diffraction (μ-XRD) gives information on the crystallographic structures of solid particles. X-ray absorption near edge structure spectrometry (μ-XANES) provides information on the oxidation state (Fig. 2).

Extended X-ray absorption fine structure analysis (EXAFS) could also potentially provide information on the coordination number and the distance to neighbouring atoms, if detailed knowledge on all atoms involved is available (Conradson 2000).

Following solid-state (non-destructive) speciation techniques, leaching experiments may provide important data on solubility and weathering rates, which can be linked to the particle characteristics. Finally, the particles may be fully dissolved, and subjected to α, β and/or mass spectrometry to determine activity concentrations and isotopic ratios to identify the source of the particle.

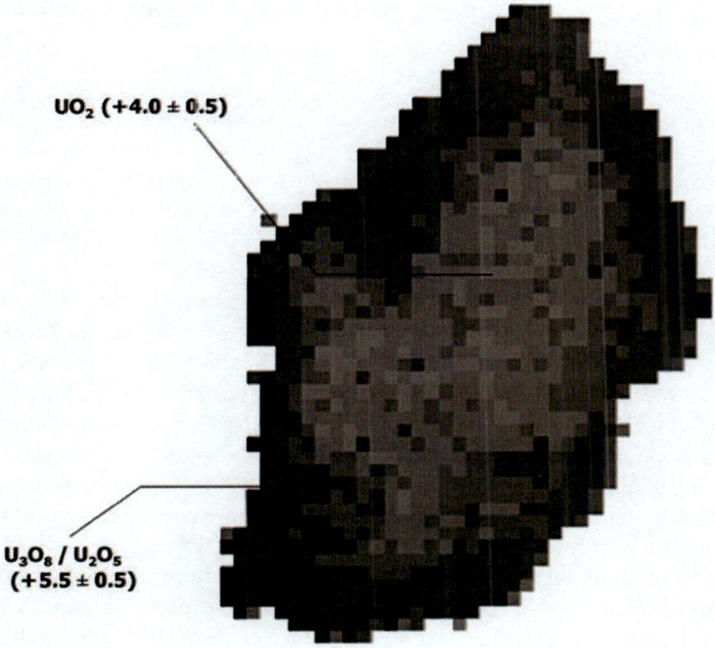

Fig. 2 Synchrotron radiation based 2D μ-XANES analysis of uranium in a particle collected to the North of the Chernobyl reactor showing the two dimensional distribution of U oxidation state. The particle has a UO_2 core, while U in the surface is oxidized to U_3O_8 (Salbu et al. 2004)

3 Sources Contributing to the Release of Radioactive Particles into the Environment

The main source of anthropogenic radionuclides in the environment is fallout from atmospheric nuclear weapon tests (UNSCEAR 1993, 2000). Other important anthropogenic sources that contribute significantly include authorized or accidental releases from the nuclear fuel cycle, especially reprocessing plants and nuclear reactors. Releases from satellite, aircraft and submarine accidents, and dumping of radioactive waste at sea have a local impact only. A significant fraction of refractory radionuclides released from these sources is present as radioactive particles (Salbu 2000a). However, radioactive particles are not exclusively of anthropogenic origin. Radionuclides in naturally occurring radioactive material (NORM) and technologically enhanced NORM (TENORM) may also be present as particles and fragments.

3.1 *Anthropogenic Radionuclides Released as Radioactive Particles*

3.1.1 Particles Released During Nuclear Weapons Tests

Fallout from more than 2,300 atmospheric, surface, underground and underwater nuclear weapon tests and sub-critical safety trials with conventional explosives are the major sources of radioactive contamination in the environment (UNSCEAR 1993).

Nuclear weapons tests have been conducted at various locations, at and under the ground surface, mounted on towers, placed on barges on the ocean surface, from balloons, dropped from airplanes, and high-altitude launchings by rockets. Depending on the location of the explosion (altitude and latitude) the radioactive debris entered the local, regional, or global environment (UNSCEAR 1993). Categorization of radioactive particles from various types of nuclear weapon tests under different conditions has been described in detail by Heft (1970). According to him, all the radionuclides (except ^{3}H, ^{14}C and the long lived rare gases) involved in a nuclear detonation are accounted for as radioactive particles. During the late 1950s and 1960s, a lot of attention was given to the local deposition of radioactive particles at test sites (Crocker et al. 1966). Reports on radioactive particles deposited within certain test sites (i.e., Marshall Islands, Nevada Test Site, Johnston Atoll, Novaya Zemlya, Maralinga, Mururoa) are available in open literature, while the presence of radioactive particles is expected, but not confirmed at several test sites (e.g. India and Pakistan). In the 1950s and 1960s, long-range transported radioactive particles originating from nuclear tests were identified and investigated in Sweden (Novaya Zemlya, Russia) and in Japan (Lop Nor, China and Semipalatinsk, Kazakhstan).

Semipalatinsk Nuclear Test Site

In the Semipalatinsk Test Site (STS), located in the north-east of Kazakhstan, a total of 456 nuclear weapon tests have been conducted during 1949–1989 (IAEA 1998). Digital phosphor imaging (Fig. 3) of dried surface soils from sites within the STS show that the radioactive contamination is highly heterogenous (Lind 2006).

Further investigations yielded the isolation of glass-like vitrified particles from the Balapan and Ground zero sites, whereas soil particles with U and Pu co-existing in highly concentrated small grains (tens of μm in size) were found at the Tel'kem craters (Lind 2006).

Novaya Zemlya

At Novaya Zemlya, 88 atmospheric, 39 underground, and at least three underwater nuclear weapon tests were carried out between 1950 and 1990. Significant

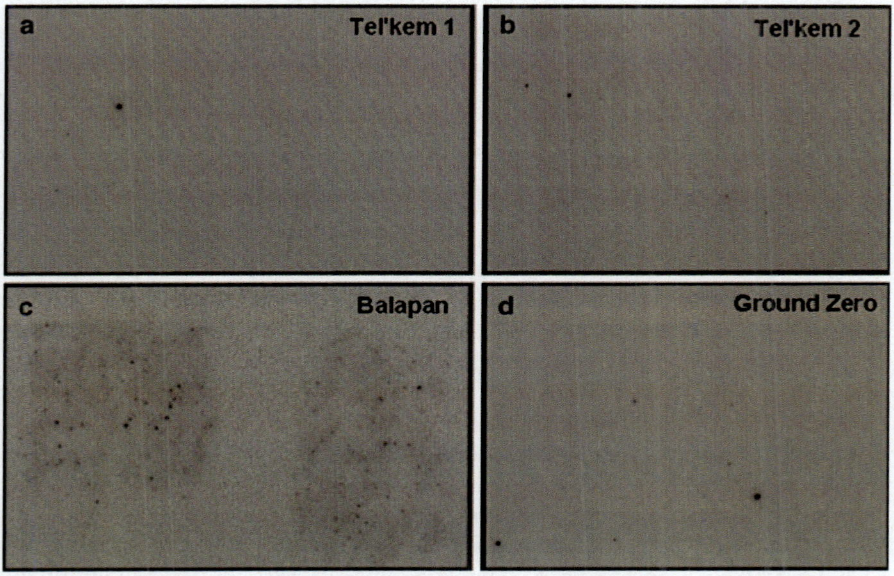

Fig. 3 Digital phosphor imaging (autoradiography) of dried surface soils (1–5 cm) from locations within the Semipalatinsk Test Site. (**a**) The outer rim (South-East) of the Tel'kem 1 crater; (**b**) Outside (north) of the Tel'kem 2 crater; (**c**) Within the fusion explosion crater of the Balapan Lake (Atomic Lake). (**d**) Ground Zero

contamination, hot spots and localised heterogeneities are observed within the three major test areas, and are also attributed to specific shot scenarios (AMAP 1997). Localised heterogeneities of radionuclides in sediments indicate the presence of radioactive particles, in particular in Chernaya Bay (Smith et al. 2000), where underwater weapon tests took place in 1955, 1957 and 1961. Information on particle characteristics is scarce. However, a large number of radioactive particles originating from atmospheric weapon tests at Novaya Zemlya were collected in October 1958 by aircraft at high altitude (up to about 13 km) over central Sweden (Sisefsky 1961). Autoradiography showed that the activity of the particles was roughly proportional to the volume. In the microscope, particles appeared spherical, translucent and colourless to reddish.

Marshall Island

At Marshall Island, the US performed 23 nuclear weapon tests in the atmosphere, and at ground surface at the Bikini and Enewetak atolls, between 1946 and 1958. Localised heterogeneities reflected particles deposited in soils (Simon et al. 1995). Based on autoradiography, large spherical particles (0.5–1 mm) with uniform distribution of radionuclides and irregular several mm-sized particles with surface contamination have been observed after ground-surface shots. According to

Crocker et al. (1966), particle characteristics (size distribution, shape, colour) depended on devices and shot conditions. Pu associated with spherical particles from high altitude detonations was inert towards leaching in water, while Pu associated with algal crust of the atoll soils and the debris from coral-surface bursts is reported to be highly soluble in water.

Nevada Test Site

In Nevada Test Site, US, 86 atmospheric weapon tests (UNSCEAR 2000) as well as 828 underground tests were conducted during 1951–1992 (Kersting et al. 1999). Hot spots and heterogeneities observed by means of autoradiography reflected the presence of radioactive particles (Anspaugh and Church 1986). A large variety of fused or partially fused Pu-particles, vesicle-shaped particles, and large agglomerates consisting of individual small particles differing in colour, specific activity, density and magnetic properties have been identified. The particle size distribution is reported to depend on device and shot conditions (Crocker et al. 1966). High altitude detonations resulted in spherical small-sized dense particles with high specific activity, while ground surface shots gave large irregular shaped particles with lower density and specific activities. The leaching of gross gamma or gross beta activity from particles depended on device and shot conditions, matrix composition, particle size and type. High-altitude shot particles were more inert than those from ground detonations. Gross-gamma and gross-beta activity in air-burst debris was dissolved in 0.1 M HCl. The majority (~98%) of the refractory radionuclides (e.g., Pu) involved in underground nuclear tests are incorporated into melt glass that forms at the bottom of the test cavities (Kersting et al. 1999).

Maralinga and Emu

During 1953–1963, UK conducted nine nuclear weapon tests and several hundred smaller-scale weapon trials at Maralinga and Emu sites, Southern Australia (Burns et al. 1994, 1995; Cooper et al. 1994). The major contamination of radioactive particles is attributed to 12 "Vixen B" safety trials at the Taranaki part of the Maralinga nuclear test site, which involved either burning or explosive dispersal of 22.2 kg Pu and 47.3 kg U (Cooper et al. 1994). Pu was predominantly present as particles within the 250–500 μm fraction, and also as finely dispersed Pu particles as well as fragments up to several hundred microns, Pu-contaminated soil particles, and coatings on device reminiscences (Cooper et al. 1994). Based on individual large particles characterised by gamma spectrometry and proton-induced X-ray emission spectroscopy, Pu and U were localised on particle surfaces. A number of quite porous Pu particles from the Taranaki site were subjected to solubility tests using simulated lung fluids and 0.16 M HCl. No significant solubility was found in simulated lung fluids, but in 0.16 M HCl the solubility varied from 1 to 96% over a period of 40 days (Burns et al. 1995). A 0.93 $\pm$ 0.12 kilotons weapon test in 1957 at the Tadje area of Maralinga contaminated an area close to the detonation site with both small sized Pu particles and glassy spherical particles containing Pu and Am (Cooper et al. 1994).

Fangataufa and Mururoa atolls

Between 1966 and 1996, France conducted 193 nuclear tests (178 nuclear detonations and 15 safety trials) above and below ground at the Mururoa and Fangataufa Atolls in French Polynesia. According to Danesi et al. (1998), the presence of radioactive particles originating from five safety test trials performed between 1966 and 1974 has been confirmed at the safety test sites at Colette, northern part of Mururoa. Despite extensive cleanup operations, Pu particles were left on the surface of the region and in the adjacent sand bank in the lagoon. Relatively large Pu-containing particles, ranging in size from 200 μm to 1 mm, have been separated from coral debris and crushed coral rocks. Particles with glassy relatively smooth compact surfaces to conglomerates of small particles with rough appearances were observed (Salbu 2001). Using μ-tomography, Eriksson and co-workers found a ~100 μm Mururoa Pu inclusion to be attached to coral matrix (Eriksson et al. 2005). The activity levels of ^{239}Pu and ^{241}Am in the particles ranged from 5 kBq to about 1 MBq and 0.2 to 5.6 kBq, respectively. Danesi also reported that 99.9% of the mass and 95.8% of the activity were present in particles larger than 250 μm. However, the possible presence of particles smaller than 10 μm with ^{239}Pu activities of several hundred Bq were not excluded (Danesi et al. 1998). In vitro dissolution studies of individual particles in simulated serum showed solubility of Pu of less than 0.07%, indicating dissolution characteristics similar to particles from the Maralinga nuclear test site (Danesi et al. 1998).

Lop Nor

The Chinese nuclear weapons testing programme included 22 atmospheric tests between 1964 and 1980 at the Lop Nor test site in Gansu, China (UNSCEAR 2000). No information on local deposition is available, however, spherical radioactive particles containing U and with optical diameters ranging from 7 to 22 μm originating from the first test on 16th October 1964 (small scale land surface burst) were identified in Japan 3–4 days later (Mamuro et al. 1965). Colour varied from particle to particle, black or reddish black being the most frequent. Spherical radioactive particles with optical diameters up to about 15 μm originating from the tests were also identified in Japan in 1961–1962 (Mamuro et al. 1962). Uniformly coloured, reddish brown or black particles were most frequent (Mamuro et al. 1966).

3.1.2 Depleted Uranium Ammunitions

Large amounts of depleted uranium (DU), a waste product of U enrichment, have been applied as armour-piercing ammunition, due to the high density, hardness of the metal and pyrophoric properties. Several reports have been published on DU contamination due to DU penetrator use during war time in Kosovo (Danesi et al. 2003a, b; Salbu et al. 2003), Serbia-Montenegro (McLaughlin et al. 2003), Bosnia and Herzegovina (UNEP 2003), Kuwait (Salbu et al. 2005) and Iraq (Gerdes et al. 2004). At all sites, except Iraq, a part of the DU contamination was identified as DU particles. DU ammunition have also contaminated military proving grounds and their surroundings (Sowder et al. 1999).

Following the impact with hard targets, DU is dispersed and ignited. Therefore, DU particles ranging from submicrons to several hundred micrometers, mostly in the respiratory fraction, have been observed in sand, soils, and in damaged vehicles.

Based on synchrotron X-ray microscopic techniques (Fig. 4), crystalline phases such as UO_2, $UO_{2.34}$, UC and metallic U or U–Ti alloy were identified in moderately soluble DU particles released during impact with tanks at sites in Kosovo and Kuwait (Lind et al. 2009). Following a fire in a DU ammunition storage facility, oxidized U(+6) was identified in large, fragile, highly soluble and bright yellow

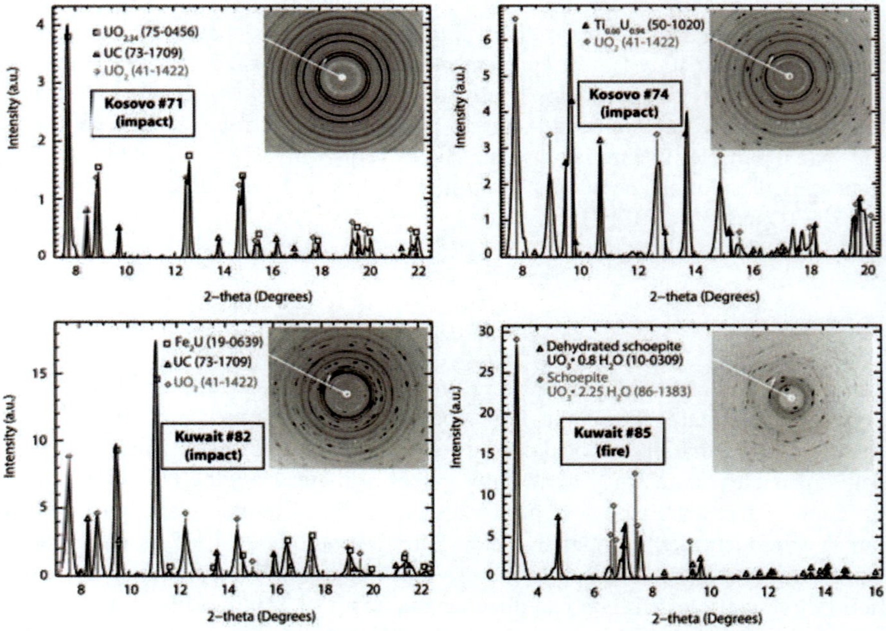

Fig. 4 Diffraction patterns collected from local spots on individual DU particles sampled in Kosovo and Kuwait with corresponding ICDD PDF-2 database entries

DU particles. Furthermore, crystalline phases such as schoepite ($UO_3 \cdot 2.25\ H_2O$), dehydrated schoepite ($UO_3 \cdot 0.75\ H_2O$) and metaschoepite ($UO_3 \cdot 2.0\ H_2O$) were identified (Lind et al. 2009). Although the source was similar (DU penetrators), the release scenario (detonation, fire, corrosion) influenced the characteristics of released particles.

3.1.3 Nuclear Reactor Accidents

Although most nuclear accidents have had a local impact only, some events (fires, explosions) have resulted in global or regional contamination. When nuclear accidents occur, releases of refractory radionuclides will indicate the presence of fuel particles (Salbu 2000b).

Windscale

Two air-cooled graphite moderated metal U reactors operated at Windscale from 1951 to 1957, when both were shut down due to the fire in Pile No1. Atmospheric particle releases were observed already in the mid 1950s. Due to low-temperature oxidation and corrosion of spent fuel elements misplaced in the air-cooled ducts leading to the discharge stack with an inefficient filtering system, fuel particles containing actinides and fission products were continuously dispersed (Salbu 2001). The total release over several years is estimated to have been about 20 kg U as large particles, up to 700 μm in length (Jakeman 1986). Based on electron microscopy, the particles had a flake-like structure, significantly different from those observed in the Chernobyl fallout (Salbu et al. 1994). The particles were inert towards leaching with 1 M HCl (Oughton et al. 1993).

Due to the graphite fire in 1957, about 10^9 Bq of ^{239}Pu, were released, probably associated with U particles observed as localised heterogeneities by autoradiography of samples from the stack filters, from contaminated surfaces from outside the reactor and from contaminated soils (Arnold 1992; Chamberlain and Dunster 1958). Particles in the size range 20–500 μm were observed up to 4 km from the site, but were considered of little relevance to public health (Appleby and Luttrell 1993).

Chernobyl

The energy generated in the Chernobyl nuclear power plant in 26th April 1986 led to a series of explosions that destroyed the outer containment, exposing the core to the environment and injecting highly radioactive debris into the atmosphere. The releases continued in the following days due to the subsequent fire in the graphite moderator.

As a result of the accident, about 6–8 tonnes of UO_2 fuel were released into the atmosphere (Victorova and Garger 1990). Large fuel particles with variable radionuclide composition deposited within a 30 km zone with respect to the plant, while small-sized particles were identified up to 2,000 km from the site (Devell et al. 1986; Kashparov et al. 1999; Kuriny et al. 1993; Salbu 1988). During the initial explosion on April 26th, mechanical destruction of the UO_2 fuel occurred under high pressure and high temperatures (>2000°C immediately prior to the explosion and reaching 2400–2600°C in local regions of the reactor), and deposition of fuel particles took place to the west of the reactor. In the period 26–30 April, volatile fission products and U fuel particles were released under moderate temperatures (330–930°C) and oxidising conditions due to the fire (Kashparov et al. 1999), and deposition of particles occurred to the north, northeast and south of the plant. From April 30th to May 6th, the temperature was lowered and, subsequently, the emission of volatiles decreased (Kashparov et al. 1999). The particle weathering rate constant (k, year^{-1}) of U fuel particles estimated from a series of NH_4Ac extraction studies on Chernobyl soils collected over a 10 years period were found to range from 0.04 to 0.4 year^{-1} (Kashparov et al. 1999). The results indicated that varying weathering rates could be related to different particle characteristics, as well as varying soil pH. The weathering rate constants (k, year^{-1}) for U particles deposited to the west of the reactor were significantly lower than for particles deposited to the north and south of the reactor (Kashparov et al. 1999). μ-XANES and μ-XRD measurements demonstrated that fuel particles released during the initial explosion were not oxidised and some particles contained apparently inert, reduced forms of U associated with Zr or carbide (Salbu et al. 2001). Furthermore, the results from μ-XANES imaging (Fig. 2) and μ-XRD demonstrated that a particle released during the reactor fire (north of the reactor) was characterised by a UO_2 core surrounded by oxidised U (a layer of U_2O_5/U_3O_8 and possibly other intermediate forms) (Salbu et al. 2004).

Dobrovolsky and Lyalko (1995) divided reported hot Chernobyl particles into four categories:

1. Fuel particles, i.e. more or less pure U oxides (UO_2 to U_3O_8/U_2O_5).
2. Fuel construction particles, i.e. particles with a matrix of nuclear fuel mixed with reactor construction material. A lot of (Zr, U)O_x particles have been found inside the 30 km exclusion zone around the Chernobyl nuclear power plant, both to the west and to the north. Particles of U–Zr composition seem to be rather inert.
3. Construction particles are composed of reactor construction materials. They are carriers of some radionuclides.
4. Particles of hybrid type. Particles that were mainly derived from interaction of fuel and reactor construction material with fire distinguishing material etc.

This categorization seems valid for close-in fall-out around Chernobyl. However, a fifth category comprising single element (e.g., Ru, Cs) particles should be added. Such particles were found as far away as Norway, 2,000 km from Chernobyl (Fig. 5).

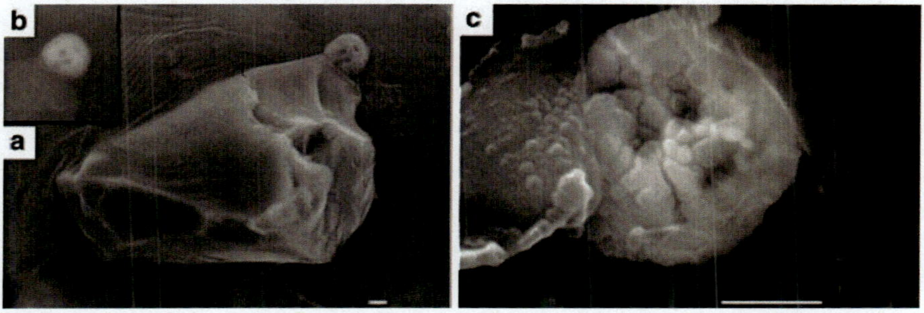

Fig. 5 Scanning electron micrograph of a Ruthenium particle found in 1987 in Valdres, Norway. (**a**) Overview of the Ru particle attached to a larger silicon particle, SEI-mode; (**b**) Close up of the Ru particle, BEI-mode. Bright areas show the presence of elements with high atomic numbers; (**c**) Higher magnification of the hot particle, BEI-mode (Salbu 1988)

3.1.4 Nuclear Accidents Involving Satellites

On at least three occasions, satellites powered by ^{238}Pu-based radioisotope thermoelectric generators (RTGs) or nuclear reactors have been destroyed upon re-entry to earth's atmosphere causing regional and global contamination of radioactive particles.

SNAP9-A

In 1964, the re-entry to the earth's atmosphere of the US SNAP9-A (satellite nuclear auxiliary power unit) containing 1 kg (629 TBq) of ^{238}Pu metal added to the global fallout of Pu. It is estimated that about 95% of Pu deposited by the end of 1970, most probable as finely dispersed particles (Krey 1967).

Cosmos 954 and 1402 Satellites

In 1978, the re-entry into the atmosphere of the Soviet satellite Cosmos 954 with a fission reactor (50 kg of 90% ^{235}U) contaminated large areas (124,000 km^2) in Canada with radioactive debris and only a fraction of the total mass was recovered (AMAP 1997). The size distribution pattern varied from submicron to fragments, and about 25% of the inventory was estimated as mm-sized particles. Surface contaminated remnants of the experimental device and soils were also observed (AMAP 1997; Krey et al. 1979). In 1983, a nuclear reactor from the Cosmos 1402 Soviet radar reconnaissance satellite re-entered the earth's atmosphere and disintegrated over the South Atlantic Ocean (Leifer et al. 1987). Information on the size distribution of debris appears to be scarce.

3.1.5 Conventional Detonation of Nuclear Weapons

Since the beginning of the nuclear age, there have been numerous accidents involving nuclear weapons. It is believed that at least 230 accidents have occurred involving the nuclear weapons of the US, the Soviet Union and the UK between 1950 and 1980 (Gregory and Edwards 1989). The US Department of Defence has confirmed the occurrence of 32 serious accidents between 1950 and 1980 (Mian et al. 2001). This includes a number of events where the high explosive (HE) in nuclear weapons has burnt or detonated, and led to contamination of fissile material (Drell and Peurifoy 1994). Information about accidents in the FSU is more difficult to obtain. However, more than 25 serious nuclear weapon accidents have been mentioned in the literature (Shaun 1990). No inadvertent nuclear weapon detonation (fission) is known to have occurred, and the risk is considered low (AMAP 1997).

Apparently, only four nuclear weapon accidents have been described in the open literature, i.e. accidents which occurred on the McGuire air force base, New Jersey and Johnston Atoll, USA and at Palomares, Spain and Thule air base, Greenland. In all four cases, the nuclear weapons were fragmented, and contamination was largely in the form of radioactive particles.

McGuire Air Force Base

On 7 June 1960, a Boeing Michigan Aeronautical Research Center (BOMARC) missile was destroyed at the former US McGuire air force base, New Jersey, when a high-pressure helium tank exploded and ruptured the fuel tank. Although safety devices prevented the HE from detonating, a fire took place and along with it the entire warhead containing Pu, enriched U and bottled tritium burned (Drell and Peurifoy 1994; Mian et al. 2001). A significant quantity of weapons-grade Pu particles with varying sizes (sub-micrometer-sized or single large particles) containing ^{239,240}Pu and ^{241}Am were inhomogeneously dispersed over a 7-acre area (Lee and Clark 2005). No further information on the physico-chemical characteristics of the particles seems to be available.

Johnston Atoll

Johnston Atoll was the launch site of atmospheric nuclear weapons tests in the late 1950s and early 1960s (Wolf et al. 1997). In 1962, three nuclear warhead-carrying THOR missiles were destructed; one on the launch pad and two others at altitudes of 9 and 33 km, and Pu- and U-containing particles were deposited throughout the atoll with a higher concentration of activity near the launch pad. Wolf et al. (1997) has reported of discrete hot particles with actinide containing inclusions exhibiting widely varying U/Pu ratios and comprising >99% of the total activity of the sample.

Thule and Palomares

In 1966 and 1968, two accidents involving USAF B-52 bombers and conventional explosions of weapon material and subsequent fires occurred in Palomares, Spain and Thule, Greenland, respectively. The surrounding areas were contaminated with submicron to mm particles containing Pu and enriched U (Eriksson et al. 2008; Jimenez-Ramos et al. 2007; Lind et al. 2007). Based on synchrotron radiation X-ray microtechniques, the Pu–U particles were oxidised (Lind et al. 2007).

3.1.6 Particles Released from Nuclear Installations

In nuclear fuel reprocessing, U and Pu from spent nuclear fuel are recovered for civil or military use. During reprocessing, the spent fuel is brought into solution thereby increasing the potential for release as liquid waste discharges. On the other hand, reports on U fuel particles found in the vicinity of reprocessing sites indicate that the dissolution of fuel may be incomplete (Bolsunovsky and Tcherkezian 2001; Maucec et al. 2004; Voillequé et al. 2002), and that residual fuel fragments and particles in the discharges represent important sources of fuel particles.

Particles Released from US Sites

According to Voillequé et al. (2002), episodic releases of physically large radioactive particles to the air occurred at the nuclear Hanford Site in the late 1940s and early 1950s. The first releases of particles were due to corrosion of the canals in the exhaust system during the early years of reprocessing plant operations. The contaminated corrosion particles were larger than those in routine releases and were deposited around and downwind of the processing plant stacks. Environmental surveys documented the presence of the particles, which contained iron and a mixture of radionuclides. The second group of episodic releases of active particles occurred in the 1950s. The key radioactive constituents were ^{103}Ru and ^{106}Ru in large particles released from the REDOX reprocessing facility. A layer of radioactive contamination formed on the lining of the stack by deposition of volatile ruthenium tetroxide and ammonium nitrate. When the layer of contaminated ammonium nitrate was disturbed, large radioactive particles were discharged. Although these releases from the stack walls were not well monitored, environmental surveys clearly showed the presence of the released particles in the vicinity and downwind of the stack.

The Columbia River has received releases of radioactivity following “a large number of ruptures of irradiated fuel elements in the single-pass production reactors” (Voillequé et al. 2002). In addition, discrete ^{60}Co particles have been found along the shorelines downstream of the Hanford reactors (Nees and Corley 1973). However, no further information appears to be available on the physico-chemical forms of the radioactive contamination in the Columbia River.

Rocky Flats

From 1952 to 1989, the Rocky Flats Nuclear Weapons Plant was a major processor of weapons grade Pu to build nuclear weapons triggers called “pits”. Rocky Flats was also responsible for recycling Pu retrieved from decommissioned nuclear warheads.

Workplace accidents, spills, fires, emissions, leakages from storage containers, and routine operations allowed Pu and many chemicals to be released from the plant site (Mongan et al. 1996a). Two major events caused the largest releases of Pu outside the Rocky Flats Plant boundaries. These events included a fire that occurred in the Pu processing building in 1957 (Mongan et al. 1996a), and wind-blown releases, mainly during 1968 and 1969, from an outdoor waste storage area called the 903 Area, located at the plant (Mongan et al. 1996b). Evidence of Pu particles being present in the surrounding area has been provided (McDowell and Whicker 1978), however, no particle characteristics except for particle size (submicron to ~7 μm) appear available.

Particles Released from Russian Sites

Nuclear weapon materials were produced at three main sites in the former Soviet Union: Mayak PA (Chelyabinsk-45 or 65), Krasnoyarsk Mining and Chemical Industrial Complex (Krasnoyarsk-26), and Siberian Chemical Combine (Tomsk-7). Relatively large routine releases occurred during the early years of operation of these facilities (UNSCEAR 1993). In addition, accidents have contributed to radioactive contamination in the vicinity of the sites. An accidental explosion occurred in the radiochemical plant at Tomsk-7 on April 6th April 1993 (AMAP 2004). As a result of the accident, the snow-covered region close to the facility was contaminated and a radioactive trace including radioactive particles spread 20 km in a North-Eastern direction. Total γ-activity of some of the hottest particles reached 12 kBq (Tcherkezian et al. 1995). Releases from Krasnoyarsk-26, including those resulting from three accidents, have contaminated the Yenisey river downstream of the site with U fuel particles (Fig. 6) and other forms of radioctive particles (Bolsunovsky and Tcherkezian 2001; Chuguevskii et al. 2010; Sukhorukov et al. 2004).

Literature on radioactive particles at Mayak PA is scarce. However, screening of contaminated soil and sediments from in and around the site using digital autoradiography indicated widespread presence of radioactive particles (Fig. 1), and radioactive, beta-emitting particles with a Sr-matrix was present in reservoir sediments (Fig. 7) (JNREG 2004).

European Reprocessing Plants

Among the three major reprocessing plants in Europe (Sellafield and Dounreay, UK, La Hague, France), Sellafield has been the major contributor of radionuclides

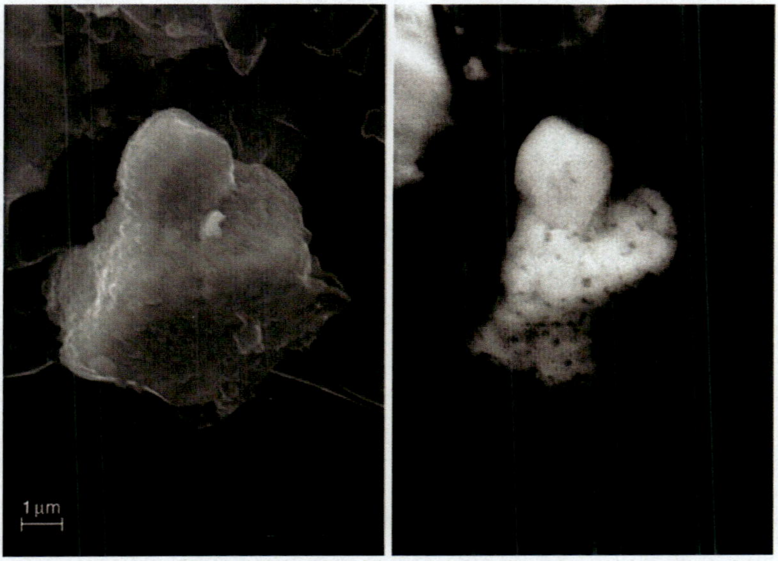

Fig. 6 Radioactive particle with a U matrix isolated from the river floodplain of the Beriozovy Island, downstream the Krasnoyarsk-26. *Left*: Image in SEI-mode, *Right*: Image in BEI-mod. Bar 1 µm

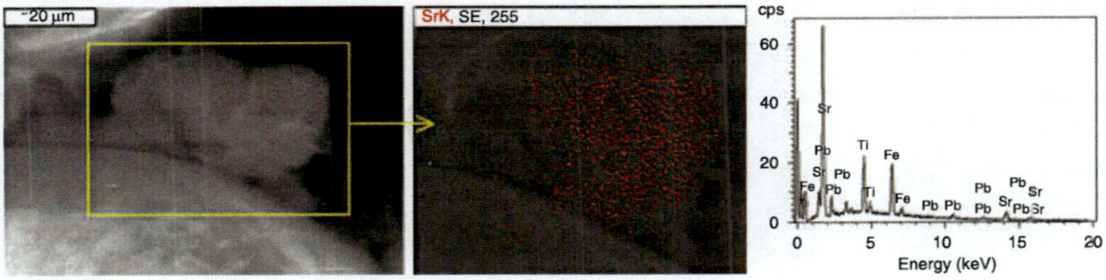

Fig. 7 Scanning electron microscopy of a particle isolated from Mayak Reservoir 10 sediment sample. Secondary electron imaging mode show the particle embedded in residual sediment material (*left*, bar 20 µm). X-ray mapping reveal that Sr is a major component of the beta-emitting particle matrix (*middle*) and X-ray microanalysis spot measurement confirm the presence of Sr, as well as Fe, Pb and Ti (*right*)

to the north European seas (e.g. McCartney et al., 1994). Radionuclides have been discharged from Sellafield to the environment, as aerosols and particles in air emissions or as ions, colloids and particles via the effluent pipeline to the Irish Sea (Leonard et al. 1995; Salbu et al. 1993). Sediments in the Irish Sea, especially outside the pipeline, i.e., in the mud patch, act as a sink and are highly contaminated partly due to sedimentation of particles. Radioactive particles with a U matrix containing about 2/3 U(IV) and 1/3 U(VI) as determined by µ-XANES have been isolated (Jernström et al. 2004). During the years, releases from Sellafield have also

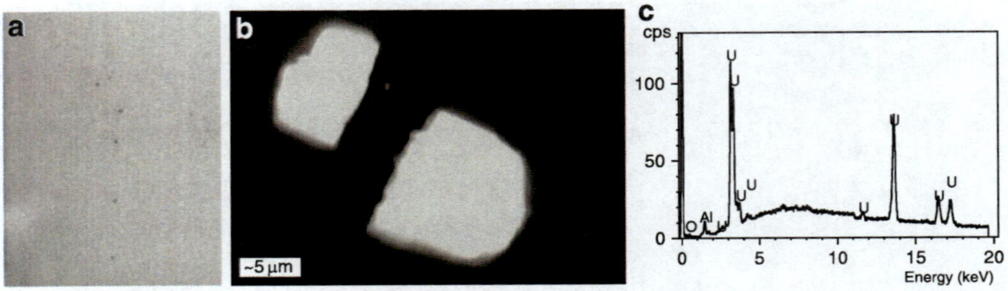

Fig. 8 Radioactive particles in Ravenglass estuary sediments (Lind 2006). (**a**) Digital autoradiography of Ravenglass surface sediments (0–5 cm). (**b**) Electron micrograph (BEI mode) of two particles with X-ray microanalysis (**c**) confirming the presence of U. Bar = 5 μm. The appearance of the particles indicates that they were actually one particle which was divided upon manipulation

contaminated the coastal zones, especially in Cumbria, UK. Hot particles (Fig. 8), including fuel fragments, have been identified in the effluent and have been shown to persist in the marine environment close to Sellafield.

For almost 20 years, highly radioactive particles and fragments have been recovered from the foreshore at the United Kingdom Atomic Energy Authority's (UKAEA) former reactor research establishment at Dounreay, Caithness, Scotland and from a neighbouring beach. It is believed that the fragments were inadvertently released into the marine environment during reprocessing operations involving irradiated Al-U fuel, which took place during the 1960s and 1970s (Maucec et al. 2004). The fragments vary substantially in size, but are most typically similar in size to grains of sand (0.2–2 mm). The principal radionuclides contained within the particles are the fission products ^{137}Cs (ranging from 10^3 to 10^8 Bq) and ^{90}Sr/^{90}Y, although they also contain small amounts of Pu isotopes and ^{241}Am (Darley et al. 2003). In addition to an alloy (Al/U ~ 7) matrix of Al and highly enriched U (82.9 ± 0.8 ^{235}U) (Tamborini 2004), DMTR fragments (Dounreay Materials Testing Reactor) contain Nd and Fe. DFR (Dounreay Fast Reactor) particles contain U, Nb and Fe. U is present as UO_2 in both DMTR and DFR particles (Salbu and Lind, unpublished).

3.1.7 Particles Associated with Dumping of Waste

During 1959–1991, radioactive waste including six reactors with fuel, 11 reactors without fuel, vessels, barges and more than 6,000 containers were dumped in the Abrosimov, Stepovogo and Tsivolky bays and in the Kara Sea Trough (AMAP 1997). Furthermore, the fuel assembly from the Lenin reactor, assumed dumped along the east coast of Novaya Zemlya, has not yet been localised. The re-estimated inventory is 37 PBq at the time of dumping, and about 4.7 PBq at present (AMAP 1997). In the close vicinities of dumped objects, especially containers, enhanced levels of Pu-isotopes and fission products in sediments were observed. Based on autoradiography, heterogeneities in the sediment samples reflected the presence of particles, and strong oxidising solutions (H_2O_2 in HNO_3) were needed to leach the Pu-isotopes from the sediments. Crud particles containing ^{60}Co were also identified using SEM (Salbu et al. 1997).

In the Kola Bay, NW Russia, two types of radioactive particles have been identified in marine sediment and in lichen samples collected in the vicinity of nuclear fuel storage and radioactive waste sites (Pöllänen et al. 2001). ^{137}Cs was present in the sediment matrix in large (~100 μm) greenish particles that were most probably pieces of paint. Although their element composition was heterogeneous, ^{137}Cs was found to be evenly distributed. ^{60}Co in the lichen matrix was present in small (~1 μm) particles. Neither U nor transuranium elements were identified in either type of particle.

3.2 *Radioactive Particles of Technologically Enhanced Naturally Occurring Radioactive Material origin*

Naturally occurring radionuclides may also occur as radioactive particles. Th, U and daughter nuclides such as ^{210}Po/^{210}Pb (Landa et al. 1994) may be heterogeneously distributed in minerals (Alsecz et al. 2007; Landa et al. 1994), thus occurring as hot spots in U mining and tailing sites. Recently, single grains of U minerals (Fig. 9) have been isolated from soil samples originating from former U mining sites in Kazakhstan and Kyrgyzstan (Lind et al. 2008). Similarly, heterogeneities (i.e., Th hot spots) have been observed on surfaces of Norwegian mineral specimens using ESEM with energy dispersive X-ray microanalysis (EDX) (O.C. Lind, pers. comm).

4 Conclusions

In all types of severe nuclear events with releases of refractory elements, a major fraction of released radionuclides will most probably be associated with particles. Thus, the phenomenon of radioactive particle should be expected to occur more frequently than previously anticipated. The identification of refractory radionuclides in releases from a nuclear source, implies the presence of particles.

Based on several years of research the particle composition will depend on the specific source, i.e., reflecting the matrix and the refractory radionuclide composition (e.g., burn-up). However, particle characteristics that are essential for transport and ecosystem behaviour such as particle size distribution, crystallographic structures and oxidation states, will be strongly influenced by the release scenarios (e.g., temperature, pressures, redox conditions).

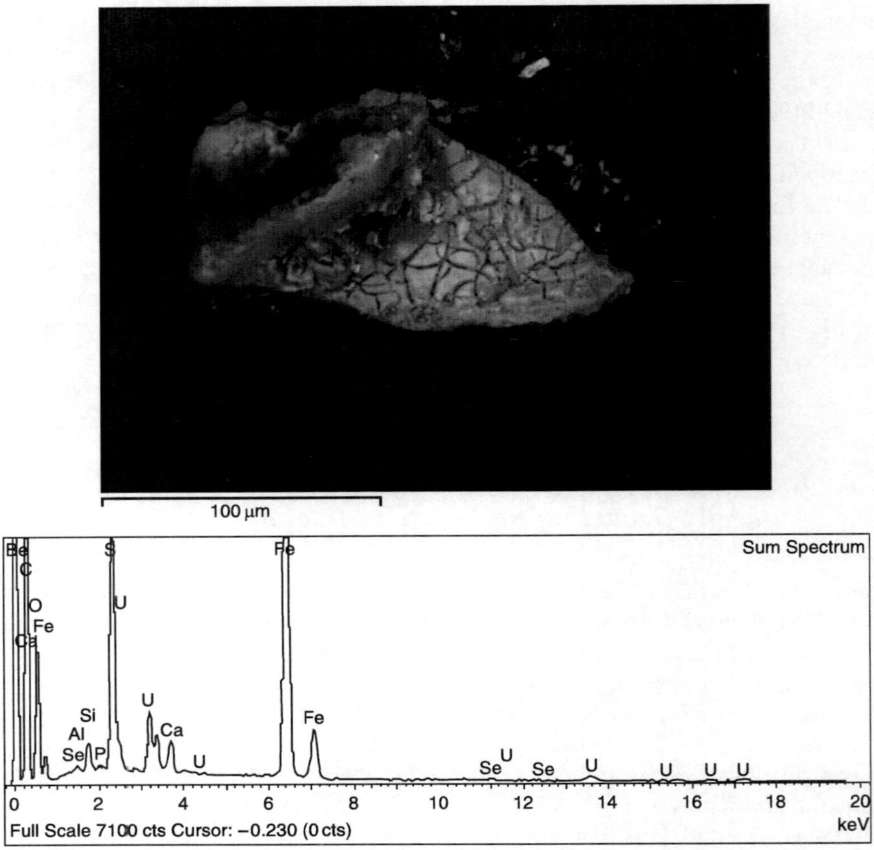

Fig. 9 Radioactive particle of TENORM origin isolated from soil collected in the vicinity of the abandoned Kadji-say U mine, Kyrgyzstan. Electron micrographs recorded in backscattered electron imaging mode (*upper*) using an environmental scanning electron microscope. Bright areas reflect high atomic number areas on the surface of the specimen. Energy dispersive X-ray microanalysis spectrum (*lower*) from the bright areas shows that U was a major element in the mineral grain and are associated with Fe, Ca, Si, S and Se

Soil and sediments can act as a sink for deposited particles, and particle contaminated soil and sediments may also act as a potential diffuse source in the future. Thus, within radioecology there is a need for knowledge with respect to particle characteristics and processes influencing particle weathering, and remobilisation of associated radionuclides to assess long-term impact from radioactive particle contamination. To obtain such information, advanced analytical methods including microanalytical techniques are needed for identification, isolation and characterisation of radioactive particles, enabling radioecologists to provide information essential to assess environmental impact of radioactive particles in contaminated areas.

Acknowledgments The authors gratefully acknowledge the support provided by EU 5. and 6. FP (RAFF, ADVANCE), IAEA (CRP, "Characterization of radioactive particles"), The Norwegian Research Council, and The Norwegian Foreign Ministry. Furthermore, the authors are indebted to close collaborating scientists: V. Kashparov, UIAR (Chernobyl), P.R. Danesi, IAEA (Kosovo, Kuwait), M. García León, University of Seville (Palomares), P. Mitchell, Univ. Coll. Dublin/N. Priest, Univ. of Middlesex (Semipalatinsk), P. Dale, SEPA (Dounrey), (Sellafield), S. Lukashenko, Kazakhstan (Semipalatinsk), H. Dahlgaard, Risoe National Laboratory (Thule), Joint Russian-Norwegian Expert Group (Kara Sea, Mayak, Kola) and K. Janssens, Univ. of Antwerp (ESRF, HASYLAB).

References

Alsecz A, Osan J, Kurunczi S, Alfoldy B, Varhegyi A, Torok S (2007) Analytical performance of different X-ray spectroscopic techniques for the environmental monitoring of the recultivated uranium mine site. Spectrochim Acta Part B At Spectrosc 62:769–776.

AMAP (1997) Arctic pollution issues: Radioactive contamination. Norwegian Radiation Protetion Authority, Østerås.

AMAP (2004) AMAP Assessment 2002: Radioactivity in the Arctic. Arctic Monitoring and Assessment Programme, Oslo, Norway.

Anspaugh LR, Church BW (1986) Historical estimates of external gamma exposure and collective external gamma exposure from testing at the Nevada Test Site. 1. Test series through Hardtack II, 1958. Health Phys 51:35–51.

Appleby LJ, Luttrell SJ (1993) Case-studies of significant radioactive releases. In: Warner F, Harrison RM, editors. Radioecology after Chernobyl. John Wiley & Sons LTD, Chichester.

Arnold L (1992) The Windscale Fire 1957. Anatomy of a Nuclear Accident. Macmillan Press, London.

Bolsunovsky AY, Tcherkezian VO (2001) Hot particles of the Yenisei River flood plain, Russia. J Environ Radioact 57:167–174.

Bunzl K (1997) Probability for detecting hot particles in environmental samples by sample splitting. Analyst 122:653–656.

Bunzl K (1998) Detection of radioactive hot particles in environmental samples by repeated mixing. Appl Radiat Isot 49:1625–1631.

Bunzl K, Tschiersch J (2001) Detection of radioactive hot particles in environmental samples using a Marinelli-beaker measuring geometry. Radiochim Acta 89:599–604.

Burns PA, Cooper MB, Johnston PN, Martin LJ, Williams GA (1994) Determination of the ratios of Pu-239 and Pu-240 to Am-241 for nuclear-weapons test sites in Australia. Health Phys 67:226–232.

Burns PA, Cooper MB, Lokan KH, Wilks MJ, Williams GA (1995) Characteristics of plutonium and americium contamination at the former UK atomic weapons test ranges at Maralinga and Emu. Appl Radiat Isot 46:1099–1107.

Chamberlain AC, Dunster HJ (1958) Deposition of radioactivity in north-west England from the accident at Windscale. Nature 182:629–630.

Chuguevskii AV, Sukhorukov FV, Mel'gunov MS, Makarova IV, Titov AT (2010) "Hot" particles of the Yenisei River: Radioisotope composition, structure, and behavior in natural conditions. Dokl Earth Sci 430:51–53.

Conradson SD (2000) XAFS – A technique to probe local structure. Los Alamos Sci 26:422–435.

Cooper MB, Burns PA, Tracy BL, Wilks MJ, Williams GA (1994) Characterization of plutonium contamination at the former nuclear-weapons testing range at Maralinga in South-Australia. J Radioanal Nucl Chem 177:161–184.

Crocker GR, Oconnor JD, Freiling EC (1966) Physical and radiochemical properties of fallout particles. Health Phys 12:1099.

Danesi PR, de Regge P, La Rosa JJ, Makarewicz M, Moreno JMB, Radecki Z, Zeiller, E. (1998) Residual plutonium isotopes and americium in the terrestrial environment at the former nuclear test sites of Mururoa and Fangataufa. 7th International Conference on Low level measurements of actinides and long-lived radionuclides in biological and environmental samples. Salt Lake City.

Danesi PR, Bleise A, Burkart W, Cabianca T, Campbell MJ, Makarewicz M, Moreno J, Tuniz C, Hotchkis M (2003a) Isotopic composition and origin of uranium and plutonium in selected soil samples collected in Kosovo. J Environ Radioact 64:121–131.

Danesi PR, Markowicz A, Chinea-Cano E, Burkart W, Salbu B, Donohue D, Ruedenauer F, Hedberg M, Vogt S, Zahradnik P, Ciurapinski A (2003b) Depleted uranium particles in selected Kosovo samples. J Environ Radioact 64:143–154.

Darley PJ, Charles MW, Fell TP, Harrison JD (2003) Doses and risks from the ingestion of Dounreay fuel fragments. Radiat Prot Dosimetry 105:49–54.

Devell L, Tovedal H, Bergstrom U, Appelgren A, Chyssler J, Andersson L (1986) Initial observations of fallout from the reactor accident at Chernobyl. Nature 321:192–193.

Dobrovolsky E, Lyalko V (1995) Acidification of soils and radioactive hot particles behavior: A macrokinetic approach. Water Air Soil Pollut 85:767–772.

Drell S, Peurifoy B (1994) Technical issues of a nuclear test ban. Annu Rev Nucl Part Sci 44:285–327.

Eriksson M, Osan J, Jernstrom J, Wegrzynek D, Simon R, Chinea-Cano E, Markowicz A, Bamford S, Tamborini G, Torok S, Falkenberg G, Alsecz A, Dahlgaard H, Wobrauschek P, Streli C, Zoeger N, Betti M (2005) Source term identification of environmental radioactive Pu/U particles by their characterization with non-destructive spectrochemical analytical techniques. Spectrochim Acta B 60:455–469.

Eriksson M, Lindahl P, Roos P, Dahlgaard H, Holm E (2008) U, Pu, and Am nuclear signatures of the Thule hydrogen bomb debris. Environ Sci Technol 42:4717–4722.

Gerdes A, Weyer S, Brey G, Durakovic A, Zimmermann I (2004) Monitoring depleted uranium contamination in the biosphere of Iraq using MC-ICP-MS. Geochim Cosmochim Acta 68: A506.

Gregory S, Edwards A (1989) The hidden cost of deterrence – Nuclear-weapons accidents 1950–88. Bulletin of Peace Proposals 20:3–26.

Heft RE (1970) Characterization of radioactive particles from nuclear weapons tests. Advances in Chemistry Series.

IAEA (1998) Radiological conditions at the Semipalatinsk test site, Kazakhstan: Preliminary assessment and recommendations for further studies. International Atomic Energy Agency, Vienna.

IAEA CRP (2001) Radiochemical, chemical and physical characterisation of radioactive particles in the environment. IAEA, Vienna.

Jakeman, D (1986) Notes of the level of radioactive contamination in the Sellafield area arising from discharges in the Early 1950s. Atomic Energy Establishment, Winfrith.

Jernström J, Eriksson M, Osan J, Tamborini G, Torok S, Simon R, Falkenberg G, Alsecz A, Betti M (2004) Non-destructive characterisation of low radioactive particles from Irish Sea sediment by micro X-ray synchrotron radiation techniques: micro X-ray fluorescence (mu-XRF) and micro X-ray absorption near edge structure (mu-XANES) spectroscopy. J Anal At Spectrom 19:1428–1433.

Jimenez-Ramos MC, Barros H, Garcia-Tenorio R, Garcia-Leon M, Vioque I, Manjon G (2007) On the presence of enriched uranium in hot-particles from the terrestrial area affected by the Palomares accident (Spain). Environ Pollut 145:391–394.

JNREG (2004) Impacts on man and the environment in Northern areas from hypothetical accidents at "Mayak" PA, Urals, Russia. Joint Norwegian-Russian expert group for investigation of radioactive contamination in the Northern areas, Østerås.

Kashparov VA, Oughton DH, Protsak VP, Zvarisch SI, Levchuk SE (1999) Kinetics of fuel particle weathering and 90Sr mobility in the Chernobyl 30 km exclusion zone. Health Phys 76:251–259.

Kersting AB, Efurd DW, Finnegan DL, Rokop DJ, Smith DK, Thompson JL (1999) Migration of plutonium in ground water at the Nevada Test Site. Nature 397:56–59.

Krey PW (1967) Plutonium-238 from Snap-9A Burnup. Trans Am Nucl Soc 10:1.

Krey PW, Leifer R, Benson WK, Dietz LA, Coluzza JL, Hendrikson HC (1979) Atmospheric burnup of the Cosmos-954 reactor. Science 205:583–585.

Kuriny VD, Ivanov YA, Kashparov VA, Loshchilov NA, Protsak VP, Yudin EB, Zhurba MA, Parshakov AE (1993) Particle-associated Chernobyl fallout in the local and intermediate zones. Ann Nucl Energy 20:415–420.

Landa ER, Stieff LR, Germani MS, Tanner AB, Evans JR (1994) Intense alpha-particle emitting crystallites in uranium mill wastes. Nucl Geophys 8:443–454.

Lee MH, Clark SB (2005) Activities of Pu and Am isotopes and isotopic ratios in a soil contaminated by weapons-grade plutonium. Environ Sci Technol 39:5512–5516.

Leifer R, Juzdan ZR, Kelly WR, Fassett JD, Eberhardt KR (1987) Detection of uranium from Cosmos-1402 in the stratosphere. Science 238:512–514.

Leonard KS, McCubbin D, Lovett MB (1995) Physico-chemical characterisation of radionuclides discharged from a nuclear establishment. Sci Total Environ 175:9–24.

Lind OC (2006) Characterisation of radioactive particles in the environment using advanced techniques. PhD thesis, Norwegian University of Life Sciences.

Lind OC, Salbu B, Janssens K, Proost K, Garcia-Leon M, Garcia-Tenorio R (2007) Characterization of U/Pu particles originating from the nuclear weapon accidents at Palomares, Spain, 1966 and Thule, Greenland, 1968. Sci Total Environ 376:294–305.

Lind, OC, Strømman, G, Skipperud, L, Rosseland, BO, Janssens, K, De Nolf, W, Jaroszewicz, J, Burkitbayev, M, Salbu, B (2008) Speciation and source identification of U from contaminated TENORM sites. European Geological Union, Vienna.

Lind OC, Salbu B, Skipperud L, Janssens K, Jaroszewicz J, De Nolf W (2009) Solid state speciation and potential bioavailability of depleted uranium particles from Kosovo and Kuwait. J Environ Radioact 100:301–307.

Mamuro T, Fujita A, Yoshikawa K, Matsunami T (1962) Microscopic examination of highly radioactive fall-out particles. Nature 196:529.

Mamuro T, Fujita A, Matsunam T (1965) Microscope examination of highly radioactive fallout particles from first Chinese nuclear test explosion. Health Phys 11:1097–1101.

Mamuro T, Yoshikaw K, Matsunami T, Fujita A (1966) Radionuclide fractionation in debris from a land surface burst. Health Phys 12:757–763.

Maucec A, de Meijer RJ, Rigollet C, Hendriks PHGM, Jones DG (2004) Detection of radioactive particles offshore by mu-ray spectrometry – Part I: Monte Carlo assessment of detection depth limits. Nucl Instr Meth A 525:593–609.

McCartney M, Kershaw PJ, Woodhead DS, Denoon DC (1994) Artificial radionuclides in the surface sediments of the Irish Sea, 1968–1988. Sci Total Environ 141:103–138.

McDowell LM, Whicker FW (1978) Size characteristics of plutonium particles in rocky flats soil. Health Phys 35:293–299.

McLaughlin JP, Vintro LL, Smith KJ, Mitchell PI, Zunic ZS (2003) Actinide analysis of a depleted uranium penetrator from a 1999 target site in southern Serbia. J Environ Radioact 64:155–165.

Mian Z, Ramana MV, Rajaraman R (2001) Plutonium dispersal and health hazards from nuclear weapon accidents. Curr Sci 80:1275–1284.

Mongan TR, Ripple SR, Brorby GP, diTommaso DG (1996a) Plutonium releases from the 1957 fire at Rocky Flats. Health Phys 71:510–521.

Mongan TR, Ripple SR, Winges KD (1996b) Plutonium release from the 903 Pad at Rocky Flats. Health Phys 71:522–531.

Nees WL, Corley JP (1973) Environmental surveillance at Hanford for CY-1973. Pacific Northwest Laboratories, Richland, WA.

Oughton DH, Salbu B, Brand TL, Day JP, Aarkrog A (1993) Under-determination of Sr-90 in soils containing particles of irradiated uranium oxide fuel. Analyst 118:1101–1105.

Pöllänen R, Klemola S, Ikaheimonen TK, Rissanen K, Juhanoja J, Paavolainen S, Likonen J (2001) Analysis of radioactive particles from the Kola Bay area. Analyst 126:724–730.
Salbu B (1988) Radionuclides associated with colloids and particles in rainwaters, Oslo, Norway. In: von Philipsborn H, Steinhäuser F, editors. Hot particles from the Chernobyl Fallout. Bergbau – und Industrimuseum, Theuern.
Salbu B (2000a) Source-related characteristics of radioactive particles: A review. Radiat Prot Dosimetry 92:49–54.
Salbu B (2000b) Speciation of radionuclides in the environment. In: R.A.Meyers, editor. Encyclopedia of analytical chemistry. John Wiley & Sons Ltd, Chichester.
Salbu B (2001) Actinides associated with particles. In: Kudo A, editor. Plutonium in the environment. Elsevier, Tokyo.
Salbu B (2009) Fractionation of radionuclide species in the environment. J Environ Radioact 100:283–289.
Salbu B, Bjornstad HE, Svaren I, Prosser SL, Bulman RA, Harvey BR, Lovett MB (1993) Size distribution of radionuclides in nuclear-fuel reprocessing liquids after mixing with seawater. Sci Total Environ 130:51–63.
Salbu B, Krekling T, Oughton DH, Ostby G, Kashparov VA, Brand TL, Day JP (1994) Hot particles in accidental releases from Chernobyl and Windscale nuclear installations. Analyst 119:125–130.
Salbu B, Nikitin AI, Strand P, Christensen GC, Chumichev VB, Lind B, Fjelldal H, Bergan TDS, Rudjord AL, Sickel M, Valetova NK, Foyn L (1997) Radioactive contamination from dumped nuclear waste in the Kara sea – Results from the joint Russian-Norwegian expeditions in 1992–1994. Sci Total Environ 202:185–198.
Salbu B, Krekling T, Lind OC, Oughton DH, Drakopoulos M, Simionovici A, Snigireva I, Snigirev A, Weitkamp T, Adams F, Janssens K, Kashparov VA (2001) High energy X-ray microscopy for characterisation of fuel particles. Nucl Instr Meth A 467:1249–1252.
Salbu B, Janssens K, Lind OC, Proost K, Danesi PR (2003) Oxidation states of uranium in DU particles from Kosovo. J Environ Radioact 64:167–173.
Salbu B, Lind OC, Skipperud L (2004) Radionuclide speciation and its relevance in environmental impact assessments. J Environ Radioact 74:233–242.
Salbu B, Janssens K, Lind OC, Proost K, Gijsels L, Danesi PR (2005) Oxidation states of uranium in depleted uranium particles from Kuwait. J Environ Radioact 78:125–135.
Shaun G (1990) The hidden cost of deterrence: nuclear weapons accidents. Brassey's, London.
Simon SL, Jenner T, Graham JC, Borchert A (1995) A comparison of macroscopic and microscopic measurements of plutonium in contaminated soil from the Republic-of-the-Marshall-Islands. J Radioanal Nucl Chem 194:197–205.
Sisefsky J (1961) Debris from tests of nuclear weapons – Activities roughly proportional to volume are found in particles examined by autoradiography and microscopy. Science 133:735.
Smith JN, Ellis KM, Polyak L, Ivanov G, Forman SL, Moran SB (2000) 239, 240Pu transport into the Arctic Ocean from underwater nuclear tests in Chernaya Bay, Novaya Zemlya. Cont Shelf Res 20:255–279.
Sowder AG, Clark SB, Field RA (1999) The transformation of uranyl oxide hydrates: The effect of dehydration on synthetic metaschoepite and its alteration to becquerelite. Environ Sci Technol 33:3552–3557.
Sukhorukov FV, Degermendzhi AG, Bolsunovsky A, Belolipetskii VM, Kosolapova LG (2004) Distribution and migration behaviours of radionuclides in the Yenisei River floodplain. SB RAS Publishers, Novosibirsk.
Tamborini G (2004) SIMS analysis of uranium and actinides in microparticles of different origin. Microchim Acta 145:237–242.
Tcherkezian V, Galushkin B, Goryachenkova T, Kashkarov L, Liul A, Roschina I, Rumiantsev O (1995) Forms of contamination of the environment by radionuclides after the Tomsk accident (Russia, 1993). J Environ Radioact 27:133–139.

UNEP (2003) Depleted uranium in Bosnia and Herzegovina: Post-conflict environmental assessment. United Nations Environment Programme, Geneva.
UNSCEAR (1993) Sources and effects of ionizing radiation. United Nations, New York.
UNSCEAR (2000) Sources and effects of ionizing radiation. United Nations, New York.
Victorova NV, Garger EK (1990) Biological monitoring of the deposition and transport of radioactive aerosol particles in the Chernobyl NEP zone of influence. Proceedings of the CEC seminar on comparative assessment of the environmental impact of radionuclides released during three major nuclear accidents; Kyshtym, Windscale, Chernobyl.
Voillequé PG, Killough GG, Rope SK (2002) Methods for estimating radiation doses from short-lived gaseous radionuclides and radioactive particles released to the atmosphere during early Hanford operations – Final report. Centers for Disease Control and Prevention Department of Health and Human Services.
Wolf SF, Bates JK, Buck EC, Dietz NL, Portner JA, Brown NR (1997) Physical and chemical characterization of actinides in soil from Johnston Atoll. Environ Sci Technol 31:467–471.

Effect of Redox Conditions on Actinide Speciation and Partitioning with Colloidal Matter

Stepan N. Kalmykov, Elena V. Zakharova, Alexander P. Novikov, Boris F. Myasoedov, and Satoshi Utsunomiya

Abstract Colloid-facilitated transport of actinides in subsurface environments depends on various geochemical and geological properties among which the redox conditions are very important. In reducing conditions, redox-sensitive actinides are stabilized in low valence states and could potentially form intrinsic colloids while in oxidizing conditions, they form pseudocolloids by sorption or coprecipitation onto natural aquatic colloids. The difference in actinides' behavior associated with colloids at two sites in Russia under different redox conditions is discussed in this chapter. Successive micro- and ultrafiltrations of groundwater samples were performed by electron microscopic methods, nanoSIMS and sequential extraction for partitioning. Under oxidizing conditions of the PA "Mayak" site (Southern Urals, Russia), U and Np are present in the higher oxidation states while Pu is mostly found in the tetravalent state. Under these conditions, actinides are bound to amorphous hydrous ferric oxide (HFO) and Mn oxides, forming pseudocolloids as demonstrated by nanoSIMS. According to nanoSIMS, at the Tomsk site (Siberian Chemical Combine, Russia) oxidizing waste solutions were mixed with

---

S.N. Kalmykov (✉)
Division of Radiochemistry, Chemistry Department, Lomonosov Moscow State University, Leninskie Gory, 119991 Moscow, Russia
and
Vernadsky Institute of Geochemistry and Analytical Chemistry, Russian Academy of Sciences, Kosygin st., 17, 119991 Moscow, Russia
e-mail: stepan@radio.chem.msu.ru

E.V. Zakharova and B.F. Myasoedov
Frumkin Institute of Physical Chemistry and Electrochemistry, Russian Academy of Sciences, Leninsky ave. 31, 119991 Moscow, Russia

A.P. Novikov
Vernadsky Institute of Geochemistry and Analytical Chemistry, Russian Academy of Sciences, Kosygin st., 17, 119991 Moscow, Russia

S. Utsunomiya
Department of Chemistry, Kyushu University, 6-10-1 Hakozaki, Higashi-ku, Fukuoka 812-8581, Japan

S.N. Kalmykov and M.A. Denecke (eds.), *Actinide Nanoparticle Research*,
DOI 10.1007/978-3-642-11432-8_13,

reducing groundwater resulting in reduction of U(VI) and formation of intrinsic U(IV) hydroxocolloids. Most of the U and Pu in a reducing environment are present in a residual fraction of low-mobility that favors slow migration rates.

1 Introduction

The development of the nuclear fuel cycle in recent decades has resulted in the accumulation of large amounts of high-level wastes (HLW) and spent nuclear fuel (SNF). The final stage of nuclear fuel cycle adopted in many countries is the disposal of SNF and HLW in geological formations. Several concepts are accepted in different countries for the disposal of SNF or HLW. In the US, Yucca Mountain was proposed as the site for SNF disposal in volcanic tuff, but the project was terminated in 2009. Another active site located in salt formations is the Waste Isolation Pilot Plant (Carlsbad, NM) used for transuranium waste disposal. In Russia, several options are being considered for HLW and SNF disposal including the proposed international repository in granite formations near Krasnoyarsk. Despite the significant differences in the national strategies of the nuclear industry and repository design, a common feature is the assessment programs required for the licensing of such repositories. For the actinides that are long-lived components of SNF and HLW, speciation data obtained from natural analog sites (e.g., Cigar Lake in Canada and Oklo in Gabon) or artificially contaminated sites (e.g., Hanford site in USA and PA "Mayak" in Russia) having different geochemical conditions are important in the development of reactive contaminant transport models.

The most challenging problem is the development of performance assessment programs that can take into account the colloid-facilitated transport of radionuclides. Three types of colloids are distinguished including aquatic colloids, intrinsic colloids (or eigencolloids) and pseudocolloids. Aquatic colloids are mostly mineral particles, precipitates of hydrolyzed metal ions, high-molecular-weight organic substances, microorganisms, etc. When radionuclides are sorbed by aquatic colloids, the so-called "pseudocolloids" are generated. The formation of intrinsic colloids are typical for highly hydrolyzable low-soluble radionuclides, e.g., tetravalent cations (Neck et al. 2002; Fanghänel and Neck 2002; Walther et al. 2009; Cho et al. 2005) and Tc(IV) (Sekine et al. 2002). They produce aggregates of hydrolyzed species through oxo or hydroxo bridges. Several studies have reported the formation of An(IV) hydroxocolloids (Neck et al. 2002; Fanghänel and Neck 2002; Walther et al. 2009) including those in simulated groundwater (Kaminski et al. 2005); however, to the best of our knowledge, "intrinsic" actinide colloids have never been reported under real environmental conditions.

The lack of knowledge on the effect of colloids on radionuclide speciation should be compensated by reliable experimental data obtained under well-defined laboratory conditions and supported by the examination of samples collected at radionuclide contaminated territories. This chapter deals with the actinides' (U, Np, Pu and Am) speciation and partitioning to colloids in groundwater samples collected from two different geochemical environments: oxidizing groundwater collected at the PA "Mayak" site and reducing groundwater collected from the clay–sand environment of the deep liquid radioactive waste injection site at Siberian Chemical Combine.

2 Description of Sampling Sites

Production Association "Mayak" is an SNF plant located in the South Urals. For several decades nitrate waste effluents containing fission products (^{90}Sr, ^{99}Tc, ^{137}Cs) and actinides (U, Pu, Np and Am isotopes) were discharged to Karachai Lake. These brine solutions had pH values between 7.9 and 9.3, average salt concentrations around 50 g/L with the major dissolved ionic species being NO_3^-, CH_3COO^-, $C_2O_4^{2-}$, SO_4^{2-}, Cl^-, Na^+, Ca^{2+}, Mg^{2+} and some surfactants. Lake Karachai is connected to a 55- to 100-m-thick groundwater zone, in which fluids flow through fractured Silurian and Devonian metavolcanic rocks with andesitic and basaltic composition (Fig. 1). The waste effluents penetrate into the groundwater system by gravity. The contamination plume has migrated as far as 4.5 km from Karachai Lake over the course of about 50 years (Novikov et al. 2006). The concentration of waste components, i.e., radionuclides, nitrates, dissolved oxygen and hydrous ferric oxide (HFO) colloids, increases with depth and decreases with

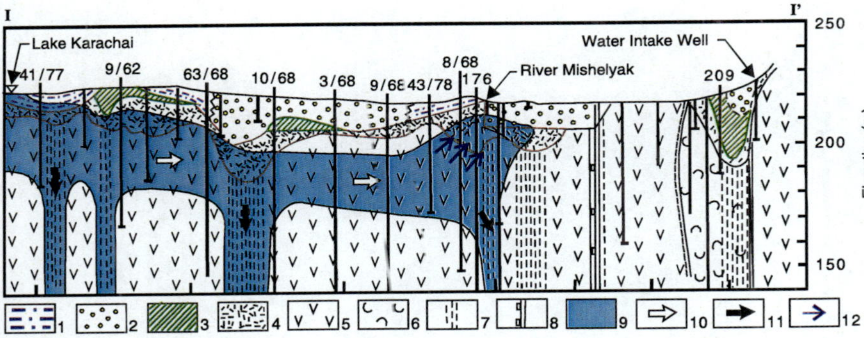

Fig. 1 Geologic cross section in a southerly direction from Karachi Lake and hydrogeochemical sections through the southern limb of the plume of contaminated water. The zone of open observation wells is delineated by a *dashed line*. Hydrogeochemical sections are prepared from hydrogeochemical logging results and interval sampling. 1 – late Miocene deluvial and proluvial loams with rare quartz pebbles, 2 – late Miocene alluvial sands and clays, 3 – loams of the weathered crust, 4 – weathered bedrocks, 5 – porphyrites, 6 – tuffs, 7 – schistose zones, 8 – a "rejuvenated" fault, 9 – plume, 10 – predominant direction of the contaminated water movement, 11 – structural "traps" (zones of accumulation) of dense brines, 12 – assumed direction of discharge of contaminated waters in the valley of the Mishelyak River (reproduced from Solodov et al. 1994)

distance. The contaminated groundwater is highly oxic with Eh between +50 and +500 mV (vs s.h.e.). A more detailed description of the geochemical and environmental conditions around Karachi Lake was published earlier (Solodov et al. 1994; Lisizin et al. 2001; Myasoedov and Novikov 1997; Drozhko et al. 1996; Samsonova and Drozhko 1996; Novikov et al. 1998).

Injection of low- and intermediate level liquid wastes containing short-lived fission products and some tracer amounts of actinides is the strategy proposed (Rybalchenko et al. 1994, 2005) at three nuclear waste sites, i.e., the Krasnoyarsk Mining and Chemical Combine, the Siberian Chemical Combine and the Research Institute of Atomic Reactors at Dimitrovgrad. At the Siberian Chemical Combine, low-level radioactive wastes are injected into the II and III geological horizons located at depths of about 370 m and 300 m, respectively (Fig. 2). Intermediate-level radioactive wastes are injected into the II horizon located at the depth of 330 m. The reservoir horizons are composed of quartz (40–70%), illite, montmorillonite and kaolinite (10–40%), feldspars (5–20%), micas, chlorites and other minerals. Weakly cemented sand–clay formations are separated by impermeable clay strata that prevent radionuclide migration away from the injection wells. Both the horizons contain groundwater with a low salt content of 0.3–0.4 g/L and low Eh values around −150 to −200 mV (vs s.h.e.) that is controlled by S^{2-} and Fe^{2+}. More detailed information about the sampling site, its geochemical and geological conditions as well as injection strategy was published earlier (Rybalchenko et al. 1994, 2005).

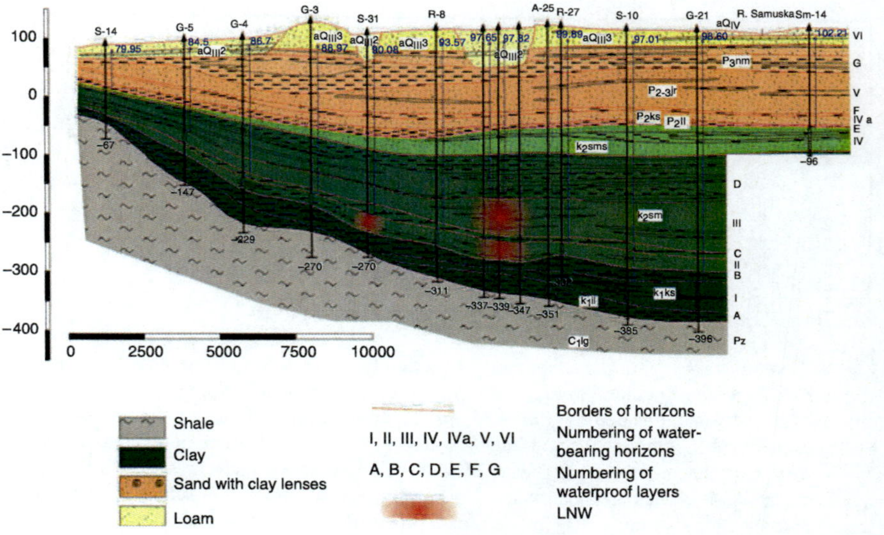

Fig. 2 Geological cross section of the deep liquid nuclear waste injection site [reproduced from (Rybalchenko et al. 1994)]

3 Groundwater Sampling and Analysis

Field work at both the sites includes in situ measurements of pH, Eh, temperature, and concentrations of dissolved O_2, H_2S, Na^+, NH_4^+, NO_3^- performed using a multichannel hydrogeochemical probe (developed in the IGEM RAS).

The groundwater properties, especially for samples from a reducing environment, can be altered significantly after sampling due to Fe(II) oxidation with precipitation of Fe(III) hydroxide colloids. Groundwater samples are pumped in situ from the well at a rate of 0.2–0.25 m^3/h, which will prevent immobilization of colloids sorbed onto rocks. Samples are collected into glass bottles, previously rinsed with high-purity nitric acid and deionized water. All the equipment is purged with nitrogen before sampling. Successive micro- and ultrafiltrations using 200 nm, 50 nm (nucleopore filters, Dubna) and 5 nm (Vladipor, Russia) are performed in an inert atmosphere immediately after the sampling. The concentration of radionuclides associated with colloids of different size is determined by complete dissolution of a known portion of the filter with colloids and subsequent ICP-MS and nuclear spectroscopy detection after chemical separation.

Aliquots of groundwater samples are taken from each sample for the determination of major and trace elements using conventional analytical techniques, e.g., ICP-MS, AAS and IC as presented in Table 1.

The salt content in the sample collected in the oxidizing conditions near Karachai Lake at the "Mayak" site is significantly higher than for samples collected in the reducing conditions of the Siberian Chemical Combine and is defined by sodium nitrates and other salts. The oxidizing groundwater is visibly oversaturated with respect to HFO and calcite. The important difference between two groundwater samples is the value of the redox potential. In the case of groundwater sampled

Table 1 Chemical composition of the groundwater samples from oxidizing and reducing environments

Component	Concentration, mg/L	
	Oxic	Anoxic
Na	750	44.4
K	9.48	16.6
Ca	2,291	28.5
Mg	181	12.5
Ba	n.d.[a]	0.15
Sr	n.d.	1.4
Mn	8.7	0.29
Al	3.2	0.25
Fe	15.3	3.30
HCO_3^-/CO_3^{2-}	266	235
Cl^-	472	7.8
SO_4^{2-}	220	27
PO_4^{3-}	0.5	0.25
NO_3^-	6,900	1.5
pH	7.8	7.0
Eh, mV	+200	−118

[a]Not determined

under reducing conditions, the low Eh value is due to the presence of S^{2-} and Fe^{2+} in the sand–clay formations and the absence of atmospheric oxygen.

4 Association of Radionuclides with Groundwater Colloids

Figure 3 depicts the association of actinides with colloids of different size for both groundwater samples. It is not surprising that for samples collected under oxidizing conditions, soluble uranium and neptunium are in the filtrate, while plutonium and americium are associated with colloids. The reverse is seen under reducing conditions where all actinides are predominantly associated with colloids. The difference in behavior of U and Np in these two samples results from their redox speciation under these conditions.

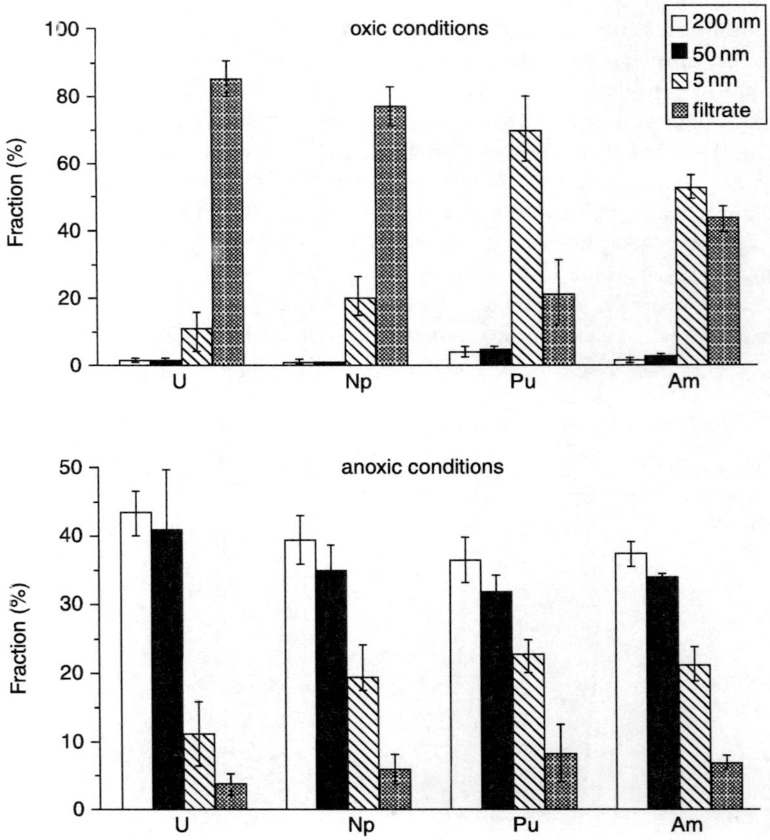

Fig. 3 Association of actinides with colloids of different size under oxidizing conditions (upper figure) and reducing conditions (lower figure)

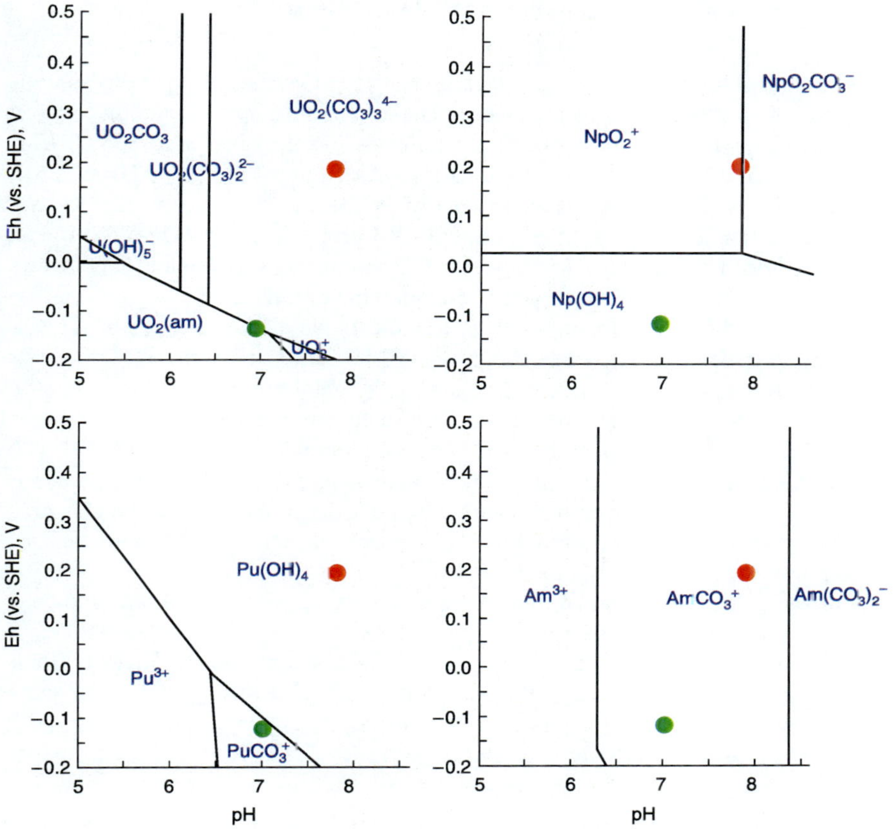

Fig. 4 Pourbaix diagrams for actinides. *Red markers* correspond to samples collected in oxidizing conditions and *green markers* correspond to samples collected in reducing conditions

The Pourbaix diagrams for actinides showing conditions for both samples are presented in Fig. 4. The important conclusion from Fig. 4 is that under reducing conditions uranium and neptunium are present in the tetravalent form while for oxidizing conditions they occur as soluble carbonate species. Tetravalent actinides form hydroxides of very low solubility which precipitate by way of the formation of intrinsic colloids at the nanoscale with further aggregation to sibmicron and micron particles. Considering that the total concentration of uranium in the sample collected in reducing conditions exceeds the solubility limit for $U(OH)_4$(am), the formation of intrinsic colloids or their aggregates is possible.

Plutonium speciation is dominated by reduced forms, Pu(IV)/Pu(III), in both cases. Despite the similarity of americium speciation in the collected samples, the fraction of americium in filtrate is higher for samples taken under oxidizing conditions. This might be caused by the presence of simple carboxylic acids and surfactants which could increase the concentration of soluble species.

5 Characterization of Groundwater Colloids

After transportation to the laboratory, colloids are characterized by various analytical techniques including different types of electron microscopy – scanning electron microscopy (SEM, Hitachi S-3200N) and transmission electron microscopy (TEM, JEOL JEM2010F). High-angle annular dark-field scanning TEM (HAADF-STEM), high-resolution (HR) TEM and energy dispersive X-ray spectroscopy (EDX) analyses are conducted using a 0.5 nm sized STEM probe. The HAADF-STEM is an option in TEM that enables HR images with Z-contrast to be obtained. This feature is important when searching for actinide-containing particles.

The ion-probe nanomapping of major colloid-borne elements (Fe, Mn, Ti, Ca, Al) and actinides is carried out using nanoSIMS (Cameca, NanoSIMS-50). For nanoSIMS measurements, aliquots containing colloidal particles are deposited on silicon chips and dried. All analyses are performed using 16 keV O^- primary ions, and detecting positive secondary ions. Mass resolution was set at $M/\mathrm{d}M = 2{,}000$ for all data. Because the contrast of elemental maps presented in this chapter is enhanced to show the distribution clearly, the intensity of the color in the chemical maps corresponds to the relative concentration for each element, but cannot be used to compare one element with another.

For samples collected in an oxidizing environment, different colloids and their aggregates are detected, although amorphous HFO that also contains admixtures of Si and Ca dominates (Novikov et al. 2006). The concentration of other colloids decreases in the following order: clays $\approx$ calcite ($CaCO_3$) > rutile (TiO_2) $\approx$ hematite (Fe_2O_3) $\approx$ barite ($BaSO_4$) $\approx$ rancieite ($CaMn_5O_9$) $\approx$ plagioclase ($NaCaAlSi_2O_8$) > monazite (REE phosphate) $\approx$ quartz (SiO_2). Typical SEM and TEM images of colloids separated from the groundwater collected in oxidizing conditions are presented in Figs. 5 and 6.

It was shown earlier (Novikov et al. 2006) that HFO and MnO_2 colloids are responsible for uranium and plutonium binding and formation of pseudocolloids.

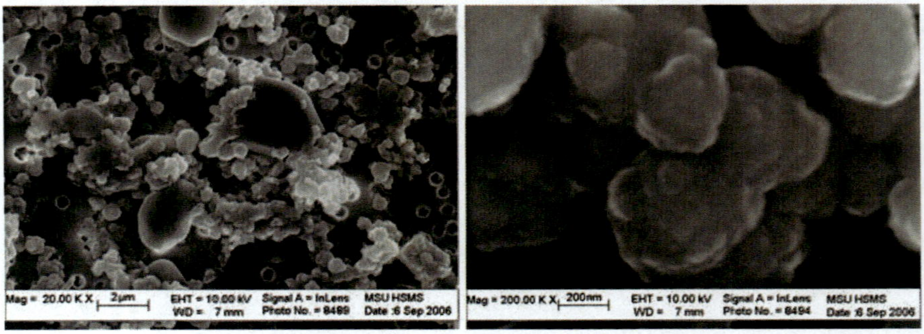

Fig. 5 Typical SEM images of colloids from groundwater collected under oxidizing conditions: *left*: general view, *right*: aggregate of HFO colloids as determined by EDX

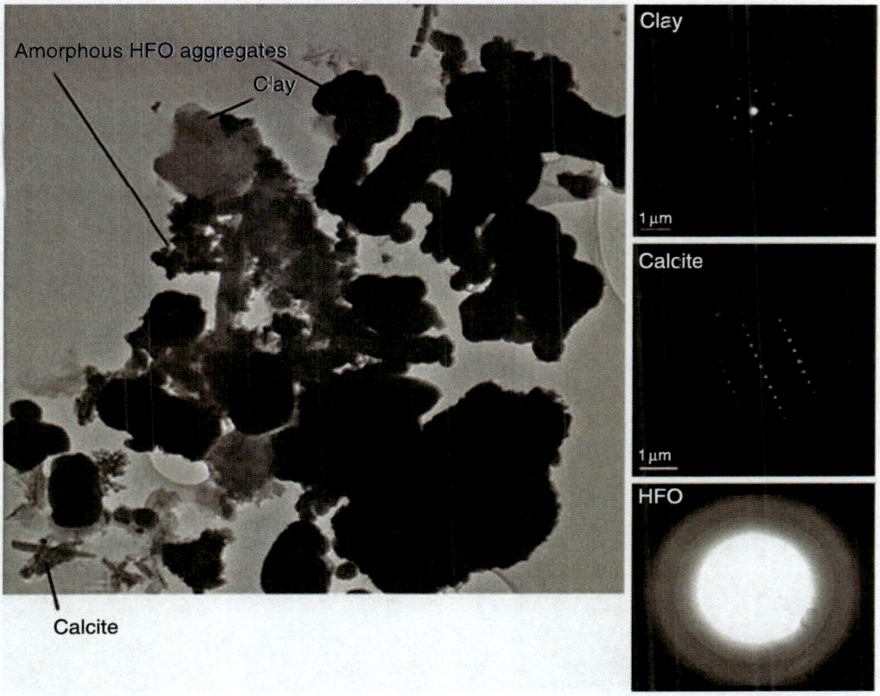

Fig. 6 TEM image and corresponding SAED patterns of different colloids from groundwater collected under oxidizing conditions

The typical element distributions of U, Pu and some major colloid borne elements as obtained by nanoSIMS are presented in Fig. 7. The corresponding normalized correlation plots of U versus Fe and U versus Mn are shown in Fig. 8.

Colloids sampled from the groundwater collected under reducing conditions are clays, pyrite, monazite, hematite, zeolites and barite, with a small fraction of other phases as demonstrated by SEM (Fig. 9) and EDX (Fig. 10).

The nanoSIMS examination of colloids from samples collected in reducing conditions does not allow the determination of the local distribution of plutonium because of its very low concentration, which is below the detection limit for nanoSIMS. However, several elemental maps for uranium were obtained as presented in Fig. 11. The main feature of element distributions in samples collected from reducing conditions is the absence of correlation of U with major colloid-borne elements, i.e., Fe, Mn, Si, Al and Ti. This could be interpreted by the formation of U(IV) hydroxide particles, whose distribution does not correlate with the distribution of other colloid particles. This is possible upon interaction of oxic liquid wastes with reducing groundwater.

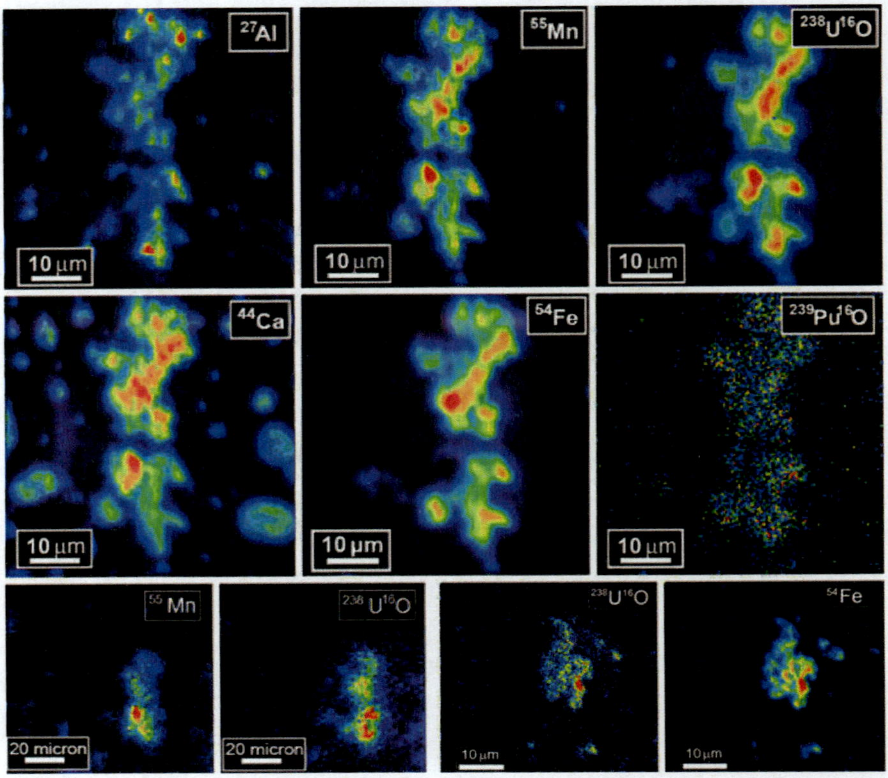

Fig. 7 NanoSIMS elemental maps showing correlation of U and Pu with Fe and Mn in colloid matter from an oxidizing environment (reproduced from Novikov et al. 2006)

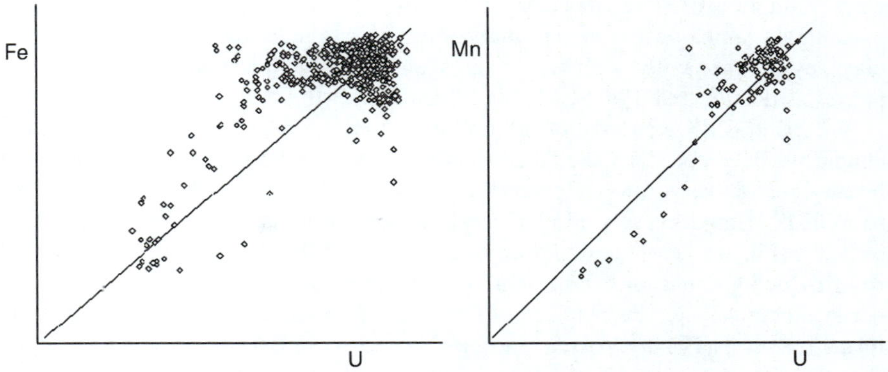

Fig. 8 Correlation plots for distributions of normalized intensities for U versus Fe and U versus Mn

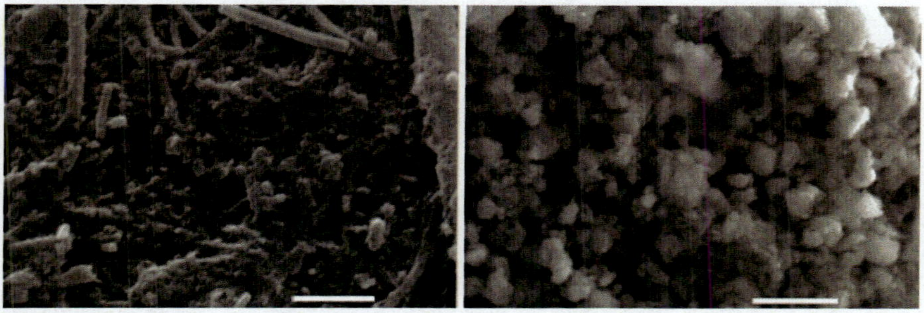

Fig. 9 Typical SEM images of colloids separated from groundwater collected in reducing conditions. *Scale bar* = 1 μm

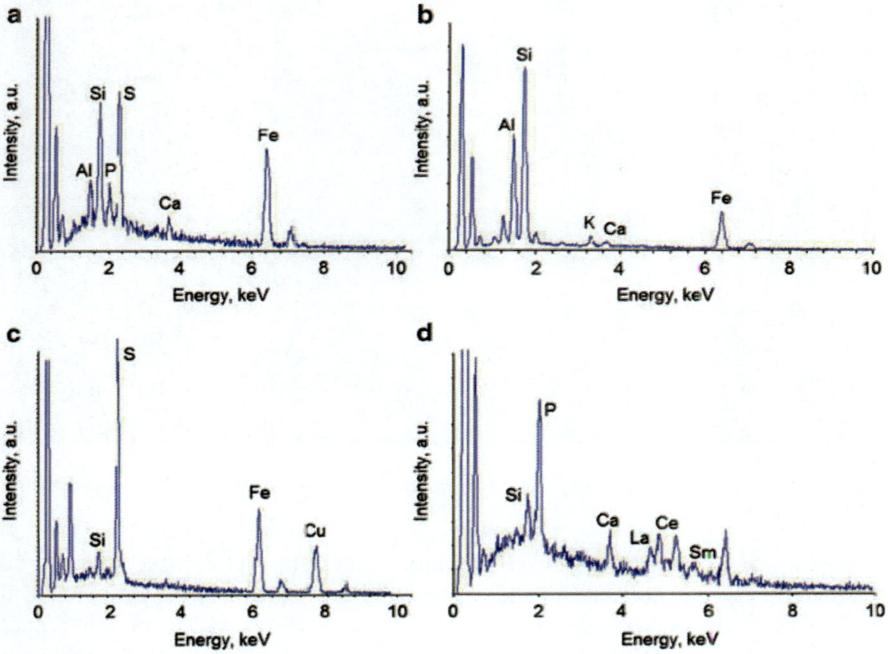

Fig. 10 Typical EDX spectra corresponding to different colloidal particles from the sample collected in reducing conditions: (**a**) pyrite and clay, (**b**) chlorite and hematite, (**c**) chalcopyrite, (**d**) monazite

6 Effect of Redox Conditions on the Actinides Partitioning to Colloids and Their Mobility

The migration rates of redox-sensitive elements such as actinides, Tc, I, Se and others are controlled by the redox conditions, the presence of complexing ligands in solution and reactive surfaces as well as geochemical properties of the site such as permeability, groundwater flow rate, etc. Hu et al. (2008) studied the effect of redox

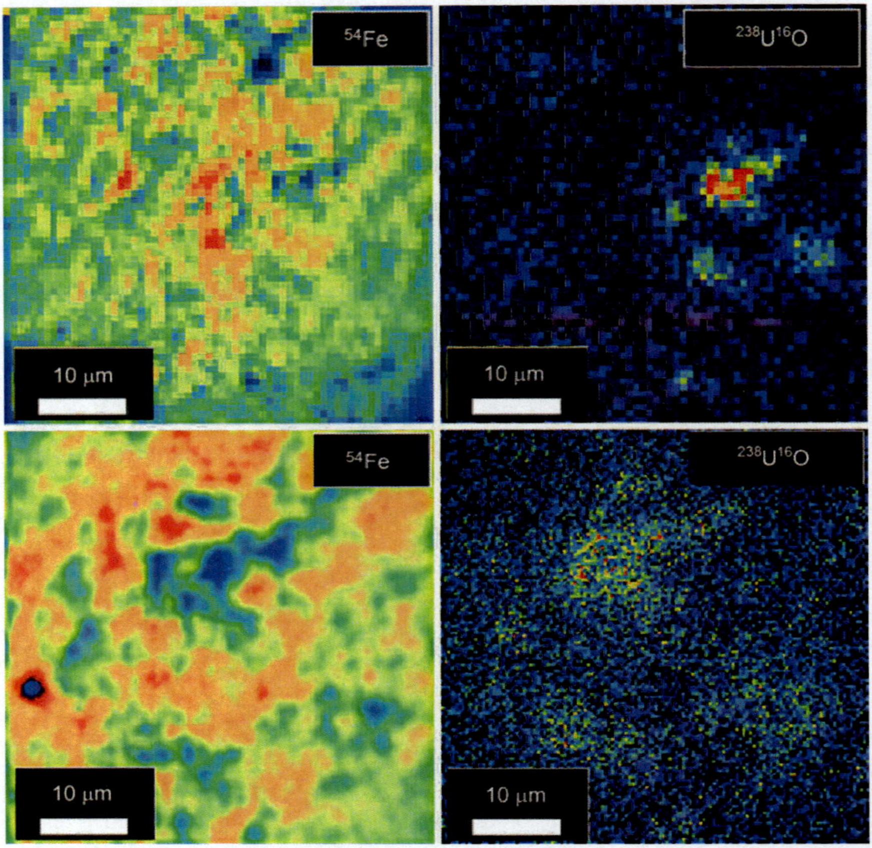

Fig. 11 NanoSIMS elemental maps showing typical distribution of Fe and U in colloid matter from reducing conditions

conditions on the sorption of different radionuclides onto solids typical for Yucca Mountain. The Kd obtained in the experiments in a glove box with a reduced oxygen level was higher than under the ambient atmosphere. However, this cannot be explained by the reduction of U(VI) to U(IV) species under the mildly reducing conditions of the experiments. In the case of deep liquid waste injection, the reduction of U(VI) occurs, stabilizing uranium in colloidal form. For plutonium the effect of Eh is much less important – it stays in the reduced Pu(III)/Pu(IV) form through a broad interval of Eh values (Fig. 12).

The formation of U(IV) hydroxocolloids would result in the decrease of uranium and other actinide migration ability by their sorption or coprecipitation with $U(OH)_4$(am). Such An(IV) hydroxocolloids are very reactive and tend to precipitate onto different surfaces.

One important conclusion about the relative mobility of actinide species under different redox environments is based on their partitioning. Sequential extraction is a technique that is commonly applied to study partitioning of toxic and radioactive

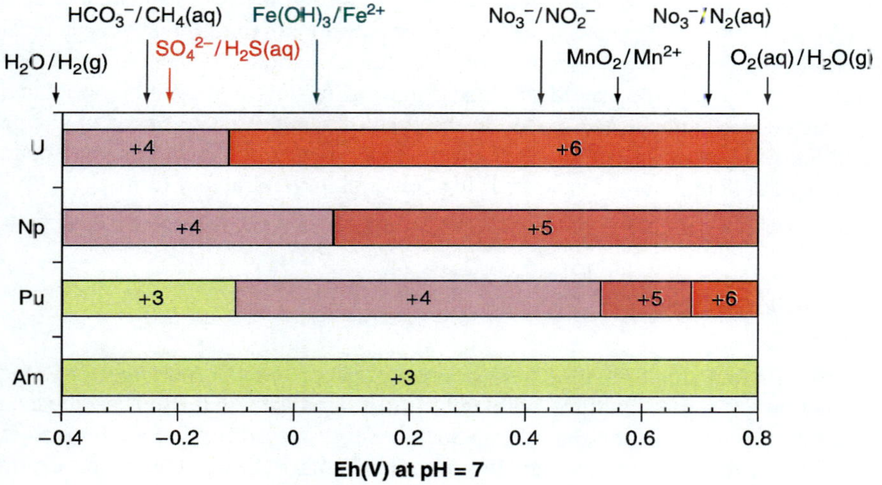

Fig. 12 Expected dominant oxidation states of actinides as a function of Eh (modified from Banaszak et al. 1999)

Table 2 Partitioning of uranium and plutonium bound to colloids separated from groundwater collected in oxidixing and reducing environments

Fraction	Reagent / pH	200 nm filter				50 nm filter			
		Oxidizing		Reducing		Oxidizing		Reducing	
		U	Pu	U	Pu	U	Pu	U	Pu
Exchangeable	0.5M $Ca(NO_3)_2$ / pH 5.5	5.4	4.2	4.5	3.2	2.9	2.7	2.0	1.5
Acid soluble	0.44M CH_3COOH, 0.5M Ca $(NO_3)_2$ / pH 2.5	19.2	17.9	16.7	15.9	11.2	14.2	13.3	10.9
Fe/Mn oxides	0.01M $NH_2OH{\cdot}HCl$ in 25% CH_3COOH / pH 2	31	30.9	16.9	17.9	33.4	28.3	6.8	7.2
Oraganics	0.1M $Na_4P_2O_7$ / pH 10	15.4	22.6	5.8	9.0	27.9	29.9	1.6	0.8
Amorphous oxides	0.18M $(NH_4)_2C_2O_4$ 0.1M $H_2C_2O_4$ / pH 3.5	18	16.9	6.5	2.2	11.6	10.9	0.9	1.1
Crystalline oxides	0.15M $Na_2C_2O_4$, 0.75 g Na ditionite / pH 5.5	5.5	2.2	18	14.6	11.1	12.4	18.0	14.6
Residual	HF + H_2SO_4 / pH<0	5.5	5.3	31.6	37.2	1.9	1.6	57.4	63.9

elements in soils and bottom sediments (Loyland Asbury et al. 2001). In this study, we used this technique to study actinide partitioning to colloids separated from both groundwater samples. Among various protocols for sequential extraction, we used those proposed by Miller et al. (1986), in which seven steps are employed. The sequential extraction is carried out as follows: the filters with colloids are sequentially washed with different reagents and the solutions collected to determine actinides. The sequential extraction can only be applied for 200 nm and 50 nm pore filters since 5 nm filters are partially or completely dissolved by acidic solutions.

The sequential extraction data are presented in Table 2 as the percentage for the total concentration of the element. Because of some experimental problems (readsorption, incomplete dissolution, etc.) typical for this method, sequential extraction could provide only an estimate of element partitioning to different geochemical fractions. However, in our case for both the studied samples, sequential extraction completely supports the nanoSIMS data; the main fraction of colloid-borne Pu and U is associated with Fe/Mn oxides in the case of samples taken under oxidizing conditions. In contrast, in samples taken under reducing conditions, the main fraction of colloidal-borne U and Pu is in a low-mobility, refractory form.

7 Conclusions

For the first time, the formation of U(IV) hydroxide colloids or their aggregates on interaction of waste solutions with reducing groundwater has been established under real environmental conditions. Such precipitates sorb other radionuclides and stabilize them in refractory and low mobility forms. In contrast, under oxidizing conditions actinides form pseudocolloids by sorption onto HFO and MnO_2.

The experimental data obtained in this study clearly indicate that, together with local hydrology and groundwater chemistry, the redox conditions of the storage site is very important in defining actinide migration.

Acknowledgments This work was supported by the President of Russian Federation Grant (to SNK, # MD- 7150.2010.3), the Federal Ministry for Education and Sciences (02.740.11.0853) and the Russian Basic Research Foundation (10-03-01029-a).

References

Banaszak JE, Rittmann BE, Reed DT (1999) Subsurface interactions of actinide species and microorganisms: Implications for the bioremediation of actinide-organic mixtures. J Radioanal Nucl Chem, 241:385–435

Cho HR, Walther C, Rothe J, Neck V, Denecke MA, Dardenne K, Fanghänel Th (2005) Combined LIBD and XAFS investigation of the formation and structure of Zr(IV) colloids. Anal Bioanal Chem, 383:28–40

Drozhko EG, Ivanov IA, Samsonova LM (1996) Retention of radionuclides by solids in geological formations near Karachai Lake. Radiat Saf, 2:22–27

Fanghänel Th, Neck V (2002) Aquatic chemistry and solubility phenomena of actinide oxides/hydroxides. Pure Appl Chem, 74:1895–1907

Hu QH, Zavarin M, Rose TP (2008) Effect of reducing groundwater on the retardation of redox-sensitive radionuclides. Geochem Trans, 9:12–36

Kaminski MD, Dimitrijevic NM, Mertz CJ, Goldberg MM (2005) Colloids from the aqueous corrosion of uranium nuclear fuel. J Nucl Mater, 347:77–87

Lisizin AK, Myuskin VI, Gunina VI, Shulik LS, Drozko EG, Ivanov IA, Aleksahkin AI (2001) Protective geochemical properties of the bedrock in the Karachai Lake area. Radiat Saf, 4:9–24

Loyland Asbury SM, Lamont SP, Clark SB (2001) Plutonium partitioning to colloidal and particulate matter in an acidic, sandy sediment: Implications for remediation alternatives and plutonium migration. Environ Sci Technol, 35:2295–2300

Miller WP, Martens DC, Zelasni LW (1986) Effect of sequence on extraction of trace metals from soils. J Soil Sci Soc Am, 50:598–601

Myasoedov BF, Novikov AP (1997) Main sources of radioactive contamination in Russia and methods for their determination and speciation. J Radioanal Nucl Chem, 229:33–38

Neck V, Müller R, Bouby M, Altmaier M, Rothe J, Denecke MA, Kim JI (2002) Solubility of amorphous Th(IV) hydroxide – application of LIBD to determine the solubility product and EXAFS for aqueous speciation. Radiochim Acta, 90:485–494

Novikov AP, Pavlotskaya FI, Goryachenkova TA (1998) Radionuclide content in groundwaters from observation wells around Karachai Lake. Russ Radiochemб, 40:484–490

Novikov AP, Kalmykov SN, Utsunomiya S, Ewing RC, Horreard F, Merkulov A, Clark SB, Tkachev VV, Myasoedov BF (2006) Colloid transport of plutonium in the far-field of the Mayak Production Association, Russia. Science, 314:638–641

Rybalchenko AI, Pimenov MK, Kostin PP, Balukova VD, Nosuchhin AV, Mikerin EI, Egorov NN, Kaimin EP, Kosareva IM, Kurochkin VM (1994) Deep injection of liquid radioactive waste in Russia. Battelle Press, Columbus Richland, WA

Rybalchenko AI, Pimenov MK, Kurochkin VM, Kamnev EK, Korotkevich VM, Zubkov AA, Khafizov RR (2005) Deep injection disposal of liquid radioactive waste in Russia, 1963–2002: Results and consequences. Dev Water Sci, 52:13–19

Samsonova LM, Drozhko EG (1996) Migration of high-density industrial waste solutions through fresh groundwaters. In: Deep injection disposal of hazardous and industrial waste: Scientific and engineering aspects, (eds. J. A. Apps and C-F. Tsang), Chap. 41, pp. 669–680. Academic Press, San Diego, CA

Sekine T, Narushima H, Kino Y, Kudo H, Lin M, Katsumura Y (2002) Radiolytic formation of Tc (IV) oxide colloids. Radiochim Acta, 90:611–616

Solodov IN, Velichkin VI, Zotov AV, Kochkin BT, Drozhko EG, Glagolev AV, Skokov AN (1994) Distribution and geochemistry of contaminated subsurface waters in fissured volcanogenic bed rocks of the Lake Karachai Area, Chelyabinsk, Southern Urals. Russian-American Center for Contaminant Transport Studies. Lawrence Berkeley Laboratory. Rept. LBL-36780 (UC-603)

Walther C, Rothe J, Brendebach B, Fuss M, Altmaier M, Marquardt CM, Büchner S, Cho HR, Yun JI, Seibert A (2009) New insights in the formation processes of Pu(IV) colloids. Radiochim Acta, 97:199–207

X-Ray Absorption Spectroscopy of Plutonium Particles at the Rocky Flats US Nuclear Weapons Production Site

Steven D. Conradson, David L. Clark, Christophe den Auwer, and Juan S. Lezama-Pacheco

Abstract The Rocky Flats Environmental Technology Site (RFETS) was a U.S. Department of Energy (DOE) environmental cleanup site located about 15 miles northwest of downtown Denver. Nearly 40 years of nuclear weapons production had left behind a legacy of contaminated facilities, soils, surface and groundwater. Ultrafiltration studies had shown that plutonium was associated with particles. We employed X-ray absorption spectroscopy to demonstrate that plutonium in soils at the Site was in the insoluble chemical form of PuO_2. This information coupled with ultrafiltration studies was used to make the case for particle transport mechanisms as the basis of plutonium and americium mobility, rather than aqueous solubility processes, and established a successful scientific basis for the dominance of physical transport processes by wind and water. This understanding allowed Site contractors to rapidly move to application of soil erosion and sediment transport models as the means of predicting plutonium and americium transport, which led to design and application of site-wide soil erosion control technology to help control downstream concentrations of plutonium and americium in streamflow. Good scientific understanding in the public interest helped the most extensive cleanup in the history of Superfund legislation to finish 1 year ahead of schedule, ultimately resulting in billions of dollars in taxpayer savings.

S.D. Conradson (✉) and D.L. Clark
Los Alamos National Laboratory, Los Alamos, NM 87545, USA
e-mail: conradson@lanl.gov

C. den Auwer
Commissariat al'Energie Atomic Marcoule, 30207, France

J.S. Lezama-Pacheco
Los Alamos National Laboratory, Los Alamos, NM 87545, USA
and
CINVESTAV-Mérida, Mérida, Yucatán 97310, México

S.N. Kalmykov and M.A. Denecke (eds.), *Actinide Nanoparticle Research*,
DOI 10.1007/978-3-642-11432-8_14, © Springer-Verlag Berlin Heidelberg 2011

1 Introduction

The Rocky Flats Environmental Technology Site (RFETS) was a U.S. Department of Energy (DOE) environmental cleanup site located about 15 miles northwest of downtown Denver (Fig. 1). From 1952 to 1989, the Rocky Flats Plant fabricated components for the nation's nuclear weapons arsenal using various radioactive and hazardous materials, including plutonium and uranium, toxic metals such as beryllium, and hazardous chemicals such as solvents and degreasers. In 1989, the Federal Bureau of Investigation (FBI) and the Environmental Protection Agency (EPA) abruptly halted nuclear production work to investigate environmental and safety concerns, and the site was added to the Superfund list later that year. In 1993, the Secretary of Energy announced the end of the Rocky Flats nuclear production mission. Nearly 40 years of nuclear weapons production had created a large legacy of contaminated facilities, soils, and surface and groundwater at the site.

Production operations at Rocky Flats generated a huge volume of wastes contaminated with radionuclides and other hazardous substances. The majority of wastes were shipped off the site for disposal, but many areas at Rocky Flats had contaminated soil and water due to the improper disposal of contaminated materials, ruptured or leaking pipes, fires, or faulty storage units. Since many of the activities at the plant involved plutonium, much of the contamination was associated with this element [1].

Fig. 1 A 1995 aerial photograph of the RFETS showing the industrial area

In March 1995, DOE estimated that the cleanup for Rocky Flats would cost in excess of $37 billion and take 70 years to complete [2]. By 1996, DOE and Kaiser-Hill initiated a massive accelerated closure effort that resulted in a plan to reach closure by December 31, 2006, at a contracted cost of $7 billion. After a troubled start, Kaiser-Hill completed the task nearly a year ahead of schedule [3].

Scientific understanding of plutonium behavior in soils and surface water played a crucial role for decision makers in cleaning up the site. The impetus to understand the science behind plutonium contamination gained momentum in 1995 when intense rainfall and wet spring conditions raised concerns among Site personnel and stakeholder groups about the potential for increased plutonium mobility and offsite transport. There was a hypothesis that plutonium was soluble in surface and ground water in order to account for increased plutonium concentrations at on-site surface water monitoring locations. Modeling efforts and site data at the time predicted no to very limited movement of plutonium. The predictions of no plutonium mobility, coupled with the observation of plutonium transport at surface water monitoring stations, led to public mistrust and lack of confidence. When coupled with other questions regarding the mobility of different actinide elements (U, Pu, Am) at different Site locations, DOE and Kaiser-Hill established the Actinide Migration Evaluation (AME) advisory group in 1995 to provide advice and technical expertise on issues of actinide behavior and mobility in the air, surface water, groundwater, and soil. In this report, we summarize how X-ray absorption spectroscopy (XAS) was used to identify the chemical form of plutonium particles and how that understanding was used to aid cleanup decisions at the Site.

2 About the Site

The RFETS occupied approximately 6,585 acres and resembled a small city, with its own fire department, medical offices, cafeterias, steam plant, and water- and sewage-treatment plants (Figs. 1 and 2). Over 800 structures were located within a centralized 385-acre Industrial Area surrounded by a 6,200-acre grassland Buffer Zone. This open space continues to serve as a buffer between Rocky Flats and the nearby, growing communities and is home to many species of animals and plants.

Water at RFETS and the surrounding area is distributed among surface water, shallow groundwater, and deep groundwater [4]. Surface water flows across RFETS primarily from west to east along North and South Walnut Creek, and Woman Creek in the Industrial Area (Fig. 2). The site had constructed several series of detention ponds along these drainages to manage wastes and surface water runoff from the industrial area of the site. The A- and B-series ponds are located on North and South Walnut Creeks, and the C-series ponds on Woman Creek. Past discharge of low-level contaminated wastes to the A- and B-series ponds resulted in the accumulation of plutonium and americium in the pond sediments.

Shallow groundwater refers to water within the alluvium and weathered bedrock geologic units to a depth of 30 m. Surface water and shallow groundwater are

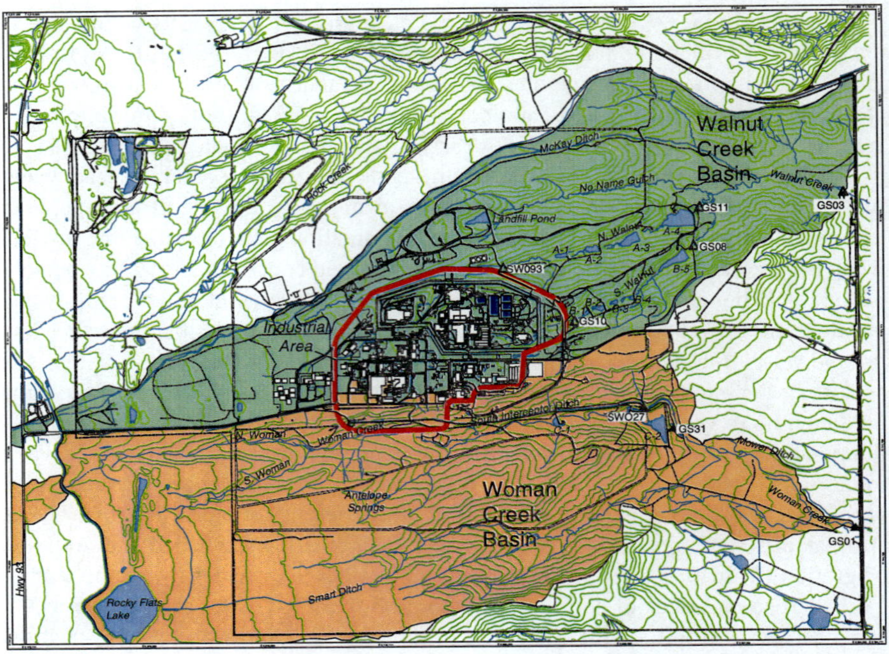

Fig. 2 The Rocky Flats site map showing the industrial area (denoted by the outline in the center), and the Walnut Creek (*green*) and Woman Creek (*brown*) drainage basins

inseparably linked. Water from stream channels, storm water, and industrial areas infiltrates downward, recharging the shallow groundwater, which, in turn, recharges the stream channels depending upon the time of the year. Beneath the alluvium is a highly impermeable bedrock layer that inhibits vertical flow. As a result, shallow groundwater flows laterally, discharging as baseflow into the streams or as hillslope springs and seeps. Approximately 200–300 m below the surface lies the Fox Hills Sandstone, where deep regional groundwater flows. Because of the intervening bedrock, this regional groundwater aquifer is hydraulically isolated from the Rocky Flats surface and shallow groundwater and actinide contaminants [5].

The climate is temperate and semiarid, characteristic of Colorado's Front Range. The average annual precipitation is approximately 37 cm, with about half occurring as rain from May to October and half as snow from late October through early April. Evapotranspiration averages over 40 cm per year, creating a water deficit in most years. Prior to the removal of the Industrial Area, much of the runoff feeding the Site's drainages occurred rapidly from the impervious Industrial Area surfaces. Winds at RFETS are predominantly from the northwest toward the southeast. The RFETS is noted for the periodic occurrence of strong, gusty winds ($\geq$100 mph) that are an important factor in the resuspension of soil and actinides. Air monitoring and calculations of the actinide loads showed that air transport has been a dominant actinide migration pathway [4].

3 The Nature of Actinide Contaminants

Production operations conducted at Rocky Flats generated a huge volume of wastes contaminated with radionuclides and other hazardous substances. The majority of wastes were shipped off the site for disposal, but many areas at Rocky Flats had contaminated soil and water due to the improper disposal of contaminated materials, ruptured or leaking pipes, fires, or faulty storage units. Since many of the activities at the plant involved plutonium, much of the contamination was associated with this element. Based on historical data, there were five plutonium accidents and incidents that resulted in the majority of accidental releases of plutonium into the environment, including two major glovebox fires that, had they not been contained within the buildings, threatened contamination of a broad area surrounding Rocky Flats. These incidents brought significant media attention to the Site.

By far the largest source of plutonium and americium contamination in soils emanated from the drum storage area known as the 903 Pad. From 1958 to 1969, drums containing plutonium-contaminated lathe coolant were stored on the Pad, located on the southeastern part of the Industrial Area (Fig. 2). These drums leaked, and wind and water erosion carried plutonium and americium in a well-defined pattern to the east and southeast (Fig. 3), past the eastern Site boundary in some cases. Based upon material balance around the drums, it was estimated that a total of 5,000 gallons containing approximately 86 g (5.3 Ci) of plutonium were released into soils [1, 4].

Between 1991 and 1999, nearly 2,500 surface soil samples from across the site were collected and analyzed to assess the level and extent of actinide contamination. A heterogeneous small-scale concentration distribution over a large spatial area at RFETS led to the use of state-of-the-art geostatistical analyses, including the techniques of variograms and Kriging [6]. Estimated spatial concentrations of $^{239/240}Pu$ in surface soil at RFETS that were calculated by Kriging analysis are shown in Fig. 3. Plutonium and americium generally exhibited the same spatial distribution in surface soils, with wide variations in activities occurring throughout the site. The highest concentrations were found at the 903 Pad and areas to the east of the Pad, and displayed a wind-driven dispersal pattern to the east of the primary source area – the 903 Pad (see Fig. 3). The plutonium and americium radioactivity in RFETS soils was highly heterogeneous, often consisting of "hot particles" [7]. Approximately 90% of the plutonium and americium inventory was in the top 12 cm of the soil.

In addition to the contamination produced during manufacturing operations at RFETS, the Closure Project was required to deal with contamination that resulted from remediation and D&D activities. Concentrations of plutonium (and americium) ranged from fractions of picocuries (pCi) in streams and pond waters, up to nanocuries (nCi) in soils and sediments, and up to curies (Ci) in contaminated building materials.

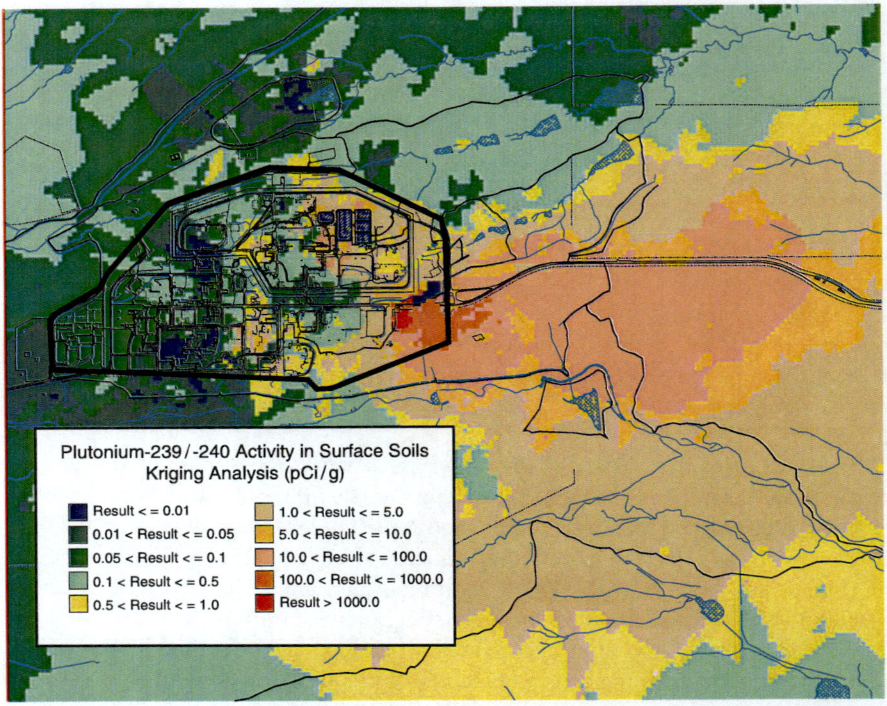

Fig. 3 Because it was not feasible to sample surface soil at every location, a geostatistical modeling technique known as Kriging was applied to the plutonium surface soil data for $^{239/240}$Pu to estimate concentrations in surface soil. The "hot spot" of $^{239/240}$Pu concentrations in excess of 1,000 pCi/g at the 903 Pad is shown in *red*. A clear plume of $^{239/240}$Pu contamination that tracks roughly with the prevailing winds from NW to SE is evident from the data. This figure represents conditions at Rocky Flats prior to soil remediation actions

4 Expectations for Plutonium Chemistry at the Site

Migration of actinides in the environment takes place within the context of physical and chemical processes [8, 9]. Chemical reactions, particularly redox reactions within soil and ponds, were and are often hypothesized to explain actinide mobility. At one extreme, the actinide(s) may react chemically along with the surrounding materials, to create soluble (dissolved) and mobile components. At the other extreme, the actinide(s) might remain unchanged at the molecular and atomic scale while the associated materials react, to create mobile and immobile components.

The great contrast between actinide solubilities – plutonium and americium have very low solubilities whereas uranium has relatively higher solubility – drove the consideration of colloidal and particulate transport processes and required careful evaluation of evidence that could distinguish solubility and redox process results. For example, actinide chemists have long recognized the preferred stability of plutonium in the form of oxides [10], and as colloidal sized particles [11], but detailed knowledge of reactivity and mobility of such materials in the environment is limited at concentrations of pCi/L in waters and pCi/g to nCi/g in soils.

Extensive field observations and research have been conducted internationally on the environmental behavior of actinide elements in diverse sets of environments over the past 30–40 years. This research has provided a good base for understanding the major types of species and their transport mechanisms in soils and natural waters. In natural waters, plutonium solubility is limited by the formation of amorphous $Pu(OH)_4$ or polycrystalline PuO_2, which provides an upper limit on the amount of *dissolved* (i.e., ionic/molecular) plutonium that can be present. PuO_2 has a laboratory-measured solubility range of 10^{-10}–10^{-13} mol/L and is limited by the formation of amorphous $Pu(OH)_4$ [11]. Due to the very low solubility, and the tendency of Pu(IV) compounds to adhere to organic and mineral particles, the primary path of plutonium transport is usually migration of fine sediments including colloidal particles. Indeed, when concentrations above fallout levels of plutonium have been investigated in detail, they have been linked to colloids and particulates [12–15]. There was a clear need to provide the stakeholders with a scientific understanding of actinide transport mechanisms in order to move forward with planning the strategy to close the Site.

5 Ultrafiltration of Field Samples

Based on the geochemical characteristics discussed above, one might conclude that insoluble oxides of plutonium and americium would be trapped by the geological matrix and remain immobile in the environment. However, there are a growing number of field studies documenting movement of low concentrations of low-solubility radionuclides in surface and groundwater [12, 16, 17]. Small concentrations of low-solubility radionuclides, such as plutonium and americium, can be transported in surface or groundwater as small-sized particulates (1 nm–1 μm) [18, 19] that remain suspended and are therefore transported in natural aquatic systems. At RFETS, the monitoring data and historical observations were that plutonium and americium could migrate in small quantities that would settle out in the A-, B-, and C-series ponds. Thus colloidal-sized particles were likely to play a role in plutonium and americium mobility at RFETS.

From 1998 until 2001, Santschi and coworkers used field studies and laboratory experiments to examine $^{239,240}Pu$ and ^{241}Am concentrations and phase speciation in the surface environment at RFETS [17]. Since the environmental forms of actinides present in the surface waters were in the 10^{-3}–10^{-1} pCi/L concentration range, filtration and tangential-flow ultrafiltration were the only methods suitable for the separation and analysis.

Measurements of total $^{239,240}Pu$ and ^{241}Am concentrations in storm runoff and pond discharge samples, collected during spring and summer in 1998–2000, demonstrated that most of the $^{239,240}Pu$ and ^{241}Am transported from contaminated soils to streams occurred in the particulate (≥0.45 μm) and colloidal (~2 nm or 3 kDa to 0.45 μm) phases. In general, most of the plutonium and americium in RFETS water was found in the particulate phase (>20 μm and 0.5–20 μm) with most of the material that passed a 0.5 μm filter being colloidal, i.e., it was filtered out by 100 kDa or 3 kDa ultrafilters, with 20% or less of 0.5 μm filter-passing plutonium and americium passing a 3 kDa ultrafilter. Based on graphite furnace atomic absorption spectroscopy, transmission electron microscope (TEM), and energy dispersive X-ray microprobe (EDAX) images, colloids in these size ranges were primarily composed of clay and organic-matter [17].

Plutonium concentrations above the action level (≥0.15 pCi/L) at RFETS historically occurred predominantly during storm runoffs in early spring and summer, at monitoring locations either within or adjacent to the Industrial Area. During these storm events, elevated concentrations of colloidal plutonium, accompanying those in the particle phase, were observed. Controlled laboratory investigations of soil resuspension, which simulated storm and erosion events, confirmed that most of the plutonium in the 0.45 μm filter-passing phase was in the colloidal phase (≥80%) and that remobilization of colloid-bound plutonium during soil erosion events can be greatly enhanced by humic and fulvic acids present in these soils. However, the fraction of remobilized colloidal plutonium, at most 1% in these soil resuspension experiments, was much smaller than what was observed during storm runoff or pond discharge, where the fraction of colloidal plutonium passing a 0.5 μm filter was as much as 66%. This showed that suspended matter and colloids were likely derived less from eroding soils directly but more likely from sediment resuspension of stream beds and banks.

Isoelectric focusing experiments of radiolabeled colloidal matter extracted from RFETS soils revealed that colloidal plutonium was in the tetravalent state and was mostly associated with a negatively charged organic colloid with molecular weight of 10–15 kDa, rather than with the more abundant inorganic (iron oxide and clay) colloids. Santschi's evidence strongly argued against the presence of mobile colloidal particles consisting mainly of PuO_2 but suggested that PuO_2 was imbedded in, or attached to, a matrix of organic matter and some iron, with organic matter determining the surface charge and migration behavior of the 10–15 kDa colloid.

Santschi's studies were of great importance in that they provided strong evidence that the low levels of plutonium and americium in surface water at RFETS were transported by the colloidal and particulate fraction and not the dissolved fraction of the water. These findings were essential for the understanding of actinide transport pathways, the confirmation that groundwater is not a major pathway for plutonium and americium transport, and the development of actinide transport models.

6 X-Ray Absorption Fine Structure Evaluation of Chemical Speciation of Pu

Corroboration of the colloid/particulate hypothesis and a deeper understanding of the history, current behavior, and predicted fate of the Pu requires information on the smaller, molecular length scale that is associated with and determines the chemical speciation. Specifically, primary colloids of Pu are most closely correlated with highly insoluble Pu(IV) in oxy-hydroxide form [15], which has been shown to be the thermodynamic well for Pu in equilibrium with air and water. Definitive experimental proof of this prediction – especially in a format easily defensible in a public forum – was critical in validating the models and their predictions for Pu. X-ray absorption fine structure (XAFS) spectroscopy, arguably the most incisive method for identifying chemical speciation of individual elements in real samples, is the only method with the requisite combination of sensitivity, selectivity, and minimal risk of artifacts required to solve this problem [20–27]. It has therefore been a key technique in the area of Molecular Environmental Sciences, which in turn is based on the premise that even in the most complex geological system, the behavior of all of the components can be reduced to the expected chemistry of their local environment – both homo- and heterogeneous – which is why chemical speciation is the essential input.

XAFS is the technique of choice in many cases because its output is the valence of the target element and the identification and characterization of its neighbor atoms in terms of the elements constituting these shells to $Z \pm 2$–5, the absorber – neighbor distances to $R \pm 0.01$–0.04 Å, and the numbers of neighbor atoms in the shells to $N \pm 15$–35% for neighbor shells out to 3–7 Å from the absorber – the parameters that, cumulatively, define the chemical speciation of the target element [26, 27]. Insofar as the precision is significantly enhanced in comparisons of related samples, XAFS is exquisitely sensitive to changes in speciation. The method's enhanced sensitivity to changes in these metrics and the associated speciation of the target element and its capabilities for obtaining this information from complicated samples with analyte concentrations down to the parts per million level – since the underlying mechanism of exciting a core-level electron is an atomic process – makes it ideally suited to the task of evaluating Pu speciation. This is especially true when the primary objective is to discriminate between typical oxyhydroxide compounds of Pu(IV) that are nonspherically symmetric and the highly oblate trans-dioxo complexes (PuO_2^{n+}, $n = 1, 2$) of Pu(V,VI), readily accomplished from the Pu $L_{II/III}$ edge XANES [21].

XAFS spectra consist of the region through the absorption edge when the absorbance undergoes a large, abrupt increase because the incident X-ray comes into resonance with the core electron shell of the target element and the extended region beyond the edge. The X-ray absorption near edge spectra (XANES), because of its origin in transitions to bound and quasi-bound states single and multiple scattering resonances, is sensitive to the valence and site symmetry of the absorber and can serve as a fingerprint for the speciation. The extended X-ray absorption fine structure (EXAFS) contains the metric information on the absorber environment, but its modulation of the absorption coefficient begins as only a few percent of the edge jump and damps out quickly with increasing energy to only a few parts per thousand, making it much more difficult to measure accurately, especially in samples with low absorber concentrations. The oscillations that compose the EXAFS are typically Fourier transformed to give a $\chi(R)$ representation (instead of the original $\chi(k)$, where k is the photoelectron wave vector magnitude) that bears a superficial resemblance to the element-specific partial pair distribution function. The metrical information is extracted via curve-fits to the EXAFS.

7 Soils Contaminated with Originally Soluble or Suspended Pu Compounds

Samples of soils contaminated with Pu were obtained from the 903 pad area and also from some additional locations where waste pipes had leaked. Larger pebbles and plant material were removed, and the dirt was lightly ground to give a more homogeneous powder that was packed into sample containers for XAFS measurements. These experiments were performed prior to the development of microfocusing capability, so the beam was typically 1 $\times$ 12 mm and the results are an average over all of the Pu present. However, within the context of particles, a highly relevant result was obtained on a concrete sample from a separations area. A relatively strong signal was obtained in one location that gave a highly anomalous, even unique for Pu XANES spectrum whose edge and peak energies were consistent with Pu(VI). Moving the 1 $\times$ 12 mm beam to other locations on the sample that was in a 2.5 $\times$ 20 mm slot gave a much weaker signal that was within the noise level of being that of PuO_2. This result did demonstrate that the Pu could be particulate, that the particles could be large enough so that even one could dominate the total spectrum, and that the separate particles could contain different Pu species.

Because the emission line from the Pu L_{III} edge overlaps with that of the Sr K edge of Sr-containing minerals in the soils, the Pu L_{II} edge spectra were measured instead with some loss of sensitivity. X-ray energies were calculated by defining the first inflection point of the XAFS spectra of an Ru metal foil as 22,116 eV. The combination of extremely low analyte concentrations – of the order of 1 ppm – and the complex soil matrix that contains dozens of elements at parts per million and higher level and is also sufficiently dense to absorb much of the beam, made these measurements exceedingly challenging with some samples requiring 16–20 h of signal averaging just for XANES. Therefore, with a single exception, only the Pu XANES could be measured from soils in this category. The other challenge was to perform the experiments in a way that could make the results accessible to the general public. This second goal was accomplished by always running a PuO_2 standard first and then both a PuO_2 and a PuO_2^{2+}-hydroxy-carbonate standard that had been mixed with soil or concrete so as to have the same background. These spectra could then be overlaid with the spectra from the contaminated soils so that direct visual comparisons could be performed, which was found to be convincing even for people with minimal scientific background.

XANES spectra were obtained for seven samples from soils, one of which gave a sufficiently strong signal to also measure its EXAFS. Samples 1–4 were from soils contaminated from leaking pipes carrying waste streams. Samples 10–12 (the sample numbers used in the original report are retained here) were soils at or in proximity to the 903 pad.

8 Soil Samples Contaminated from Leaking Pipes

The source term for these samples would most likely have been Pu dissolved in aqueous solution. Precipitation would then have occurred via modification of the solution parameters (pH, Eh, etc.) on contact with the soil that would have caused the Pu to become insoluble. Precipitation may also have been mediated by mineral surfaces and the formation of surface complexes and other soil components. These types of chemical reactions could also have promoted the formation of secondary or pseudocolloids.

For our first measurements on samples 1 and 2 (Fig. 4), only Pu(IV) standards were measured, so the Pu(VI) standard's spectrum is that measured at a later date (dotted red line, identical to its solid red line counterpart in Fig. 5). For both samples 1 and 2, despite the relatively noisy spectra, the highest groups of points are below the peak of the Pu(VI) spectrum and at or below that of Pu(IV). The peak position of sample 1 as judged from the first-derivative spectrum (Fig. 4, inset) almost exactly coincides with that of the Pu(IV) (PuO_2) standard. In addition, the inflection points of 1 and 2, although shifted somewhat higher than those from the

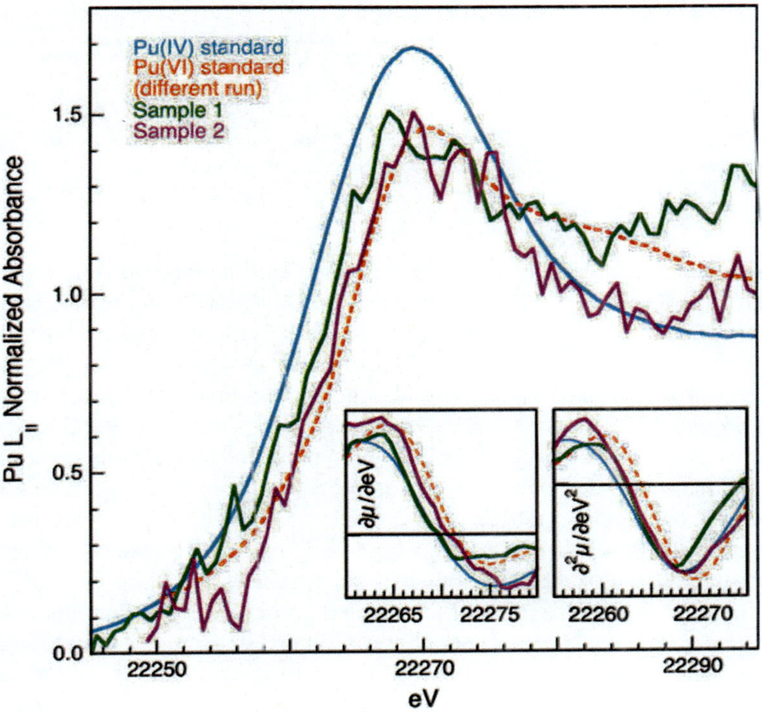

Fig. 4 Plutonium L_{II} XANES of soil samples 1 and 2; *left* and *right insets* show, respectively, first (zeros = peak energies) and second (zeros = inflection point energies) derivatives

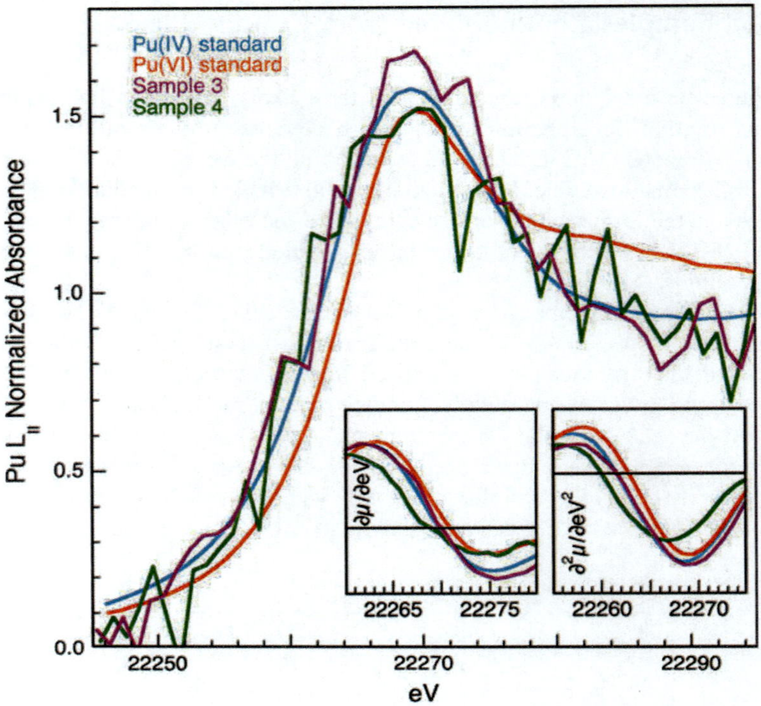

Fig. 5 Plutonium L_{II} XANES of soil samples 3 and 4; *insets* show first (zeros = peak energies), *left*, and second (zeros = inflection point energies), *lower right*, derivatives

Pu(IV) standard for this run, are within the range of this standard from other runs. The same is true of sample 2's inflection point, although its peak position is slightly higher than the (IV) standard. Inspection of the spectra, however, shows that the energy found for the peak of sample 2 is distorted by the flat top of the spectrum that resembles neither of the standard spectra and that is primarily the result of the two points around 22,274 eV that create an upward spike. These observations suggest that more credence be placed on the inflection point and shape of its spectrum that parallels that of the Pu(IV) standard over the entire range of energies, lacking the post-edge shoulder characteristic of Pu(VI) spectra. The addition of what appears to be a positively sloped line to the spectrum of 1, most likely because of an overlapping fluorescence peak whose intensity increased as the beam position shifted very slightly to illuminate more of the mineral grain containing that other element in the inhomogeneous sample, could not cause the spectrum of 2 to exhibit the Pu (VI) shoulder because, as with that of 1, it would raise the spectrum more at 22,295 eV than around 22,280 eV. Such an effect would shift the peak position to slightly higher energy but have no effect on the energy of the inflection point. The XANES spectral evidence is therefore consistent with the conclusion that plutonium in sample 1 occurs predominately as Pu(IV). Even if sample 2 contains a Pu(VI) component affecting peak energy, the edge energy implies a significant fraction of Pu(IV), which would be consistent with a conversion, over time, of the higher-valence species to the expected more stable lower one that is still at an intermediate point in the process.

The analysis of soil samples 3 and 4 (Fig. 5) is more straightforward. Both peak energies and inflection points are either coincident with the Pu(IV) standard (sample 3) or slightly lower than standard (sample 4). Inspection of the spectrum of 4 shows an anomaly at 22,266–22,268 eV where the spectrum becomes flat before resuming its ascent to give a slightly higher peak. Having the beam position shift slightly with respect to a particle that contains the Pu could again cause this effect. It would result in the observed reduction of the spectral energies, a small change for the peak since this error is relatively far away from it (note the anomalous change in slope in 3 where the derivative goes from positive to negative indicating the peak position and its crossing of the axis, relative to sample 2), but a large shift for the inflection point because of its proximity. Both spectra also show the distinctive shape of the Pu(IV) standard at higher energy, descending from the peak into the minimum near 22,290 eV, with no sign of the shoulder characteristic of the Pu(VI) standard. Comparison with the Pu(VI) standard provides no evidence for Pu(VI) speciation. Hence, these two soils meet the criteria for containing plutonium of predominately or exclusively Pu(IV) speciation.

9 Contaminated Soils from the 903 Pad Area

The source term for these samples was largely through leakage of waste-storage drums containing plutonium metal fines in cutting fluids from machining operations. In contrast to the contamination from leaking pipes, the Pu here began in pure, solid form so that subsequent reactions – including any that might have produced soluble Pu species – occurred on and via the surfaces of macroscopic particles. Similarly, formation of colloids would likely result from fragmentation of these original particles into nanoparticulates, with dissolution and reprecipitation occurring not at all or at least much less frequently unless a highly unusual set of local conditions occurred. It was therefore of substantial interest to determine if the Pu speciation of these 903 pad and the other samples from leaking pipes had converged despite their antithetical sources. This would not only validate the thermodynamic prediction of the importance of PuO_2 in the environment but also demonstrate that convergence can occur on the decades or shorter time scale even with radically different sources.

The spectral shapes, peak positions, and close coincidence of inflection points with the Pu(IV) standard, all support the conclusion that samples 10, 11, and 12 from the 903 pad area contain plutonium of primarily Pu(IV) speciation (Fig. 6). Two standards of Pu(IV) are included because these spectra were obtained in two

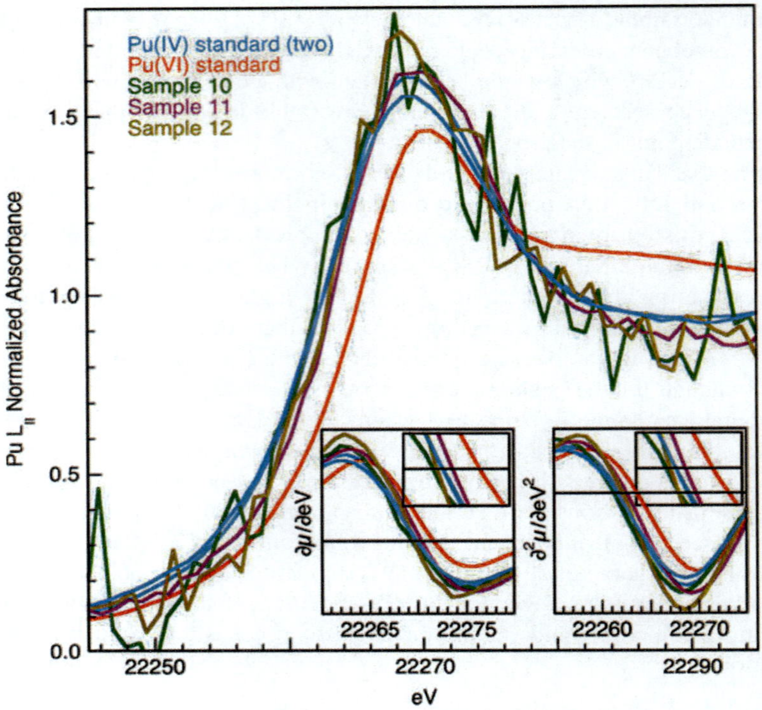

Fig. 6 Plutonium L_{II} XANES of Pad 903 soil samples 10, 11, and 12; *insets* show first (zeros = peak energies), *left*, and second (zeros = inflection point energies), *lower right*, derivatives. The boxes at extreme *top-right* of the insets show a higher magnification of the first derivatives' crossing of the zero line, and the second derivative at the zero line, illustrating the close coincidence of peak positions and inflection points of all spectra with that of the Pu(IV) standard

separate runs. Only the PuO_2 standard was available for elucidating the XANES spectra of samples 10 and 11; the Pu(VI) standard is hence associated with the run that generated spectrum 12. All spectra closely parallel that of the Pu(IV) standard within the context of their different experimental uncertainties, and these uncertainties are also much less than the differences between the Pu(IV) and Pu(VI) standard spectra. Small disparities in peak amplitudes attributable to background differences are observed here, mitigated in the derivatives. None of the samples show evidence for the higher-energy shoulder indicative of the trans-dioxo moiety of Pu(V) and Pu(VI) [21]. The conclusion is therefore that Pu is predominantly, if not entirely, Pu(IV) and exhibits the XANES signatures consistent with PuO_2. This conclusion is reinforced by analysis of the EXAFS from 903 Pad sample 11, which gave a sufficiently strong signal so that the spectrum could be measured to energies well past the absorption edge.

10 Pu(IV)-Oxide Speciation Modes Derived from EXAFS

For sample 11 (903 Pad), the quality of the data was sufficiently high that in addition to the XANES, the EXAFS region of the spectrum was obtained up through $k = 11$ Å^{-1}. This allows specific metrical parameters of the speciation to be extracted. This is important because, although the XANES of both samples points to Pu(IV), recent work has demonstrated that the PuO_2 system is complicated by two factors [23, 24, 28, 29]. Chemisorption and subsequent hydrolysis of H_2O results in stable OH^- complexation and associated distortions in the lattice. Partial oxidation can form superstoichiometric mixed-valence compounds in which the adventitious O occurs as Pu(V)-like oxo groups that localize the charge. Analysis of the EXAFS can elucidate where these samples reside within this range or if they exhibit novel behaviors.

PuO_2-type materials display distinctive patterns in both $\chi(k)$ and $\chi(R)$ representations (Fig. 7) [23, 24, 28]. The switch from domination of the spectrum by the low-frequency wave at lower k to the high-frequency one at higher k as the Pu–O amplitude diminishes and the Pu–Pu amplitude increases from its initially small contribution is easily observed in both spectra, even in that from sample 11 with its higher noise level. These two features subsequently dominate the Fourier transforms as well, as the large, pointed, single peak from the O contribution near $R\text{-}\phi = 1.9$ Å and the Pu/O second/third nearest neighbor peak near $R\text{-}\phi = 3.8$ Å with its low R shoulder caused by the complicated Pu wave. The limited range of the data in combination with the disorder in the Pu shell result in the spectra being dominated by the nearest-neighbor O contribution. These characteristics nevertheless easily differentiate it from the *trans*-dioxo cation moiety that is the norm for Pu(V/VI) that only displays a single broad peak around $R\text{-}\phi = 1.6$ Å that splits into two overlapping ones as the data range is increased and the contributions from the oxo and equatorial ligands are resolved [21]. The spectrum from sample 11 also displays a large feature between these two, which is consistent with other PuO_2-type samples where it indicates relatively large numbers of O atoms located in noncrystallographic shells more likely associated with hydration and hydrolysis [23–25, 28, 30].

More quantitative characterization of the local ordering of the Pu–O/Pu shells is obtained from the metrical results calculated by the EXAFS curve-fits (Table 1). These fits utilizing the PuO_2 structure as an organizing metric for the numbers and types of neighbor shells exhibit correspondence between data and fit that is reasonable considering the noise levels, demonstrating that the assignment of both samples to a PuO_2-like class of material is correct. The correspondence between both moduli and real components of the fits and the data clearly establish the origin of the features as O at lower R and Pu for the higher-R feature. Only Pu–O/Pu distances are reported because correlation between the numbers of atoms and the Debye–Waller factors is exacerbated by the short extent of the data and the noise level, causing them to be unrealistic and misleading. The relative numbers of atoms in a shell are probably best estimated from the area of the modulus of its individual

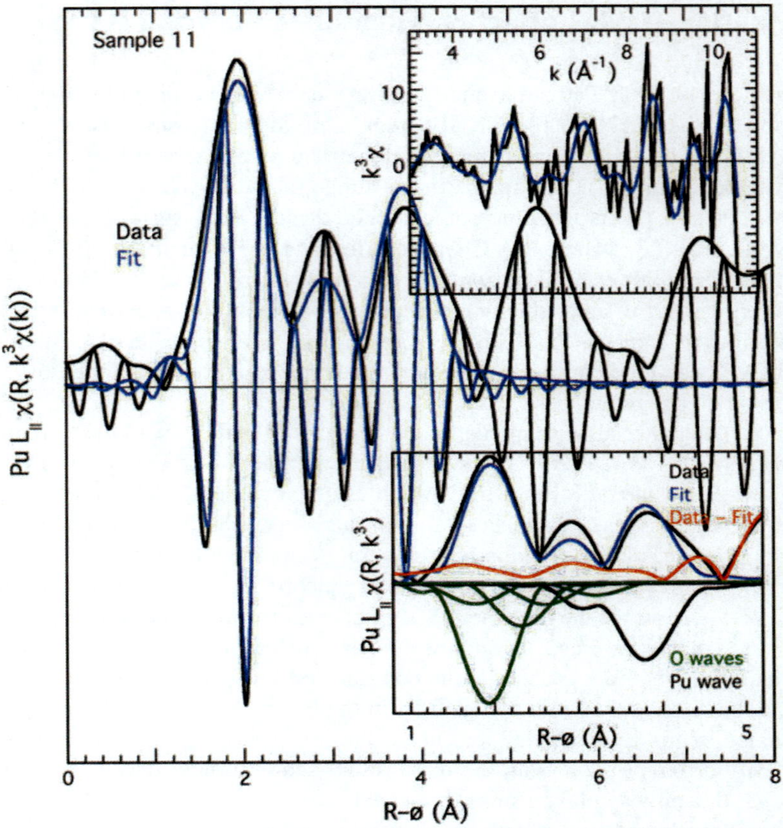

Fig. 7 $\chi(R)$ EXAFS spectrum of sample 11. This is from the pad 903 sample. The higher noise levels are apparent in the amplitudes of the features beyond 5 Å and in the relatively poor fit, which nevertheless correctly located the PuO_2-like attributes of the Pu environment. Data are shown in black and a fit to the data are shown in blue. Top inset shows the EXAFS data and fit in k-space. Bottom inset shows the data and fit with positive amplitudes, the individual shells that make up the fit with negative amplitudes, and the difference between the data and fit in red

Table 1 Radial Pu–O/Pu distances for Pu(IV) compounds

Sample	O at 2.34	O at 2.8	O at 3.0	O at 3.5	O at 4.47	Pu
PuO_2[a]	2.34				4.47	3.82
PuO_2 standard	2.33				4.48	3.85
Sample 11	2.38 ± 0.02	2.74 ± 0.03	3.00 ± 0.02	3.65 ± 0.02	[b]	3.83 ± 0.015

Blank field – no evidence for presence of this PuO_{2+x} shell in fit of the spectrum from this sample
[a]Crystallographic structure
[b]Feature in spectrum but extra shell not included in fit

contribution in the fit. The distances found in these fits, which model the nearest-neighbor distribution with only a single shell because of the limited range of the data, are quite close to those that are common among a large number of PuO_2-type compounds prepared by different methods that are indicative of specific moieties and bonding modes in the material [20, 23].

This similarity confirms the identification of this sample as conforming to the speciation found in this class of materials. The deviations from the crystallographic structure in terms of the extra O shells are very similar to what are found in laboratory produced materials, indicating that the chemical reactivity of the Pu in soils at the RFETS site is similar to that under much more carefully controlled conditions.

The remaining issue in the speciation is demonstrating whether this sample is partly oxidized by the mechanism found for laboratory samples, the oxidation of PuO_2 to $PuO_{2.25}$ via the formation of Pu(V)-oxo moieties [23, 24, 28]. As pointed out in the previous section, however, there are precedents for multisite Pu–O distributions that do not include a Pu-oxo shell, implying hydrolysis types of reactions but not oxidation. The evidence for mixed valence is best found by identifying the signature of the oxo group in the EXAFS as a component around $R\text{-}\phi = 1.3$ Å that is well fit by an O shell at a Pu–O distance at 1.8–1.9 Å. This contribution is isolated by subtracting from the data a fit containing the waves from all of the near neighbors except this one, which can then be fit separately to

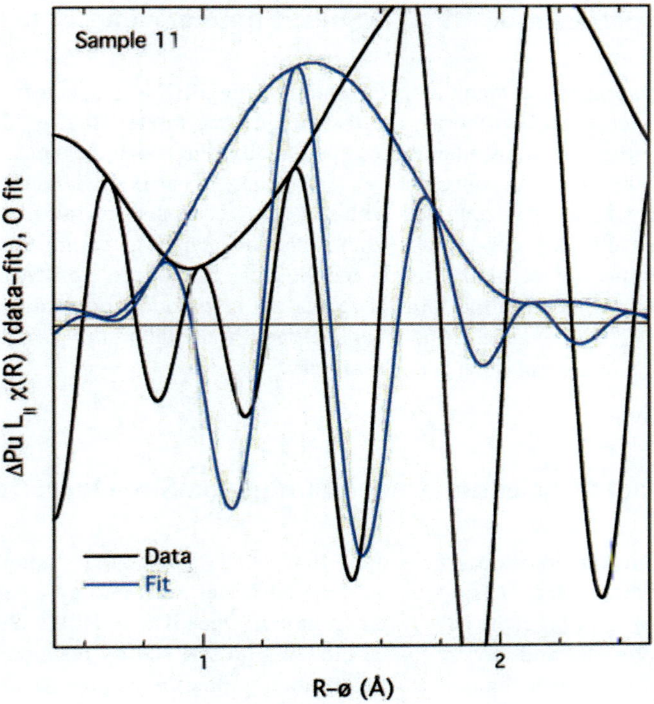

Fig. 8 $\chi(R)$ fit of an O shell (*blue*) to the residual in the pad 903 data (*black*) obtained by subtracting the fit with the shells listed in Table 1 from the data. The poor correspondence between the data and fit reduces the possibility that the oxo-type shell found in some PuO_{2+x} materials exists in this sample to any significant extent

establish the correspondence between any residual remaining after subtraction and a Pu–O wave (Fig. 8). The contribution of the O shell at 1.83 Å that is the best fit to this large residual for the spectrum from sample 11 is not associated with any particular feature and does not correspond well to the data. It is therefore unlikely that sample 11 has been subjected to any significant degree of oxidation beyond PuO_2. The non-PuO_2 shells at distances >2.34 Å must therefore demonstrate that the Pu has undergone the other type of reaction associated with chemisorption of H_2O with retention of the H^+.

The available EXAFS spectra therefore provide convincing evidence in support of the conclusion that the Pu in sample 11 (pad 903) is not only Pu(IV) but also resides in a PuO_2-type of compound of the type we have found before in laboratory prepared materials. This corroborating analysis provides an additional degree of confidence in the conclusions about Pu speciation derived solely from XANES, affirming that the XANES criteria used to assess speciation of the other samples are largely valid.

11 Pu Speciation in RFETS Soils: Summary and Conclusions

The combination of the many XANES and the single EXAFS spectrum indicating PuO_2-like speciation corroborate the thermodynamic models that predict that Pu (IV) is the expected valence, and more specifically, PuO_2 is the expected chemical species in soils in equilibrium with air and water. What is of interest is that this equilibrium was already obtained within a few decades from radically different source terms and local conditions. What is also of interest is that the small particle sizes determined by the ultrafiltration and the concomitant probability that some of the Pu may have been in the form of secondary or pseudocolloids did not disrupt this pattern. The relevance of models developed in the laboratory to the complexities of real world conditions is encouraging.

12 Impact of Scientific Understanding on Site Operations

The scientific understanding developed through the integrated studies described above provided clarity and focus on the real issues surrounding plutonium and americium migration in the RFETS environment. Once Kaiser-Hill, DOE, EPA, the State of Colorado, and the concerned citizen's groups had all reached a common understanding of the technical issues surrounding plutonium and americium migration at the Site, these groups were able to reach long sought-after agreements on how to proceed with cleanup. The common understanding that plutonium and americium were predominantly in particulate and colloidal forms led to the recognition that environmental migration occurs through sedimentation and resuspension of small particles by action of wind and surface water at the Site. This helped all parties focus remediation efforts on surface contamination and wind and surface water transport pathways that posed the greatest risk to human health and the environment. It helped guide selection of surface-specific removal technologies and future land configuration strategies.

In recognition of the new understanding, Site operators responded with a major emphasis shift to erosion and the need to control it. The recognized need for erosion controls "close in space and close in time" helped to prevent movement of contaminants during Site remediation activities and reduced the transport of plutonium and americium to the Site's stream channels and ultimately off site. The additional protection provided by soil erosion control measures allowed Site remediation to proceed rapidly and thus meet or exceed the project deadlines.

In 1996, the Rocky Flats Cleanup Agreement (RFCA) radionuclide soil action level for plutonium cleanup was 651 pCi/g, and it was based on dose. In 2002, armed with improved understanding of plutonium behavior, the DOE, Colorado Department of Public Health (CDPHE), and EPA released a series of reports that formed the basis for a new surface soil action level of 50 pCi/g that was based on risk and resulted from unprecedented community involvement. Since plutonium contamination was generally confined to surface soils, the greatest risk to public health was from dispersal due to action of wind and surface water erosion processes. This new risk-based agreement focused on the removal of surface soil contamination at a more aggressive 50 pCi/g standard to three feet below the surface, with the tradeoff that contaminated soil below three feet could remain in place at higher concentrations.

The Site developed a Storm Water Pollution Prevention Plan (SWPPP) that encouraged minimum soil disturbance, which resulted in control and minimization of erosion and sedimentation, and minimization of runoff across the Site. Each project was reviewed for impacts to surface water with a specifically designed control system. Erosion control measures included straw bales and wattles, straw crimping, silt fences, mats, hydromulch and Flexterra™, and rip-rap lining of drainage channels (Fig. 9). In addition, some new wetland areas were installed.

In actual decontamination, demolition, and remediation work, the Site employed a combination of tents, comprehensive dust and erosion control measures, and general environmental protection during cleanup activities. As a result, surface water and air monitoring stations at the Site boundary showed little change in actinide migration as a result of the site cleanup activities.

13 Conclusions

Making the case for particle transport mechanisms as the basis of plutonium and americium mobility, rather than aqueous sorption–desorption processes, established a successful scientific basis for the dominance of physical transport processes by wind and water. The scientific basis was successful because it was in agreement

Fig. 9 The reconfigured Walnut Creek basin functional channel shows the heavily armored channel and extensive erosion-control measures on stream banks. The measures included straw bales, wattles, and crimping; silt fences; mats; hydromulch and Felxterra^{TM}, and rip-rap (rock) lining of drainage channels

Fig. 10 Aerial photo of the remediated Rocky Flats site, October 2005

with general theory on insolubility of PuO_2 in oxidation state IV, results of ultrafiltration analyses of field water/sediment samples, XAFS analyses of soil, sediment, and concrete samples, and was also in general agreement with on-site monitoring data. This understanding allowed Site contractors to rapidly move to application of soil erosion and sediment transport models as the means of predicting plutonium and americium transport, which led to design and application of site-wide soil erosion control technology to help control downstream concentrations of plutonium and americium in streamflow. Finally, good scientific understanding in the public interest helped bring clarity and focus to real issues of actinide migration at RFETS. This in turn helped to develop a more defined scope with a clearer endpoint that allowed the most extensive cleanup in the history of Superfund legislation to finish 1 year ahead of schedule, ultimately resulting in billions of dollars in taxpayer savings (Fig. 10).

Acknowledgments The authors acknowledge Kaiser-Hill LLC and DOE for their support of AME studies, and we acknowledge the Office of Basic Energy Sciences, and the Office of Biological and Environmental Research at the U.S. Department of Energy for their support of actinide science and the Stanford Synchrotron Radiation Lightsource that assisted RFETS cleanup activities.

References

1. DOE, "Final Phase II RCRA Facility Investigation/Remedial Investigation Report, 903 Pad, Mound, and East Trenches Area, Operable Unit No. 2" *Tech. Report No. RF/ER-95-0079.UN* (US Department of Energy, Rocky Flats Plant, 1995).
2. DOE, "Baseline Environmental Management Report" (US Department of Energy, Environmental Management, 1995).
3. GAO, "Nuclear Cleanup of Rocky Flats. DOE Can Use Lessons Learned to Improve Oversight of Other Sites' Cleanup Activities" *Tech. Report No. GAO-06-352* (US Government Accountability Office, 2006).
4. "Actinide Migration Evaluation Pathway Analysis Summary Report" *Tech. Report No. ER-108* (Kaiser-Hill, Rocky Flats Environmental Technology Site, 2002).
5. Hurr, R. T., "Hydrology of a Nuclear-Processing Plant Site, Rocky Flats" *Tech. Report No. Open File Report 76-268* (U.S. Geological Survey, 1976).
6. Myers, J. C., *Geostatistical error management; quantifying uncertainty for environmental sampling and mapping*, Van Nostrand Reinhold, New York, 1997.
7. Litaor, M. I., Barth, G. R., and Zika, E. M., *J. Environ. Qual.*, 1996. **25**(4): p. 671–683.
8. Runde, W., and Neu, M. P.; in *The chemistry of the actinide and transactinide elements*, L.R. Morss, N.M. Edelstein, and J. Fuger, Editors. Volume 6, Springer, Heidelberg, 2010.
9. Langmuir, D., *Aqeuous environmental geochemistry*, Prentice Hall, New Jersey, 1997, 600 pp.
10. Harnish, R. A., McKnight, D. M., and Ranville, J. F. *Particulate, colloidal, and dissolved-phase associations of plutonium and americium in a water sample from well 1587 at the Rocky Flats Plant, Colorado*, Geological Survey, Denver, CO, USA., Report 1994, 32 pp.
11. Knopp, R., Neck, V., and Kim, J. I., *Radiochim. Acta*, 1999. **86**(3–4): pp. 101–108.
12. Kersting, A. B., Efurd, D. W., Finnegan, D. L., Rokop, D. J., Smith, D. K., and Thompson, J. L., *Nature (London)*, 1999. **397**(6714): pp. 56–59.
13. Salbu, B., *J. Environ. Radioact.*, 2007. **96**: pp. 47–53.
14. Lind, O. C., Salbu, B., Janssens, K., Proost, K., García-León, M., García-Tenorio, R., *Sci. Total Environ.*, 2007. **376**: pp. 294–305.
15. Hayden, J., DeHerrera, M., and Stewart, C., *Particle size distribution of Plutonium on soil in Rocky Flats East buffer zone*, Rockwell International, Report 1975.
16. Buddemeier, R. W., and Hunt, J. R., *Appl.Geochem.*, 1988. **3**(5): pp. 535–548.

17. Santschi, P. H., Roberts, K. A., and Guo, L., *Environ. Sci. Technol.*, 2002. **36**(17): pp. 3711–3719.
18. Shaw, D. J., *Colloid and surface chemistry*, Butterworth-Heinemann, Oxford, 1992.
19. Stumm, W., *Chemistry of the solid-water interface: processes at the mineral-water and particle-water interface in natural systems*, Wiley, New York, 1992.
20. Clark, D. L., Janecky, D. R., and Lane, L. J., *Phys. Today*, 2006. **59**(9): pp. 34–40.
21. Conradson, S. D., *Appl. Spectrosc.*, 1998. **52**(7): pp. 252A–279A.
22. Conradson, S. D., Abney, K. D., Begg, B. D., Brady, E. D., Clark, D. L., den Auwer, C., Ding, M., Dorhout, P. K., Espinosa-Faller, F. J., Gordon, P. L., Haire, R. G., Hess, N. J., Hess, R. F., Keogh, D. W., Lander, G. H., Lupinetti, A. J., Morales, L. A., Neu, M. P., Palmer, P. D., Paviet-Hartmann, P., Reilly, S. D., Runde, W. H., Tait, C. D., Veirs, D. K., and Wastin, F., *Inorg. Chem.*, 2004. **43**(1): pp. 116–131.
23. Conradson, S. D., Begg, B. D., Clark, D. L., den Auwer, C., Ding, M., Dorhout, P. K., Espinosa-Faller, F. J., Gordon, P. L., Haire, R. G., Hess, N. J., Hess, R. F., Keogh, D. W., Morales, L. A., Neu, M. P., Paviet-Hartmann, P., Runde, W., Tait, C. D., Veirs, D. K., and Villella, P. M., *J. Am. Chem. Soc.*, 2004. **126**(41): pp. 13443–13458.
24. Conradson, S. D., Begg, B. D., Clark, D. L., den Auwer, C., Espinosa-Faller, F. J., Gordon, P. L., Hess, N. J., Hess, R., Keogh, D. W., Morales, L. A., Neu, M. P., Runde, W., Tait, C. D., Veirs, D. K., and Villella, P. M., *Inorg. Chem.*, 2003. **42**(12): pp. 3715–3717.
25. Ding, M., Conca, J. L., den Auwer, C., Gabitov, R. I., Hess, N. J., Paviet-Hartmann, P., Palmer, P. D., LoPresti, V., and Conradson, S. D., *Radiochim. Acta*, 2006. **94**(5): pp. 249–259.
26. Koningsberger, D. C., Prins, R., *X-ray absorption: principles and techniques of EXAFS, SEXAFS AND XANES*, Wiley, New York, 1988.
27. Teo, B.T., *EXAFS: basic principles and data analysis*, Springer-Verlag, New York, 1986, 349 pp.
28. Conradson, S. D., Begg, B. D., Clark, D. L., den Auwer, C., Mei, D., Dorhout, P. K., Espinosa-Faller, F. J., Gordon, P. L., Haire, R. G., Hess, N. J., Hess, R. F., Keogh, D. W., Lander, G. H., Manara, D., Morales, L. A., Neu, M. P., Paviet-Hartmann, P., Rebizant, J., Rondinella, V. V., Runde, W., Tait, C. D., Veirs, D. K., Villela, P. M., and Wastin, F., *J. Solid State Chem.*, 2005. **178**(2): pp. 521–535.
29. Martin, P., Grandjean, S., Ripert, M., Freyss, M., Blanc, P., and Petit, T., *J. Nucl. Mater.*, 2003. **320**(1–2): pp. 138–141.
30. Rothe, J., Walther, C., Denecke, M. A., and Fanghanel, T., *Inorg. Chem.*, 2004. **43**(15): pp. 4708–4718.

Colloid-Facilitated Transport of Plutonium at the Nevada Test Site, NV, USA

Annie B. Kersting and Mavrik Zavarin

Abstract Increasing observations of subsurface actinide transport has reinforced the need for a basic geochemical and hydrologic understanding of how these contaminants are transported in groundwater. At the Nevada Test Site, NV, USA, low-levels of plutonium have been shown to migrate on the scale of kilometers. At all but one sampling location the Pu is associated with the colloidal fraction (<1 μm) consisting predominantly of clays and zeolites. The majority of the Pu ($\sim$70%) is associated with the smallest nanoparticle (e.g., colloid) size fraction of 10–100 nm and in one case was identified as associated with a clay colloid. At one unique sampling site, Pu is predominantly associated with dissolved organic matter. Monitoring at over 20 contaminated groundwater sites at the NTS suggests that low-levels of Pu can be mobilized either via colloid-facilitated transport in fractured rock or as a result of aqueous Pu stabilization by anthropogenic levels of dissolved organic matter. Nevertheless, activities above the EPA's Maximum Contaminant Level (MCL) for alpha-emitting radionuclides (15 pCi/L, 0.56 Bq/L) have been measured at only two locations at the NTS.

1 Introduction

Over 800 underground nuclear tests were detonated from 1951 to 1992 at the Nevada Test Site (NTS), as part of the United States' nuclear testing program (DOE 2000) (Fig. 1). Approximately one third of the nuclear tests were detonated at or below the standing water table (Laczniak et al. 1996). An estimated 1.3×10^8 curies (4.8×10^{18} Bq) of radioactivity (calculated to 1992) was deposited and remains in the subsurface at the NTS, consisting of tritium, fission products, actinides, and activation products (Bowen et al. 2001).

A.B. Kersting (✉) and M. Zavarin
Glenn T. Seaborg Institute, Lawrence Livermore National Laboratory, P.O. Box 808, Livermore, CA 94550 , USA
e-mail: kersting1@llnl.gov; zavarin1@llnl.gov

S.N. Kalmykov and M.A. Denecke (eds.), *Actinide Nanoparticle Research*,
DOI 10.1007/978-3-642-11432-8_15, © Springer-Verlag Berlin Heidelberg 2011

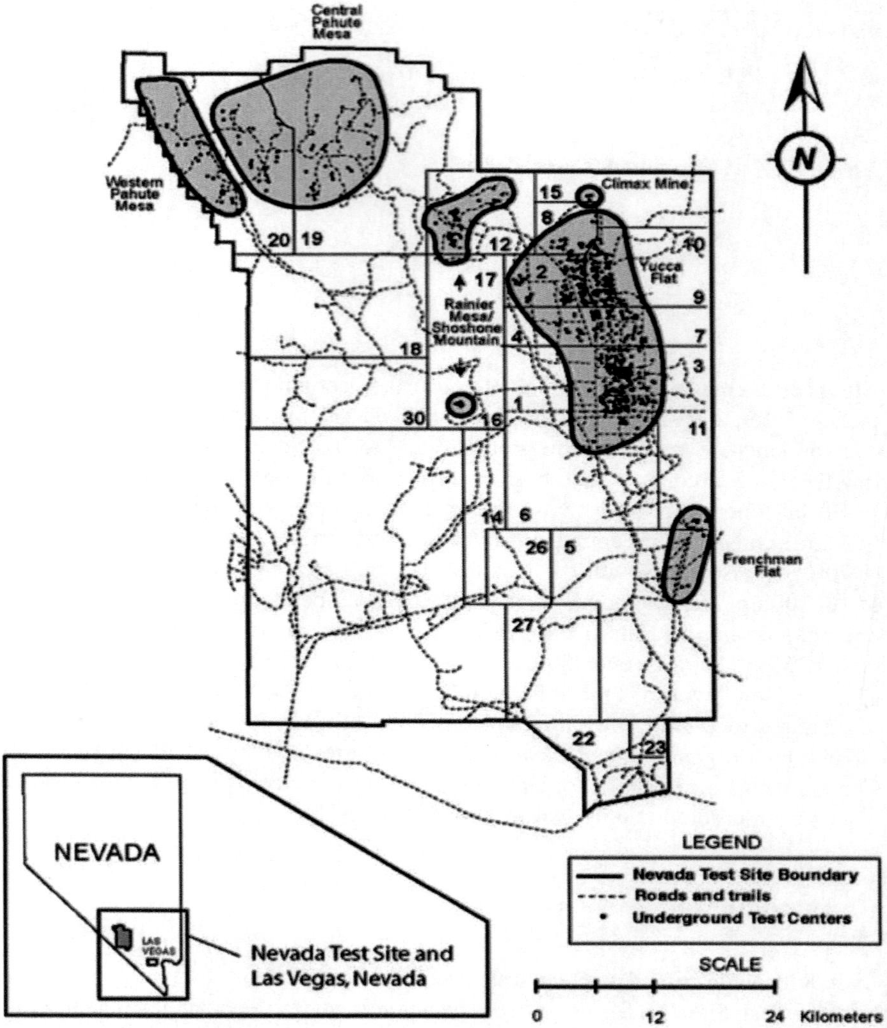

Fig. 1 The NTS showing the location of each of the 828 underground nuclear tests. *Circles* denote geographic regions

The radionuclide inventory at the NTS is similar in magnitude to the inventory at the Hanford Reservation (~1.0×10^{19} Bq) and approximately two orders of magnitude less than estimates for the proposed Yucca Mountain Repository (~1.8×10^{21} Bq). Although the majority of the NTS inventory is currently tritium, approximately 2,775 kg (3.1×10^{16} Bq) is plutonium (Pu) (Bowen et al. 2001). As a result, the NTS contains a large inventory of Pu in the subsurface and thus provides a unique opportunity for studying its fate and transport.

Details of the potential transport of Pu have been a contentious issue especially in the area of nuclear energy and waste disposal. Plutonium, with its low solubility and high surface reactivity, was thought to remain immobile in the subsurface, adsorbed to minerals that constitute the soil and bedrock. Buddemeier and Hunt (1988) implicated colloids in the transport of transition element and lanthanide radionuclides in groundwater at the NTS. In a later field study, Kersting et al. (1999) documented the unexpected appearance of Pu downgradient from its known source and showed that Pu was transported in association with the colloidal fraction consisting of clays and zeolites. Although not a new idea (Kim 1991; McCarthy and Zachara 1989; Santschi and Honeyman 1989; Ryan and Elimelech 1996), colloid-facilitated transport has slowly gained acceptance for the transport of low solubility contaminants. Recent work by Novikov et al. (2006) and Xu et al. (2006) has further advanced our understanding of this process by showing that Pu transported from its source was adsorbed to Fe-oxide and organic colloids, respectively.

These field studies indicate that the Pu associated with mobile colloids can move on the scale of kilometers. Colloidal particles are ubiquitous in both groundwaters and surface waters (Degueldre et al. 1989; McCarthy and Degueldre 1993; Ryan and Elimelech 1996) and are chemically, structurally, and temporally complex organic and inorganic species of both natural and anthropogenic origin. Plutonium can adsorb to colloidal particles, be incorporated within them, or form its own colloidal polymer. Yet, fundamental details of the surface mechanisms that control this process are not well understood and continue to hinder both the conceptualization and predictive capability of colloid-facilitated actinide transport. One of the difficulties is that, in general, Pu measured in field samples is at extremely low levels (e.g., femtomolar Pu concentrations), and this has limited the application of spectroscopic techniques to identify the characteristics of Pu–colloid interactions. Without knowledge of how Pu moves in groundwater, transport models fail to reliably predict the subsurface behavior of Pu. The current inability to effectively predict Pu transport in the subsurface presents a serious scientific challenge.

Information presented in this chapter is based on data accumulated by a number of programs at the NTS that have examined the initial distribution of radionuclides in the subsurface after a nuclear test and its subsequent transport. Most recently, the effort has been led by the Underground Test Area project, Department of Energy, National Nuclear Security Administration, Nevada Operations Office, U.S.A. Details regarding sampling programs and groundwater characterization can be found in Smith (1995), Tompson et al. (2002), and Hu et al. (2008). The focus of this chapter is the observation and characterization of plutonium transport mechanisms at the NTS. We evaluate the Pu detected in several of wells at the NTS, characterize the colloidal fraction of the groundwater, provide a conceptual model for Pu transport at the NTS, and extend this model to other locations.

2 Geology and Hydrology of the NTS

The NTS occupies 2,173 km^2 in south-central Nevada (Fig. 1) and lies within the southern Great Basin region that is characterized by low rainfall and sparsely distributed intermittent streams (Laczniak et al. 1996). The groundwater is deep, roughly more than 250 m below ground surface and flows south–southwest through thick sequences of carbonate and volcanic rocks and discharges at Ash Meadows and Death Valley. Most of the nuclear tests were detonated in either the valley-fill alluvium or volcanic rock. Groundwater near each underground test generally moves laterally through these units and, in some cases, may contact the underlying regional carbonate aquifer. In general, the groundwater in alluvium and volcanic rock is dominantly Na-bicarbonate type with a pH of ~8 and low ionic strength, while the carbonate aquifer groundwater is enriched in Ca relative to the overlying volcanic waters.

3 Phenomenology of Underground Nuclear Tests

During the detonation of a nuclear device, the rock immediately surrounding the device vaporizes and expands as a plasma, temperatures exceed 1×10^6 K locally, and pressures exceed 7 Mbar (Fig. 2a). Approximately 70 metric tons of rock is vaporized and another 700 metric tons of rock is melted for every kiloton (kt) of

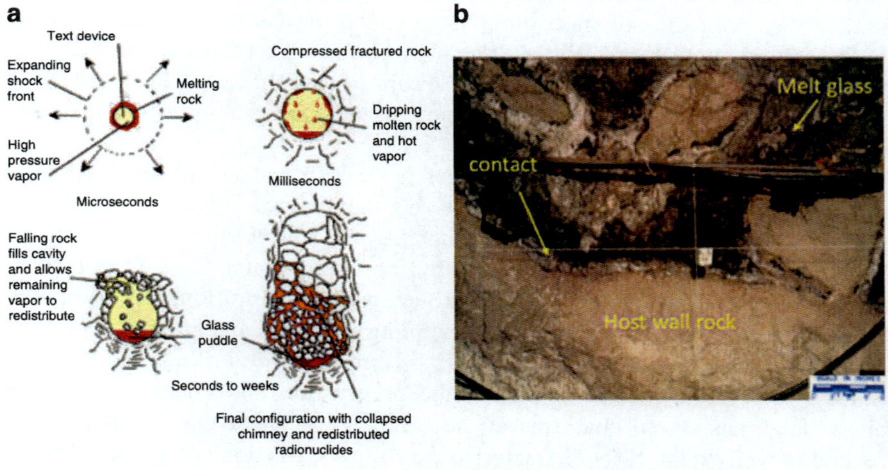

Fig. 2 (**a**) Phenomenology of underground nuclear test, showing nuclear melt glass (puddle glass) coalescing at the bottom of the cavity below the chimney rubble zone. (**b**) Photograph of the Rainier nuclear test (1957) showing the contact between the volcanic host rock below (*pink*) and the nuclear melt glass (*dark brown*) at the bottom of the cavity (Wadman and Richards 1961)

yield (Smith 1993). The majority of refractory radionuclide species [e.g., rare earths (Ce, Sm, Eu), alkaline earths (Mg, Sr, Ba), Zr, and actinides (Am, Pu)] coalesce in the nuclear melt glass at the bottom of the cavity (Fig. 2b). In addition, many of the fission and neutron activation products (^{54}Mn, ^{60}Co, ^{99}Tc, ^{106}Ru, ^{125}Sb, ^{137}Cs, and ^{147}Pm) are also found in the melt glass (Borg 1975). These radionuclides are structurally incorporated into the glass matrix and are not immediately available for transport with groundwater. As the groundwater reinfiltrates the cavity, the melt glass begins to alter, resulting in the slow release of radioactivity into the groundwater. The volatile species (e.g., alkali metals, U, Sb, I, Ru and gases Ar, Kr, Xe) are more broadly distributed in the cavity and overlying rubble chimney created directly above the cavity (Borg et al. 1976). For a more detailed explanation of the nuclear test phenomenology, see: Office of Technology Assessment (1989) and Smith (1995).

Underground nuclear tests ranged in yield from less than 1 kt to detonations in excess of 1,000 kt (DOE 2000) and varied in design and performance. The volume of rock affected, extent of fracturing, thermal perturbations, and most importantly, the residual inventory of radioactivity and its initial distribution in the subsurface are unique to each test. The radionuclides deposited after a test are heterogeneously distributed throughout the cavity, chimney area, and within several cavity radii of the emplacement location. Although the amount of radioactivity at each test may be unique, once the contaminants are deposited, the processes that control their transport in groundwater are similar.

4 Plutonium Detected in NTS Groundwater

Currently, there are approximately 20 wells that can be used to collect contaminated groundwater from either inside or outside a nuclear test cavity (Fig. 3). In addition to high levels of tritium (up to 5×10^7 Bq/L), low-levels of other radionuclides have been detected (Kersting et al. 1999; Hu et al. 2008). Low concentrations of Pu were detected in several of these wells (Table 1), which will be described in some detail below.

5 Well ER-20-5 #1 and #3: Down-Gradient from the Benham Site

The ER-20-5 well cluster consists of two near-field monitoring wells (#1 and #3) located on Pahute Mesa, northwest of the NTS (Fig. 3). These wells were drilled in 1995–1996 to look for down-gradient transport of radionuclides from the Tybo nuclear test. The ER-20-5 well cluster is located 280 m southwest of the Tybo test site and 1.3 km south of the Benham test site (Kersting et al. 1999). The ER-20-5 #1

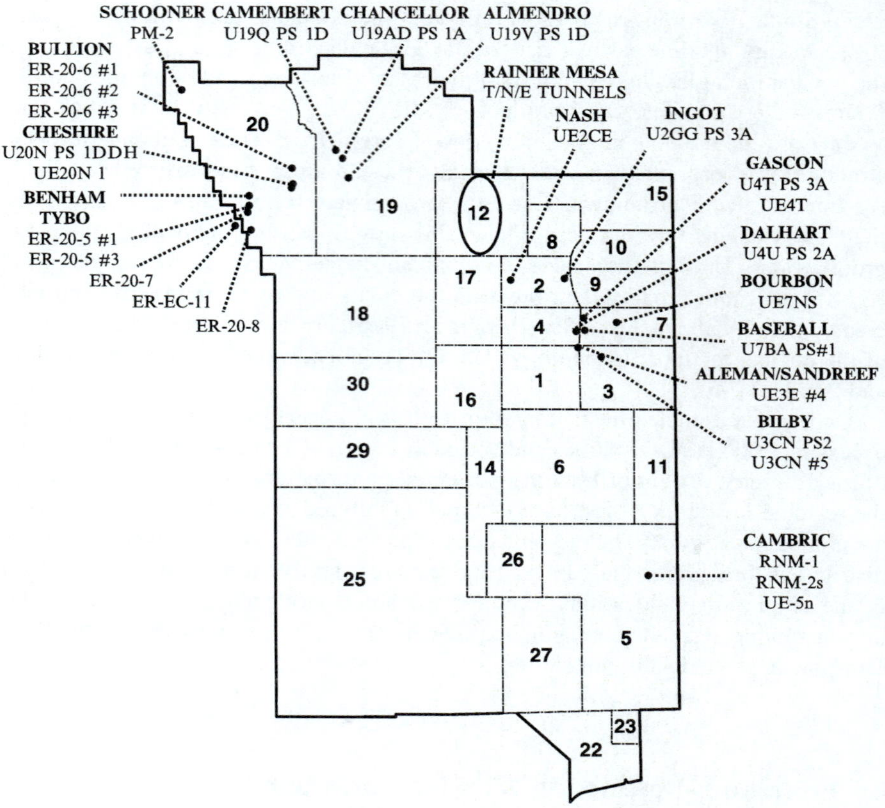

Fig. 3 Map of the NTS showing both the nuclear test locations and the well locations where wells are currently available for groundwater sampling

and #3 wells were screened at a depth of 701–784 m and 1,046–1,183 m, respectively. The Tybo nuclear test was detonated on May 14, 1975, with a yield range of 200–1,000 kt (DOE 2000). The Benham test was detonated on December 19, 1968, with an announced yield of 1.15 Mt.

Groundwater was collected on three different occasions from 1996 through 1997. Plutonium was detected at low concentrations in both wells. In 1997, the ^{239,240}Pu concentration in the ER-20-5 #1 and #3 wells was 0.63 pCi/L (2.3×10^{-2} Bq/L) and 1.13×10^{-2} pCi/L (4.2×10^{-4} Bq/L), respectively (Thompson 1998). Subsequent sampling efforts have measured Pu activity similar to these initial values. The ^{240}Pu/^{239}Pu isotopic ratio of the groundwater from each well was identical and matched that of the Benham nuclear test (Kersting et al. 1999). Despite the close proximity of the ER-20-5 well cluster to the Tybo test, plutonium isotope measurements indicate that the Benham test was the sole source of the Pu activity in both of the ER-20-5 wells.

Table 1 Plutonium concentrations in NTS hot well groundwater samples

Test	Well name	Distance from detonation point, m[b]	Sampling date	^{3}H pCi/L	^{3}H Bq/L	Pu, total pCi/L	Pu, total Bq/L
Bilby	U-3cn PS#2	210 (vert.)	2007	7.7×10^{6}	2.8×10^{5}	0.06	2.2×10^{-3}
Camembert	U19q PS1d	462 (vert.)	2003	1.1×10^{7}	4.1×10^{5}	<0.02	$<7.4 \times 10^{-4}$
Cambric	RNM-2S	91	2007	1.0×10^{5}	3.7×10^{3}	<0.02	$<7.4 \times 10^{-4}$
Cambric	RNM-1	6	2007	8.7×10^{2}	3.2×10^{1}	0.006	2.2×10^{-4}
Cambric	UE-5n	600	2004	1.5×10^{5}	5.6×10^{3}	<0.02	$<7.4 \times 10^{-4}$
Schooner	PM-2	270 (hor.) 490 (vert.)	2005	2.6×10^{5}	9.6×10^{3}	0.006	2.2×10^{-4}
Dalhart	U4u PS2a	140 (vert.)	2008	2.4×10^{7}	8.9×10^{5}	0.44	1.6×10^{-2}
Bourbon	UE-7ns	137	2005	1.3×10^{2}	4.8	<0.04	$<1.5 \times 10^{-3}$
Nash	UE-2ce	183	2008	2.7×10^{5}	1.0×10^{4}	<0.008	$<3.0 \times 10^{-4}$
Gascon	U-4t PS3a	54 (vert.)	2008	3.0×10^{3}	1.1×10^{2}	<0.0015	$<5.6 \times 10^{-5}$
Gascon	UE-4t #1	168	2008	6.8×10^{1}	2.5	[a]	
Gascon	UE-4t #2	168	2008	1.7×10^{3}	6.3×10^{1}	[a]	
Aleman/Sandreef	UE-3e #4	60/280	2009	1.5×10^{6}	5.6×10^{4}	<0.002	$<7.4 \times 10^{-5}$
Tybo/Benham	ER-20-5 #1	280/1,310	1997	6.9×10^{7}	2.6×10^{6}	0.63	2.3×10^{-2}
Tybo/Benham	ER-20-5 #3	290/1,340	1997	1.4×10^{5}	5.3×10^{3}	1.13×10^{-2}	4.2×10^{-4}
Cheshire	U20n PS1 DDh		2005	3.3×10^{7}	1.2×10^{6}	0.46	1.7×10^{-2}
Almendro	U19v PS1ds	125 (vert.)	1999–2006	1.1×10^{8}	4.1×10^{6}	<0.004–9.5.	0–0.35
Chancellor	U19ad PS1A	48 (vert.)	2008	1.3×10^{7}	4.8×10^{5}	47	1.7
U12n tunnel	U12n extension drift gas sealed plug	450–1,600	2006	1.2×10^{6}	4.4×10^{4}	$1.2\ (16)^{2}$	4.4×10^{-2}
U12n tunnel	U12n main drift gas sealed plug	450–1,600	2006	1.8×10^{6}	6.7×10^{4}	$1.1\ (16)^{2}$	4.1×10^{-2}
U12n.10	Tunnel vent	120	2008	6.3×10^{6}	2.3×10^{5}	0.80	3.0×10^{-2}
U12t tunnel	U12t gas sealed door	450–1,000	2006	9.4×10^{3}	3.5×10^{2}	$0.009\ (7)^{2}$	3.3×10^{-4}
U12t tunnel	U12t gas sealed plug	450–1,000	2006	4.8×10^{7}	1.8×10^{6}	$54\ (21)^{2}$	2.0

Note: each well has been sampled more than once, but only one value is listed. For a complete listing of all radionuclides measured at the NTS wells, see Stoller-Navarro (2007)

[a]Not measured

[b]Distance is horizontal unless otherwise noted

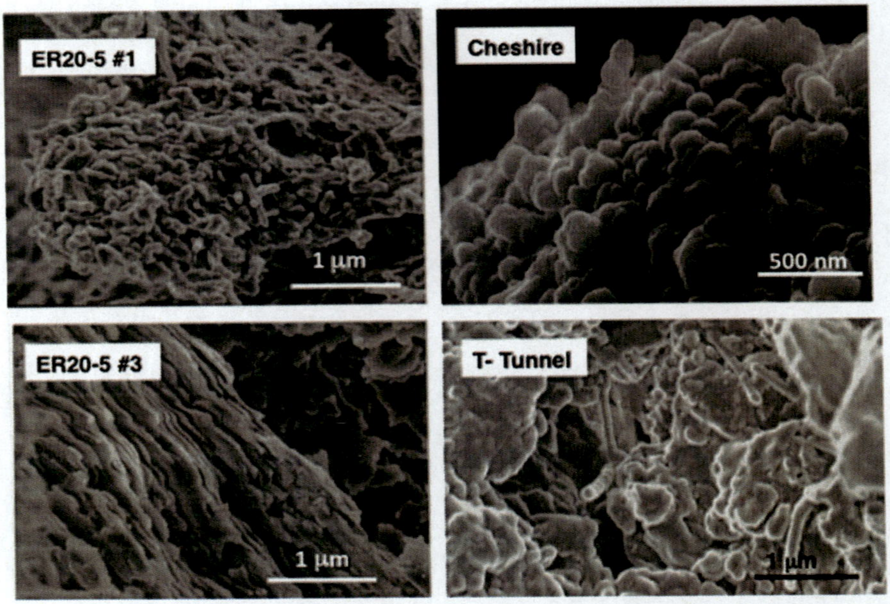

Fig. 4 SEM photomicrographs of colloids filtered from groundwater at the NTS. ER-20-5#1: *rod shaped* colloids are mordenite, ER-20-5#3 platy minerals are stacks of clays

The groundwater from each well was filtered to 7 nm and the colloid fraction analyzed by X-ray diffraction (XRD). The colloidal fraction from each well was the same and composed of clays (illite and smectite), zeolites (mordenite and clinoptilolite), and cristobalite (Fig. 4) (Kersting 2003). There was no difference in mineralogy between the different colloid fractions (>1 μm, 1,000–50 nm, and 50 to ~7 nm). Over 95% of the Pu was associated with the colloidal fraction of the groundwater.

6 Almendro (Well U-19v PS#1ds)

The Almendro nuclear test was conducted on 6 June, 1973, at a vertical depth of 1,063 m beneath the surface of Pahute Mesa. The postshot hole (U-19v PS#1ds) was later developed for use as a near-field monitoring well (Thompson 2001). Groundwater characterization samples were first collected at this site in 1993 at a vertical depth of 942 m; 125 m above the working point. The well was not pumped and sampling was always conducted using a wireline bailer due to the narrow diameter of the borehole. A borehole temperature log run in 1996 recorded a maximum value of 157°C at the bottom of the cavity. The persistence of elevated temperatures more than 30 years after the test suggests that Almendro cavity fluids may be effectively isolated from the surrounding groundwater.

Over the last 10 years, Pu concentrations in groundwater from the Alemendro well have been measured five times, each time producing a different concentration, ranging from nondetect (<0.004 pCi/L) to 9.5 pCi/L $^{239,240}Pu$ (Rose et al. 2003). The samples were collected 125 m above the working point which is near the edge of the cavity and may not be representative of the average cavity conditions (cavity radius was estimated using the empirical equation reported in Pawloski (1999) with the maximum announced yield from DOE (2000), a depth of burial from DOE (1997), and an approximate overburden density of 2.0 g/cm^3). The samples contained substantial amounts of drilling fluids and as a result, no filtering of the colloids or further characterization was attempted.

7 Dalhart (Well U-4u PS#2)

The Dalhart test (U−4u) was conducted in 1988 with an announced yield of <150 kt (DOE 2000). The working point was at 640 m below ground surface and located in zeolitized tuff. The water level was at a depth of 508 m. In 1990, the postshot drillback U−4u PS#2 was drilled approximately 224 m due south from the working point and at an angle (Mathews et al. 1994; Thompson 1995). The sampling interval is 140–170 m above the working point (2.25–2.75 cavity radii based on estimation methods described in the previous section).

Water samples were collected intermittently at this site since 1992. However, early samples had a significant particulate load resulting from drilling mud (Thompson 1995, 1999). The well was not thoroughly cleaned out until 2003 when 7,058 gallons (26.7 m^3) of water was pumped prior to sampling.

In 2008, samples were collected from Dalhart and the concentration of Pu was 0.44 pCi/L. Groundwater was further separated into colloidal (>40 nm) and aqueous fractions (<40 nm). The colloidal fraction contained approximately 65% of the Pu. It may be surprising that ~35% of the Pu appears to be aqueous (i.e., not associated with colloids). However, it is important to remember that the partitioning between aqueous and colloidal fractions is method-dependent. In this case, the colloidal fraction was only separated to a particle size of >40 nm. Thus, it is quite possible that the colloidal fraction between 10 and 40 nm contributed to the apparently high "aqueous" Pu as noted in the Chancellor groundwater sampling described below. The apparent colloidal Pu fraction should be considered a minimum value based on the size exclusion criterion used.

8 Cheshire (Well U-20n PS1 DDh)

The Cheshire test was detonated on 14 February, 1976, at a depth of 1,167 m beneath the surface of Pahute Mesa with an announced range in yield between 200 and 500 kt (DOE 2000). A postshot hole (U-20n PS1-DDh) was slant-drilled soon after the test and later was converted to a monitoring well for radionuclide migration investigations (Sawyer et al. 1999). The well has been modified several times over its history, but still provides access to the Cheshire cavity and/or chimney region for groundwater sampling.

Groundwater samples were collected in 2005 after the groundwater was pumped from the slant-depth interval 1,250–1,253 m (4,100–4,110 ft) below the surface, within the Cheshire test cavity. Unfiltered groundwater had a total $^{239}Pu + ^{240}Pu$ activity of 0.46 pCi/L. This result compares very well with earlier Pu measurements and suggests little or no change in the Cheshire cavity/chimney Pu concentrations over time.

Filtration performed on the 1998 and 2003 samples showed that >90% of the plutonium was associated with the colloidal fraction (>20 nm). Colloids were characterized by XRD and consist mostly of clays and minor zeolites (Brachmann and Kersting 2003).

9 Chancellor (Well U-19ad PS#1A)

The Chancellor test was detonated on 1 September, 1983, at a vertical depth of 2,047 ft (624 m) below the surface of Pahute Mesa. It had an announced yield of 143 kt (DOE 2000). The regional water table lies at a depth of 644 m below the surface, located approximately 20 m below the working point of the test (Rose et al. 2008). Groundwater flow in this area occurs principally along interconnected faults and fractures within the thick sequence of Tertiary volcanic rocks that underlie Pahute Mesa (Blankennagel and Weir 1973; Laczniak et al. 1996).

In 2004, a postshot well (U-19ad PS#1A) was drilled in order to retrieve core and groundwater samples in the vicinity of the original test cavity (Rose et al. 2008). Prior to sampling, 69.7 m^3 of groundwater was purged from the well at a rate of 0.09 $m^3\ min^{-1}$. Temperatures were recorded downhole and reached a maximum of 97°C. The high temperatures were observed again in 2008. Persistent high temperatures imply that a significant amount of residual heat remains from the nuclear detonation, even after 25 years. The elevated temperatures appear to have modified the major ion chemistry of the water, resulting in high concentrations of dissolved F^-, Cl^-, and SO_4^{2-} relative to environmental groundwater samples from this same area.

Groundwater collected from the Chancellor postshot well contains one of the highest plutonium concentrations yet observed from a NTS test cavity. The total $^{239,240}Pu$ activity is 47 pCi/L (~1.74 Bq/L). This is above the Maximum Contaminant Level (MCL) set by the U.S. Environmental Protection Agency for alpha particle emitting radionuclides (15 pCi/L or 0.56 Bq/L). Groundwater was further filtered to determine if the Pu was associated with the colloidal fraction of the groundwater. The colloids were centrifuged and size fractioned at >1,000 nm, 100–1,000 nm and 10–100 nm, and <10 nm (dissolved fraction). Pu was extracted from each size fraction and analyzed using mass spectrometry.

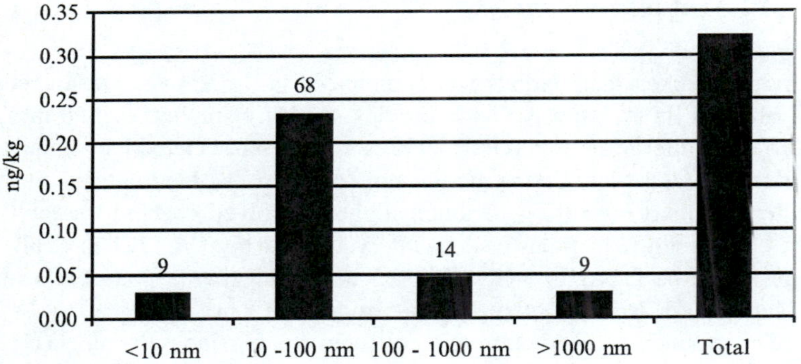

Fig. 5 Plot of total Pu in Chancellor groundwater sample compared to the concentration of Pu in each colloidal size fraction. *Numbers* denote the percentage of Pu in that given size fraction

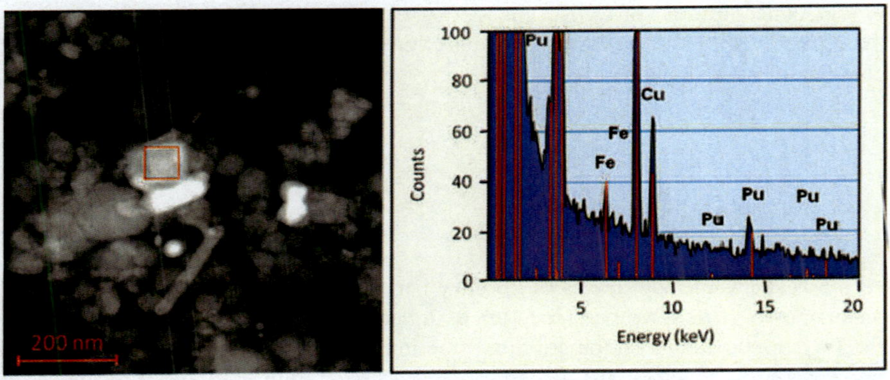

Fig. 6 STEM image (*left*) and EDX spectrum (*right*) of natural colloids from groundwater collected from the Chancellor test cavity. *Box* region shows Pu associated with a clay colloid

Figure 5 shows the total Pu and the distribution in each colloidal fraction. The majority of the Pu (>70%) was in the smallest (10–100 nm) fraction. Less than 10% of the Pu was in the dissolved fraction. Approximately 84% of the total Pu was associated with the colloidal fraction of the water (>20 nm) and when the size cutoff was reduced to 10 nm, the colloidal fraction increased to 91%.

The colloids were further analyzed by XRD and scanning transmission electron microscopy (STEM). All three size fractions consisted of clays, zeolites, and feldspar (Fig. 4). There was no significant difference in the mineralogy between the different fractions. Figure 6 shows a STEM image of Pu associated with a clay colloid and an accompanying EDX analysis of the Pu and clay colloid. Concentrations of Pu in the sample were approximately 10^{-15} M.

10 U12 Tunnel System

A total of 62 nuclear detonations were conducted in the Rainier Mesa area of the NTS, primarily in an extensive and complex system of tunnels drilled into zeolitized tuff of Rainier Mountain (DOE 2000). Activities associated with underground nuclear testing and tunnel construction on Rainier Mesa have altered the tunnel water geochemistry, resulting in much higher dissolved organic matter (DOM) content (15–19 mg C/L) than local spring water (0.2 mg C/L) (Zhao et al. 2011). The source of the DOM is believed to be tunnel lagging, wood debris, drilling fluids, and their microbial decomposition products.

In 2006, groundwater samples were collected from inside two tunnel (U12n and U12t) complexes and analyzed for radionulcides. In contrast to the other groundwater samples collected at the NTS, over 65% of the Pu passed through the 20 nm filter (dissolved fraction). More than 90% of the DOM also remained in solution after filtration. Pu batch sorption experiments were carried out and K_ds calculated for Pu adsorbed to tuff volcanic rock and tunnel fluid with and without the presence of high concentrations of DOM. The K_ds were two orders of magnitude lower when the groundwater contained high concentrations of DOM, suggesting that the Pu is complexed with DOM.

11 Conclusions

At the NTS, where low levels of Pu have been detected in both cavity waters and waters downgradient from the location of underground nuclear tests, the majority of the Pu is associated with the colloidal fraction and not the dissolved fraction. The one exception is when the groundwater has been impacted by anthropogenic increases in DOM, where the majority of the Pu is associated with the dissolved fraction (>20 nm). The majority of groundwaters at the NTS are near neutral pH, have low ionic strength and are naturally low in organic matter. When associated with colloids, the majority of Pu is found in the smallest, 10–100 nm colloidal fraction. If filtration cutoff values are larger than 10–20 nm, caution should be used when quantifying the colloidal vs. dissolved fractions as colloidal fraction may only represent a minimum value. At the NTS, the groundwater colloids have a strikingly consistent mineralogy of clays and zeolites with minor cristobalite and feldspar. In groundwater from the Chancellor test, Pu detected with a clay nanoparticle using STEM. Currently there is no evidence that the Pu is transported as its own PuO_2 colloid. The colloidal fraction of the groundwater is able to transport Pu significant distances in the subsurface. Given similar groundwater and hydrologic conditions, colloid-facilitated transport of low-levels of Pu should be expected. This mechanism must be taken into account if reliable transport models are to be developed for Pu and other low-solubility, high reactive contaminants.

Acknowledgments Prepared by LLNL under Contract DE-AC52-07NA27344. This work was supported by the Subsurface Biogeochemical Research Program of the U.S. Department of Energy's Office of Biological and Environmental Research.

References

Blankennagel RK, Weir JE Jr (1973) Geohydrology of the Eastern Part of Pahute Mesa, Nevada Test Site, Nye County, Nevada, U.S. Geological Survey, Geological Survey Professional Paper 712-B

Borg IY (1975) Radioactivity Trapped in Melt Produced by a Nuclear Explosion. Nucl Technol, 26:88–100

Borg IY, Stone R, Levy HB, Ramspott LD (1976) Information Pertinent to the Migration of Radionuclides in Ground Water at the Nevada Test Site, In: Part 1: Review and Analysis of Existing Information. Lawrence Livermore National Laboratory. UCRL-52078

Bowen SM, Finnegan DL, Thompson JL, Miller CM, Baca PL, Olivas LF, Geoffrion CG, Smith DK, Goishi W, Esser BK, Meadows JW, Namboodiri N, Wild JF (2001) Nevada Test Site Radionuclide Inventory, 1951–1992. Los Alamos National Laboratory. LA-13859

Brachmann A, Kersting AB (2003) Colloid-Facilitated Transport of Low-Solubility Radionuclides: A Field, Experimental, and Modeling Investigation. In: Kersting AB, Reimus PW (eds) Lawrence Livermore National Laboratory. UCRL-ID-149688

Buddemeier RW, Hunt JR (1988) Transport of Colloidal Contaminants in Groundwater: Radionuclide Migration at the Nevada Test Site. Appl Geochem, 3:535–548

Degueldre C, Baeyens B, Goerlich W, Riga J, Verbist J, Stadelmann P (1989) Colloids in Water from a Subsurface Fracture in Granitic Rock, Grimsel Test Site, Switzerland. Geochim Cosmochim Acta, 53:603–610

Department of Energy (1997) Shaft and Tunnel Nuclear Detonations at the Nevada Test Site: Development of a Primary Database for the Estimation of Potential Interactions with the Regional Groundwater System. DOE/NV-464

Department of Energy (2000) United States Nuclear Tests: July 1945 through September 1992. DOE/NV-209 Rev. 15

Hu Q H, Rose TP, Zavarin M, Smith DK, Moran JE, Zhao PH (2008) Assessing Field-Scale Migration of Radionuclides at the Nevada Test Site: "Mobile" Species. J Environ Radioact, 99:1617–1630

Kersting AB (2003) Colloid-Facilitated Transport of Low-Solubility Radionuclides: A Field, Experimental, and Modeling Investigation. In: Kersting AB, Reimus PW (eds) Lawrence Livermore National Laboratory. UCRL-ID-149688

Kersting AB, Efurd DW, Finnegan DL, Rokop DJ, Smith DK, Thompson JL (1999) Migration of Plutonium in Ground Water at the Nevada Test Site. Nature, 397:56–59

Kim JI (1991) Actinide Colloid Generation in Groundwater. Radiochim Acta, 52:71–81

Laczniak RJ, Cole JC, Sawyer DA, Trudeau DA (1996) Summary of hydrogeologic controls on ground-water flow at the Nevada Test Site, Nye Country, Nevada. United States Geological Survey, Geological Survey Professional Paper 96-4109

Mathews M, Hahn K, Thompson J, Gedeken L, Madigen W (1994) Subsurface Radionuclide Investigation of a Nuclear Test. J Appl Geophys, 32:279–291

McCarthy JF, Degueldre C (1993) Chapter 6. In: Buffle J, van Leeuwen HP (eds) Environmental Particles, Vol. 2. Lewis Publisher, Chelsea, MI

McCarthy JF, Zachara JM (1989) Subsurface Transport of Contaminants – Mobile Colloids in the Subsurface Environment May Alter the Transport of Contaminants. Environ Sci Technol, 23:496–502

Novikov AP, Kalmykov SN, Utsunomiya S, Ewing RC, Horreard F, Merkulov A, Clark SB, Tkachev VV, Myasoedov BF (2006) Colloid Transport of Plutonium in the Far-Field of the Mayak Production Association, Russia. Science, 314:638–641

Office of Technical Assessment (1989) The Containment of Underground Explosions, U.S. Congress, Office of Technology Assessment, Washington D.C., OTA-ISC-414

Pawloski GA (1999) Development of Phenomenological Models of Underground Nuclear Tests on Pahute Mesa, Nevada Test Site – BENHAM and TYBO. Lawrence Livermore National Laboratory. UCRL-ID-136003

Rose TP, Eaton GF, Kersting AB (2003) Hydrologic Resources Management Program and Underground Test Area Project FY 2001-2002 Progress Report. Lawrence Livermore National Laboratory, UCRL-ID-154357

Rose TP, Hu Q, Zhao P, Conrado CL, Dickerson R, Eaton GF, Kersting AB, Moran JE, Nimz G, Powell BA, Ramon EC, Ryerson FJ, Williams RW, Wooddy PT, Zavarin M (2008) Radionuclide Partitioning in an Underground Nuclear Test Cavity. Lawrence Livermore National Laboratory, UCRL-TR-409817

Ryan JN, Elimelech M (1996) Colloid Mobilization and Transport in Groundwater. Colloids Surf A, 107:1–56

Santschi PH, Honeyman BD (1989) Radionuclides in Aquatic Environments. Radiat Phys Chem, 34:213–240

Sawyer DA, Thompson JL, Smith DK (1999) The CHESHIRE migration experiment: A summary report. Los Alamos National Laboratory, LA-13555-MS

Smith DK (1993) A Review of Literature Pertaining to the Leaching and Sorption of Radionuclides Associated with Nuclear Explosive Melt Glasses, Lawrence Livermore National Laboratory, UCRL-ID-113370

Smith DK (1995) Characterization of Nuclear Explosive Melt Debris. Radiochim Acta, 69:157–167

Stoller-Navarro JV (2007) Geochem07.mdb and A User's Guide to the Comprehensive Water Quality Database for Groundwater in the Vicinity of the Nevada Test Site. S-N/99205-059-Rev.2, Stoller-Navarro Joint Venture, Las Vegas

Thompson JL (1995) Laboratory and Field Studies Related to the Hydrologic Resources Management Program, October 1, 1993 – September 30, 1994. Los Alamos National Laboratory, LA-12917-PR

Thompson JL (1998) Laboratory and Field Studies Related to Radionuclide Migration at the Nevada Test Site, October 1, 1996 – September 30, 1997. Los Alamos National Laboratory, LA-13419-PR

Thompson JL (1999) Laboratory and Field Studies Related to Radionuclide Migration at the Nevada Test Site, October 1, 1997 – September 30, 1998. Los Alamos National Laboratory, LA-13576-PR

Thompson JL (2001) Laboratory and Field Studies Related to Radionuclide Migration at the Nevada Test Site, October 1, 1999 – September 30, 2000. Los Alamos National Laboratory

Tompson AFB, Bruton CJ, Pawloski GA, Smith DK, Bourcier WL, Shumaker DE, Kersting AB, Carle SF, Maxwell RM (2002) On the Evaluation of Groundwater Contamination from Underground Nuclear Tests. Environ Geol, 42:235–247

Wadman RE, Richards WD (1961) Postshot Geologic Studies of Excavations Below Rainier Ground Zero. Lawrence Livermore National Laboratory, UCRL-6586

Xu C, Santschi PH, Zhong JY, Hatcher PG, Francis AJ, Dodge CJ, Roberts KA, Hung CC, Honeyman BD (2006) Colloidal Cutin-Like Substances Cross-Linked to Siderophore Decomposition Products Mobilizing Plutonium from Contaminated Soils. Environ Sci Technol, 42:211–216

Zhao P, Zavarin M, Leif RN, Powell BA, Singleton MJ, Lindvall RE, Kersting AB (2011) Mobilization of actinides by dissolved organic compounds at the Nevada Test Site, Appl Geochem (in press)

Speciation of Actinides in Granite Subjected to Tracer Studies

Melissa A. Denecke, Wout de Nolf, Alexander Rack, Remi Tucoulou, Tonya Vitova, Gerald Falkenberg, Sousan Abolhassani, Peter Cloetens, and Bernhard Kienzler

Abstract Tomographic studies of granite from the Äspö Underground Laboratory in Sweden are performed to understand the observed release of natural uranium in column studies upon application of groundwater flow. X-ray phase-sensitive tomography images reconstructed from in-line X-ray phase contrasted radiographs were compared with scanning fluorescence tomography reconstructions. The latter are based on sinograms of X-ray emission lines recorded with spatial resolution on a nanometer scale for a granite rock containing U after being subjected to a radionuclide tracer experiment. The results show that the U distribution follows microfissures or pores in the granite. Naturally occurring Th is revealed to be intimately associated with what appears to be feldspar, suggesting its being present as a dopant ion in the mineral structure. Neptunium tracer was not found in a sample prepared using FIB lift-out techniques, although the presence of Np was identified in the original granite piece during screening experiments prior to FIB-ing. The Np-containing part of the sample broke off during the preparation. Although this case study was limited to only three

M.A. Denecke (✉), T. Vitova, and B. Kienzler
Karlsruhe Institute of Technology, Institute for Nuclear Waste Disposal, P.O. Box 3640 76021 Karlsruhe, Germany
e-mail: melissa.denecke@kit.edu, tonya.vitova@kit.edu, bernhard.kienzler@kit.edu

W. de Nolf
Department of Chemistry, University of Antwerp, Universiteitsplein 1, 2610 Antwerp, Belgium
e-mail: wout.denolf@ua.ac.be

A. Rack, R. Tucoulou, and P. Cloetens
European Synchrotron Radiation Facility, 6 rue Jules Horowitz, 38043 Grenoble Cedex, France
e-mail: alexander.rack@esrf.fr, tucoulou@esrf.fr, cloetens@esrf.fr

G. Falkenberg
HASYLAB at DESY, Notkestr. 85, 22607 Hamburg, Germany
e-mail: gerald.falkenberg@desy.de

S. Abolhassani
Laboratory for Materials Behaviour, Paul-Scherrer-Insitute, CH-5232 Villigen PSI Villigen, Switzerland
e-mail: sousan.abolhassani@psi.ch

S.N. Kalmykov and M.A. Denecke (eds.), *Actinide Nanoparticle Research*,
DOI 10.1007/978-3-642-11432-8_16,

samples, initial interpretation of results confirm that Th is less likely than U to become mobile in groundwater upon drilling and excavation of emplacement caverns and shafts for a nuclear waste repository in granite bedrock.

1 Introduction

Granite is considered a potential host rock for the disposal of radioactive wastes notably in the Nordic countries Sweden and Finland. The retention properties of pristine and altered granite are to this end of utmost importance. Granite generally contains varying natural amounts of uranium and thorium. The natural background of these elements poses a potential source for these actinide elements, should they dissolve and enter groundwater upon drilling and excavation of shafts and deposition tunnels for emplacement of the waste packages in the granite host rock for a nuclear waste repository. Elution of natural uranium (nat-U) from the host granite rock during migration experiments performed on granite drill cores from the Äspö Hard Rock Laboratory (HRL) in Sweden has been reported to be significant [1]. The concentrations of nat-U in the granite rock bore cores at Äspö HRL are on the order of around 5 μg/g and associated groundwater nat-U borehole concentration on the order of 1 E–10 mol/L. The concentrations of natural ^{238}U eluted during tracer migration experiments in the Äspö granite experiment was reported to exceed the ^{233}U tracer concentrations significantly. Apart from experimentally induced breakthrough peaks, the ^{238}U concentration was in the range of 2–4 $\times$ 10^{-7} mol/l. These values surmount the U concentration of the groundwater as well as the calculated solubility of U(IV) minerals. The nat-U initial breakthrough was eluted faster than the inert tritiated water (HTO) in the tracer cocktail. This indicates that the nat-U was already present in the groundwater, filling the granite fractures in significant amounts prior to the tracer experiments. The nat-U released exhibits repeated breakthrough-like elution curves following interruption and subsequent restarting of granite core column flushing. The amount nat-U released is greater than expected from simple dissolution of low soluble U(IV) species. The interpretation of these results is that after drilling and during storage, the cores were in contact with air, which led to oxidation of naturally occurring U(IV) species in the granite to more soluble U(VI) species.

Granite also contains natural levels of thorium in amounts similar to that for U (4–8 μg/g) at Äspö. Thorium is redox insensitive and is present in tetravalent form under natural conditions with a stable Rn electron shell configuration. No breakthrough for nat-Th was observed in column migration studies. For this reason, we performed a comparative study of the distribution of uranium and thorium in granite pieces stemming from a bore core column migration study, in order to ascertain if differences in their spatial distributions relative to micropores of the granite exist. No ^{233}U was in the tracer cocktail introduced onto this particular column in the experiment. The tracer cocktail did contain ^{237}Np(V) and ^{241}Am, and previous X-ray fluorescence and X-ray absorption near edge structure spectroscopy investigations performed using an X-ray beam with micrometer dimensions and confocal excitation/detection geometry (confocal μ-XRF/μ-XANES) revealed neptunium to be immobilized in the granite as Np(IV) localized within microfissures having dimensions on the order of tens of microns [2]. Due to the small size of fissures where neptunium was immobilized, we selected to perform the comparative investigations presented here using a submicron spatial resolution. Our original intention was also to determine if the immobilized Np(IV) "hot spots" are present as precipitated hydrolysis (particulate) products or if they are present as sorbed Np localized at regions on the fissure walls. This difference in species, particulate versus sorbed species, is crucial for predicting potential remobilization of Np into flowing groundwater. However, during focused ion beam (FIB) preparation of a piece of granite, which exhibited Np Lα1 and Lβ1 intensity in μ-XRF screening experiments to select appropriate pieces for the investigation (see below), the sample broke at a fissure seam. As no Np was found remaining on the sample, we assume that the Np was lost with the piece that broke off.

The chapter is structured as follows. Details of the sample preparation and experimental details of the screening and tomography studies, including their reconstruction using measured sinograms of X-ray emission lines and radiograms are presented first. The results of the studies are then presented and discussed in detail. In the concluding remarks, we discuss the lessons learned from this investigation and the implications of the results for nuclear waste disposal in thorium/uranium-containing granite.

2 Experimental

2.1 *Sample Preparation*

The granite pieces investigated stem from the same Äspö granite bore core section, where previous investigations identified Np(IV) in a posttracer experiment mentioned above [2]. Initially, the granite section was cut into small pieces and mounted onto a quartz slide using an acrylic resin. By means of repeated grinding, recutting, and mounting, pieces with dimensions less than 1 mm were obtained. These pieces were then screened for their elemental content using scanning μ-XRF at Hamburg Synchrotron Radiation Laboratory (HASYLAB) experimental station BL-L. Three granite pieces, which were observed to exhibit X-ray emission lines for Np, Th, and U, were selected for further study at the ID22NI station at the European Synchrotron Radiation Facility (ESRF) and designated as sample 1, sample 3, and sample 6, respectively. Sample 1 was downsized further using an FIB.

2.1.1 Sample 1

An optical microscopic image of sample 1 before FIB extraction of a suitable sized piece, its Fe, Np, and Zn distributions, the FIB-cut sample 1 mounted on its copper

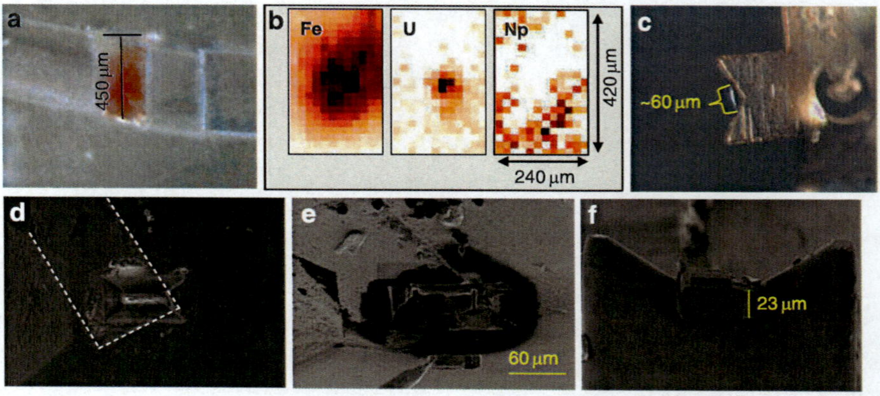

Fig. 1 Sample 1 screening and preparation. (**a**) Microscopic image of an approximately 250 × 450 µm^2 granite piece before FIB lift-out to prepare sample 1. (**b**) 2D elemental distribution screening results. The maps are made from Fe Kα, U Lα_1, and Np Lα_1 intensities measured with scanning µ-XRF (step size = 20 µm; counting time = 7 s). Np appears along a diagonal across the *lower right corner*, likely localized in a micro fissure. (**c**) Micrograph of sample 1 pillar approximately 60 µm long, after FIB lift-out and mounting on a *copper holder*. SEM images of: (**d**) FIB trenches during cutting of the sample 1 pillar. The sample silhouette is indicated by *dotted lines*. (**e**) Lift-out using a W tip micromanipulator attached to the sample with sputtered carbon. (**f**) The final 60 × 23 µm^2 pillar shaped sample 1 after transferring and mounting onto a *copper sample holder*

sample holder, and the key stages of FIB preparation are depicted in Fig. 1a–f, respectively. A Zeiss NVision40 instrument was used for FIB sample preparation, equipped with a Ga liquid metal ion source and a field emission electron source; the scanning electron microscope (SEM) column is a GEMINI high resolution column. Sample 1 was freed from embedding resin prior to FIB milling and subsequently mounted in the correct orientation for milling. As large amounts of material had to be removed, the sample was milled at 30 kV with a beam current of 27 nA and 3 nA. An approximately 60 × 25 µm^2 pillar-shaped section (selected from the region at right angles where a diagonal line of Np Lα intensity was measured near the corner of the granite piece, see Fig. 1b) was cut using the lift-out FIB technique. The FIB lift-out technique is well known for sample preparation for transition electron microscope (TEM) studies [3]. In this technique, the sample is cut directly from the "bulk" specimen using FIB (Fig. 1d), lifted out with a micromanipulator (the cut sample and attached micromanipulator are visible at the bottom of Fig. 1e) and transferred to a copper sample holder (Fig. 1f). The copper holder is mounted onto a standard SEM holder, which fits onto the ID22NI sample holder.

2.1.2 Sample 3

Sample 3 was used without any further treatment than the repeated cutting and grinding (Fig. 2a), by simply removing the resin-mounted sample of approximately

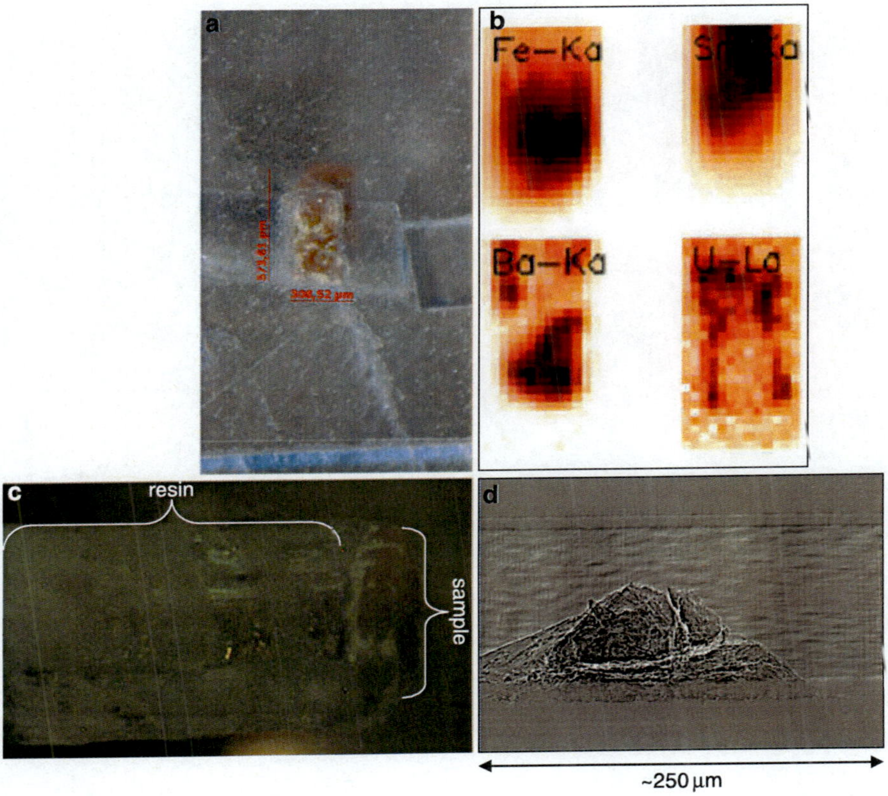

Fig. 2 (**a**) Micrograph of prethinned sample 3 embedded in resin on a glass slide mount. (**b**) 2D distribution scanning μ-XRF results for Fe, Sr, Ba, and U (500 × 340 μm^2, step size = 20 μm; counting time = 7 s). (**c**) Profile of sample removed from slide mount. (**d**) Projection radiograph of the corner of sample 3 placed in the beam path for the tomographic scan. The field of view is $(250\ \mu m)^2$. The embedding resin used for cutting and grinding the sample is visible on the *left*

500 × 340 μm^2 dimensions from the glass slide (Fig. 2c). The sample on the resin was mounted on a capillary tip, which was placed on an IUCr standard brass pin. The μ-XRF screening showed sample 3 to contain U, especially near one corner (Fig. 2b). This corner of the sample was placed in the beam path for the tomography studies (Fig. 2d) by mounting and aligning the capillary plus sample onto an IUCr standard eucentric goniometer head. This goniometer head can be directly mounted onto the sample positioning system at ID22NI.

2.1.3 Sample 6

Screening results of a rectangular section of the precut and ground sample 6 (Fig. 3a) using μ-XRF showed this section to contain a significant amount of Th

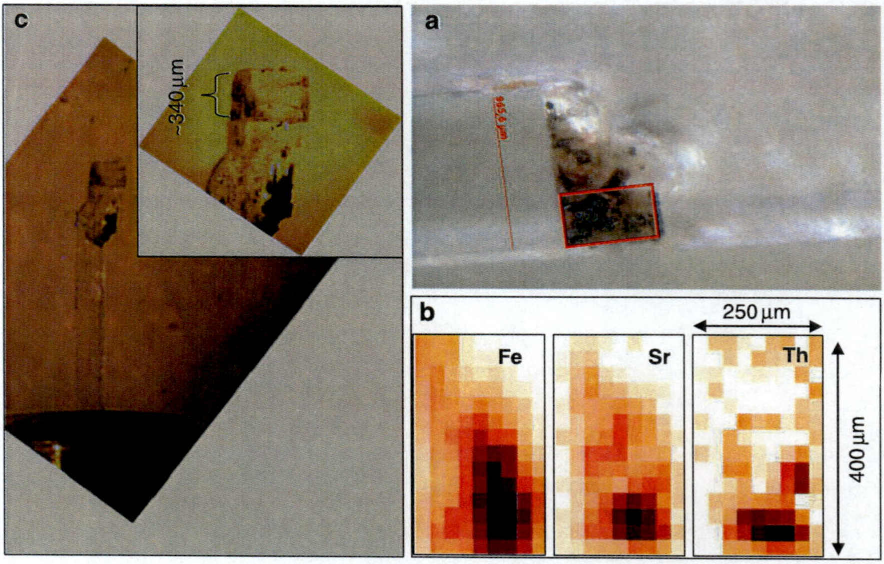

Fig. 3 (**a**) Micrograph of cut and milled sample 6 embedded in resin on glass slide mount. The *rectangle* marks the approximate area screened using μ-XRF. (**b**) 2D distribution scanning μ-XRF results for Fe, Sr, and Th (400 × 270 μm^2, step size = 30 μm; counting time = 10 s). (**c**) Micrograph of triangular sample cut from (**a**) and mounted on a glass capillary. The *inset* shows a zoom of the sample 6 piece investigated of approximately 340 μm height. The resin remaining on the sample is evident on the *right side*

(Fig. 3b). This part of was cut off from the rest of the sample using a razor blade, then mounted on a glass capillary (Fig. 3c), and aligned on a eucentric goniometer head as for sample 3. The sample had an arrow head shape and retained resin attached to its back side (Fig. 3c inset).

2.2 Experimental Setup

Scanning X-ray fluorescence screening with a focused beam of the original granite pieces obtained by repeated cutting and grinding experiments were performed at Beamline L of the DORIS III synchrotron ring, which is operated by the HASYLAB at the Deutsches (German) Elektronen-Synchrotron DESY (Hamburg, Germany). A band pass of energies with a narrow wavelength range centered at 0.34 Å (36.5 keV) was selected from the white beam from a bending magnet using a NiC multilayer (ML) monochromator. The incident radiation was focused with an elliptical single-bounce monocapillary to deliver a footprint on the sample approximately 15 μm diameter at 45° orientation of the sample to the incident beam [4, 5].

The sample was mounted on an *x*, *y* positioning stage to scan in a step-by-step fashion (i.e., pixel by pixel in generally 20 μm steps) over selected sample areas. The fluorescence emitted by the sample was detected at an angle of 90° to the incoming beam using either a Si-drift detector (Vortex, SII Technologies Inc.) or a Si(Li) detector (Gresham Scientific Instruments Ltd.). Two-dimensional X-ray intensity maps were obtained using a 7–10 s acquisition time for each pixel. X-ray fluorescence spectra of each pixel were evaluated by nonlinear least-square fits using the AXIL software package [6]. Elemental distribution maps were obtained by plotting background-corrected fluorescence intensities for selected emission lines as a function of pixel position and scaling intensities, with white being assigned the pixel with the smallest fluorescence intensity signal and black the pixel with the highest intensity of 100%; the remaining pixel intensities were scaled gray shades accordingly on a relative linear scale.

High resolution experiments were performed at the ID22NI station (NI = nanoimaging) at the ESRF. ID22NI is a synchrotron hard X-ray nanoprobe recently built at the ESRF and mostly dedicated to 2D and 3D multimodal analysis (fluorescence, diffraction, imaging) of heterogeneous materials. It offers an accurate and powerful analytical tool coupling noninvasive properties, high detection limits, and deep submicron spatial resolution. A wide range of scientific communities including Earth and Planetary Sciences [7], Biology and Medicine [8], and Material Sciences [9] have already shown a strong interest in this new end-station supplementing existing methods for sample characterization.

ID22NI has been developed as the continuation of the existing ID22 beamline at the ESRF sharing the same X-ray source and some of the main optical elements, previously described in [10]. The X-ray beam focusing is performed with Kirkpatrick-Baez (KB) ML mirrors, in our case W/B_4C as ML, which delivered a focal spot size of approximately 127 nm (H) × 147 nm (V) at a nominal incident energy of 23 keV. A double crystal [Si(111)] monochromator is also available but offers lower flux, down from around 10^{12} ph/s with the ML to 10^{10} ph/s, which is less desirable for the trace concentrations in our system of interest.

A photograph of the sample stage setup is shown in Fig. 4. The sample was placed at the focal spot on an *x,y,z* positioning system plus rotation stage. In contrast to most of the existing hard X-ray microprobes measuring XRF, the sample was placed in normal incidence (and not at 45°), thus avoiding any lateral resolution degradation due to beam projection on the sample surface. Two silicon drift detectors (Vortex 50 mm^2, SII Nanotechnology USA Inc., Northridge, CA) mounted on each side of the sample were used simultaneously to measure the fluorescence signal. In our case, these were placed as near ±90° from the incoming beam as possible. The sample stage was on a motorized ramp to move it downstream from the focal spot, where magnified radiographs using the ESRF FReLoN CCD camera in combination with visible light microscopy optics and a scintillator [11] were recorded, in order to align the sample along the axis of rotation. Following alignment, a series of radiographs was recorded from 0° to 180° in 0.1° steps for samples 3 and 6 (24 μm physical pixel size of the FReLoN CCD,

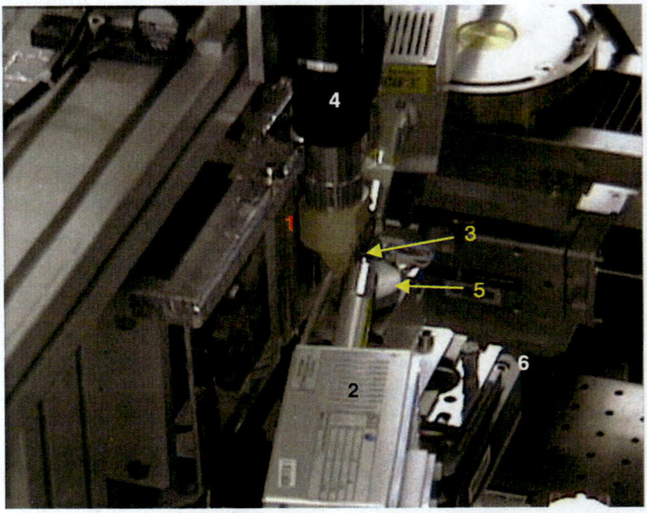

Fig. 4 Photograph of the experimental set-up. (1) 127 nm (H) × 147 nm (V) focused beam delivered by the KB mirror system (multilayer = W/B_4C, nominal energy = 23 keV). (2) Solid state Si drift fluorescence detector (Vortex, SII Instruments). (3) Sample appearing as a *white spot* due to the illumination by the in-line microscope. (4) In-line microscope. (5) Sample rotation with fine positioning piezo stages. (6) Positioning stage (x, y, z). The latter can also be ramped downstream in order to record radiographic projection images with a high resolution indirect X-ray image detector (FReLoN CCD camera with microscope visible light optics and scintillator, ESRF in-house design)

1 μm effective detector pixel size, 200 nm estimated pixel size in the images). The sample stage was then returned to the beam focus, and two dimensional (2D) XRF images were recorded in fast scanning mode by continuously translating the sample through the beam using piezo-driven motors and reading out integrated detector counts at specified time intervals, corresponding to recording fluorescence spectra at pixels of defined size. For samples 1 and 6, a 400 μm and 200 μm thick Al foil filter, respectively, was placed in front of the detector to reduce the high incoming count rate and associated dead time.

Horizontal line scans of interest were selected from 2D elemental distributions constructed from the fast XRF scan data. Fluorescence tomographic sinograms were recorded at horizontal sections of interest selected from 2D images (indicated as dotted lines in Figs. 6, 10 and 14) by repeatedly registering fast line scans and changing the sample orientation angle in incremental steps from 0° to 180°. Because two fluorescence detectors were used, it was not necessary to rotate the sample the full 360°. Reconstruction of emission sinograms to obtain elemental distributions in sample cross sections was performed by means of the maximum likelihood expectation maximization (MLEM [12]) algorithm implemented in the XRDUA software version 5.1.12.1 [13].

2.3 Data Analysis

Elemental distributions of 2D scanned areas were obtained by plotting measured emission intensities registered at each pixel. The emission intensities are obtained from fits to each pixel fluorescence spectrum using the program PyMCA [14], available as freeware from the ESRF beamline instrumentation software support group (BLISS). Three dimensional images were reconstructed from the radiographic projection images using the ESRF software package PyHST, saved as binary volumes with 32-bit resolution. Classical X-ray microtomography measures the 3D distribution of the linear attenuation coefficient by assigning different gray values to different constituents in the image. Due to the use of partial coherent synchrotron light for imaging, diffraction of X-rays at interfaces within the sample contributes to the contrast in our case as well, the so-called in-line X-ray phase contrast (edge enhancement). Better visibility of features within the tomographic image with respect to the in-line X-ray phase contrast was obtained after applying a phase-retrieval algorithm. The latter was performed by means of the code ANKA-phase, based on the approach by Paganin et al., to get a phase map using only a single sample-to-detector distance (18 cm) [15]. Commonly, holotomography is performed at ID22NI in order to derive quantitative phase maps while conserving the high spatial resolution [16]. The latter requires scanning the sample at different distances from the detector. Due to stability issues with the mounted sample (cf. motion artifacts in Figs. 11 and 12), the single-distance approach was chosen. It does not allow for a quantitative study but delivers qualitative information about the morphology of the main constituents present in the sample while introducing an acceptable image blurring with respect to holotomography. A comparison of the reconstructed in-line X-ray phase contrast tomogram and the tomogram after phase retrieval is shown in Fig. 5. The phase shift of short wavelength radiation when it

Fig. 5 Tomographic reconstructions using plain in-line X-ray phase contrast (edge-enhancement, *left*) and the corresponding image after phase-retrieval has been performed (*right*). The resin mount attached to the left flank of the sample is more evident in the in-line X-ray phase contrast image

transverses material comprised of light elements is greater than their attenuation or absorption. This is especially evident in the organic resin attached to the sample, which appears nearly transparent in the plain in-line X-ray phase contrast tomographic reconstruction, as it only weakly attenuates and hence is only visible due to its edges. In contrast, the resin is clearly evident in the reconstruction after phase retrieval.

3 Results and Discussion

3.1 Sample 6

The element distribution maps for major elements with atomic numbers $Z \geq 19$ present in sample 6 and extracted from 2D fast XRF scans measured over a $184 \times 298\ \mu m^2$ (2 μm steps) area are shown in Fig. 6. It is evident from the distribution maps that Y and Nb are concentrated on the left hand of the area, the transition metals, the alkaline metals, and the alkaline earth metals are concentrated to the right, whereby Rb dominates the top of the sample and Sr the bottom. The Zr is found intermittently as localized hot spots. Thorium is concentrated at the bottom of the area scanned and near, but not on, one of the Zr hot spots; it is rather associated with areas with significant but lower Zr Kα intensity. No correlation between elements is particularly noticeable, but a negative correlation between the Zr hot spots and a number of elements (Th, Nb, Y, Sr, Rb, Fe) exists. The dotted line in the Nb distribution indicates the sample position where the emission tomographic cross section shown in Fig. 7 was measured.

In the element distributions extracted from the emission sinograms and displayed in Fig. 7, Th is observed to be concentrated in the upper corner of the triangular shaped sample in an egg-like form. The Th distribution is depicted as an overlay image to the corresponding phase-retrieval tomographic cross section and also as a RGB overlay for Fe, Sr, and Th. Inspection of these images reveals that Th is associated with an area of higher contrast (lighter gray scale, i.e., higher electron density) and positively correlated to the emission tomogram for Sr Kα fluorescence. The Th hot spot is relatively large, around 35 μm across, and exhibits fine fissures running through it with dimensions of a few micrometers. It is not only associated directly with Sr but also has a Sr layer on one side and both ends. The fact that Th is not associated with Fe or Mn points to it likely not being associated with biotite (or associated altered phases such as clinochlore) in sample 6. If the observed fissures are considered indications of cleavage planes at different angles, then this Th-containing part of the granite rock might be a tectosilicate. As it is not apparently associated with K and only small Ca K intensity, this Th-containing grain may be a plagioclase feldspar comprised of majority sodium endmember, albite (Na K emission at ~1.05 eV is energetically too low to be observed here). The U distribution is shown in Fig. 7, too. However, the counts associated with the darkest

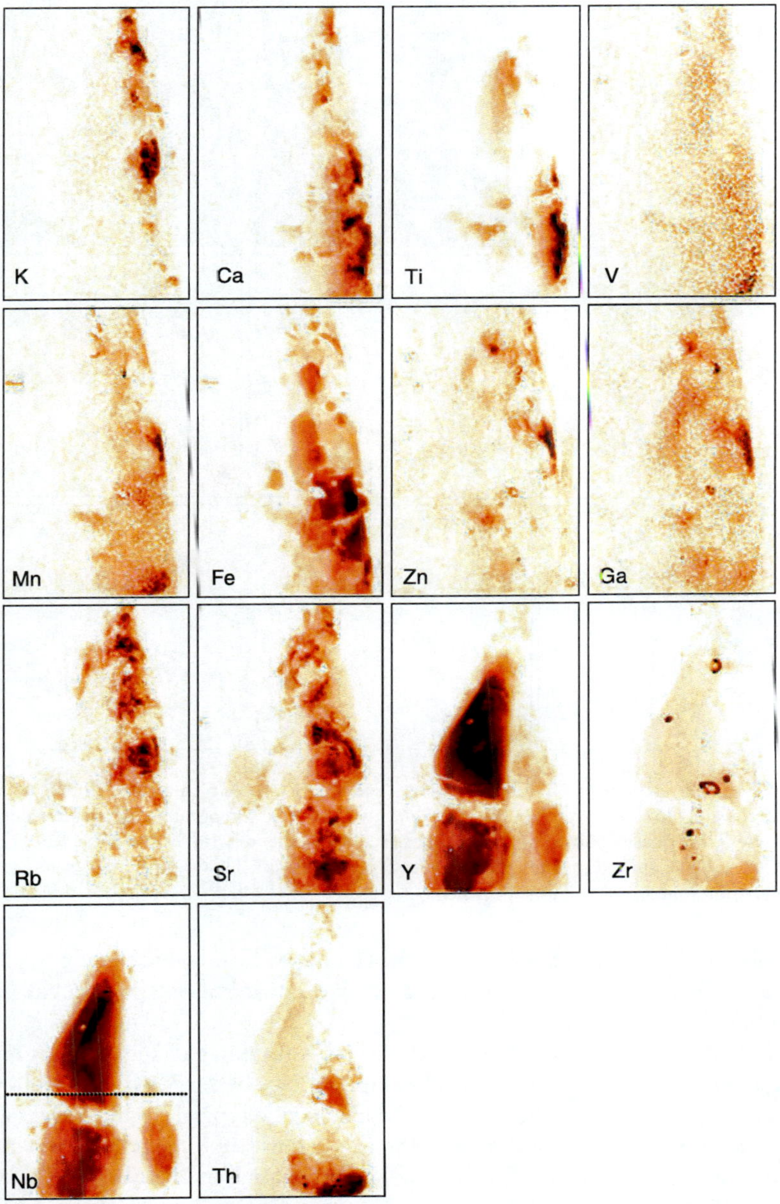

Fig. 6 Element distributions for major elements with $Z < 19$ present in sample 6 extracted from y, z scans measured over a $184 \times 298\ \mu m^2$ (2 μm steps) area. The *dotted line* in the Nb distribution indicates the sample position where the emission tomographic cross section shown in Fig. 7 was measured

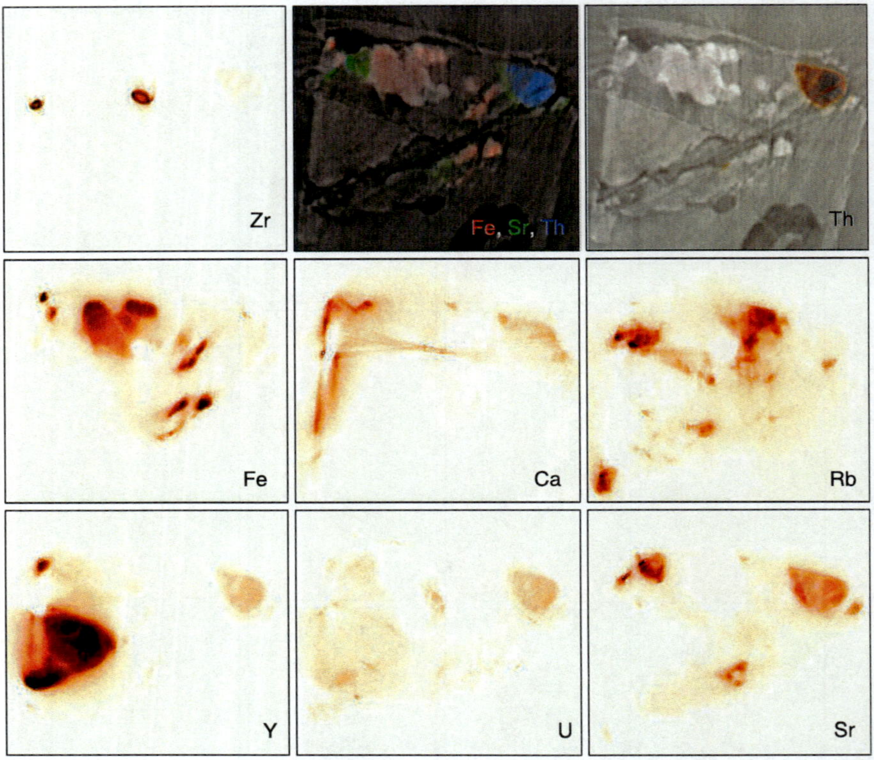

Fig. 7 Element distributions in a selected cross section of sample 6 (indicated in Fig. 6). Image dimensions are 215 × 193 μm^2. In the *top right*, two full-field tomographic images are depicted: the in-line X-ray phase contrast tomogram cross section is shown with Fe, Sr, Th as RGB overlay printed on *top* while the slice after phase-retrieval is shown with Th only. All images are scaled to 80% maximum intensity

U pixels are around 50 times less than the Th counts. If the same scaling were used in both distribution maps, instead of a relative linear scale, the U would not be visible.

The Y distribution is localized in the lower left corner of Fig. 7, as is the Nb distribution (not shown). This part of sample 6 may be fergusonite, $YNb0_4$ [17]. Some Ce was detected in this area, too, which is quite typical for this mineral. Iron is associated with the lightest areas in the phase-retrieval tomogram. Areas rich in Rb are situated between the Y- and Fe-containing grains and between an Fe feature and the Th hot spot. Calcium is not always associated with Fe and is generally found in the same pixels as Ti (not shown). Zirconium is observed to have high fluorescence intensity in the pixels indicated. These might be $ZrSiO_4$ and are visible in the phase-retrieval tomogram as round structures at the top left border of the Y-containing grain and the lower right edge of the irregularly shaped large Fe grain. Neither U nor Th is associated with the Zr hot spots.

3.2 Sample 3

Two-dimensional element distributions for a 100 × 100 μm^2 area in sample 3 are shown in Fig. 8. The area at the top of the image is dominated by K, Rb, and, interestingly, some Ba. The concentrated Ba and K pixels in the upper left quadrant of the images outline the protruding feature in the radiogram displayed in Fig. 2d. One of the near vertically oriented fissures visible in Fig. 2d can also be discerned in the K and Ba maps as white areas. Strontium is concentrated at the bottom two-thirds of the image, as are Ca, Mn, Fe, Ga, and Y. The distribution of Y is remarkable, being concentrated along lines running in different directions, possibly outlining grain interfaces. The Fe maps shows Fe hot spots to be embedded in a diffuse Fe distribution, likely marking accessory or secondary minerals in the

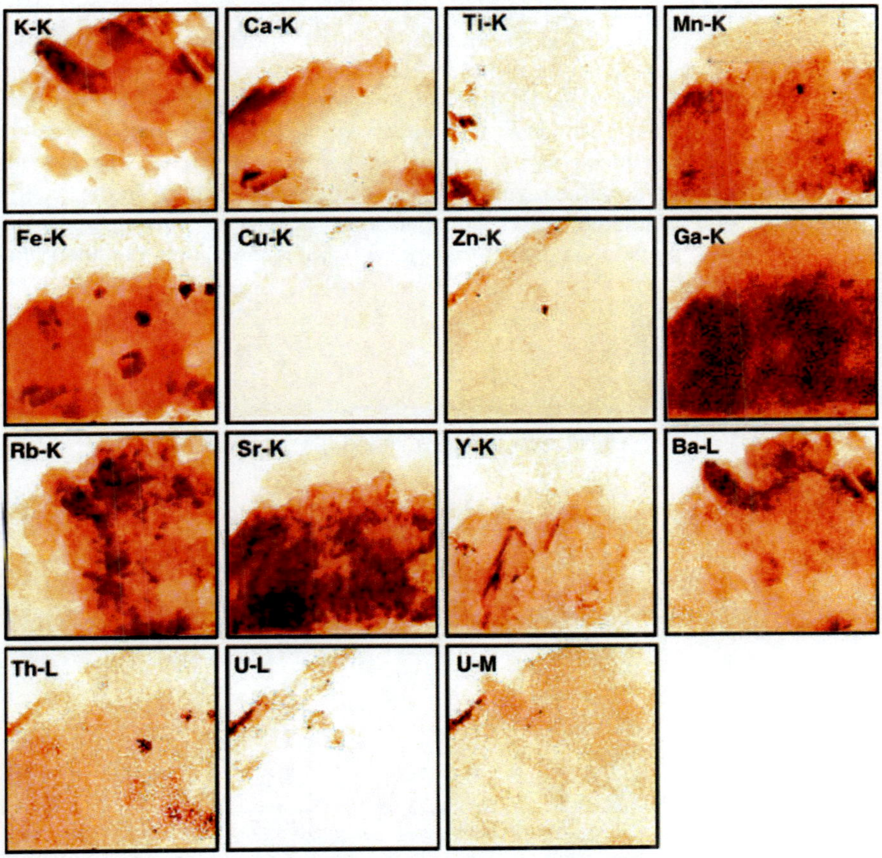

Fig. 8 Element distributions for major elements with $Z < 19$ present in sample 3 extracted from y, z scans measured over a 100 × 100 μm^2 (1 μm steps) area. The U distribution extracted from both L and M emission lines are shown

surrounding rock forming aluminosilicates. Three of these Fe hot spots are associated with relatively high Th Lα intensity. The U distribution extracted from L and M emission lines are both shown. The U is concentrated at the top left of both these images. There is also some Th intensity in the same area, but the Th fluorescence counts are around 60 times less than the U counts in these pixels. The Th appears more concentrated in the hot spots to the right in Fig. 8. The U appears to follow a fissure at the left side of the sample. The distribution of Zn is at first glance similar to that of U. As the Np distribution in the granite in the previous post-tracer μ-XRF study was also observed to be positively correlated to that for Zn [2], a correlation plot of the U versus the Zn distribution for the top half of the images in Fig. 8 was created (see Fig. 9). There is no strong correlation evident between these elements; there is no U present in pixels with the highest Zn Kα intensity (data points on the abscissa in Fig. 9). The normalized amount of Zn in pixels exhibiting significant U Lα intensity is clustered at around 20%.

A 2D scan with smaller step size over a 280 × 50 μm^2 stripe across the entire width of sample 3 was recorded, in order to select appropriate horizontal slices to record emission sinograms. The resulting 2D element distributions are shown in Fig. 10. The dashed line in the Fe distribution shows the position where the first emission cross section depicted in Fig. 11 was recorded. This plane transects two of the three prominent Fe hot spots and another smaller one to the left, as well as the near vertically oriented fissures, which appear white in the images. In the reconstructed emission tomogram, the Fe hot spots are observed to be associated with pores; they appear as light areas next to dark spots, indicating pores in underlying

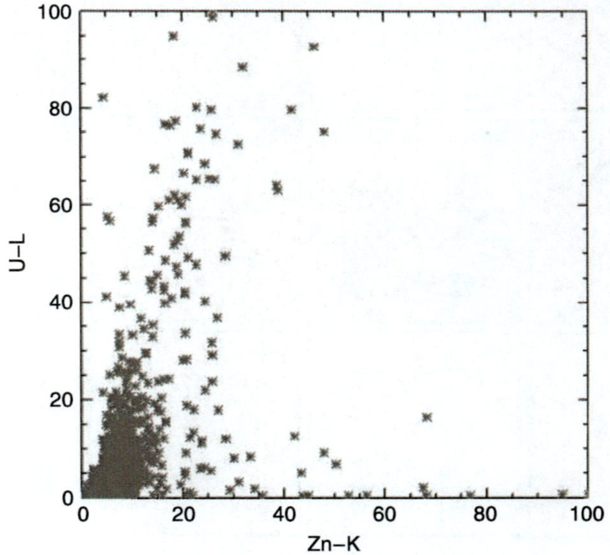

Fig. 9 Correlation plot for distributions of Zn Kα versus U Lα (normalized) intensities for the *upper half* of the images shown in Fig. 8

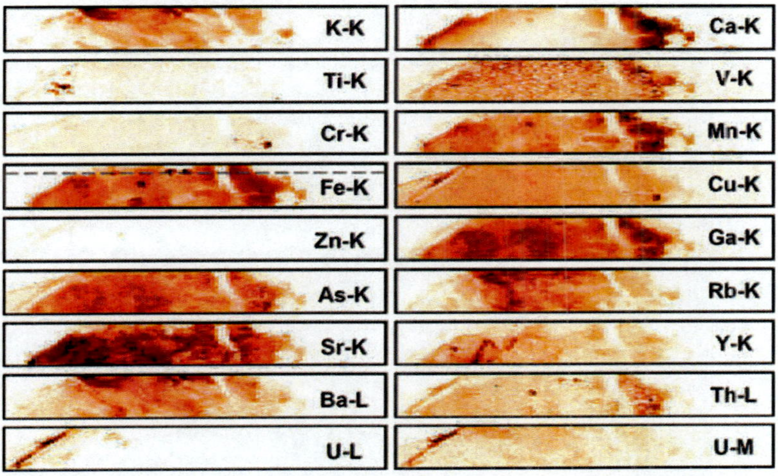

Fig. 10 Element distributions for sample 3 extracted from y, z scans measured over a 280 (2.8 μm steps) × 50 (0.5 μm steps) μm^2 area located near the *top* of the radiographic image shown in Fig. 2d. The *dashed line* in the Fe distribution shows the position where the emission cross section depicted in Fig. 11 was recorded

phase-retrieval tomogram in Fig. 11. The pores are embedded in the areas of high Sr concentration. Unlike the Th–Sr association observed in a grain for sample 6, the U in sample 3 is not correlated with Sr at all. Uranium is concentrated in the area of sample 3, which is dominated by pores and fissures in the upper left of the cross section (corresponding to far left of Figs. 8 and 10) and pores at the protrusion at the bottom of Fig. 11. Note that U is not associated with the pores, where Fe is concentrated.

Another emission sinogram was measured at a position on sample 3, which was 6 μm below that for Fig. 11. The reconstructed tomogram is shown in Fig. 12, also as an overlay of the phase-retrieval tomogram. The Fe in this cross section is not associated with any hot spots, and its distribution is generally similar to the Sr distribution, concentrated at the bottom of the image and with Ca on the edge of the sample. This is evident in the correlation plots for Fe versus Sr and Ca shown in Fig. 13. Calcium might have been in the areas inside the sample, too, but was not detected due to self-absorption of the low energy emission. This might explain the deviation of the Fe–Ca correlation curve in Fig. 13 from a straight line to one having more like two different slopes. The Sr is found concentrated at boundaries between grains with different densities, as well as at pores in sample 3. Herein might lie the reason for the strange spikes in the Sr Kα intensity observed at Fe Kα relative intensities of around 14%, 22%, and 35% and indicated in Fig. 13 as dotted lines; Sr is concentrated in the pores, but Fe distribution remains more homogeneous. We do not interpret these spikes as indicative of the presence of any true solid solution (Fe_xSr_{1-x}) phase. The U is not associated with these Sr-rich pores but solely with those at the top left of the image, where the pores are near, but not on

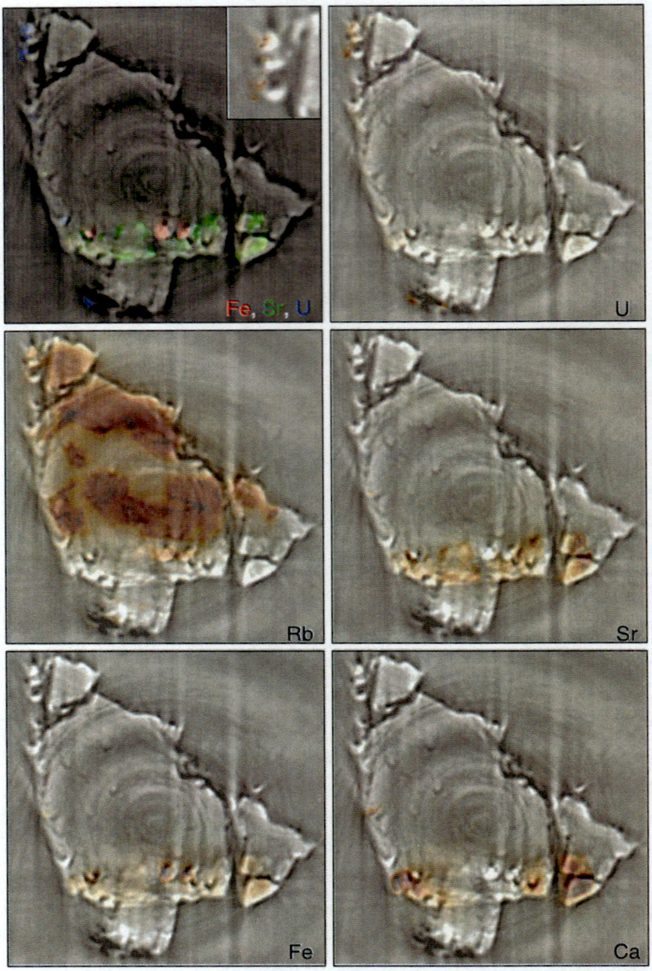

Fig. 11 Element distributions in a selected cross section of sample 3 (indicated by the *dashed line* in Fig. 10) and plotted together with the (partially transparent) phase-retrieval tomographic cross section (motion artifacts are present as well). Image dimensions are 183 × 190 μm^2 (0.2 μm steps). In the *top left* image, the tomographic cross section is shown together with Fe, Sr, U as RGB overlay. The *inset* in the *upper right* of the overlay is a zoom of the U distribution (without RGB color coding). All images are scaled to 80% maximum intensity, except for the RGB image, which is scaled to 30% maximum intensity for Fe and Sr and to 10% for U

top, an area of high Rb concentration. The U is observed to be not only adjacent to (or lining) pore walls in the upper left part of sample 3, but some U is also found in the lighter area of the phase-retrieval tomogram between pores. This may indicate diffusion from pores into the surrounding matrix. This matrix is associated with some K (not shown) and not associated with high concentrations of any $Z > 19$

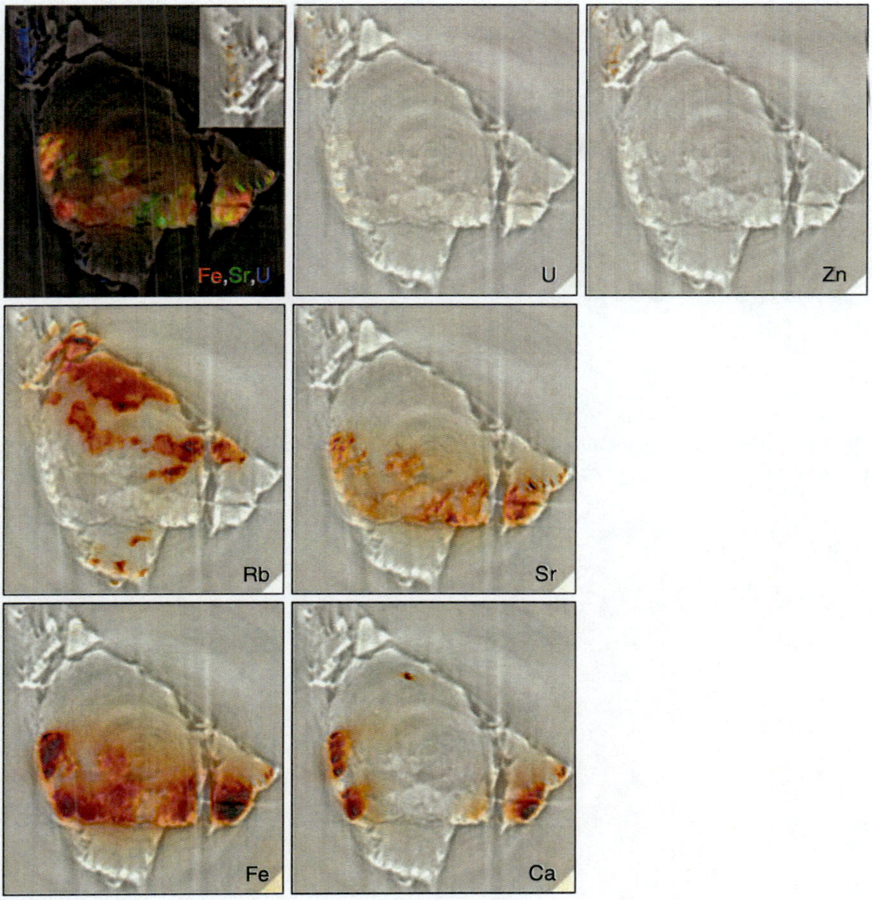

Fig. 12 Element distributions in a cross section of sample 3 reconstructed from fluorescence sinograms recorded 6 μm below the slice shown in Fig. 10. Image dimensions are 199 × 114 μm^2 (0.2 μm steps). A partially transparent phase-retrieval tomographic slice is depicted under each image (motion artifacts are present as well). In the *top left* image, Fe, Sr, U distributions are shown as an RGB overlay. The *inset* in the *upper right* of the overlay is a zoom of the U distribution. All images are scaled to 80% maximum intensity, except for the RGB, which is scaled to 60% maximum intensity for Fe and Sr and to 20% for U

elements. This may indicate that it is an Na–Al–Si mineral such as Na-rich alkali feldspar or albite, $NaAlSi_3O_8$.

3.3 *Sample 1*

The Ga distribution measured in a 2D scan over sample 1 is shown at the left of Fig. 14. Sputtered Ga atoms not evacuated during the FIB process were redeposited

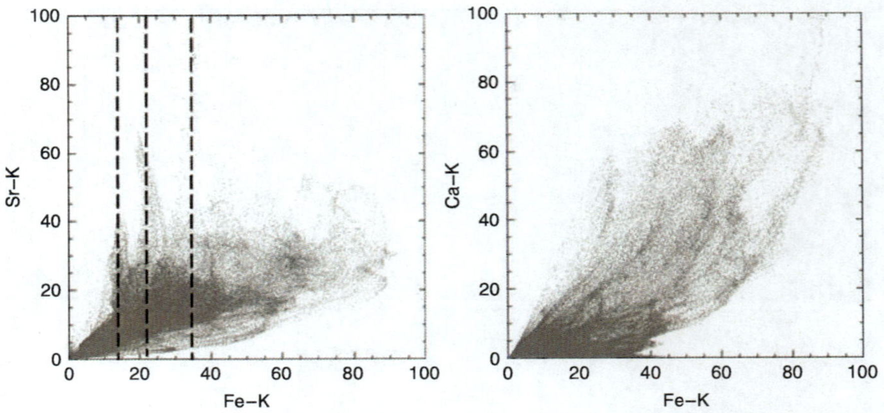

Fig. 13 Correlation plots for distributions of normalized intensities for Fe Kα versus Sr Kα (*left*) and versus Ca Kα (*right*) for the emission tomographic cross sections shown in Fig. 12

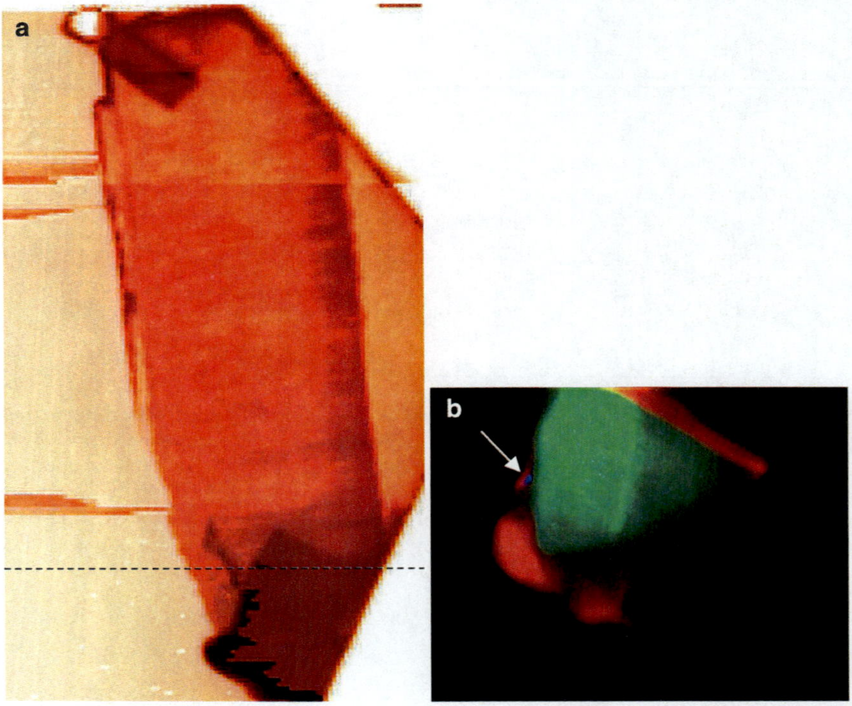

Fig. 14 (**a**) Silhouette of sample 1 as measured Ga distribution. The image height is ~65 μm. The *dark diagonal stripe* at the *top* marks where the W-manipulator was attached for lift-out FIB. The *dotted line* in the *lower half* marks the position where the emission tomographic cross section shown in (**b**) was measured. (**b**) Tomogram RGB overlay for Ga, Rb, U in a 37.25 × 28.5 μm^2 (0.125 μm steps) cross section. A U hot spot 2–3 steps in width, sandwiched between the Ga coating from FIB preparation and Rb from the sample corpus is indicated

at the attacked areas. This led to a high amount of Ga deposited on the surface of sample 1 to such an extent that it caused saturation of the detectors during XRF scanning. In addition, the Cu sample holder also caused large numbers of counts and therefore deadtime in the detector. We had to use a 400 μm thick Al filter in front of the detectors for these measurements. In fact, the Ga image in Fig. 14 is a composite of a number of smaller 2D scans, from which bad lines of data were removed (remeasured and "pasted" in). Figure 15 shows the elemental distributions extracted from data of one of these small area scans. In addition to the Ga from the FIB process, we observed W emission in the XRF spectra as an artifact from the micromanipulator used in the FIB lift-out. Unlike the other two samples, this small granite piece did not exhibit any measurable amounts of metals others than those shown in Fig. 15 (except a small U hot spot, see below). Part of the area where a piece of the sample pillar broke off during FIB preparation is in the lower part of these images. In this region, we find small globular Fe structures 400 nm and less in size. This likely marks the fissure, where the Np was observed to be localized in the scanning XRF studies (Fig. 1). We did not find, however, any emission signal for Np any more in this FIB-cut sample. The Np was presumably lost with the piece that broke off.

In the 2D scan of the bottom of the sample pillar, a single hot spot of U emission was observed. An emission tomogram of the cross section at the pillar height containing U was recorded. The dotted line in the lower half of the Ga distribution in Fig. 14a left this position. The resulting emission tomogram is displayed in Fig. 14b as an RGB overlay for Ga, Rb, and U. A U hot spot 2–3 steps in width or 250–375 nm in size is found sandwiched between the Ga coating from FIB preparation and Rb from the sample corpus is indicated.

4 Conclusions

One of the lessons we learned from this investigation is that care must be used when applying FIB technology to prepare samples for subsequent synchrotron-based XRF studies. It is well-established that FIB will induce damage and a large deposition of Ga ions on the surface of the material [18], the extent of which depends on the beam current and the nature of the sputtered material. It is therefore necessary to remove damaged areas by several milling steps, using smaller beam currents, and at a final stage using a low voltage milling (usually 2 kV). This was not done in this case, and the results indicate that the amount of deposited Ga was very large. However, in our FIB-cut sample 1, the Ga defines the sample silhouette remarkably. This could potentially be used to facilitate creating exact overlays of phase-sensitive tomograms with emission tomograms. By leaving Ga-markers on a sample corpus, one could more easily align cross section image overlays. To avoid detector deadtimes, the marking could be left as thin lines with varying pattern. Our sample 3 and 6 had a pyramidal or arrow shape. This shape naturally tapers towards the sample top, which was helpful in aligning tomographic cross sections with the

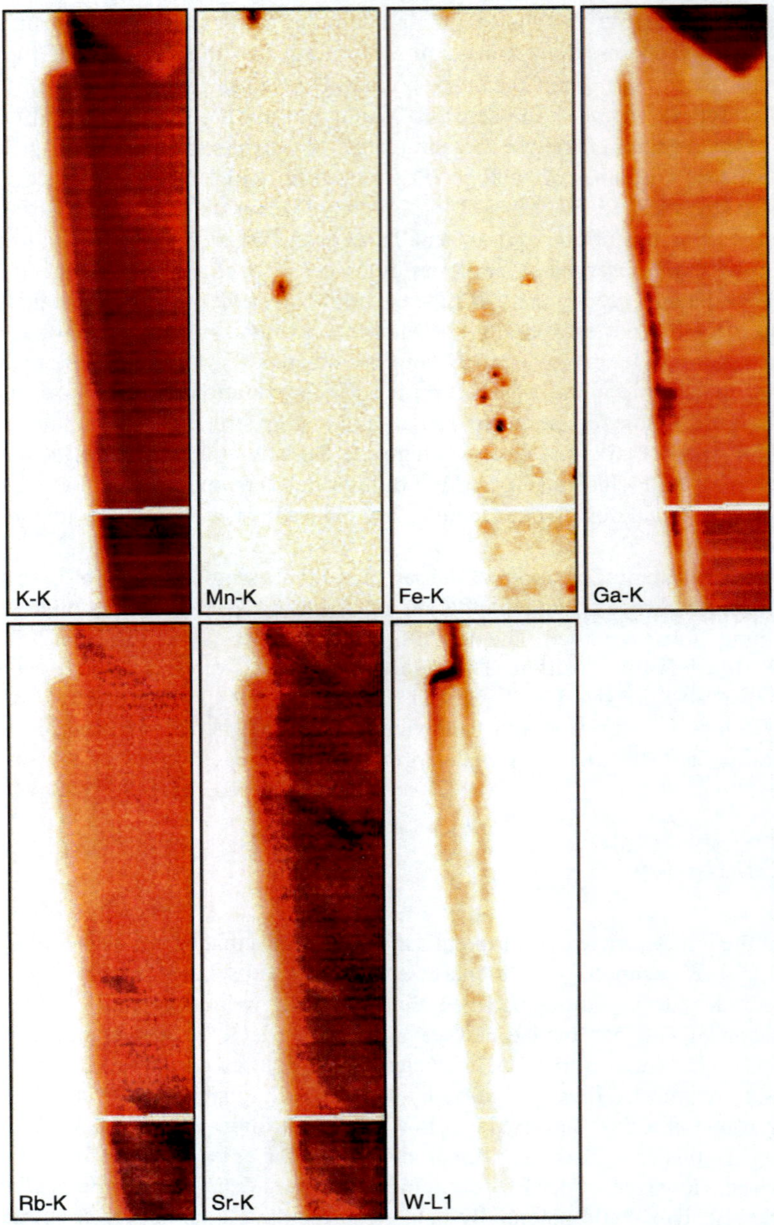

Fig. 15 Element distributions from a 2D scan of sample 1. The image dimensions are $10.75 \times 37.375\ \mu m^2$ (125 nm steps). Recorded with a 200 μm Al filter in front of the detector

correct emission ones. Having element markers available such as lines and patterns of a deposited element drawn with FIB or other techniques and defining the sample contours, however, would have been advantageous and facilitated the image overlay alignment.

We were not able to directly address our original question if the observed Np(IV) hot spots immobilized in the granite during a tracer study [2] are present as precipitated hydrolysis particulate products or if they are present as sorbed Np localized at regions on the fissure walls. This is because no Np was observed in the nano-XRF images of sample 1, the only granite piece originally observed to exhibit Np fluorescence in the μ-XRF screening studies prior to FIB preparation of the sample. Apparently, the sample broke off at the fissure during the lift-out procedure and the Np was lost. Note, however, that nanosized Fe structures were observed on the remaining fissure wall, but no Np was found there. It appears unlikely that the Np was exclusively sorbed onto the lost fissure wall and not on the remaining one, meaning that Np was unlikely to be present as a sorbed species on the fissure walls. This is an indirect indication that the Np was likely present as particulate hydrolysis products formed when the Np(V) of the tracer solution was reduced to Np(IV), which then subsequently hydrolyzed to particles large enough to be occluded within the fissure. The Fe nanoparticulate structures on the fissure wall might have been involved in the process responsible for Np immobilization. To corroborate this interpretation, TEM/EDX investigations of selected pieces from the remaining bulk sample 1 and neighboring sections of the original granite bore core used in the tracer study are planned in hopes of finding Np and evaluating its correlation to Fe. These results can hopefully be compared to new data collected at the Hard X-Ray Micro/Nanoprobe beamline P06 [19] now in its commissioning phase at the new light source PETRAIII in Hamburg, Germany.

The sort of problem we faced here of samples breaking or crumbling, where the area of interest consists of fissures, pores, or grain boundaries, may pose a general challenge for any investigation with nanometer resolution. Preparation of heterogeneous samples to obtain dimensions small enough for a reasonable field of view (<~150 μm) in such nanotomography studies may destroy the fissures, pores, or grain boundaries of original interest, as preferential cleavage at these areas during sample preparation using methods such as cutting, grinding, or FIB-ing of the sample will likely occur. Using impregnation methods to "glue" the heterogeneous parts of samples together prior to cutting might not help for micropores and grain boundaries, as even with vacuum impregnation these are difficult to fill with resin. A different strategy using a technique that does not require reducing samples to such small dimensions such as confocal irradiation-detection geometry [20] might be advantageous. However, this will not offer the resolution attainable in a similar emission tomography experiment. Typical resolution of confocal μ-XRF is on the order of 10–20 μm in the lateral and depth dimensions.

What we also learned about the samples is that the Th and U in these granite pieces was found to be likely generally associated with feldspar, in case of Th, and with micropores in the case of U. That Th is intimately associated with mineral phases suggests that it is present as a trace structural component. Although this case study was limited to only three samples and the Th-mineral association should be corroborated by, e.g., XRD studies, initial interpretation of results suggest that Th is less likely than U to become mobile in groundwater upon drilling and excavation emplacement caverns and shafts for a nuclear waste repository in granite. This is because Th is redox stable and expected to be relatively less soluble than U as a structural component of the host rock, isolated from micropores. The U that was observed to be associated with granite pores may be U that became mobilized upon eluting the granite bore core with groundwater and transported via the pores. We cannot know this for sure without extensive further study. However, it does not matter if the U observed was originally associated with the pores in the granite, or if it was mobilized from another part of the granite and transported to there, we may still conclude from these results that the U in the granite pieces studied would present a higher potential contamination source in groundwater upon drilling activities in granite bed rock than Th in these samples. If the U was originally associated with the pores, it has more contact with water than the Th, which appears isolated from pores. If the U was transported to the granite piece studied, this already attests its mobility.

In closing, a short comment on the relevance of such investigation to nuclear waste disposal is in order. Studies of the association of U and Th in granite components using microscopic and alpha track imaging were performed in the 1980s (see, e.g., [21]). From these and other investigations, insight into the general modes of occurrence of these elements in granite were identified: (1) associated with accessory minerals, (2) along grain boundaries, (3) in microfissures and pores, and (4) associated with quartz and feldspar grains. Indeed, our study reported here did not discover anything otherwise. However, it is known that these occurrences can be influenced by, e.g., alteration. Drilling and excavating of the granite also affects changes, the extent and mechanism of which is not known well-enough to be quantified. Further study using this type of investigation, in combination with migration studies such as the one reported in [1] for characterization of proposed nuclear repository sites in granite host rock, would be necessary to be able to successfully include these in models and codes used for safety cases.

Acknowledgments We thank HASYLAB and the ESRF for awarding us beamtime. We also are grateful to E. Sobolla (KIT-INE) and Julijana Krbanjevic (PSI) for preparation of the ground and FIB-ed samples and J. Göttlicher (KIT) for expert improvement of this manuscript. We acknowledge financial support as a Joint Research Group between the German Helmholtz Society and the Russian Basic Research Foundation (HRJRG-011 and 07-03-92280-SIG_a).

References

1. Kienzler, B, Vejmelka, P, Römer, J, Jansson, M (2009) Actinide migration in fractures of granite host rock: laboratory and in situ investigations. Nucl. Technol. 165:223–240
2. Denecke, MA, Brendebach, B, De Nolf, W, Falkenberg, G, Janssens, K, Simon, R (2009) Spatially resolved μ-XRF and μ-XAFS study of a fractured granite bore core following a radiotracer experiment. Spectrochim. Acta B 64:791–795

3. Overwijk, MHF, van den Heuvel, FC, Bulle-Lieuwma, CWT (1993) Novel scheme for the preparation of transmission electron microscopy specimens with a focused ion beam. J. Vac. Sci. Technol. B11:2021–2024
4. Falkenberg, G, Rickers, K, Bilderback, DH, Huang, R (2003) A Single-bounce Capillary for Focusing of Hard X-rays. http://hasyweb.desy.de/science/annual_reports/2003_report/part1/intern/11062.pdf
5. Huang, R, Bilderback, DH (2006) Single-bounce monocapillaries for focusing synchrotron radiation: modeling, measurements and theoretical limits. J. Synchrotron Radiat. 13:74–84
6. Vekemans, B, Janssens, K, Vincze, L, Adams, F, Van Espen, P (1994) Analysis of X-ray spectra by iterative least squares (AXIL): new developments. X-ray Spectrom. 23:278
7. Bleuet, P, Simionovici, A, Lemelle, L, Ferroir, T, Cloetens, P, Tucoulou, R, Susini, J (2008) Hard X-rays nanoscale fluorescence imaging of earth and planetary science samples. Appl. Phys. Lett. 92:213111
8. Ortega, R, Cloetens, P, Devès, G, Carmona, A, Bohic, S (2007) Iron storage within dopamine neurovesicles revealed by chemical nano-imaging. PLoS ONE 2(9):e925
9. Bleuet, P, Cloetens, P, Gergaud, P, Mariolle, D, Chevalier, N, Tucoulou, R, Susini, J, Chabli, A (2009) A hard X-ray nanoprobe for scanning and projection nanotomography. Rev. Sci. Instrum. 80:056101
10. Somogyi, A, Tucoulou, R, Martinez-Criado, G, Homs, A, Bleuet, P, Simionovici, A (2005) ID22: a multitechnique hard X-ray microprobe beamline at the European Synchrotron Radiation Facility. J. Synchrotron Radiat. 12:208–215
11. Labiche, J-C, Mathon, O, Pascarelli, S, Newton, MA, Ferre, GG, Curfs, C, Vaughan, G, Homs, A, Carreiras, DF (2007) The fast readout low noise camera as a versatile X-ray detector for time resolved dispersive extended X-ray absorption fine structure and diffraction studies of dynamic problems in materials science, chemistry, and catalysis. Rev. Sci. Instrum. 78: 0901301
12. Lange, K, Carson, R (1984) EM reconstruction algorithms for emission and transmission tomography. J. Comput. Assist. Tomogr. 8(2):306
13. De Nolf, W, Janssen, K (2010) Micro X-ray diffraction and fluorescence tomography for the study of multilayered automotive paints. Surf. Interf. Anal. 42:411–418
14. Sole, VA, Papillon, E, Cotte, M, Walter, P, Susini, J (2007) A multiplatform code for the analysis of energy-dispersive X-ray fluorescence spectra. Spectrochim. Acta B62:63
15. Paganin, D, Mayo, SC, Gureyev, TE, Miller, PR, Wilkins, SW (2002) Simultaneous phase and amplitude extraction from a single defocused image of a homogeneous object. J. Microsc. 206:33–40
16. Cloetens, P, Ludwig, W, Baruchel, J, Van Dyck, D, Van Landuyt, J, Guigay, JP, Schlenker, M (1999) Holotomography: quantitative phase tomography with micrometer resolution using hard synchrotron radiation X-rays. Appl. Phys. Lett. 75:2912–2914
17. Weitzel, H, Schrocke, H (1980) Kristallstrukturverfeinerungen von euxenit, $Y(Nb_{0.5}Ti_{0.5})_2O_6$, und M-fergusonit, $YNbO_4$. Zeitschrift für Kristallographie 152:69–82
18. Abolhassani, S, Gasser, P (2006) Preparation of TEM samples of metal-oxide interface by focused ion beam. J. Microsc. doi: 10.1111/j.1365-2818.2006.01599
19. Schroer, CG, Boye, P, Feldkamp, JM, Patommel, J, Samberg, D, Schropp, A, Schwab, A, Stephan, S, Falkenberg, G, Wellenreuther, G, Reimers, N (2010) Hard X-ray nanoprobe at beamline P06 at PETRA III. Nucl. Instrum. Methods Phys. Res. A 616:93–97
20. Denecke, MA, Janssens, K, Proost, K, Rothe, J, Noseck, U (2005) Confocal micro-XRF and micro-XAFS studies of uranium speciation in a tertiary sediment from a waste disposal natural analogue site. Environ. Sci. Technol. 39(7):2049–2058
21. Schoch, AE, Scheepers, R (1990) The distribution of uranium and thorium in the cape columbine granite from the southwestern cape province, South Africa. Ore Geol. Rev. 5:223–246

Index

S.N. Kalmykov and M.A. Denecke (eds.), *Actinide Nanoparticle Research*,
DOI 10.1007/978-3-642-11432-8,